Interdisciplinary Applied Mathematics

Volume 59

Problems in engineering, computational science, and the physical and biological sciences are using increasingly sophisticated mathematical techniques. Thus, the bridge between the mathematical sciences and other disciplines is heavily traveled. The correspondingly increased dialog between the disciplines has led to the establishment of the series: Interdisciplinary Applied Mathematics.

The purpose of this series is to meet the current and future needs for the interaction between various science and technology areas on the one hand and mathematics on the other. This is done, firstly, by encouraging the ways that mathematics may be applied in traditional areas, as well as point towards new and innovative areas of applications; and secondly, by encouraging other scientific disciplines to engage in a dialog with mathematicians outlining their problems to both access new methods as well as to suggest innovative developments within mathematics itself.

The series will consist of monographs and high-level texts from researchers working on the interplay between mathematics and other fields of science and technology.

Luis Manuel Cruz-Orive

Stereology

Theory and Applications

Springer

Luis Manuel Cruz-Orive
Universidad de Cantabria
Santander, Spain

ISSN 0939-6047 ISSN 2196-9973 (electronic)
Interdisciplinary Applied Mathematics
ISBN 978-3-031-52450-9 ISBN 978-3-031-52451-6 (eBook)
https://doi.org/10.1007/978-3-031-52451-6

This Springer imprint is published by the registered company Springer Nature Switzerland AG
The registered company address is: Gewerbestrasse 11, 6330 Cham, Switzerland

Paper in this product is recyclable.

To Maya and Emma

Preface

The Beginnings

As a PhD student at the Department of Probability & Statistics, University of Sheffield (UK), in 1972, I came across the book of Kendall and Moran (1963). Some results on integral geometry aroused my curiosity – for instance, curve length could be expressed in terms of the number of intersections with a random plane! I discovered more wonders in the masterpiece of Santaló (1953). At that time, Terry M. Mayhew, a PhD student from the Department of Anatomy, came to see me seeking help on a stereological problem involving macrophages. He let me know that there was a discipline called stereology, explained his problem, and passed me some reprints. My first papers were coauthored with him, and we started a friendship lasting to this day.

I then learned that the term stereology (from the Greek 'στερεός' = solid), was coined in 1961, when the International Society for Stereology (ISS, now ISSIA) was founded in Germany by a group of scientists interested in the "3D interpretation of 2D images". The nexus between stereology and integral geometry was clear, but I noticed that, with rare exceptions, theoretical geometricians and stereologists did not seem to know much of each other!

Our early publishing correspondence with Ewald R. Weibel, then Editor of the *Journal of Microscopy*, and his interest in stereology – Weibel (1963) was a remarkable book – motivated his inviting me to join his research group at the Institute of Anatomy, University of Bern (CH) in 1976. In 1994 I moved to the University of Cantabria in Santander (E), where I continued my research activity along similar lines.

Development

In March 1977, Ewald Weibel organized a stereology course inviting Roger E. Miles, who showed us that stereology was a blend between integral geometry, probability and statistics. It emerged that, in essence, stereology is geometric sampling, inspired by applications in microscopy, radiology and image analysis. Hans Jørgen G. Gundersen attended that course, and between 1979 and 1998 we shared teaching with various staff at many ISS Stereology Courses around the world. The interaction between teachers and participants (mainly from biomedical sciences) contributed to the development of stereology. Nowadays stereology provides valuable tools, easy to apply, to estimate geometric parameters (such as volume, surface area, length, number, or connectivity), defined on spatial structures with no shape assumptions, based on sparse (usually systematic) sampling rather than reconstructions.

Why Stereology?

It is often argued that new imaging tools based on fast computing, combined with novel techniques in radiology and microscopy, may render stereology obsolete. If the purpose of a scientific project is to visualize shape, then reconstructing a few blocks of material may help. Examples may be the interconnection among neurons, percolation processes, etc. The scenario drastically changes, however, if the target is the number of neurons in a brain compartment, the total length of their axons, the area of a surface of interest (e.g., a fracture) per unit volume of a material, etc. In this scenario, take a few blocks from the target object, and reconstruct them into a high resolution, 3D computer rendering. Then:

1. Suppose that (a) the target object is not homogeneous (that is, the mean of the relevant measure within a block varies with the location of the block), and (b) the blocks are not properly sampled. Then, no matter the measurement precision, the scientific value of the result risks being nil. Quantifying is a committing task!
2. Suppose instead that the target object is not homogeneous (common in biomedical sciences), but sampling is correct. Or, the object is homogeneous (common in materials sciences), in which case block sampling is not that important. Then, the problem is transferred into measuring the blocks! What do we do with the neurons, or grains, hit by the boundary of the block? Do we count them, or not? Also, volume may be accessible by direct pixel counting, but curve length, or surface area, will usually not.

It turns out, therefore, that if quantifying is the task, then stereology (ideally combined with modern technologies) is the choice. This book is mainly about principles of geometric sampling, and principles are timeless.

This Book

Mathematicians willing to understand and expand stereology find that it is a combination of often unfamiliar disciplines, requires some familiarity with the applications, and the existing theory and methods are rather disperse in the literature. On the other hand, natural scientists, usually interested in the applications, lack detailed case studies grounded on solid principles, also dispersed in specialized journals, and usually with an incomplete description of the methods. To write a book covering these deficiencies had been brewing in my mind for decades, but I found it difficult to decide how to organize it. In my lectures I used different approaches to justify the methods, often resorting solely to geometrical probability and sampling theory, but I gradually realized that elementary integral geometry (Chapter 1) opened the door to stereological methods (Chapter 4) in a natural manner by incorporating probability (Chapter 2).

Chapters 1 and 2 are written in textbook style, and may be used for separate courses. Chapter 1 does not invoke probability, but it is a prerequisite for Chapter 2, which in turn deals with stereological identities involving probability and mean values, but no estimators. I decided to treat sampling designs and estimation by way of example using case studies with real data in Chapter 4.

While Chapters 1, 2 and 4 focus on design stereology, in which the target object is bounded and non-random, Chapter 3 deals with model stereology, in which the object is modelled by a stationary (or 'homogeneous') random process. On the other hand, Chapter 5 contains nearly all the (usually non-trivial) variance predictors currently used for systematic sampling, and their derivation. Finally, the Appendix contains basic concepts and tools of probability and statistics for Chapters 2 and 4, plus some mathematics for Chapter 5.

Navigating Through the Book

The flavour of stereology is probably best appreciated by first having a look at Chapter 4. This is especially recommended to readers who are mainly interested in the applications. Most section headings bear the name of the pertinent designs (Cavalieri to estimate volume from parallel systematic sections; isotropic, or vertical, designs to estimate curve length and surface area; the disector, the nucleator, and the invariator for the stereology of particles, etc.). For a finer search there is a Subject Index at the end of the book.

Chapter 4 contains the main stereological estimators, illustrated with real data. Important methods are illustrated first with toy examples (point counting with a leaf, Cavalieri and vertical sections with a banana, the isotropic Cavalieri with spaghetti), followed by more serious, usually published case studies.

The remaining Chapters (1, 2, 3, and 5) may be regarded either as auxiliary to Chapter 4, or as interesting topics in themselves. In the latter case I was thinking of undergraduate students of engineering, mathematics, or statistics. In either case, researchers in biomedicine, physics, materials sciences, or even laymen, willing to know more about the mathematical basis of any concrete estimator from Chapter 4, will be directed into a chain of results leading back to Chapter 2, and eventually to Chapter 1.

Chapter 5 is devoted to variance predictors which include those used in Chapter 4. The explicit prediction formulae, ready for application, are given under section headings reading "Practical variance prediction formulae". The mathematical details are given under "Derivations". Apparent paradoxes arising in geometric sampling are also considered.

Chapter 3 deals with model stereology, which is not the main focus of the book. Its main motivation is to show – by means of J. Mecke's theorem – that the ratio estimators used in design stereology formally extend to model stereology, and to describe tools (notably the K-function) for the analysis of spatial patterns. In addition, a few elementary concepts of stochastic geometry (which is a well-developed discipline, with a broad literature) are briefly described.

Remarks

The mathematics used in the book is free of technicalities, and should be accessible to undergraduate students and numerate readers in general. Nearly all results are proved from scratch, with the purpose that the book should serve for self-tuition. The basic ingredients introduced in Chapter 1, namely motion-invariant densities associated with moving geometric objects, are intuitive and easy to understand. Their mathematical foundations lie beyond the scope of this book, however, and are not needed for the intended purposes.

With few exceptions that should not affect understanding, the treatment is intuitive, restricted to 1D, 2D, and 3D, namely to $\mathbb{R}^d$, $d \leq 3$, and assisted by line drawings. The notation has been simplified as much as possible, but it is easy to interpret in each context, and consistent throughout. For instance, the q-dimensional measure of a target set $Y \subset \mathbb{R}^d$, $q \leq d$, is denoted by $\gamma(Y)$. For a probe T_r of dimension $r \geq d - q$ hitting the set Y, the $(q + r - d)$-dimensional measure of the intersection $Y \cap T_r$ is denoted by $\alpha(Y \cap T_r)$.

To save space, a formula of interest is often given by the last equation of a chain used to derive it. For instance, the target formula corresponding to Eq. (1.5.8) is $\int I(Y \cap L_1^2)\mathrm{d}L_1^2 = 2B$.

To avoid distraction, reference quotations, historical details, etc., are deferred to notes at the end of the pertinent sections. The list of references is far from exhaustive, mainly related to the contents of the book – it contains original sources, a subjective choice of applications, and recent developments not included in the book.

Acknowledgements

My vague idea of contributing to extend the classical sampling theory on discrete spaces to continuous ones for stereology started to take shape when I joined Ewald R. Weibel's group in 1976, and met influential visitors like Roger E. Miles and Hans Jørgen G. Gundersen. My colleagues in Bern sought my help with challenging projects that inspired new methods. Some of the material used in Chapter 4 was kindly provided by them, and used in stereology courses. In the early 1980s, Eva B. Vedel Jensen and Hans J. G. Gundersen organized stimulating stereology courses and workshops at the Institute of Mathematics of the University of Aarhus (DK), attended by leading specialists. The participants attending the International ISS courses around the world were also a source of motivation and inspiration, not to say the many members of the teaching staff, who included Adrian J. Baddeley, Hans J. G. Gundersen, Vyvyan Howard, Terry M. Mayhew, Bente Pakkenberg, and Ruth Østerby.

My scientific collaborators also contributed to shape the concepts and applications in this book over the years. I thank Ximo Gual-Arnau for his help with integral geometry questions, two anonymous Springer reviewers for a number of corrections and constructive remarks, and Barnaby Sheppard for editing the original manuscript.

The care and support to make everything possible has always come from my wife Soledad.

E-Santander, May 2023 *Luis Manuel Cruz-Orive*

Contents

Chapter 1
Basic Results of Integral Geometry

1.1 Introduction

1.1.1 Motion-invariant densities: concept

The results in this chapter are of a geometric nature, and they emanate from the basic concept of motion-invariant measure associated with a mobile geometric object.

Classical sampling theory deals mainly with discrete spaces – usually the object of interest is a discrete population $Y = \{y_1, y_2, \ldots, y_N\}$ of a fixed number N of sampling units. To each item we may assign a constant measure equal to 1, namely the element of counting measure, in which case the total measure of Y is $\mu(Y) = \sum_{i=1}^{N} 1 = N$, which does not depend on the location and orientation of the population Y – we then say that the counting measure is motion-invariant. Geometric sampling, however, works mainly on continuous spaces. What is the measure $\mu(Y)$ of the total number of points in a bounded region Y? We cannot assign a measure 1 to each point, because the result would be infinity. Rather, we need a motion-invariant density $\mu(\mathrm{d}x)$ for points $x \in \mathbb{R}$, whose integral over Y will be a motion-invariant quantity defined on Y. Further, what is the measure of the intersection points determined in the boundary of Y by all possible straight lines hitting it? Now we need a motion-invariant density for straight lines. Note that $\mu(Y)$ is a finite sum in the discrete, and an integral in the continuous case – hence the term 'integral geometry'.

As a simple illustration, consider an interval $Y = (a, b) \subset \mathbb{R}$. The problem is to find a weight function $w : \mathbb{R} \to \mathbb{R}_+$ such that the density $\mu(\mathrm{d}x) = w(x)\mathrm{d}x$ of a mobile point x, and thereby the integral

$$\mu(Y) = \int_a^b \mu(\mathrm{d}x) = \int_a^b w(x)\mathrm{d}x, \tag{1.1.1}$$

are translation-invariant (that is, they do not depend on the location of the interval Y on its support axis or, equivalently, they do not depend on the choice of origin in that axis). Thus, for any real constant c we must have,

L. M. Cruz-Orive, *Stereology*, Interdisciplinary Applied Mathematics 59,
https://doi.org/10.1007/978-3-031-52451-6_1

$$\int_{a}^{b} w(x)\mathrm{d}x = \int_{a+c}^{b+c} w(x)\mathrm{d}x. \tag{1.1.2}$$

In the left-hand side (lhs) set $x = y - c$. Then, Eq. (1.1.2) becomes

$$\int_{a+c}^{b+c} w(y-c)\mathrm{d}y = \int_{a+c}^{b+c} w(x)\mathrm{d}x, \tag{1.1.3}$$

and this implies that for all $c \in \mathbb{R}$ we must have $w(x-c) = w(x)$, whereby $w(x) =$ constant for all $x \in \mathbb{R}$. The constant is a scale factor that may be taken to be equal to 1, and therefore the translation-invariant density for points in $\mathbb{R}$ is the length element,

$$\mu(\mathrm{d}x) = \mathrm{d}x. \tag{1.1.4}$$

In integral geometry, densities are taken in absolute value because they must be positive.

1.1.2 Outline of this chapter

In design stereology, a fixed and bounded geometric object Y, with an unknown geometric measure $\gamma(Y)$, called the target parameter, is hit by a mobile geometric object T, called a probe. The problem is to express $\gamma(Y)$ in terms of the integral of the intersection measure $\alpha(Y \cap T)$ with respect to the motion-invariant density $\mathrm{d}T$ of T. Alternatively, the probe T may be fixed and the target object Y may be mobile, hitting T with motion-invariant density $\mathrm{d}Y$. For instance, if $Y \subset \mathbb{R}^3$ is a domain of volume V, and T is a plane probe, then the relevant intersection measure is the area $A(Y \cap T)$. If the plane probe T has a fixed orientation, and its density is translation-invariant, then the relevant result is Eq. (1.4.5) below, see also Fig. 1.4.1(b). If the probe density is motion-invariant, then see Eq. (1.5.2).

The choice of motion-invariant densities is convenient because (a) the relevant integrals are independent from the adopted reference frame, and (b) the results are compatible with experiment – see Section 2.4.8, Note 2, for a description of the famous Buffon's needle problem.

The purpose of this chapter is to derive special cases of geometric integrals which are combined with probability tools in Chapter 2 in order to elaborate the basic methods of geometric sampling.

1.2 Points and Unbounded Probes

1.2.1 Points

For coherence with the sequel, a point of abscissa p on an arbitrary axis (i.e. on $\mathbb{R}$) may be denoted by $L_0^1 \equiv L_0^1(p)$. The sub- and the superscript (0 and 1 in this case, respectively) are the dimensions of the probe and of the containing space, respectively. The translation-invariant density of the point is the length element

$$\mathrm{d}L_0^1 = \mathrm{d}p, \tag{1.2.1}$$

see Fig. 1.2.1(a). As a natural extension, the translation-invariant density of a point L_0^d in d-dimensional Euclidean space $\mathbb{R}^d$ is the corresponding volume element (namely the Lebesgue measure element), that is,

$$\mathrm{d}L_0^d = \mathrm{d}x = \mathrm{d}x_1\mathrm{d}x_2 \cdots \mathrm{d}x_d, \tag{1.2.2}$$

where $x \in \mathbb{R}^d$ represents a point of Cartesian coordinates $(x_1, x_2, \ldots, x_d)$, see Fig. 1.2.2(a).

In general an invariant density is essentially a wedge or exterior product. For instance, if $x \in \mathbb{R}^2$ then the expression $\mathrm{d}x = \mathrm{d}x_1\mathrm{d}x_2$ is correct only if the coordinate axes are mutually orthogonal, see Fig. 1.2.2(a). If the angle between the coordinate axes is α, then $\mathrm{d}x$ is not an ordinary product of length elements, but

$$\mathrm{d}x = \mathrm{d}x_1 \wedge \mathrm{d}x_2 = \sin\alpha \, \mathrm{d}x_1 \, \mathrm{d}x_2, \tag{1.2.3}$$

namely the area of an oblique box of side lengths $\mathrm{d}x_1, \mathrm{d}x_2$ at an angle α. An alternative notation, occasionally used in the sequel, is $\mathrm{d}x = [\mathrm{d}x_1 \, \mathrm{d}x_2]$.

A test probe equipped with the motion-invariant density is called an invariant probe, for short. Henceforth the pertinent invariant densities are given without proof, which can be found in the books mentioned in Section 1.2.10.

1.2.2 Invariant straight lines in the plane

Fix a rectangular frame Ox_1x_2 with origin O in the plane $\mathbb{R}^2$. An axis $L_{1[0]}^2 \equiv L_1^2(0, \phi)$ of direction $\phi \in [0, \pi)$ is an unoriented straight line through O (hence the subscript '[0]') making an angle ϕ with the positive half axis of abscissas. This is equivalent to joining O with a point ϕ of the unit semicircle $\mathbb{S}_+^1$. The rotation-invariant density of an axis is

$$\mathrm{d}L_{1[0]}^2 = \mathrm{d}\phi, \quad \phi \in [0, \pi), \tag{1.2.4}$$

namely the length element in $\mathbb{S}_+^1$, see Fig. 1.2.1(b).

A straight line $L_1^2 \equiv L_1^2(p,\phi)$ in $\mathbb{R}^2$ is normal to an axis $L_{1[0]}^2$ of direction ϕ, called the orthogonal complement of the line, at a signed distance $p \in (-\infty,\infty)$ from O. The pair (p,ϕ) are called the normal coordinates of L_1^2, see Fig.1.2.1(c). The motion-invariant density of L_1^2, namely the density invariant with respect to rotations and translations, is

$$\mathrm{d}L_1^2 = \mathrm{d}p\,\mathrm{d}\phi. \tag{1.2.5}$$

It is equivalent to take $p \in [0,\infty)$ and $\phi \in [0,2\pi)$.

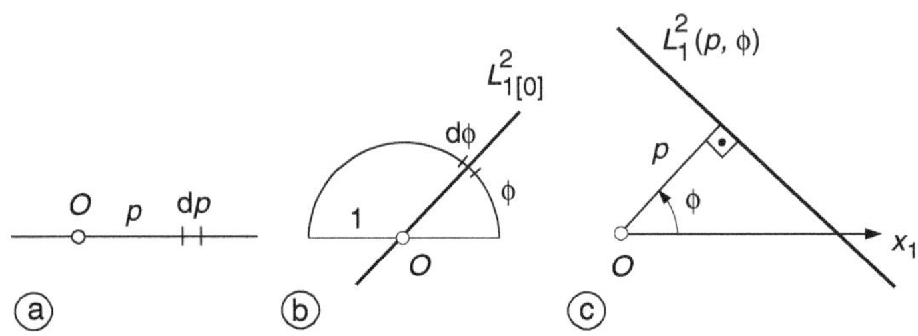

Fig. 1.2.1 (a) The translation-invariant density for a point p on the real axis is the length element dp. (b) The rotation-invariant density for a point ϕ on the unit semicircle, or equivalently for an axial direction in the plane, is the arc element dϕ. (c) A straight line in the plane with its normal coordinates (p,ϕ).

1.2.3 Invariant straight lines and planes in space

Fix an orthogonal trihedron $Ox_1x_2x_3$ with origin O in space $\mathbb{R}^3$. An axis $L_{1[0]}^3 \equiv L_1^3(0,u)$ of direction $u \equiv u(\phi,\theta) \in \mathbb{S}_+^2$ is an unoriented straight line joining O with a point u of the unit hemisphere $\mathbb{S}_+^2$, see Fig. 1.2.2(b). The angles (ϕ,θ) are the spherical polar coordinates of u, namely the longitude $\phi \in [0,2\pi)$, and the colatitude $\theta \in [0,\pi/2]$. The rotation-invariant density of an axis is

$$\mathrm{d}L_{1[0]}^3 = \mathrm{d}u = \sin\theta\,\mathrm{d}\phi\,\mathrm{d}\theta, \tag{1.2.6}$$

namely the area element in $\mathbb{S}_+^2$, see Fig. 1.2.2(c).

The rotation-invariant density of a plane $L_{2[0]}^3 \equiv L_2^3(0,u)$ through O is equal to du, namely the same as the invariant density of its normal axis $L_1^3(0,u)$.

A plane $L_2^3 \equiv L_2^3(p,u)$ is parallel to $L_{2[0]}^3$ at a distance $p \in (-\infty,\infty)$ along its normal axis $L_1^3(0,u)$, which is called the orthogonal complement of the plane, see Fig. 1.2.3(a). The motion-invariant density of L_2^3 is

$$\mathrm{d}L_2^3 = \mathrm{d}p\,\mathrm{d}u. \tag{1.2.7}$$

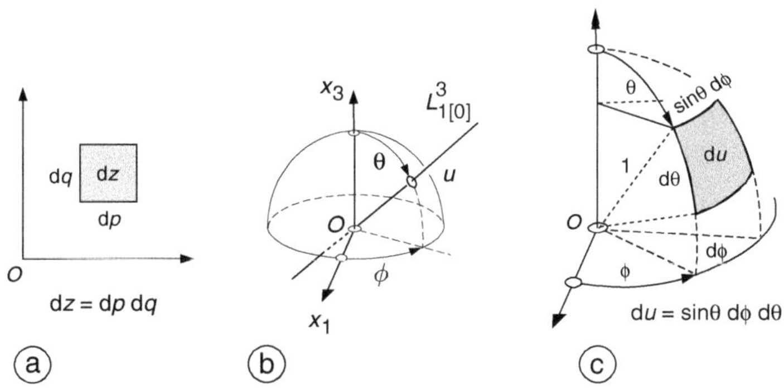

Fig. 1.2.2 (a) The motion-invariant density for a point z in the plane is the area element dz. (b) Axial direction through a point u of the unit hemisphere. (c) The rotation-invariant density for a direction u in three-dimensional space is the area element du on the unit hemisphere.

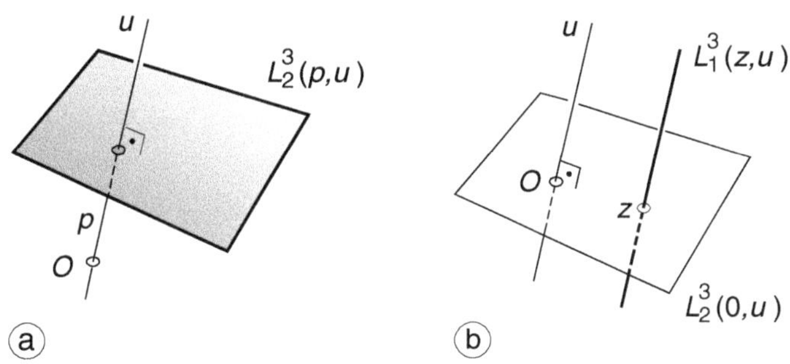

Fig. 1.2.3 (a) Parametrization of (a), a plane, and (b), a straight line in space.

A straight line $L_1^3 \equiv L_1^3(z, u)$ of direction u is a translate of the axis $L_1^3(0, u)$ to a point z in the orthogonal complement of the line, namely in the perpendicular plane $L_2^3(0, u)$, see Fig. 1.2.3(b). The motion-invariant density of L_1^3 is

$$\mathrm{d}L_1^3 = \mathrm{d}z\,\mathrm{d}u. \tag{1.2.8}$$

In turn, if (p, q) denote the Cartesian coordinates of z in $L_2^3(0, u)$, then

$$\mathrm{d}z = \mathrm{d}p\,\mathrm{d}q, \tag{1.2.9}$$

see Fig. 1.2.2(a), namely the motion-invariant density for points in the plane.

1.2.4 Indirect generation of invariant probes

The following results are based on factorizations of invariant densities – see also Section 1.2.10, Note 1.

Pivotal planes and lines in space

The following two generation procedures are equivalent.

1. Generate a plane $L^3_{2[x]}$ with rotation-invariant density $\mathrm{d}L^3_{2[x]} = \mathrm{d}u$ through a point $x \equiv L^3_0$ which is itself translation-invariant, namely $\mathrm{d}L^3_0 = \mathrm{d}x$. The plane $L^3_{2[x]}$ is called a pivotal plane with respect to the (pivotal) point x, because it is free to rotate restricted to contain the 'pivot' x.
2. Generate a plane L^3_2 with invariant density $\mathrm{d}L^3_2 = \mathrm{d}p\,\mathrm{d}u$, and then a point $L^2_0 \in L^3_2$ with translation-invariant density $\mathrm{d}L^2_0 = \mathrm{d}p\,\mathrm{d}q$ within that plane, where the linear elements of lengths $\mathrm{d}p$ and $\mathrm{d}q$ are mutually perpendicular. Then the pairs $(L^3_2, L^2_0 \in L^3_2)$ and $(x, L^3_{2[x]} \ni x)$, are equivalent.

In fact, consider a linear element of length $\mathrm{d}p$ along $u \in \mathbb{S}^2_+$. Then,

$$[\mathrm{d}L^3_{2[x]}][\mathrm{d}L^3_0] = [\mathrm{d}u][\mathrm{d}p\,\mathrm{d}L^2_0] = [\mathrm{d}p\,\mathrm{d}u][\mathrm{d}L^2_0] = [\mathrm{d}L^3_2][\mathrm{d}L^2_0]. \qquad (1.2.10)$$

In a similar way, the pairs $(L^3_1, L^1_0 \in L^3_1)$ and $(x, L^3_{1[x]} \ni x)$ are equivalent. The planar case is analogous.

Lines contained in planes in space

A pair consisting of a rotation-invariant axis $L^3_{1[0]}$ and a rotation-invariant plane $L^3_{2[1]}$ containing the axis – so that the angle $\varphi \in [0, \pi)$ of rotation of the plane around the axis has the invariant density $\mathrm{d}\varphi$ – is equivalent to a rotation-invariant axis $L^2_{1[0]}$ within a rotation-invariant plane $L^3_{2[0]}$, see Fig.1.2.4(a,b). In fact,

$$[\mathrm{d}L^3_{1[0]}][\mathrm{d}L^3_{2[1]}] = [\mathrm{d}u][\mathrm{d}\varphi] = [\mathrm{d}L^3_{2[0]}][\mathrm{d}L^2_{1[0]}]. \qquad (1.2.11)$$

A pair consisting of an invariant line L^3_1 and a rotation-invariant plane $L^3_{2[1]}$ containing the line with density $\mathrm{d}\varphi$, as above, is equivalent to an invariant line L^2_1 within an invariant plane L^3_2, see Fig. 1.2.4(c,d). In fact, with the preceding definition of $L^3_{2[1]}$, we have

$$[\mathrm{d}L^3_1][\mathrm{d}L^3_{2[1]}] = [\mathrm{d}z\,\mathrm{d}u][\mathrm{d}\varphi] = [\mathrm{d}p\,\mathrm{d}u][\mathrm{d}q\,\mathrm{d}\varphi] = [\mathrm{d}L^3_2][\mathrm{d}L^2_1]. \qquad (1.2.12)$$

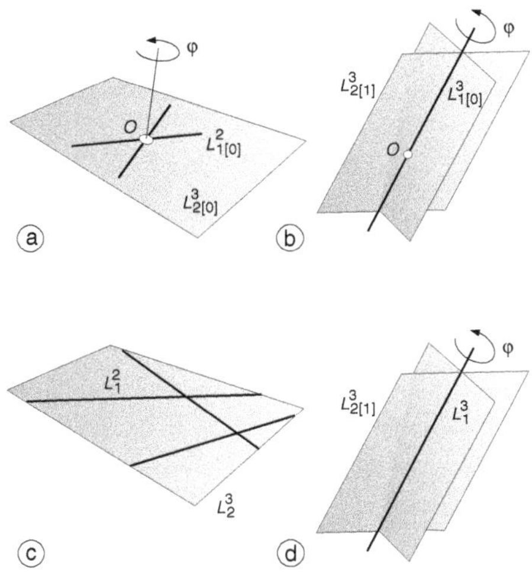

Fig. 1.2.4 Illustration of the equivalences described in Section 1.2.4.

1.2.5 Invariant stripes in the plane and slabs in space

A stripe $L^2_{1,t} \equiv L^2_{1,t}(p, \phi)$ of thickness $t > 0$ is the portion of the plane between two parallel straight lines $L^2_1(p, \phi)$ and $L^2_1(p+t, \phi)$ at a distance t, see Fig. 1.2.5(a). The former line, called the reference face, is included in the stripe, the latter is not. The motion-invariant density of $L^2_{1,t}$ is that of its reference face, namely,

$$dL^2_{1,t} = dp\, d\phi. \tag{1.2.13}$$

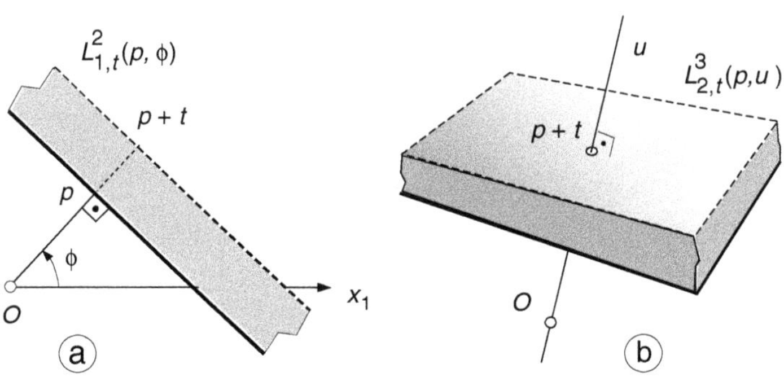

Fig. 1.2.5 Parametrization of (a), a stripe in the plane, and (b), a slab in space.

Analogously, a slab $L^3_{2,t} \equiv L^3_{2,t}(p,u)$ of thickness $t > 0$ is the portion of space between two parallel planes, namely $L^3_2(p,u)$, which is the reference face, and $L^3_2(p+t,u)$, see Fig. 1.2.5(b). The motion-invariant density of $L^3_{2,t}$ is that of its reference face, namely,

$$\mathrm{d}L^3_{2,t} = \mathrm{d}p\,\mathrm{d}u. \tag{1.2.14}$$

1.2.6 Invariant straight line in space contained in a vertical plane

Here we show that an invariant line L^3_1 can be generated within a vertical plane, namely a plane normal to a horizontal plane which may be fixed as convenient. In some applications the latter construction is easier to implement than that of invariant lines within invariant planes (Fig. 1.2.4(c)).

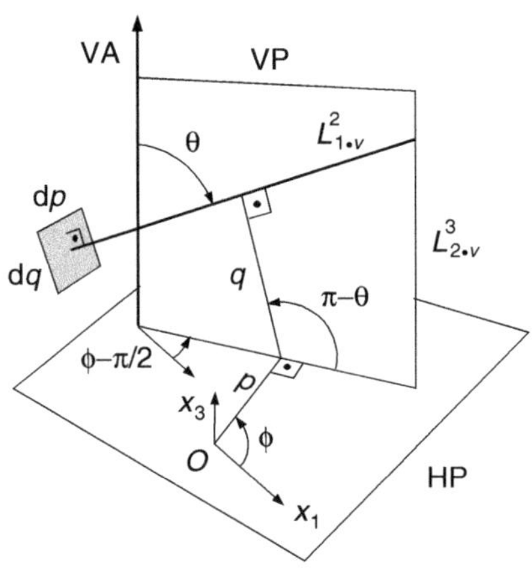

Fig. 1.2.6 Parametrization of a motion-invariant straight line in space contained in a vertical plane. VA: vertical axis, VP: vertical plane, HP: horizontal plane.

In an arbitrarily fixed orthogonal trihedron $Ox_1x_2x_3$, adopt Ox_1x_2 as the horizontal plane (HP), and the axis Ox_3 as the vertical axis (VA). A vertical plane $L^3_{2\cdot v}$ is perpendicular to the HP, and their intersection or 'trace' is a straight line $L^2_1 \equiv L^2_1(p,\phi)$, $p \in \mathbb{R}$, $\phi \in [0,\pi)$, with motion-invariant density $\mathrm{d}L^2_1 = \mathrm{d}p\,\mathrm{d}\phi$, whereby the density of the vertical plane is also $\mathrm{d}L^3_{2\cdot v} = \mathrm{d}p\,\mathrm{d}\phi$. Within the vertical plane consider a line $L^2_{1\cdot v} \subset L^3_{2\cdot v}$ making an angle $\theta \in [0,\pi)$ with the VA, whereby $L^2_{1\cdot v} \equiv L^2_{1\cdot v}(q,\pi-\theta)$ in the usual notation, see Fig. 1.2.6 for details.

In order that $L^2_{1\cdot v}$ is in fact equivalent to a line with invariant density $\mathrm{d}L^3_1 = \mathrm{d}z\,\mathrm{d}u$, the density $\mathrm{d}L^2_{1\cdot v}$ may be obtained as follows,

$$\begin{aligned} \mathrm{d}L_1^3 &= [\mathrm{d}z][\mathrm{d}u] = [\mathrm{d}p\,\mathrm{d}q][\sin\theta\,\mathrm{d}\phi\,\mathrm{d}\theta] \\ &= [\mathrm{d}p\,\mathrm{d}\phi][\mathrm{d}q\sin\theta\,\mathrm{d}\theta] = [\mathrm{d}L_{2\cdot v}^3][\mathrm{d}L_{1\cdot v}^2], \end{aligned} \tag{1.2.15}$$

that is,

$$\mathrm{d}L_{1\cdot v}^2 = \sin\theta\,\mathrm{d}q\,\mathrm{d}\theta, \tag{1.2.16}$$

namely the density $\mathrm{d}q\,\mathrm{d}\theta$ of an invariant line in the plane, weighted by the sine of the angle θ between the line and the VA.

1.2.7 Invariant densities for local stereology

The classical invariant test probes L_1^2, L_1^3, L_2^3 considered so far are free to translate and rotate in their containing space. In contrast, local stereology deals with probes that are restricted to contain a fixed point or a fixed axis, namely a fixed subspace. Often, local probes are the easier to implement in practice. The axes $L_{1[0]}^2$ and $L_{1[0]}^3$, and the plane $L_{2[0]}^3$, as described above, may be regarded as local probes. In addition, the following local probes are considered.

Ray emanating from a fixed point

A rotation-invariant ray $L_{1+}^2 \equiv L_{1+}^2(\omega)$ in the plane is a half axis emanating from a fixed origin O in a direction $\omega \in [0, 2\pi)$. The corresponding density is the element of arc of the unit circle $\mathbb{S}^1$, namely

$$\mathrm{d}L_{1+}^2 = \mathrm{d}\omega, \quad \omega \in [0, 2\pi). \tag{1.2.17}$$

A rotation-invariant ray $L_{1+}^3 \equiv L_{1+}^3(u_2)$ in space is a half axis emanating from a fixed origin O in a vector direction $u_2 = u_2(\phi, \theta) \in \mathbb{S}^2$. The corresponding density is the area element on the unit sphere,

$$\mathrm{d}L_{1+}^3 = \mathrm{d}u_2 = \sin\theta\,\mathrm{d}\phi\,\mathrm{d}\theta, \tag{1.2.18}$$

namely the same expression as $\mathrm{d}L_{1[0]}^3$, with the important difference that, unlike u, which varies on the unit hemisphere, now u_2 is a point on the unit sphere, so that $\phi \in [0, 2\pi)$, as usual, but $\theta \in [0, \pi]$ instead of $\theta \in [0, \pi/2]$.

Plane, or slab, through a fixed point

In local stereology, a pivotal plane $L_{2[0]}^3 \equiv L_2^3(0, u)$, see Section 1.2.4, plays an important role. If such a plane is the reference face or, more conveniently in this case, the midplane of a slab $L_{2[0],t}^3 \equiv L_{2,t}^3(0, u)$, then the latter is a slab through a fixed point O with the same density. Thus,

$$\mathrm{d}L^3_{2[0],t} = \mathrm{d}L^3_{2[0]} = \mathrm{d}u, \quad u \in \mathbb{S}^2_+. \tag{1.2.19}$$

Plane, or slab, containing a fixed axis

Consider a rotation-invariant plane $L^3_{2[1]} \equiv L^3_{2[1]}(\phi)$, as defined in Section 1.2.4, containing an arbitrarily fixed axis Ox_3. The intersection between $L^3_{2[1]}$ and the plane Ox_1x_2 normal to Ox_3 is a rotation-invariant axis $L^2_{1[0]} \equiv L^2_1(0, \phi)$ with density $\mathrm{d}\phi$. If the plane $L^3_{2[1]}$ is the midplane of a slab $L^3_{2[1],t} \equiv L^3_{2[1],t}(\phi)$, then the latter is also rotation-invariant containing Ox_3, namely,

$$\mathrm{d}L^3_{2[1],t} = \mathrm{d}L^3_{2[1]} = \mathrm{d}\phi, \quad \phi \in [0, \pi). \tag{1.2.20}$$

Instead of an entire plane $L^3_{2[1]}$ it is often convenient to consider a half-plane $L^3_{2+} \equiv L^3_{2+}(\omega)$ emanating from a fixed axis Ox_3. The trace determined by $L^3_{2+}(\omega)$ in the plane Ox_1x_2 is a ray $L^2_{1+}(\omega)$. A half slab $L^3_{2,t+}(\omega)$ is defined analogously, and its rotation-invariant density is

$$\mathrm{d}L^3_{2,t+} = \mathrm{d}L^3_{2+} = \mathrm{d}\omega, \quad \omega \in [0, 2\pi). \tag{1.2.21}$$

1.2.8 Invariant straight line in space contained in a pivotal plane: the invariator principle

We have seen that an invariant line L^3_1 may be generated in an invariant plane, or in a vertical plane. The invariator is a technique to generate L^3_1 in a pivotal plane $L^3_{2[0]}$ through a fixed pivotal point O (Section 1.2.4).

In the pivotal plane, consider a straight line $L^2_{1(z)}$ through an invariant point $z \in L^3_{2[0]}$ and perpendicular to the axis Oz, see Fig. 1.2.7(a), so that the normal coordinates of $L^2_{1(z)}$ are the polar coordinates (r, ω) of the point z. Thus, the density of $L^2_{1(z)}$ is in fact the density $\mathrm{d}z$ of the invariant point z, and the notation $L^2_{1(z)}$ is intended to avoid confusion with $L^2_{1[z]}$. Thus,

$$\mathrm{d}L^2_{1(z)} = \mathrm{d}z = r\,\mathrm{d}r\,\mathrm{d}\omega, \tag{1.2.22}$$

and therefore,

$$\mathrm{d}L^3_1 = \mathrm{d}z\,\mathrm{d}u = [\mathrm{d}L^2_{1(z)}][\mathrm{d}u], \tag{1.2.23}$$

which means that $L^2_{1(z)} \subset L^3_{2[0]}$ is in fact equivalent to an invariant line L^3_1.

Alternatively,

$$dL_1^3 = [r\,dr\,d\omega][du] = [r\,dL_1^2][du], \tag{1.2.24}$$

which means that an ordinary invariant line $L_1^2(r,\omega) \subset L^3_{2[0]}$ is equivalent to an invariant line L_1^3 provided that its density is multiplied by r, see Fig. 1.2.7(b).

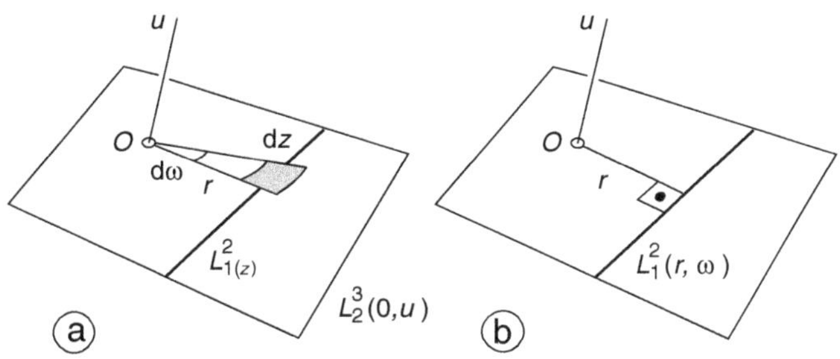

Fig. 1.2.7 The invariator principle states that a motion-invariant straight line in space may be generated as a properly weighted straight line in a pivotal plane. (a) A priori weighting, (b) a posteriori weighting. See Section 1.2.8.

Eq. (1.2.23) corresponds to a priori weighting because $dL^2_{1(z)}$ incorporates the weight r by construction, see Eq. (1.2.22). In contrast, Eq. (1.2.24) corresponds to a posteriori weighting (by the factor r) of the ordinary invariant density dL_1^2.

1.2.9 Invariant densities for sets of points: Blaschke–Petkantschin formulae

A special class of Blaschke–Petkantschin formulae express the product of invariant point densities in terms of the invariant densities of straight lines, or planes, containing the points. The different variants yield important applications in stereology.

An invariant point in the plane and in space

An invariant point $x \equiv L_0^2$ in the plane is equivalent to an invariant point L_0^1 within an invariant line L_1^2, see Section 1.2.4. However, L_0^2 is not equivalent to an invariant point within a rotation-invariant axis $L^2_{1[0]}$. In fact, if (r,ω) are the polar coordinates of x, then

$$dx = [r\,dr][d\omega] = [r\,dr][dL^2_{1[0]}], \tag{1.2.25}$$

which means that the density of x within an axis through the origin O is proportional to the distance r of x from O.

Likewise, if $x = L_0^3$ is an invariant point in space, then

$$dx = [r^2\,dr][du] = [r^2\,dr][dL^3_{1[0]}]. \tag{1.2.26}$$

A pair of invariant points in the plane

Let $\{x_0, x_1\} \in \mathbb{R}^2$ and $L_1^2(p,\phi)$ denote a pair of invariant points and the line joining them, respectively. Fix an origin O' of polar coordinates (p,ϕ), and let x_0, z_1 denote the abscissas of z_0, z_1, respectively, within the line, see Fig. 1.2.8. We want to express the product $[dx_0\,dx_1]$ in terms of the coordinates p, ϕ, z_0, z_1. We have $dx_0 = dp\,dz_0$ and, keeping z_0 fixed, we also have $dx_1 = |z_1 - z_0|\,d\phi\,dz_1$. Thus,

$$\begin{aligned} dx_0\,dx_1 &= [dp\,dz_0][|z_1 - z_0|\,d\phi\,dz_1] \\ &= |z_1 - z_0|\,dz_0\,dz_1\,dL_1^2, \end{aligned} \tag{1.2.27}$$

which is the Blaschke–Petkantschin formula in the plane. Note that the straight line L_1^2 joining the two invariant points turns out to be invariant, namely $dL_1^2 = dp\,d\phi$.

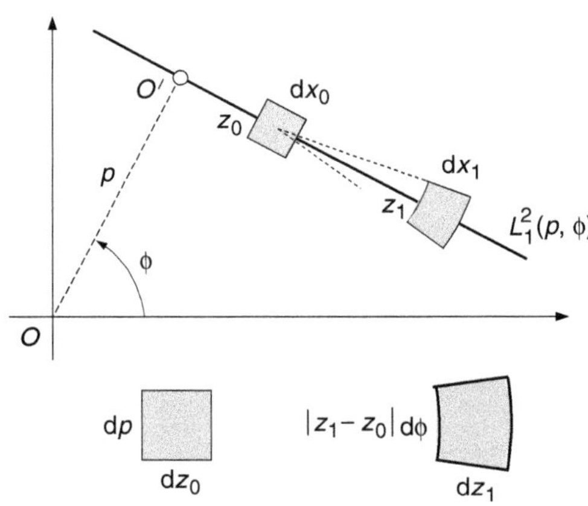

Fig. 1.2.8 A pair of invariant points in the plane may be generated as a pair of translation-invariant points within a motion-invariant straight line in the plane, in which case their joint density is given by Eq. (1.2.27). The polar angle element in the expression of dx_1 does not depend on the position of the pair – if z_0 is at O', then we see that the angle element is $d\phi$.

A pair of invariant points in space referred to the straight line joining them

For a pair of invariant points $\{x_0, x_1\} \in \mathbb{R}^3$, consider the straight line $L_1^3(z, u)$ joining them. Let $z_0, z_1 \in L_1^3$ denote the points x_0, x_1, respectively, within L_1^3 with origin z. By a similar argument as above, we have

$$\begin{aligned} \mathrm{d}x_0\,\mathrm{d}x_1 &= [\mathrm{d}z\,\mathrm{d}z_0][|z_1 - z_0|^2\,\mathrm{d}u\,\mathrm{d}z_1] \\ &= |z_1 - z_0|^2\,\mathrm{d}z_0\,\mathrm{d}z_1\,\mathrm{d}L_1^3. \end{aligned} \tag{1.2.28}$$

A pair of invariant points in space referred to the pivotal plane containing them

Fix an origin $O \in \mathbb{R}^3$, and consider a pair of invariant points $\{x_1, x_2\} \in \mathbb{R}^3$. We want to express the product $[\mathrm{d}x_1\,\mathrm{d}x_2]$ with reference to the unique pivotal plane $L_{2[0]}^3$ containing them.

Let $L_{1[0]}^3 \equiv L_1^3(0, u)$ denote the axis joining the points O and x_1, see Fig. 1.2.9(a). Using spherical polar coordinates, the volume element at the site x_1 reads

$$\mathrm{d}x_1 = r_1^2\,\mathrm{d}u\,\mathrm{d}r_1 = r_1^2\,\mathrm{d}r_1\,\mathrm{d}L_{1[0]}^3, \tag{1.2.29}$$

see Fig. 1.2.9(b), where r_1 is the distance of x_1 from O. On the other hand, adopting cylindrical coordinates (r_2, φ) with respect to the axis $L_{1[0]}^3$, the volume element at x_2 may be expressed as follows,

$$\mathrm{d}x_2 = r_2\,\mathrm{d}\varphi\,\mathrm{d}z_2 = r_2\,\mathrm{d}z_2\,\mathrm{d}L_{2[1]}^3, \tag{1.2.30}$$

see Fig. 1.2.9(b), where $\mathrm{d}z_2$ represents the area element at the site x_2 within the containing plane $L_{2[1]}^3 \equiv L_{2[1]}^3(\varphi)$. Note that the latter plane contains the points $\{O, x_1, x_2\}$, and by Eq. (1.2.11) it is therefore equivalent to the unique pivotal plane $L_{2[0]}^3$ containing these points. Referred to this plane, the point x_1 is denoted by z_1 and, by Eq. (1.2.11), the axis $L_{1[0]}^3$ is equivalent to $L_{1[0]}^2$ within $L_{2[0]}^3$, whereby,

$$\mathrm{d}z_1 = r_1\,\mathrm{d}r_1\,\mathrm{d}L_{1[0]}^2. \tag{1.2.31}$$

The corresponding Blaschke–Petkantschin formula becomes,

$$\begin{aligned} \mathrm{d}x_1\,\mathrm{d}x_2 &= [r_1^2\,\mathrm{d}r_1][r_2\,\mathrm{d}z_2][\mathrm{d}L_{1[0]}^3\,\mathrm{d}L_{2[1]}^3] \\ &= [r_1^2\,\mathrm{d}r_1][r_2\,\mathrm{d}z_2][\mathrm{d}L_{2[0]}^3\,\mathrm{d}L_{1[0]}^2] \\ &= r_1 r_2 [r_1\,\mathrm{d}r_1\,\mathrm{d}L_{1[0]}^2]\,\mathrm{d}z_2\,\mathrm{d}L_{2[0]}^3 \\ &= 2\nabla_{2[0]}\,\mathrm{d}z_1\,\mathrm{d}z_2\,\mathrm{d}L_{2[0]}^3, \end{aligned} \tag{1.2.32}$$

where $\nabla_{2[0]}$ is the area of the triangle of vertices $\{O, z_1, z_2\}$ within the pivotal plane $L_{2[0]}^3$, whose base and height are r_1 and r_2 respectively, see Fig. 1.2.9(c).

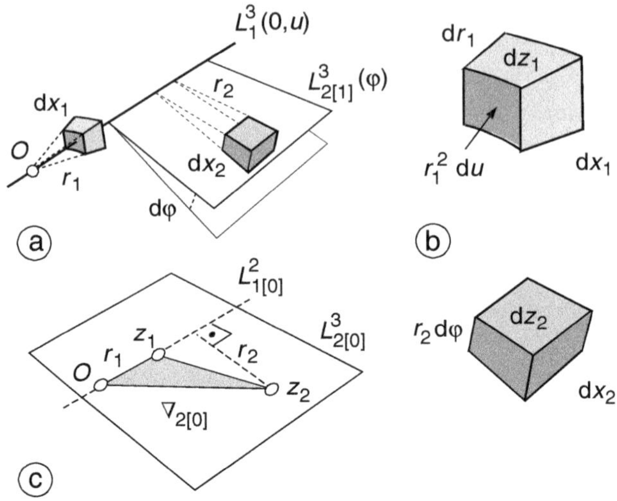

Fig. 1.2.9 A pair of invariant points in space may be generated as a pair of invariant points within a pivotal plane, in which case their joint density is given by Eq. (1.2.32).

Three invariant points in space referred to their containing plane

Fix an origin $O \in \mathbb{R}^3$, and consider a triplet $\{x_0, x_1, x_2\} \in \mathbb{R}^3$ of invariant points. We want to express the product $[\mathrm{d}x_0\, \mathrm{d}x_1\, \mathrm{d}x_2]$ with reference to the unique plane $L_2^3 \equiv L_2^3(p, u)$ containing the triplet. We may express the volume element at x_0 as follows,

$$\mathrm{d}x_0 = \mathrm{d}z_0\, \mathrm{d}p, \tag{1.2.33}$$

where $\mathrm{d}z_0$ represents the area element in the containing plane L_2^3 at the site x_0. The required product is $\mathrm{d}x_0$ times the joint density $\mathrm{d}x_1\, \mathrm{d}x_2$ given by Eq. (1.2.32), in which the origin O is now replaced with x_0. Further, $\mathrm{d}p\, \mathrm{d}L_{2[0]}^3 = \mathrm{d}p\, \mathrm{d}u = \mathrm{d}L_2^3$, whereby the corresponding Blaschke–Petkantschin formula becomes

$$\mathrm{d}x_0\, \mathrm{d}x_1\, \mathrm{d}x_2 = 2\nabla_2\, \mathrm{d}z_0\, \mathrm{d}z_1\, \mathrm{d}z_2\, \mathrm{d}L_2^3, \tag{1.2.34}$$

where ∇_2 is the area of the triangle of vertices $\{z_0, z_1, z_2\}$ within the containing plane L_2^3.

1.2.10 Notes

1. Motion-invariant densities

The motion-invariant density of a geometric probe is a basic concept of integral geometry. The underlying theory falls outside the scope of this book – for a complete treatment, using a differential geometry approach, see Santaló (1976). Integral geometry emergeds in the early 20th century as a solid foundation of geometric probability. After the seminal work of Élie J. Cartan (1869–1951), the Hamburg school of Wilhelm J. E. Blaschke (1885–1962) exerted a great influence through its disciples, among which there was the Spanish-Argentinian mathematician Luis Antonio Santaló Sors (1911–2001), who showed an interest in stereology in later years. The Swiss mathematician Hugo Hadwiger (1908–1981), was also a key contributor (Debrunner, Mani, Meyer, & Rätz, 1982). Classic books are Blaschke (1936–1937), who introduced the term 'integral geometry', Hadwiger (1957), Santaló (1953, 1976) – see also Naveira and Reventós (2009) – and the more recent Schneider and Weil (2008).

2. Vertical plane

The embedding of an invariant test line in space into a vertical plane was first proposed by A. J. Baddeley (personal communication in a letter of 21st March 1982), see also Baddeley (1984). The problem was motivated by a talk of E. Hasselager entitled "Stereological studies on the porcine placenta", delivered at the 1st International Workshop in Stereology, Institute of Mathematics, University of DK-Aarhus, in November 1981.

3. Local stereology

The term was coined by Jensen (1998). For early surveys see Cruz-Orive (1987d), and Jensen and Gundersen (1989).

4. The invariator

The idea in Section 1.2.8 was inspired by a result of Varga (1935), see Cruz-Orive (2005); the term "invariator" was coined in Cruz-Orive (2009b). For general proofs of the decomposition in Eq. (1.2.24), see Gual-Arnau and Cruz-Orive (2009), Auneau and Jensen (2010) and Jensen and Kiderlen (2017). According to Schneider and Weil (2008, p. 285) the decomposition is implicit in a general formula of Petkantschin (1936). For a survey, see Cruz-Orive and Gual-Arnau (2015).

5. The Blaschke–Petkantschin formula

The original references are Blaschke (1935) and Petkantschin (1936), see also Miles (1971). For a general proof, see Santaló (1976), Eq. 12.22. The elementary derivations given in Section 1.2.9 follow the approach of Jensen (1998), who provides general proofs and stereological applications.

1.3 Unbounded Probes Hitting a Target Set

1.3.1 Purpose and preliminaries

Consider a fixed geometric object hit by an invariant probe, e.g. a bounded planar curve of finite length hit by a straight line. Here the intersection will typically consist of a finite set of points. Our purpose is to compute the invariant integral of the total number of such intersection points for 'all possible test lines' hitting the object. Below it is shown that, if the density of the lines is motion-invariant, then the result is twice the length of the curve: this opens the possibility of estimating curve length by intersection counting.

A geometric object will be modelled by a subset Y of Euclidean space. This chapter is concerned with fixed and bounded sets – random sets are considered in Chapter 3. A set is bounded if it can be contained in a ball of finite radius. Curves and surfaces are assumed to be piecewise smooth, which means that they have a unique tangent at almost every point. 'Almost' means that the condition holds with the possible exception of a finite number of vertices, or edges. A set Y is closed if it is the union of its boundary ∂Y and its interior Y°, and a set is compact if it is both closed and bounded. We will usually be concerned with compact sets with piecewise smooth boundary. A real object such as a neuron is the union of the neuronal body, which may be modelled by a three-dimensional compact set, and a number of dendrites, which are dealt with separately. Thus, the compact sets considered here do not include lower-dimensional sets, such as curves, or even surface fragments in the three-dimensional case, attached to the boundary of the compact set. With this restriction, the number of intersection points determined by a straight line in the boundary of a full dimensional compact set is typically even. Further shape assumptions are generally not necessary.

A compact set is convex if, for any two points contained in it, the straight line segment joining them is entirely contained in the set.

A compact set will be assumed to have a finite measure – for instance a curve, a surface, a planar (2D) domain, or a three-dimensional (3D) domain, will have a positive finite length, surface area, planar area, or volume, respectively. In this book a full-dimensional compact set, namely a non-void domain with boundary, will be called a 'domain', for short.

1.3.2 Dimensional relations

Consider a compact set $Y \subset \mathbb{R}^d$, $(d = 1, 2, \ldots)$, and a test probe T equipped with the motion-invariant density. A basic prerequisite is that, whenever $Y \cap T \neq \emptyset$, the intersection must consist of points, curves, etc., that are observable for measurement. Thus,

$$\dim(Y \cap T) \geq 0, \tag{1.3.1}$$

where $\dim(\cdot)$ denotes the dimension of a set, namely the number of parameters, or coordinates, required to identify the location of a point within the set. For instance, the dimension of a finite aggregate of points, curves, surfaces, or domains in $\mathbb{R}^3$ is 0, 1, 2, 3, respectively. Then, the identity

$$\dim(Y \cap T) = \dim(Y) + \dim(T) - d \tag{1.3.2}$$

holds up to a set of positions of the probe (e.g. tangential positions) of total zero measure. Together with Eq. (1.3.1), the preceding identity implies that, given the dimension of the object, the dimension of the probe must satisfy the following inequality,

$$\dim(T) \geq d - \dim(Y). \tag{1.3.3}$$

For instance, if $Y \subset \mathbb{R}^2$ represents a curve, then $d = 2$ and $\dim(Y) = 1$, whereby $\dim(T) \geq 1$, which means that, in the present context, a motion-invariant test point is not suitable to probe a curve, whereas a straight line is.

1.3.3 Intersection formulae for unbounded probes

Bounded and local probes are treated in Sections 1.13–1.18 and 1.7–1.8, respectively. Sections 1.4–1.6 and 1.11–1.12 are devoted to integral formulae for unbounded probes, more precisely for the intersection $Y \cap L_r$, where $L_r \equiv L_r^d$, $r = 0, 1, \ldots, d$, represents an r-plane in $\mathbb{R}^d$ – for instance a test point, a line, a plane, or a slab for $r = 0, 1, 2, 3$, respectively. The formulae have the following form,

$$\int \alpha(Y \cap L_r)\mathrm{d}L_r = c_1 \cdot \gamma(Y), \tag{1.3.4}$$

where

- Y is a compact set with $\dim(Y) = q$, and q-measure $\gamma(Y)$. For instance, if $Y \subset \mathbb{R}^3$, then $\gamma(Y)$ may represent number of subsets, curve length, surface area, or volume, according to whether $q = 0, 1, 2, 3$, respectively.
- $\alpha(Y)$ is the geometric measure of a compact set Y, thus $\alpha(\emptyset) = 0$. If $Y \subset \mathbb{R}^3$, then $\alpha(Y \cap L_r)$ will represent number, length, area, or volume, according to whether $q + r - d = 0, 1, 2, 3$, respectively.

- $\mathrm{d}L_r$ denotes the motion-invariant density of the r-plane probe L_r.
- $c_1 \equiv c_1(q, r, d)$ is a known constant, see Section 1.5.5.

Eq. (1.3.4) – or Eq. (1.14.1) for bounded probes – is useful because its right-hand side (rhs) is linear in the relevant geometric measure $\gamma(Y)$, thus establishing a connection between object measures and integrated section measures and facilitating the estimation of the former (Chapter 4). Classical formulae involving motion-invariant probes are known as Crofton formulae (after Morgan W. Crofton, see Section 1.5.6, Note 1). Formulae involving rotation-invariant probes, namely probes restricted to contain a fixed point, or a fixed line, however, are the object of local stereology (Sections 1.7–1.8).

Whether a general probe is unbounded or bounded, the following two cases may be distinguished.

1. At least one of the identities $\dim(Y) = d$ or $\dim(T) = d$ holds. Then, in order to establish the aforementioned connection it suffices that the probe is translation-invariant: orientation does not matter, and it may be fixed as convenient.
2. The inequalities $\dim(Y) < d$, $\dim(T) < d$ and $\dim(Y) + \dim(T) \geq d$ hold. In this case orientation matters, and the probe density must be both translation and rotation-invariant, namely motion-invariant.

As a simple illustration, if $Y \subset \mathbb{R}^2$ and $\dim(Y) = 2$, then the domain Y can be swept by a straight line L_1 in any direction, and the total area scanned will always be equal to the area of Y. However, if Y is a straight line segment, ($\dim(Y) = 1$), then the total distance swept by L_1 while hitting Y will depend on the angle between L_1 and the segment Y – the swept distance may even be equal to zero if the line is parallel to the segment. Thus, in this case orientation matters.

As indicated in Section 1.3.1, the shape assumptions required to derive the formulae are not too stringent. For the case in which $\dim(Y) < d$ and $\dim(T) < d$, the derivations essentially share a common technique. For instance, the Crofton formula involving the finite length B of a bounded planar curve $Y \subset \mathbb{R}^2$ is obtained by initially approximating Y by a polygonal Y_N consisting of N straight line segments or 'links' whose endpoints lie in the curve. Thus,

$$Y_N = \bigcup_{i=1}^{N} y_i, \tag{1.3.5}$$

where y_i denotes the ith link. The total length of Y_N is

$$B_N = \sum_{i=1}^{N} b_i, \tag{1.3.6}$$

where b_i denotes the length of y_i. We assume that Y is rectifiable, that is, there is a mechanism to reduce the lengths of the links, thereby increasing their number, in such a way that the sequence $\{B_N\}$ is non-decreasing, and its supremum exists and defines the length B of Y. Thus, in the limit B_N may be replaced with a curvilinear integral, namely,

$$B = \int_{y \in Y} \mathrm{d}b(y), \tag{1.3.7}$$

where $\mathrm{d}b(y)$ represents the length of an essentially linear curve element δy at a point $y \in Y$. In shorthand notation we write

$$B = \int_{Y} \mathrm{d}b. \tag{1.3.8}$$

A piecewise smooth, one-sided surface Y of area S is initially approximated by a polyhedral surface Y_N consisting of N essentially planar polygonal faces. Each polygon is the orthogonal projection of a surface element at a point $y \in Y$ onto the tangent plane at y. Then S is defined as the limit of the polyhedral surface area S_N as N increases and the areas of the polygonal elements decrease. We may then express S as a surface integral, namely,

$$S = \int_{Y} \mathrm{d}s, \tag{1.3.9}$$

where $\mathrm{d}s$ is shorthand for the area $\mathrm{d}s(y)$ of an essentially planar surface element δy at a point $y \in Y$ of the surface.

For details on the mathematical definition of the relevant geometric measures, see Section 1.4.4, Note 2.

Definitions

For a domain $Y \subset \mathbb{R}^d$, the intersection $Y \cap L_1^d$ is a (linear) transect, which may consist of several connected segments called (linear) intercepts. For a surface $\partial Y \subset \mathbb{R}^d$, the intersection $\partial Y \cap L_1^d$ is a set of points called intersections. For a domain $Y \subset \mathbb{R}^3$, the intersection $Y \cap L_2^3$ is a (planar) section, whereas $\partial Y \cap L_2^3$ is a (curve) trace. A particle is a connected compact set separated from other particles. A particle consisting of a point is called a point particle. If Y is a non-void particle, then $Y \cap L_2^3$ is often called a transect, which may consist of several separated parts, called profiles, if the particle is not convex. If Y is a curve, then $Y \cap L_2^3$ is a set of points also called transects. Further, the intersection $Y \cap L_{2,t}^3$ of a domain Y with a slab is called a slice. All the intersections are assumed to be non-empty.

1.4 Crofton Formulae for Unbounded Probes of a Fixed Orientation

1.4.1 Volume from point probes

The simplest Crofton formula arises when the set $Y \subset \mathbb{R}^d$ of interest is a domain, so that $\dim(Y) = d$, whereas the probe is a translation-invariant point $L_0^d \in \mathbb{R}^d$. The domain Y has a positive volume V_d. In particular, we write $V_2 = A$, the area of a planar domain, and $V_3 = V$, the volume of a three-dimensional domain. The pertinent Crofton formula for a point probe trivially reads as follows,

$$\int_{\mathbb{R}^d} P(Y \cap L_0^d)\mathrm{d}L_0^d = \int_{\mathbb{R}^d} 1_Y(x)\mathrm{d}x = \int_Y \mathrm{d}x = V_d, \tag{1.4.1}$$

where $P(\cdot)$ denotes the number of point probes hitting a domain. In the present case,

$$P(Y \cap L_0^d) = \begin{cases} 1, & \text{if } Y \cap L_0^d \neq \emptyset, \\ 0, & \text{if } Y \cap L_0^d = \emptyset, \end{cases} \tag{1.4.2}$$

which is the same as the indicator function of the set Y, namely,

$$1_Y(x) = \begin{cases} 1, & \text{if } x \in Y, \\ 0, & \text{if } x \notin Y. \end{cases} \tag{1.4.3}$$

1.4.2 Planar area and volume from lower-dimensional probes of a fixed orientation

Planar area from test lines of a fixed orientation

Consider a fixed planar domain $Y \subset \mathbb{R}^2$ of area A, and a test line $L_1^2(p, \phi)$ with an arbitrarily fixed orientation $\phi \in [0, \pi)$, which is free to move parallel to itself with translation-invariant density $\mathrm{d}L_1^2(p, \phi) = \mathrm{d}p$. In Fig. 1.4.1(a) the linear transect $Y \cap L_1^2(p, \phi)$ is marked in red. The area element associated with the transect is a stripe element of total length $L(Y \cap L_1^2(p, \phi))$ and thickness $\mathrm{d}p$, and the pertinent Crofton formula is

$$\int_{\mathbb{R}} L(Y \cap L_1^2(p, \phi))\, \mathrm{d}p = A, \quad \phi \in [0, \pi). \tag{1.4.4}$$

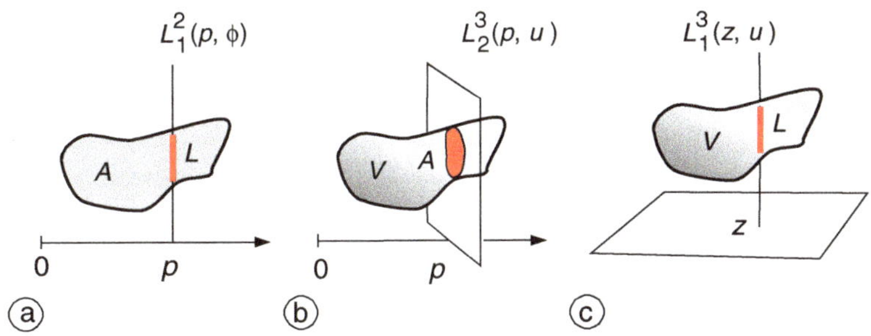

Fig. 1.4.1 (a–c), illustration of Eqs. (1.4.4)–(1.4.6), respectively.

Volume from test planes of a fixed orientation

For a fixed domain $Y \subset \mathbb{R}^3$ of volume V, hit by a test plane $L_2^3(p, u)$ which is free to move parallel to itself with a fixed orientation $u \in \mathbb{S}_+^2$ and a translation-invariant density $\mathrm{d}L_2^3 = \mathrm{d}p$, the development is analogous. The volume element corresponding to the planar transect $Y \cap L_2^3(p, u)$, see Fig. 1.4.1(b), is $A(Y \cap L_2^3(p, u))\,\mathrm{d}p$, and therefore,

$$\int_{\mathbb{R}} A(Y \cap L_2^3(p, u))\,\mathrm{d}p = V, \quad u \in \mathbb{S}_+^2. \tag{1.4.5}$$

Volume from test lines of a fixed orientation

If the domain $Y \subset \mathbb{R}^3$ is hit by a test line $L_1^3(z, u)$ of fixed orientation $u \in \mathbb{S}_+^2$ with translation-invariant density $\mathrm{d}L_1^3 = \mathrm{d}z$, then the transect $Y \cap L_1^3(z, u)$, see Fig. 1.4.1(c), may consist of several connected linear intercepts of total length $L(Y \cap L_1^3(z, u))$, whose associated volume element is the volume of a cylinder element of length $L(Y \cap L_1^3(z, u))$ and cross sectional area $\mathrm{d}z$. In this case the Crofton formula is given by the following double integral,

$$\int_{\mathbb{R}^2} L(Y \cap L_1^3(z, u))\,\mathrm{d}z = V, \quad u \in \mathbb{S}_+^2. \tag{1.4.6}$$

1.4.3 Measures of arbitrary dimension from stripe and slab probes of a fixed orientation

The set $Y \subset \mathbb{R}^d$ of interest has dimension $\dim(Y) = q \in \{0, 1, \ldots, d\}$. For instance, if $q = 0$ then Y is a bounded and finite set of N point particles. If $q = 1$, then Y is a set of curves of total length B, (or L), etc. On the other hand the probe is a stripe in $\mathbb{R}^2$, or a slab in $\mathbb{R}^3$, namely a probe of full dimension. Thus, to write the relevant intersection formulae it suffices that the probes are translation-invariant and of arbitrary fixed orientation, as in the preceding cases. For instance, we may take a stripe $L^2_{1,t}(p,\phi)$ in the plane, where $\phi \in \mathbb{S}^1_+$ is an arbitrarily fixed orientation, and the relevant invariant density is $\mathrm{d}L^2_{1,t}(p,\phi) = \mathrm{d}p$, see Fig. 1.4.2, or a slab $L^3_{2,t}(p,u)$ of a fixed orientation $u \in \mathbb{S}^2_+$, in which case $\mathrm{d}L^3_{2,t}(p,u) = \mathrm{d}p$ as well.

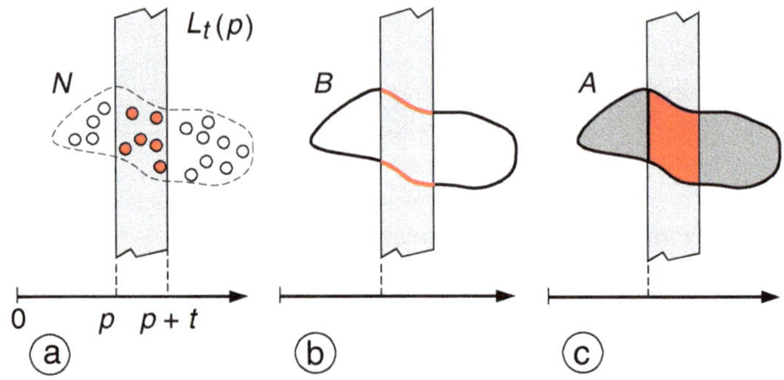

Fig. 1.4.2 (a–c), illustration of Eqs. (1.4.8), (1.4.10), and (1.4.11), respectively, for a translation-invariant stripe in the plane.

Point particles, curves, and domains in the plane intersected by a stripe

Consider first a bounded and finite union of N separate point particles in the plane, namely

$$Y = \bigcup_{i=1}^{N} y_i, \tag{1.4.7}$$

where y_i represents the ith point particle, hit by a translation-invariant stripe $L_t(p) \equiv L^2_{1,t}(p,\phi)$ of thickness $t > 0$ with a fixed orientation $\phi \in \mathbb{S}^1_+$, see Fig. 1.4.2(a). Let $N(Y \cap L_t(p))$, $p \in \mathbb{R}$, represent the total number of point particles captured by the stripe. Then the relevant intersection formula becomes

$$\begin{aligned}\int_{\mathbb{R}} N(Y \cap L_t(p))\,\mathrm{d}p &= \int_{\mathbb{R}} \mathrm{d}p \sum_{i=1}^{N} 1_{L_t(p)}(y_i)\\ &= \sum_{i=1}^{N} \int_{\mathbb{R}} 1_{L_t(0)}(y_i - p)\,\mathrm{d}p\\ &= Nt,\end{aligned} \tag{1.4.8}$$

because the integral is equal to t.

Now let $Y \subset \mathbb{R}^2$ represent a bounded planar curve of total length B, see Fig. 1.4.2(b), and recall that $\mathrm{d}b(y)$ represents the length of an essentially linear curve element δy at a point $y \in Y$. Further, let

$$B(Y \cap L_t(p)) = \int_{y \in Y \cap L_t(p)} \mathrm{d}b(y), \quad p \in \mathbb{R}, \tag{1.4.9}$$

denote the total curve length captured by the stripe. Then the intersection formula becomes,

$$\begin{aligned} \int_{\mathbb{R}} B(Y \cap L_t(p))\,\mathrm{d}p &= \int_{\mathbb{R}} \mathrm{d}p \int_Y 1_{L_t(p)}(y)\,\mathrm{d}b(y) \\ &= \int_{\mathbb{R}} \mathrm{d}p \int_Y 1_{L_t(0)}(y-p)\,\mathrm{d}b(y) \\ &= \int_Y \mathrm{d}b(y) \int_{\mathbb{R}} 1_{L_t(0)}(y-p)\,\mathrm{d}p \\ &= Bt. \end{aligned} \tag{1.4.10}$$

Finally, let $Y \subset \mathbb{R}^2$ represent a domain of total area A, see Fig. 1.4.2(c), and let $L(y) \equiv L_1^2(y, \phi)$ denote a straight line parallel to the stripe $L_t(p)$. Then, by virtue of Eq. (1.4.4),

$$\begin{aligned} \int_{\mathbb{R}} A(Y \cap L_t(p))\,\mathrm{d}p &= \int_{\mathbb{R}} \mathrm{d}p \int_p^{p+t} L(Y \cap L(y))\,\mathrm{d}y \\ &= \int_{\mathbb{R}} \mathrm{d}p \int_0^t L(Y \cap L(x+p))\,\mathrm{d}x \\ &= \int_0^t \mathrm{d}x \int_{\mathbb{R}} L(Y \cap L(x+p))\,\mathrm{d}p \\ &= tA. \end{aligned} \tag{1.4.11}$$

Generalization

Likewise, any intersection formulae involving a set $Y \subset \mathbb{R}^d$ of dimension $q \in \{0, 1, \ldots, d\}$ and measure $\gamma(Y)$, hit by a translation-invariant slab $L_t(p) \subset \mathbb{R}^d$ of an arbitrary fixed orientation and thickness $t > 0$, has the following form,

$$\int_{\mathbb{R}} \gamma(Y \cap L_t(p))\,\mathrm{d}p = t \cdot \gamma(Y). \tag{1.4.12}$$

1.4.4 Notes

1. Dimensionality of sections

In the stereology context, Eq. (1.3.2) was proposed for $d = 3$ by Weibel (1967) on an intuitive basis, see also Weibel (1980). The identity was presented as an axiom in Sommerville (1958, p. 10). For a proof, see Gual-Arnau, Cruz-Orive, and Nuño-Ballesteros (2010, Appendix A).

2. Geometric measures

The arguments outlined in Section 1.3.3 to define curve length and surface area are formalized in Sections 1–3 and 2–8, respectively, from Do Carmo (1976).

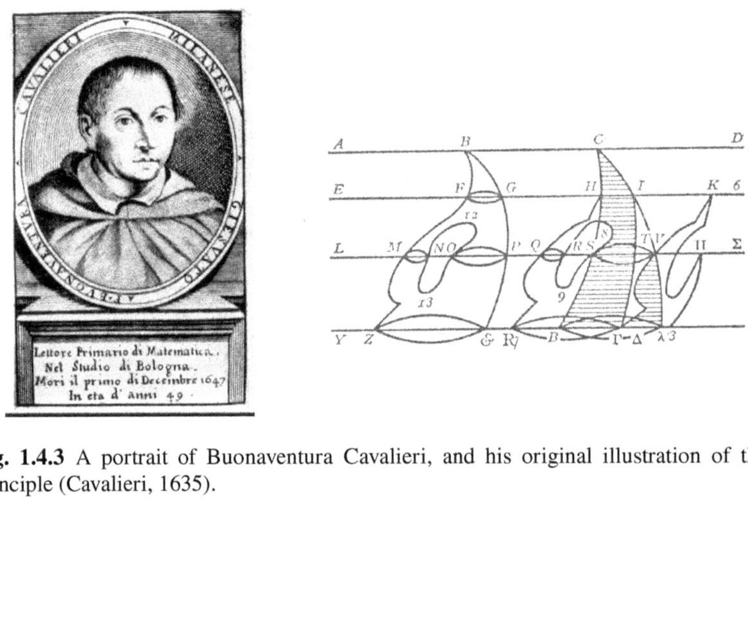

Fig. 1.4.3 A portrait of Buonaventura Cavalieri, and his original illustration of the Cavalieri principle (Cavalieri, 1635).

3. Planar area and volume. The Cavalieri principle

In elementary calculus, Eqs. (1.4.4)–(1.4.6) express area and volume by lower-dimensional integrals. The discretized version of Eq. (1.4.5), namely the decomposition of a solid into thin slices to compute its volume, was known to Greek mathematicians, notably Archimedes of Syracuse (c. 287–c. 212 BC), who applied it to regular solids, see for instance Tobias (1981). Eqs. (1.4.4), (1.4.5) are known as Cavalieri integrals in stereology, in honour of B. Cavalieri, see Cruz-Orive (1987c). The Cavalieri principle (Cavalieri, 1635), formulated by the Italian mathematician

Bonaventura Cavalieri (1598–1647), a disciple of Galileo, may be stated as follows. In Eq. (1.4.5) set $A(p) \equiv A(Y \cap L_2^3(p, u))$, for short. Consider two solids of volumes V_1, V_2. If $A_1(p) = A_2(p)$ for all $p \in (-\infty, \infty)$, then $V_1 = V_2$. Unlike ancient mathematicians, the real merit of Cavalieri is that he considers arbitrarily shaped objects. Unconsciously, one often overlooks the fact that section area and object volume have nothing to do with shape.

4. Stripes and slabs

Integral formulae for stripes and slabs were developed by Santaló (1936b), see also Rey-Pastor and Santaló (1951), Santaló (1976) and Matheron (1976).

1.5 Crofton Formulae for Motion-Invariant Unbounded Probes

1.5.1 Planar area and volume from invariant test lines and planes

If the probes are motion-invariant, then the corresponding orientation parameters may be integrated out, and Eqs. (1.4.4)–(1.4.6) become,

$$\int L(Y \cap L_1^2)\, \mathrm{d}L_1^2 = \pi A, \tag{1.5.1}$$

$$\int A(Y \cap L_2^3)\, \mathrm{d}L_2^3 = 2\pi V, \tag{1.5.2}$$

$$\int L(Y \cap L_1^3)\, \mathrm{d}L_1^3 = 2\pi V, \tag{1.5.3}$$

respectively. The adjustments for motion-invariant stripes and slabs are analogous.

1.5.2 Planar curve length from invariant test lines

Let $Y \subset \mathbb{R}^2$ denote a curve of finite length B. The latter notation is chosen because the symbol "L" is reserved for intercept length, or for curve length in $\mathbb{R}^3$, and also because in stereology a planar curve often represents the boundary of a planar set. If a test line $L_1^2 \equiv L_1^2(p, \phi)$ hits the curve, then it will determine a finite number $I(Y \cap L_1^2)$ of intersections (equal to 3 in Fig. 1.5.1(a)). Our purpose is to compute the following integral,

$$\int I(Y \cap L_1^2)\, \mathrm{d}L_1^2, \tag{1.5.4}$$

namely the total measure of the number of intersections determined in Y by all the motion-invariant test lines hitting it. The preceding integral is really a double integral extended to the following domain of variation,

$$\{(p, \phi) : p \in (-\infty, \infty),\ \phi \in [0, \pi)\}. \tag{1.5.5}$$

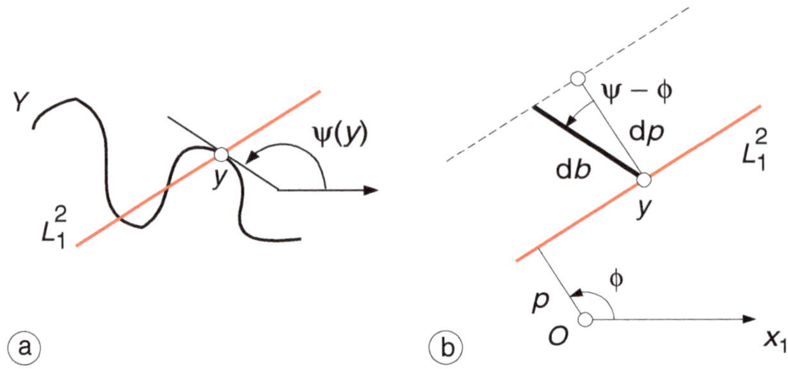

Fig. 1.5.1 (a) A test line L_1^2 determines three intersections in a curve Y. (b) Geometry for Eq. (1.5.6).

In the sequel, analogous domains of integration will be omitted whenever they can be easily deduced from the context. Note that the integral is always finite because Y is bounded and $I(\emptyset) = 0$. Let δy represent an essentially linear curve element of length $\mathrm{d}b$ at a point $y \in Y$, and let $\psi \in [0, \pi)$ denote the angle of δy with the axis of abscissas Ox_1. Whenever the test line hits the curve element we have $\mathrm{d}p = \mathrm{d}b|\cos(\psi - \phi)|$, see Fig. 1.5.1(b), and therefore,

$$\mathrm{d}L_1^2 = \mathrm{d}p\, \mathrm{d}\phi = \mathrm{d}b|\cos(\psi - \phi)|\, \mathrm{d}\phi. \tag{1.5.6}$$

Because

$$I(\delta y \cap L_1^2) = \begin{cases} 1, & \text{if } \delta y \cap L_1^2 \neq \emptyset, \\ 0, & \text{if } \delta y \cap L_1^2 = \emptyset, \end{cases} \tag{1.5.7}$$

the required invariant measure becomes,

$$\begin{aligned} \int I(Y \cap L_1^2)\, \mathrm{d}L_1^2 &= \int_Y \int I(\delta y \cap L_1^2)\, \mathrm{d}L_1^2 \\ &= \int_Y \int_{\delta y \cap L_1^2 \neq \emptyset} \mathrm{d}L_1^2 \\ &= \int_Y \mathrm{d}b \int_0^\pi |\cos(\psi - \phi)|\, \mathrm{d}\phi \\ &= 2B, \end{aligned} \tag{1.5.8}$$

namely twice the length of the curve, with no special assumptions about its shape other than piecewise smoothness.

Note that the set of local orientation angles $\{\psi(y),\ y \in Y\}$ is fixed for each $Y \subset \mathbb{R}^2$ and it does not influence the relevant integral because the probe density is motion-invariant. Thus, local orientation parameters will not be considered in the sequel.

1.5.3 Surface area from invariant lines and planes

Consider a surface $Y \subset \mathbb{R}^3$ of finite area S. If a test line $L_1^3 \equiv L_1^3(z, u)$ hits the surface, then it will determine a finite number $I(Y \cap L_1^3)$ of intersections (equal to 2 in Fig. 1.5.2(a)). Similarly as above we want to compute the measure of the total number of intersections determined in Y by all the motion-invariant test lines hitting it. Let δy represent an essentially planar surface element of area $\mathrm{d}s$ at a point $y \in Y$. Because the density $\mathrm{d}L_1^3$ is motion-invariant, the polar axis Ox_3 may be chosen to be the normal to the surface element δy, whereby $\mathrm{d}z = \mathrm{d}s|\cos\theta|$, see Fig. 1.5.2(b), and therefore,

$$\mathrm{d}L_1^3 = \mathrm{d}z\,\mathrm{d}u = \mathrm{d}s|\cos\theta|\,\mathrm{d}u. \tag{1.5.9}$$

The required invariant measure is the following triple integral,

$$\begin{aligned} \int I(Y \cap L_1^3)\,\mathrm{d}L_1^3 &= \int_Y \int I(\delta y \cap L_1^3)\,\mathrm{d}L_1^3 \\ &= \int_Y \int_{\delta y \cap L_1^3 \neq \emptyset} \mathrm{d}L_1^3 \\ &= \int_Y \mathrm{d}s \int_0^{2\pi} \mathrm{d}\phi \int_0^{\pi/2} \cos\theta \sin\theta\,\mathrm{d}\theta \\ &= \pi S. \end{aligned} \tag{1.5.10}$$

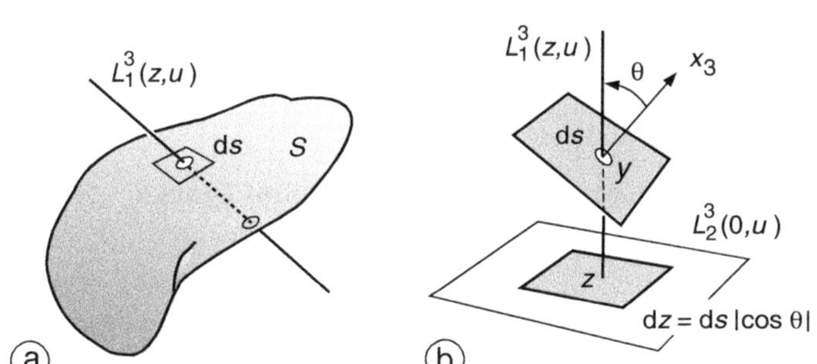

Fig. 1.5.2 (a) A test line L_1^3 determines two intersections in a surface. (b) Geometry for Eq. (1.5.9).

If the surface $Y \subset \mathbb{R}^3$ is hit by a test plane $L_2^3 \equiv L_2^3(p, u)$, then the corresponding intersection is a curve trace of total length $B(Y \cap L_2^3)$, and we want to compute the total measure of this trace length for all the motion-invariant test planes hitting the surface. As before, the polar axis Ox_3 may be chosen to be normal to δy. By Eq. (1.4.4) we have

$$\int_{\mathbb{R}} B(\delta y \cap L_2^3)\, \mathrm{d}p = \mathrm{d}s|\sin\theta|, \tag{1.5.11}$$

irrespective of the longitude angle ϕ, see Fig. 1.5.3(b). Therefore,

$$\begin{aligned} \int B(Y \cap L_2^3)\, \mathrm{d}L_2^3 &= \int_Y \int B(\delta y \cap L_2^3)\, \mathrm{d}L_2^3 \\ &= \int_Y \int_0^{2\pi} \mathrm{d}\phi \int_0^{\pi/2} \sin\theta\, \mathrm{d}\theta \int_{-\infty}^{\infty} B(\delta y \cap L_2^3)\, \mathrm{d}p \\ &= \int_Y \mathrm{d}s \int_0^{2\pi} \mathrm{d}\phi \int_0^{\pi/2} \sin^2\theta\, \mathrm{d}\theta \\ &= \frac{\pi^2}{2} S. \end{aligned} \tag{1.5.12}$$

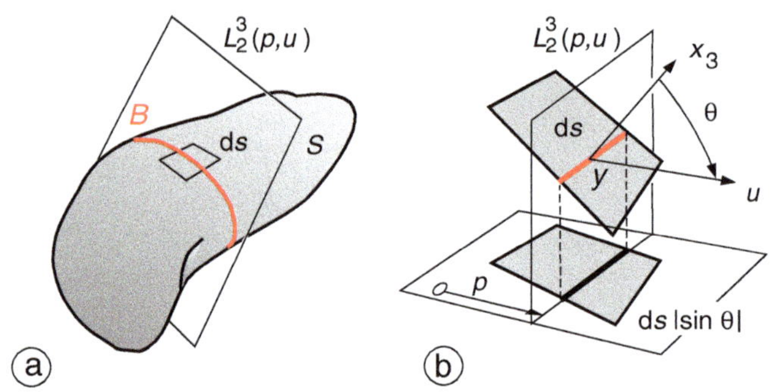

Fig. 1.5.3 (a) A test plane L_2^3 determines a trace curve of length B in a surface. (b) Geometry for Eq. (1.5.11).

1.5.4 Length of a spatial curve from invariant planes

Consider a curve $Y \subset \mathbb{R}^3$ of finite length L. A suitable probe is an invariant test plane $L_2^3 \equiv L_2^3(p, u)$, and the corresponding transect will consist of $Q(Y \cap L_2^3)$ intersection points. The notation $Q(\cdot)$ was adopted in stereology from the German term '*Querschnitt*' (namely planar cross section). We want to compute the measure of the total number of intersections determined in Y by all the motion-invariant test planes hitting it. Let δy represent an essentially linear curve element of length $\mathrm{d}l$ at a point $y \in Y$. The element δy may be taken to be parallel to the polar axis Ox_3, whereby $\mathrm{d}p = \mathrm{d}l|\cos\theta|$. Therefore,

$$\mathrm{d}L_2^3 = \mathrm{d}p\,\mathrm{d}u = \mathrm{d}l|\cos\theta|\,\mathrm{d}u. \tag{1.5.13}$$

Thus, the required invariant measure becomes

$$\begin{aligned}\int Q(Y \cap L_2^3)\,\mathrm{d}L_2^3 &= \int_Y \int Q(\delta y \cap L_2^3)\,\mathrm{d}L_2^3 \\ &= \int_Y \int_{\delta y \cap L_2^3 \neq \emptyset} \mathrm{d}L_2^3 \\ &= \int_Y \mathrm{d}l \int_0^{2\pi} \mathrm{d}\phi \int_0^{\pi/2} \cos\theta\sin\theta\,\mathrm{d}\theta \\ &= \pi L. \end{aligned} \tag{1.5.14}$$

It is not surprising that the factor 'π' appears here as it did in Eq. (1.5.10). In the latter case a planar surface element was hit by an invariant line, whereas now a linear curve element is hit by an invariant plane. In such cases one says that either result is dual of the other.

1.5.5 General Crofton formula for unbounded probes

The foregoing Crofton formulae are special cases of Eq. (1.3.4), see also Note 2 below. The relevant coefficient is

$$c_1(q, r, d) = c_{10}(r, d) \cdot \frac{O_d O_{q+r-d}}{O_q O_r}, \tag{1.5.15}$$

where

$$O_k = \frac{2\pi^{(k+1)/2}}{\Gamma((k+1)/2)}, \qquad k = 0, 1, \ldots, d, \tag{1.5.16}$$

is the surface area of the k-dimensional unit sphere $\mathbb{S}^k$, e.g., $O_0 = 2$, $O_1 = 2\pi$, $O_2 = 4\pi$, $O_3 = 2\pi^2$. On the other hand,

$$c_{10}(r, d) \equiv \int_{G_{r,d-r}} \mathrm{d}L_{r[0]}^d = \frac{O_{d-1}O_{d-2}\cdots O_{d-r}}{O_{r-1}O_{r-2}\cdots O_0}, \tag{1.5.17}$$

where $G_{r,d-r}$, called the Grassmannian in integral geometry, is the space of all non-oriented r-subspaces $L_{r[0]}^d \equiv L_r^d(0, u)$ acted upon by the group of rotations of $\mathbb{R}^d$. For instance, the coefficients in Eqs. (1.5.1)–(1.5.3) are

$$\begin{aligned} c_{10}(1,2) &= \pi, \\ c_{10}(2,3) &= c_{10}(1,3) = 2\pi, \end{aligned} \tag{1.5.18}$$

respectively.

The relevant coefficients obtained in Sections 1.5.2–1.5.4 are

$$\begin{aligned} c_1(1,1,2) &= 2, \\ c_1(2,1,3) &= c_1(1,2,3) = \pi, \\ c_1(2,2,3) &= \pi^2/2. \end{aligned} \tag{1.5.19}$$

1.5.6 Notes

1. Crofton formulae for unbounded probes

As indicated in Section 1.3.3, the term honours Morgan W. Crofton (1826–1915), who was the first to assign densities like $\mathrm{d}L_1^2 = \mathrm{d}p\,\mathrm{d}\phi$ to unbounded probes, thereby obtaining remarkable intersection formulae for convex sets with an emphasis on geometrical probabilities, see for instance Crofton (1868). More general results such as those given by Eqs. (1.5.8), (1.5.10), (1.5.12) and (1.5.14) were obtained independently by different authors during the 19th century – for a review, see for instance Hykšová, Kalousková, and Saxl (2012) and Cruz-Orive (2017).

2. Generalizations

The formal development of measure theory and integral geometry in the early 20th century revealed that the densities proposed by M. W. Crofton were indeed motion-invariant. General Crofton intersection formulae were thereby obtained for higher dimensions, and in spaces of constant curvature. For instance, the numerical coefficient in Eq. (1.5.15) is that of Eq. (14.69) from Santaló (1976), or of Eq. (6.7.11) from De-lin (1994).

1.6 Surface Area From Vertical Sections

1.6.1 The representation

Fix a vertical axis and hit a surface $Y \subset \mathbb{R}^3$ of area S with a vertical plane $L_{2\cdot v}^3(p,\phi)$, see Fig. 1.6.1(a). Then, hit the vertical curve trace $Y \cap L_{2\cdot v}^3$ with a sine-weighted straight line making an angle $\theta \in [0,\pi)$ with the vertical axis – namely with a line $L_{1\cdot v}^2(q,\pi-\theta) \subset L_{2\cdot v}^3(p,\phi)$ with density $\mathrm{d}L_{1\cdot v}^2 = \sin\theta\,\mathrm{d}q\,\mathrm{d}\theta$, see Eq. (1.2.16). Then, applying Eq. (1.2.15) to the Crofton Eq. (1.5.10), we obtain the following identity

$$\begin{aligned} \pi S &= \int I(Y \cap L_1^3)\, \mathrm{d}L_1^3 \\ &= \int \mathrm{d}L_{2\cdot v}^3 \int I\{(Y \cap L_{2\cdot v}^3) \cap L_{1\cdot v}^2\}\, \mathrm{d}L_{1\cdot v}^2, \end{aligned} \tag{1.6.1}$$

which is consistent with the fact that a sine-weighted test line$L_{1\cdot v}^2$ in a vertical plane is effectively an invariant test line in $\mathbb{R}^3$. It is useful, however, to express S as an integral with respect to the density $\mathrm{d}L_{2\cdot v}^3 = \mathrm{d}p\, \mathrm{d}\phi$ of the vertical plane. Let δy represent an essentially linear element of length $\mathrm{d}y$ at a point y of the vertical trace curve $Y \cap L_{2\cdot v}^3$, and let $\psi \equiv \psi(y) \in [0, \pi)$ denote the angle of δy with the vertical axis Ox_3, see Fig. 1.6.1(b). Then,

$$\mathrm{d}q = |\sin(\psi - \theta)|\, \mathrm{d}y, \tag{1.6.2}$$

see Fig. 1.6.1(c). Therefore, Eq. (1.6.1) becomes

$$\begin{aligned} \pi S &= \int \mathrm{d}L_{2\cdot v}^3 \int_{y \in Y \cap L_{2\cdot v}^3} \int I(\delta y \cap L_{1\cdot v}^2) \sin\theta\, \mathrm{d}q\, \mathrm{d}\theta \\ &= \int \mathrm{d}L_{2\cdot v}^3 \int_{y \in Y \cap L_{2\cdot v}^3} \mathrm{d}y \int_0^\pi |\sin(\psi - \theta)| \sin\theta\, \mathrm{d}\theta \\ &= \int W(Y \cap L_{2\cdot v}^3)\, \mathrm{d}L_{2\cdot v}^3, \end{aligned} \tag{1.6.3}$$

where

$$W(Y \cap L_{2\cdot v}^3) = \int_{y \in Y \cap L_{2\cdot v}^3} (\sin\psi + (\frac{\pi}{2} - \psi) \cos\psi)\, \mathrm{d}y(\psi), \tag{1.6.4}$$

is a functional defined on the vertical trace curve, namely the curvilinear integral of a function of the angle ψ between the tangent to the trace at a point y and the vertical axis, as the point y describes the trace.

1.6.2 Examples

As an illustration of Eq. (1.6.3) we consider the following two special cases.

Sphere

If Y is a sphere of radius R centred at the origin, then a vertical section $Y \cap L_{2\cdot v}^3(p, \phi)$ is a circle of radius $r(p) = \sqrt{R^2 - p^2}$, $p \in [-R, R]$. Moreover, the arc element of this circle is $\mathrm{d}y(\psi) = r(p)\, \mathrm{d}\psi$, whereby

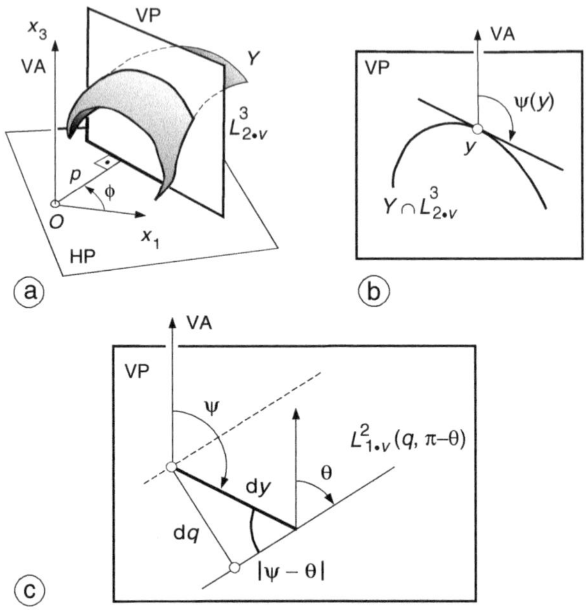

Fig. 1.6.1 (a) A vertical plane (VP) determines a trace curve, see (b), in a surface Y. (c) Geometry for Eq. (1.6.2).

$$W(Y \cap L^3_{2\cdot v}) = 2r(p) \int_0^\pi (\sin \psi + (\frac{\pi}{2} - \psi) \cos \psi)\, \mathrm{d}\psi = 8r(p), \tag{1.6.5}$$

the factor 2 stemming from the fact that the range of ψ is $[0, \pi)$ for the upper and the lower half circle. Now the rhs of Eq. (1.6.3) yields

$$8 \int_0^\pi \mathrm{d}\phi \int_{-R}^R \sqrt{R^2 - p^2}\, \mathrm{d}p = 4\pi^2 R^2 = \pi S, \tag{1.6.6}$$

as expected.

Curtain surface

A 'curtain surface' is a cylindrical surface generated by a vertical straight line segment (called the generator), of length L, as it moves along a curve Y_0 (called the directrix), of length B, lying in the horizontal plane Ox_1x_2. Thus the surface area of Y is $S = BL$. The trace $Y \cap L^3_{2\cdot v}$ typically consists of a finite union of say $I(Y_0 \cap L^3_{2\cdot v})$ vertical straight line segments of the same length L. Therefore, at any point $y \in Y \cap L^3_{2\cdot v}$ we have $\psi(y) = 0$, whereby

$$W(Y \cap L^3_{2\cdot v}) = \frac{\pi}{2} \int_{Y \cap L^3_{2\cdot v}} \mathrm{d}y = \frac{\pi}{2} L \cdot I(Y_0 \cap L^3_{2\cdot v}). \tag{1.6.7}$$

By virtue of the Crofton formula (1.5.8) for a planar curve hit by an invariant test line (namely by the straight line trace determined by $L^3_{2\cdot v}$ in the HP), Eq. (1.6.7) and Eq. (1.6.3) yield $\pi LB = \pi S$, as required.

1.6.3 Note

Eq. (1.6.3) and Eq. (1.6.4) are due to Baddeley (1985), who also considered curtain surfaces.

1.7 Formulae of Local Stereology

1.7.1 Volume from rays: the nucleator formulae

The direct nucleator

Consider a domain $Y \subset \mathbb{R}^2$ of area A in the plane. We want to represent A by means of a rotation-invariant ray probe $L^2_{1+}(\omega)$, $\omega \in [0, 2\pi)$ emanating from a fixed point O, see Eq. (1.2.17). Similarly as in Eq. (1.2.25), the area element in the plane is

$$\mathrm{d}x = [r\,\mathrm{d}r][\mathrm{d}\omega] = [r\,\mathrm{d}r][\mathrm{d}L^2_{1+}]. \tag{1.7.1}$$

To start with the simplest case, suppose that $O \in Y$ and Y is a star set with respect to O (this means that a ray joining O with any point in the boundary of Y is always connected). Define $l_+(\omega) = L(Y \cap L^2_{1+})$, the length of the radius vector of Y at an angle ω. Then, with the notation shown in Fig. 1.7.1(a),

$$\begin{aligned} A &= \int_Y \mathrm{d}x \\ &= \int_0^{2\pi} \mathrm{d}\omega \int_0^{l_+(\omega)} r\,\mathrm{d}r \\ &= \frac{1}{2} \int_0^{2\pi} l_+^2(\omega)\mathrm{d}\omega, \end{aligned} \tag{1.7.2}$$

which is the planar nucleator representation of A. Note that the preceding identity trivially holds for a disk centred at O.

If the set Y is not necessarily star-shaped, see Fig. 1.7.1(b), and the pivotal point O is not necessarily interior to Y, then whenever the ray $L^2_{1+}(\omega)$ hits Y, the corresponding intersection will in general consist of say $m(\omega) \geq 1$separate intercept

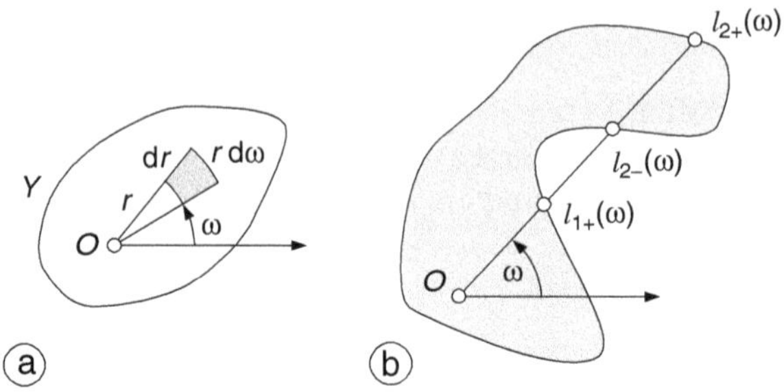

Fig. 1.7.1 (a) Notation used to derive the planar nucleator Eq. (1.7.2). (b) Case of a planar set which is not star-shaped with respect to O, see Eq. (1.7.4).

segments. The distances of the end points of these intercepts from O, arranged in increasing order of magnitude, may be denoted as follows,

$$\{l_{i-}(\omega), l_{i+}(\omega);\ i = 1, 2, \ldots, m(\omega)\}. \tag{1.7.3}$$

Then,

$$\begin{aligned} A &= \int_0^{2\pi} \mathrm{d}\omega \sum_{i=1}^{m(\omega)} \int_{l_{i-}(\omega)}^{l_{i+}(\omega)} r\mathrm{d}r \\ &= \int_0^{2\pi} f(\omega)\mathrm{d}\omega, \end{aligned} \tag{1.7.4}$$

where

$$f(\omega) = \frac{1}{2} \sum_{i=1}^{m(\omega)} (l_{i+}^2(\omega) - l_{i-}^2(\omega)), \tag{1.7.5}$$

with $f(\omega) = 0$ if the ray misses Y. Note that, if $O \in Y$, then $l_{1-}(\omega) = 0$ for all ω. If Y is star-shaped, then $m(\omega) = 1$.

For a domain $Y \subset \mathbb{R}^3$ of volume V we adopt a rotation-invariant ray probe $L_{1+}^3(u_2)$, $u_2 \in \mathbb{S}^2$ emanating from a fixed point O, see Eq. (1.2.18) and Fig. 1.7.2(a). Similarly as in Eq. (1.2.26), the volume element is,

$$\mathrm{d}x = [r^2\,\mathrm{d}r][\mathrm{d}u_2] = [r^2\,\mathrm{d}r][\mathrm{d}L_{1+}^3]. \tag{1.7.6}$$

In general, the transect $Y \cap L_{1+}^3$ may consist of $m(u_2) \geq 1$ connected intercepts. The distances of the end points of these intercepts from O, arranged in increasing order of magnitude, may be denoted as follows

$$\{l_{i-}(u_2), l_{i+}(u_2);\ i = 1, 2, \ldots, m(u_2)\}, \tag{1.7.7}$$

and analogously as in Eq. (1.7.4) the general three-dimensional nucleator identity reads as follows,

$$V = \int_{\mathbb{S}^2} f(u_2)\, \mathrm{d}u_2, \tag{1.7.8}$$

where

$$f(u_2) = \frac{1}{3} \sum_{i=1}^{m(u_2)} (l_{i+}^3(u_2) - l_{i-}^3(u_2)), \tag{1.7.9}$$

with $f(u_2) = 0$ if the ray misses Y. In Fig. 1.7.2(a), $O \in Y$ and $l_{1-}(u_2) = 0$.

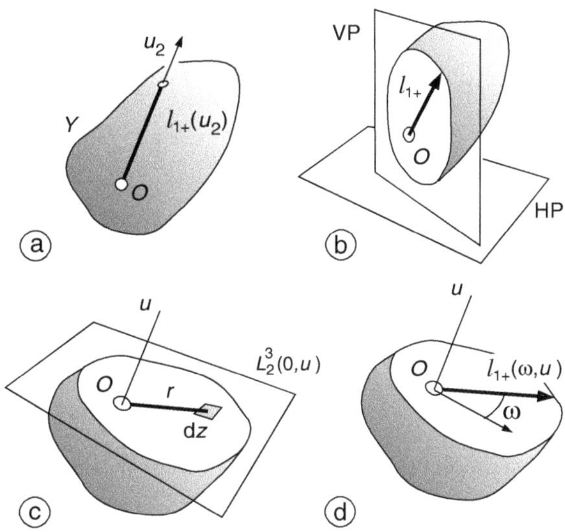

Fig. 1.7.2 (a) The direct nucleator, Eq. (1.7.8). (b) The vertical nucleator. (c) The integrated nucleator, Eq. (1.7.13). (d) The pivotal nucleator, Eq. (1.7.14).

The vertical nucleator

The nucleator Eq. (1.7.8) also holds if the ray $L_{1+}^3(u_2)$ is generated as a ray $L_{1\cdot v+}^2(\theta)$ emanating from a fixed point O within a vertical plane at an angle θ with the VA. In this case, recalling Eq. (1.2.16), $\mathrm{d}L_{1\cdot v+}^2 = \sin\theta\, \mathrm{d}\theta$. In Fig. 1.7.2(b), $l_{1+} = L(Y \cap L_{1\cdot v+}^2)$.

The integrated nucleator

Instead of generating a rotation-invariant ray probe L_{1+}^3 directly in space, as above, in practice it is convenient to generate first a pivotal plane $L_{2[0]}^3$ through O, and then a rotation-invariant ray L_{1+}^2 emanating from O in that plane. Based on Eq. (1.2.11), next we show that both constructions are equivalent, giving rise to the integrated and the pivotal nucleator, as alternatives to the direct nucleator, see Fig. 1.7.2(c),(d), respectively.

Consider a rotation-invariant plane $L_{2[1]}^3 \equiv L_{2[1]}^3(\varphi)$ containing the support axis $L_{1[0]}^3$ of the ray $L_{1+}^3(u_2)$ used above for the direct nucleator, and an invariant point $x \in \mathbb{R}^3$. For the domain $Y \subset \mathbb{R}^3$ of volume V consider the following identity,

$$\pi V = \int_0^{\pi} \mathrm{d}\varphi \int_Y \mathrm{d}x. \tag{1.7.10}$$

The application of Eq. (1.2.26), and then Eq. (1.2.11), yields

$$\begin{aligned} \mathrm{d}\varphi\, \mathrm{d}x &= \mathrm{d}L_{2[1]}^3\, \mathrm{d}L_{1[0]}^3 r^2\, \mathrm{d}r \\ &= \mathrm{d}L_{2[0]}^3\, \mathrm{d}L_{1[0]}^2 r^2\, \mathrm{d}r. \end{aligned} \tag{1.7.11}$$

In this new scenario, $L_{l[0]}^2 \equiv L_1^2(0, \omega)$, say, is a rotation-invariant axis within a pivotal plane $L_{2[0]}^3 \equiv L_2^3(0, u)$. Thus,

$$\begin{aligned} \mathrm{d}\varphi\, \mathrm{d}x &= [r\, \mathrm{d}u]\, [r\, \mathrm{d}r\, \mathrm{d}\omega] \\ &= r\, \mathrm{d}u\, \mathrm{d}z, \end{aligned} \tag{1.7.12}$$

where z is an invariant point in the pivotal plane because $\mathrm{d}z = r\, \mathrm{d}r\, \mathrm{d}\omega$. The integrated nucleator identity becomes

$$V = \frac{1}{\pi} \int_{\mathbb{S}_+^2} \mathrm{d}u \int_{Y \cap L_2^3(0,u)} r\, \mathrm{d}z, \tag{1.7.13}$$

see Fig. 1.7.2(c).

The pivotal nucleator

For simplicity assume that the domain $Y \subset \mathbb{R}^3$ is star-shaped with respect to a fixed origin $O \in Y$. In the integrated nucleator Eq. (1.7.13), return to the expression $\mathrm{d}z = r\, \mathrm{d}r\, \mathrm{d}\omega$ and set $l_+(\omega, u) \equiv L(Y \cap L_{1+}^2(\omega))$, where $L_{1+}^2(\omega)$ is a rotation-invariant ray emanating from O within the pivotal plane $L_2^3(0, u)$. Then,

$$V = \frac{1}{\pi}\int_{\mathbb{S}_+^2} \mathrm{d}u \int_0^{2\pi} \mathrm{d}\omega \int_0^{l_+(\omega,u)} r^2\,\mathrm{d}r$$
$$= \frac{1}{3\pi}\int_{\mathbb{S}_+^2} \mathrm{d}u \int_0^{2\pi} l_+^3(\omega,u)\,\mathrm{d}\omega, \qquad (1.7.14)$$

which is the pivotal nucleator representation of V, see Fig. 1.7.2(d). Thus, while the direct nucleator is based on a ray $L_{1+}^3(u_2)$, $u_2 \in \mathbb{S}^2$, generated directly in space, see Eq. (1.7.8), the pivotal nucleator is based on a two-step procedure; in the first step a pivotal plane is generated, and in the second a rotation-invariant ray is generated within the pivotal plane. The generalization of the pivotal nucleator to an arbitrarily shaped set Y, with the pivotal point not necessarily contained in Y, is immediate on replacing $l_+^3(\omega,u)$ with a summation analogous to that in Eq. (1.7.9).

1.7.2 Volume from coaxial half-planes: the Pappus–Guldin formula

Definitions

The target is the volume V of a domain $Y \subset \mathbb{R}^3$. Fix an orthogonal frame $Ox_1x_2x_3$. A point $x \in \mathbb{R}^3$ not belonging to the reference axis Ox_3 is contained in a unique half-plane $L_{2+}^3 \equiv L_{2+}^3(\phi)$, $\phi \in [0,2\pi)$ emanating from this axis, see Fig. 1.7.3. The cylindrical coordinates of x are

$$(r,\phi,x_3),\ r \in [0,\infty),\ \phi \in [0,2\pi),\ x_3 \in (-\infty,\infty), \qquad (1.7.15)$$

where x_3 denotes the ordinate of x, and r the distance of x from the reference axis Ox_3. The support of this distance is a unique ray $L_{1+}^2(x_3,\phi) \subset L_{2+}^3(\phi)$.

The intercepts representation

The volume element at the point x may be expressed as follows,

$$\mathrm{d}x = r\,\mathrm{d}\phi\,\mathrm{d}r\,\mathrm{d}x_3, \qquad (1.7.16)$$

see Fig. 1.7.3(a), whereby

$$V = \int_Y \mathrm{d}x$$
$$= \int_0^{2\pi} \mathrm{d}\phi \int_{-\infty}^{\infty} \mathrm{d}x_3 \int_{Y\cap L_{1+}^2(x_3,\phi)} r\,\mathrm{d}r$$
$$= \frac{1}{2}\int_0^{2\pi} \mathrm{d}\phi \int_{-\infty}^{\infty} \sum_{i=1}^{m} (l_{i+}^2 - l_{i-}^2)\,\mathrm{d}x_3, \qquad (1.7.17)$$

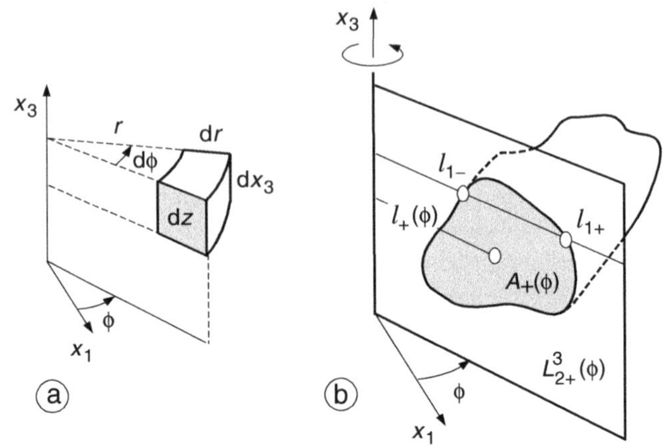

Fig. 1.7.3 (a) Cylindrical coordinates, see Eq. (1.7.15). (b) Geometry of a coaxial plane, used to obtain Eq. (1.7.17) and Eq. (1.7.19).

where $m \equiv m(x_3, \phi)$ denotes the number of connected intercept segments of the transect $Y \cap L^2_{1+}(x_3, \phi)$, whereas $l_{i+} \equiv l_{i+}(x_3, \phi)$ and $l_{i-} \equiv l_{i-}(x_3, \phi)$ are defined similarly as in the nucleator case.

The Pappus–Guldin formula

The volume element at the point x may also be expressed as follows,

$$\mathrm{d}x = r\,\mathrm{d}\phi\,\mathrm{d}r\,\mathrm{d}x_3 = r\,\mathrm{d}\phi\,\mathrm{d}z, \tag{1.7.18}$$

where $\mathrm{d}z = \mathrm{d}r\,\mathrm{d}x_3$ is the area element in the half-plane $L^3_{2+}(\phi)$, because the length elements $\mathrm{d}r$ and $\mathrm{d}x_3$ are mutually orthogonal, see Fig. 1.7.3(a). Then the volume of Y may be expressed by the general Pappus–Guldin formula, namely,

$$\begin{aligned} V &= \int_Y \mathrm{d}x \\ &= \int_0^{2\pi} \mathrm{d}\phi \int_{Y \cap L^3_{2+}(\phi)} r\mathrm{d}z \\ &= \int_0^{2\pi} l_+(\phi) \cdot A_+(\phi)\,\mathrm{d}\phi, \end{aligned} \tag{1.7.19}$$

where $A_+(\phi) \equiv A(Y \cap L^3_{2+}(\phi))$, with $A(\emptyset) = 0$, denotes the area of the intersection between Y and the half-plane $L^3_{2+}(\phi)$, whereas

$$l_+(\phi) = \frac{1}{A_+(\phi)} \int_{Y \cap L^3_{2+}(\phi) \neq \emptyset} r \, \mathrm{d}z \tag{1.7.20}$$

is the distance of the centroid (namely of the centre of mass) of a non-void intersection from the reference axis Ox_3, see Fig. 1.7.3(b). Note that, unless Y lies entirely in the half-space $x_2 > 0$, say, the mentioned intersection does not need to be a complete planar transect of Y.

Solids and surfaces of revolution

The classical Pappus–Guldin formula for volume refers to a solid of revolution Y_3 generated by a planar set $Y_2 \subset L^3_{2+}(0)$ of area A, as $L^3_{2+}(\phi)$ revolves around Ox_3 from $\phi = 0$ to $\phi = 2\pi$, in which case Eq. (1.7.19) simplifies into

$$V = 2\pi l A, \tag{1.7.21}$$

where V represents the volume of Y_3 and l the distance of the centroid of Y_2 from the axis of revolution Ox_3.

Similarly, let ∂Y_3 represent a surface of revolution – such as the boundary of the solid of revolution Y_3 – generated by a planar curve $\partial Y \subset L^3_{2+}(0)$ of length B. Then,

$$S = 2\pi l B, \tag{1.7.22}$$

where l now represents the distance of the centroid of ∂Y_2 from the axis of revolution Ox_3. In fact, the area $\mathrm{d}s(y)$ of a surface element δy at a point $y \in \partial Y_3 \cap L^3_{2+}(\phi)$ may be expressed as follows,

$$\mathrm{d}s(y) = r \, \mathrm{d}\phi \, \mathrm{d}b(y), \tag{1.7.23}$$

where $\mathrm{d}b(y)$ is the length of the trace element $\delta y \cap L^3_{2+}(\phi)$ (which does not depend on ϕ). Therefore,

$$S = \int_{y \in \partial Y_3} \mathrm{d}s(y) = \int_0^{2\pi} \mathrm{d}\phi \int_{y \in \partial Y_2} r \, \mathrm{d}b(y) = 2\pi l B, \tag{1.7.24}$$

where

$$l = \frac{1}{B} \int_{y \in \partial Y_2} r \, \mathrm{d}b(y). \tag{1.7.25}$$

No general Pappus–Guldin formula analogous to Eq. (1.7.19) exists for the surface area of the boundary ∂Y of Y, however, because the surface element can generally not be expressed in terms of parameters defined solely on the intersection $\partial Y \cap L^3_{2+}(\phi)$.

1.7.3 Surface area with the surfactor

Consider a closed smooth surface $\partial Y \subset \mathbb{R}^3$ of area S and a fixed pivotal point O. We first assume that ∂Y represents the boundary of a smooth convex set $Y \subset \mathbb{R}^3$, and also that $O \in Y^\circ$, see Fig. 1.7.4(a). A candidate probe is a ray $L_{1+}^3(u_2)$ emanating from O in a vector direction $u_2 \in \mathbb{S}^2$. This ray meets the surface at a point $y \in \partial Y$ and makes an angle $\psi \in [0, \pi/2)$ with the normal to ∂Y at y, see Fig. 1.7.4(a). Let $\mathrm{d}s$ denote the area of the surface element $\delta y \subset \partial Y$ at y. The orthogonal projected area of δy onto a plane normal to the ray is

$$\mathrm{d}s \cos\psi = l_+^2(u_2)\,\mathrm{d}u_2, \tag{1.7.26}$$

where $l_+(u_2) = L(Y \cap L_{1+}^3(u_2))$. Therefore,

$$S = \int_{\partial Y} \mathrm{d}s = \int_{\mathbb{S}^2} \frac{l_+^2(u_2)}{\cos\psi}\,\mathrm{d}u_2. \tag{1.7.27}$$

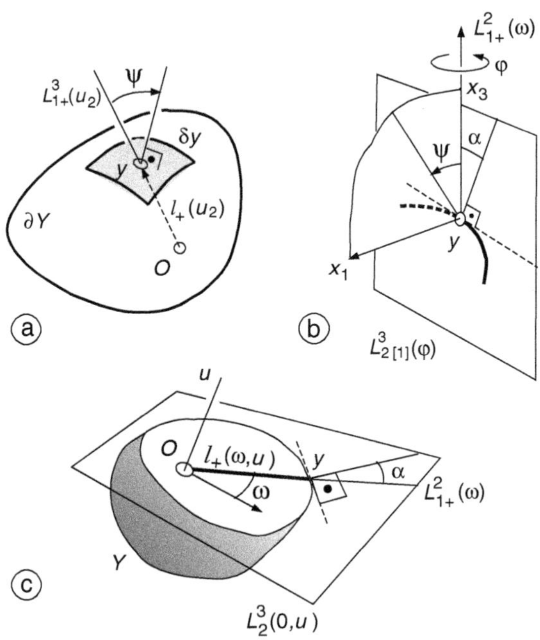

Fig. 1.7.4 (a) Geometry of the surfactor principle, Eq. (1.7.27). (b) Notation used in Eq. (1.7.28). (c) Definitions used in the surfactor Eq. (1.7.36).

Unfortunately, the angle ψ is not directly observable. To proceed, we use the fact that the ray L_{1+}^3 is equivalent to a rotation-invariant ray $L_{1+}^2(\omega)$ emanating from O within a pivotal plane $L_{2[0]}^3 \equiv L_2^3(0,u)$, as we have shown – see Eq. (1.7.14) and the ensuing text. Then,

$$S = \int_{\mathbb{S}_+^2} \mathrm{d}u \int_0^{2\pi} l_+^2(\omega,u) f(\alpha)\,\mathrm{d}\omega, \tag{1.7.28}$$

where $l_+(\omega,u) = L(Y \cap L_{1+}^2(\omega))$ and $\alpha \in (-\pi/2, \pi/2)$ denotes the angle between the ray $L_{1+}^2(\omega)$ and the normal to the pivotal trace $\partial Y \cap L_{2[0]}^3$ at the point $y = \partial Y \cap L_{1+}^2(\omega)$, see Fig. 1.7.4(b). On the other hand, $f(\alpha)$ is a function that, for each fixed pair (ω, u), satisfies the following identity,

$$\int_0^{\pi} f(\alpha(\varphi))\,\mathrm{d}\varphi = \frac{1}{\cos\psi}, \tag{1.7.29}$$

where φ is the angle of a rotation-invariant plane $L_{2[1]}^3(\varphi)$ containing the ray $L_{1+}^2(\omega)$. In order to identify $f(\alpha)$ we adopt a local orthogonal trihedron $yx_1x_2x_3$ with centre at the point y. The axis yx_3 is taken along the ray $L_{1+}^2(\omega)$, whereas the plane yx_1x_3 contains the normal to the surface at y. The unit vector of the latter normal is $(\sin\psi, 0, \cos\psi)$, whereas the unit vector of the tangent to the pivotal trace curve at the point y is $(\cos\alpha\cos\varphi, \cos\alpha\sin\varphi, \sin\alpha)$. Because both vectors are mutually perpendicular their dot product must be zero, whereby we obtain the identity

$$|\tan\alpha| = |\cos\varphi|\tan\psi. \tag{1.7.30}$$

Set $t = |\tan\alpha|$ and $x = \tan\psi$, whereby the preceding identity becomes

$$t = x|\cos\varphi|, \tag{1.7.31}$$

where x is a constant for each fixed pair (ω, u). Now Eq. (1.7.29) becomes an integral equation of the following form,

$$\int_0^{\pi} h(t(\varphi))\,\mathrm{d}\varphi = g(x), \tag{1.7.32}$$

where $g(x) = \sqrt{1+x^2}$. Thus, recalling Eq. (1.7.31),

$$\begin{aligned} \int_0^x h(t)\left|\frac{\mathrm{d}\varphi}{\mathrm{d}t}\right|\mathrm{d}t &= g(x), \\ \int_0^x \frac{h(t)}{\sqrt{x^2-t^2}}\,\mathrm{d}t &= g(x), \end{aligned} \tag{1.7.33}$$

and the latter is an Abel integral equation whose solution is

$$
\begin{aligned}
h(t) &= \frac{2}{\pi} \cdot \frac{\mathrm{d}}{\mathrm{d}t} \int_0^t \frac{x g(x)}{\sqrt{t^2 - x^2}} \, \mathrm{d}x \\
&= \frac{2}{\pi} \left(1 + t \sin^{-1} \left(t / \sqrt{1 + t^2} \right) \right).
\end{aligned}
\tag{1.7.34}
$$

Recalling that $t = |\tan \alpha|$, the required function is

$$
f(\alpha) = \frac{1}{\pi} (1 + \alpha \tan \alpha), \quad \alpha \in (-\pi/2, \pi/2),
\tag{1.7.35}
$$

which substituted into the rhs of Eq. (1.7.28) yields

$$
S = \frac{1}{\pi} \int_{\mathbb{S}_+^2} \mathrm{d}u \int_0^{2\pi} l_+^2(\omega, u)(1 + \alpha \tan \alpha) \, \mathrm{d}\omega,
\tag{1.7.36}
$$

namely the (pivotal) surfactor representation of S for a smooth convex set, see Fig. 1.7.4(c). Note that the angle α depends on (ω, u). As in the pivotal nucleator case, the extension to a non-convex set with smooth boundary is straightforward.

1.7.4 Point particle number from local slabs

Let

$$
Y = \bigcup_{i=1}^N y_i \subset \mathbb{R}^3,
\tag{1.7.37}
$$

denote a finite union of N separate point particles in a bounded region of space. Here we derive preliminary results, to be exploited in Section 2.18.5, for a slab containing a fixed axis, or a fixed point.

Slab with its midplane containing a fixed axis

Consider a slab probe $L^3_{2[1],t} \equiv L_t(\phi)$, say, of thickness $t > 0$, whose midplane contains a fixed axis, e.g. the vertical polar axis Ox_3, with the rotation-invariant density given by Eq. (1.2.20). Let $Q(Y \cap L_t(\phi))$ represent the number of point particles captured by the slab. We have,

$$
\begin{aligned}
\int_0^\pi Q(Y \cap L_t(\phi)) \, \mathrm{d}\phi &= \sum_{i=1}^N \int_0^\pi Q(y_i \cap L_t(\phi)) \, \mathrm{d}\phi \\
&= \sum_{i=1}^N \alpha_i,
\end{aligned}
\tag{1.7.38}
$$

where

$$\alpha_i = \begin{cases} 2\sin^{-1}\left(\dfrac{t}{2h_i}\right) & \text{if } t < 2h_i, \\ \pi & \text{if } t \geq 2h_i, \end{cases} \tag{1.7.39}$$

and h_i is the (horizontal) distance of the ith point particle y_i from the polar axis, see Fig. 1.7.5. Unlike Eq. (1.4.8), the preceding identity cannot be exploited because it involves all the unknown distances $\{h_i\}$. An effective twist is described in Section 2.18.5.

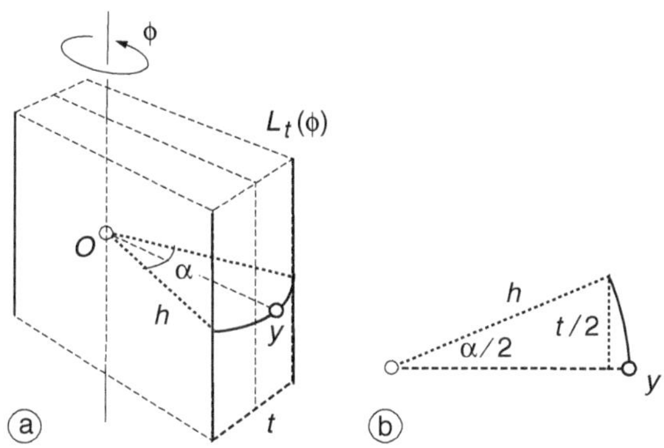

Fig. 1.7.5 (a) A slab free to rotate about a fixed axis, used as a probe to capture point particles such as the one denoted by y. (b) Geometry used in Eq. (1.7.39).

Fixed point slab

Consider an oriented slab probe $L^*_{2[0],t}(u_2) \equiv L^*_t(u_2)$, say, of thickness $t > 0$, with its midplane containing a fixed origin $O \in \mathbb{R}^3$, and with the rotation-invariant density given by Eq. (1.2.18). We have,

$$\begin{aligned} \int_{\mathbb{S}^2} Q(Y \cap L^*_t(u_2))\, du_2 &= \sum_{i=1}^{N} \int_{\mathbb{S}^2} Q(y_i \cap L^*_t(u_2))\, du_2 \\ &= \sum_{i=1}^{N} \omega_i, \end{aligned} \tag{1.7.40}$$

where

$$\omega_i = \begin{cases} 2\pi t/d_i & \text{if } t < 2d_i, \\ 4\pi & \text{if } t \geq 2d_i, \end{cases} \tag{1.7.41}$$

and d_i is the distance of the ith point particle y_i from the origin O, see Fig. 1.7.6. To show the preceding result, note that the slab $L^*_t(u_2)$ captures the point particle

y_i, namely $Q(y_i \cap L_t^*(u_2)) = 1$, whenever the midplane of the slab hits a ball of diameter t centred at y_i. This hitting event takes place whenever the vector (d_i, u_2), of modulus d_i, normal to the slab, describes a spherical zone of height t and area $2\pi t d_i$ or, equivalently, when the direction u_2 varies in a solid angle of measure $2\pi t d_i / d_i^2 = 2\pi t / d_i$ if $t < 2d_i$, and 4π if $t \geq 2d_i$. Again, the preceding identity cannot be exploited either because it involves all the unknown distances $\{d_i\}$, but see Section 2.18.5.

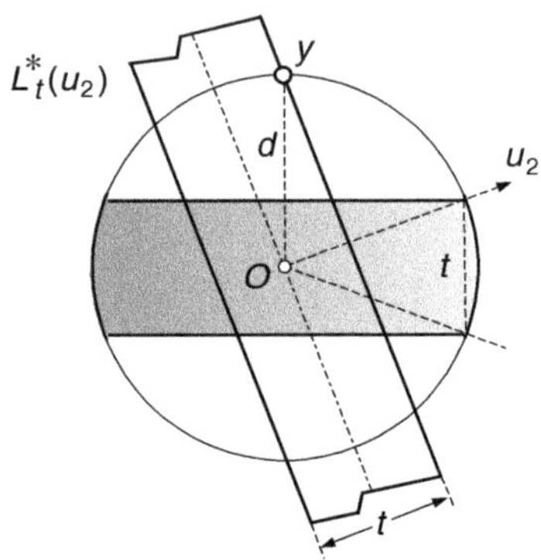

Fig. 1.7.6 (a) A slab free to rotate about a fixed point, illustrating the geometry used to obtain Eq. (1.7.41).

1.7.5 Notes

1. The direct nucleator

The nucleator Eq. (1.7.8) is a classical result of elementary calculus. The practical implementation, and the name, are due to Gundersen (1988). For a convex set the result had long been known, see for instance Santaló (1976, Eq. (12.63)). For a simple derivation of Eq. (1.7.9) see Cruz-Orive (1987a).

2. The integrated nucleator

The representation given by Eq. (1.7.13) was proposed in Jensen (1991, 1998) and applied by Hansen, Nyengaard, Andersen, and Jensen (2011).

3. The pivotal nucleator

This nucleator version, see Eq. (1.7.14), was the one adopted by Gundersen (1988), see Cruz-Orive and Gual-Arnau (2015).

4. Volume from coaxial planes

Eq. (1.7.17) and Eq. (1.7.19) were published independently by Cruz-Orive and Roberts (1993) and by Jensen and Gundersen (1993).

5. The surfactor

The surface area representation of Eq. (1.7.36) is due to Jensen and Gundersen (1987, 1989).

1.8 Surface Area and Volume With the Invariator

1.8.1 Invariator equations with a posteriori weighting

Consider a domain $Y \subset \mathbb{R}^3$ of volume V, with piecewise smooth boundary ∂Y of area S. Hit Y first with a pivotal plane $L^3_{2[0]}$, and then hit the resulting trace curve $\partial Y \cap L^3_{2[0]}$ with a motion-invariant test line $L^2_1(r, \omega) \subset L^3_{2[0]}$, $r > 0$, $\omega \in [0, 2\pi)$, whose density $\mathrm{d}L^2_1 = \mathrm{d}r\,\mathrm{d}\omega$ is weighted a posteriori by the factor r, see Fig. 1.8.1(a). Then, replacing the density $\mathrm{d}L^3_1$ with the rhs of Eq. (1.2.24) in the Crofton Eq. (1.5.10) and Eq. (1.5.3) we obtain

$$S = \frac{1}{\pi} \int \mathrm{d}L^3_{2[0]} \int I\{(\partial Y \cap L^3_{2[0]}) \cap L^2_1\} r\,\mathrm{d}L^2_1, \tag{1.8.1}$$

$$V = \frac{1}{2\pi} \int \mathrm{d}L^3_{2[0]} \int L\{(Y \cap L^3_{2[0]}) \cap L^2_1\} r\,\mathrm{d}L^2_1, \tag{1.8.2}$$

respectively, because the r-weighted test line L^2_1 is effectively an invariant test line in $\mathbb{R}^3$. The outer integrals are over the unit hemisphere $\mathbb{S}^2_+$, whereas, for each direction $u \in \mathbb{S}^2_+$ of the pivotal plane $L^3_{2[0]} \equiv L^3_2(0, u)$, the inner integrals are extended to the parameter space

$$\{(r, \omega) : \; r \in [0, \infty), \; \omega \in [0, 2\pi)\}. \tag{1.8.3}$$

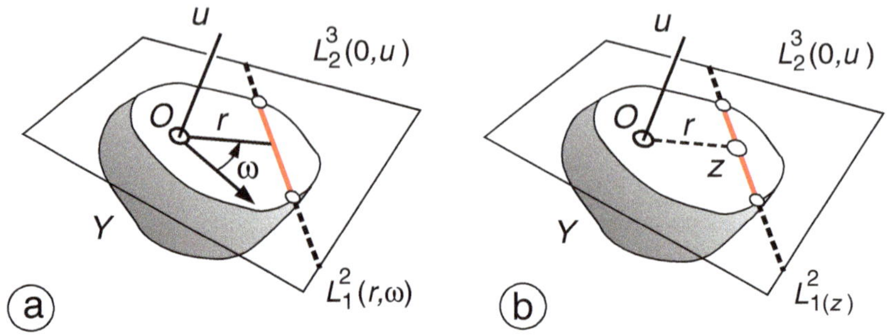

Fig. 1.8.1 (a) Invariator test line, which is effectively motion-invariant in space, generated as a motion-invariant test line $L_1^2(r, \omega)$ in a pivotal plane, whose density $\mathrm{d}L_1^2$ is weighted a posteriori by the factor r. (b) Invariator test line $L^2_{1(z)}$, also equivalent to a motion-invariant test line in space, weighted a priori by the factor r because it is drawn through a translation-invariant point z in the pivotal plane.

1.8.2 Invariator equations with a priori weighting

Choose a point $z \in L^3_{2[0]}$ with invariant density dz in a pivotal plane, and draw a line $L^2_{1(z)} \subset L^3_{2[0]}$ through z and perpendicular to the axis joining z with the fixed pivotal point O, see Fig. 1.8.1(b). Substitution of $\mathrm{d}L_1^3$ with Eq. (1.2.23) into the Crofton Eq. (1.5.10) and Eq. (1.5.3) yields,

$$S = \frac{1}{\pi} \int \mathrm{d}L^3_{2[0]} \int I\{(\partial Y \cap L^3_{2[0]}) \cap L^2_{1(z)}\} \,\mathrm{d}L^2_{1(z)}, \tag{1.8.4}$$

$$V = \frac{1}{2\pi} \int \mathrm{d}L^3_{2[0]} \int L\{(Y \cap L^3_{2[0]}) \cap L^2_{1(z)}\} \,\mathrm{d}L^2_{1(z)}, \tag{1.8.5}$$

respectively. The outer integrals are over the unit hemisphere $\mathbb{S}^2_+$, whereas, for each direction $u \in \mathbb{S}^2_+$ of the pivotal plane $L^3_{2[0]} \equiv L^3_2(0, u)$, the inner integrals are extended to $z \in L^3_{2[0]}$.

1.8.3 Surface area of a convex set: the flower formula

As in the vertical sections representation, for the applications it is useful to elaborate the inner integral in Eq. (1.8.4) into convenient functionals defined on the pivotal section.

The flower formula requires that the set $Y \subset \mathbb{R}^3$ is convex. The pivotal point O, however, does not need to be interior to Y. If a weighted test line $L^2_{1(z)} \subset L^3_2(0, u)$ hits the pivotal trace curve $C_u \equiv \partial Y \cap L^3_2(0, u)$ then, typically, $I(C_u \cap L^2_{1(z)}) = 2$ – otherwise $I(C_u \cap L^2_{1(z)}) = 0$, see Fig. 1.8.2(a). Further, recall that the invariant density of $L^2_{1(z)}$ is that of the point z, namely $\mathrm{d}z$. Thus, if Y is convex then Eq. (1.8.4) simplifies into

$$S = \frac{2}{\pi} \int_{\mathbb{S}^2_+} \mathrm{d}u \int_{C_u \cap L^2_{1(z)} \neq \emptyset} \mathrm{d}z. \tag{1.8.6}$$

The inner integral in the rhs of the preceding identity is the area of the geometric locus of the point z as the test line $L^2_{1(z)}$ hits the pivotal section, which in this case is a planar convex set. This geometric locus is generally a domain called the support set (because it is the region enclosed by the support function) or the 'flower' of the pivotal section, see Fig. 1.8.2(b). The graph of the support function $h(\omega)$ of a planar convex set K is the locus of the foot of the perpendicular from O to a variable tangent to the boundary ∂K of the set, namely the pedal curve of ∂K with respect to O. More precisely,

$$h(\omega) = \max(x \cdot e_\omega : x \in K), \quad \omega \in [0, 2\pi), \tag{1.8.7}$$

where e_ω represents a unit vector making an angle ω with the Ox_1 axis, x is a vector joining the origin O with a point of K, and '$\cdot$' denotes the scalar or dot product. Thus, the length of the radius vector of the flower H_K of K is precisely $h(\omega)$. With the flower concept, Eq. (1.8.6) becomes

$$S = \frac{2}{\pi} \int_{\mathbb{S}^2_+} A(H_{C_u})\, \mathrm{d}u, \tag{1.8.8}$$

which is the integral-geometric version of the flower formula for the surface area of a convex set.

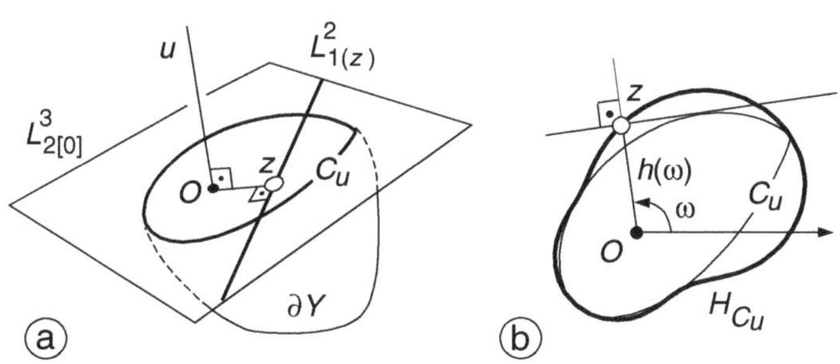

Fig. 1.8.2 (a) A priori weighted invariator test line $L^2_{1(z)}$ hitting the convex pivotal trace C_u of the boundary ∂Y of a convex body. (b) Construction of the flower of C_u.

Additional geometry for the flower of a convex set K

If a planar convex set K contains a fixed point O, namely if $O \in K^\circ \subset \mathbb{R}^2$, then the support function of ∂K satisfies $h(\omega) \geq 0$ for all ω, and the corresponding flower area is

$$A(H_K) = \frac{1}{2}\int_0^{2\pi} h^2(\omega)\,\mathrm{d}\omega, \tag{1.8.9}$$

because the area element of the flower is the area of a right triangle of leg lengths $h(\omega)$ and $h(\omega)\,\mathrm{d}\omega$ respectively, namely $\frac{1}{2}h^2(\omega)$. On the other hand, if $O \notin K$, then

$$h(\omega) = x_0 \cos\omega + y_0 \sin\omega + h_0(\omega), \quad \omega \in [0, 2\pi), \tag{1.8.10}$$

where $(x_0, y_0) \in K^\circ$ are the Cartesian coordinates (with respect to the origin O) of an arbitrary point interior to K, and $h_0(\omega)$ is the support function of K with respect to that point. Now the flower area is

$$A(H_K) = \frac{1}{2}\int_0^{2\pi} h(\omega)|h(\omega)|\,\mathrm{d}\omega, \tag{1.8.11}$$

because, in this case, $h(\omega)$ will in general be negative in some interval of ω.

Suppose that $O \subset K^\circ$, and that the boundary ∂K is of class C^2, admitting the following parametric equations,

$$\partial K = \{(X,Y) \in \mathbb{R}^2 : X = X(t),\ Y = Y(t),\ (0 \le t < 2\pi)\}. \tag{1.8.12}$$

Note that the parameter t does not need to be the polar angle of the point (X,Y). Then, the parametric equations of the support function h of ∂K, namely of the flower boundary ∂H_K, are the following,

$$\begin{cases} x(t) = -\dfrac{Y'(t)}{X'(t)} \cdot y(t), \\ y(t) = \dfrac{X'(t)Y(t) - X(t)Y'(t)}{X'^2(t) + Y'^2(t)} \cdot X'(t), \quad (0 \le t < 2\pi). \end{cases} \tag{1.8.13}$$

To see this, recall that ∂H_K is the geometric locus of the foot $(x(t), y(t))$ of the perpendicular N from O to the tangent T to ∂Kat the point $(X(t), Y(t))$. Thus,

$$\begin{cases} T:\ y(t) - Y(t) = \dfrac{Y'(t)}{X'(t)}(x(t) - X(t)), \\ N:\ y(t) = -\dfrac{X'(t)}{Y'(t)}x(t), \quad (0 \le t < 2\pi). \end{cases} \tag{1.8.14}$$

Solving for $(x(t), y(t))$, Eq. (1.8.13) is obtained.

In the first Eq. (1.8.14) replace $(x(t), y(t))$ with (x, y), and $Y'(t)/X'(t)$ with $-x/y$ by the second Eq. (1.8.14). We obtain,

$$x^2 + y^2 - X(t) \cdot x - Y(t) \cdot y = 0, \tag{1.8.15}$$

which is the equation of a family of circles depending on a parameter t, the endpoints of their diameters being O and $(X(t), Y(t)) \in \partial K$. This shows that the flower may be represented as the union of the disks of diameter Oz, $z \in \partial K$, namely

$$H_K = \bigcup_{z \in \partial K} D_2(z/2, \|z\|/2), \quad (O \in K^\circ), \tag{1.8.16}$$

where $D_2(x, r)$ denotes a disk of centre x and radius r, see Fig. 1.8.3.

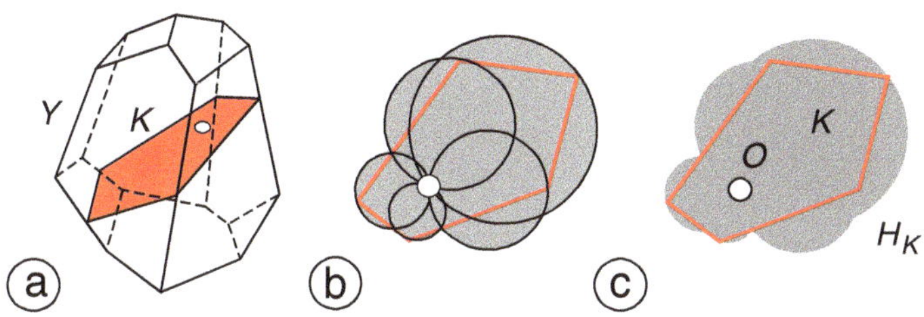

Fig. 1.8.3 (a) A pivotal section K of a convex polyhedron Y is a convex polygon. (b), (c) Construction of the flower H_K using Eq. (1.8.16).

The area of the support set H_K of a convex set K with respect to an interior origin can be expressed in terms of the parametric equations of ∂K, see Eq. (1.8.12), as follows,

$$\begin{aligned} A(H_K) &= \frac{1}{2}\int_0^{2\pi} \frac{P^2(t)Q(t)}{[X'^2(t) + Y'^2(t)]^2}\, \mathrm{d}t, \\ P(t) &= X'(t)Y(t) - X(t)Y'(t), \\ Q(t) &= X'(t)Y''(t) - X''(t)Y'(t). \end{aligned} \tag{1.8.17}$$

To show this, write Eq. (1.8.9) as

$$A(H_K) = \frac{1}{2}\int_0^{2\pi} h^2(t)\omega'(t)\, \mathrm{d}t, \tag{1.8.18}$$

with $h^2(t) = x^2(t) + y^2(t)$, $\tan(\omega(t)) = y(t)/x(t)$, and use Eq. (1.8.13).

Special case in which Y is an arbitrary triaxial ellipsoid

If $T \subset \mathbb{R}^3$ is a triaxial ellipsoid, then any pivotal trace curve $C_u \equiv \partial Y \cap L_2^3(0, u)$ is typically an ellipse, see Fig. 1.8.4(a). Suppose that $O \in Y^\circ$, and let $(M(u), m(u))$ denote the major and minor principal semiaxes of C_u. Further, let $r(u)$, $(0 \le r(u) \le M(u))$, denote the distance of the centre of C_u from the pivotal point O, see Fig. 1.8.4(b). Then,

$$A(H_{C_u}) = \frac{\pi}{2}[M^2(u) + m^2(u) + r^2(u)], \tag{1.8.19}$$

irrespective of the orientation of the axis joining the centre of the pivotal ellipse with O, see Fig. 1.8.4(c, d). By Eq. (1.8.8) the surface area of a triaxial ellipsoid containing the pivotal point is

$$S = \int_{\mathbb{S}^2_+} [M^2(u) + m^2(u) + r^2(u)]\, du. \tag{1.8.20}$$

If Y is a sphere of radius R centred at O, then $M(u) = m(u) = R$ and $r(u) = 0$ for all $u \in \mathbb{S}^2_+$, whereby $S = 4\pi R^2$, as expected.

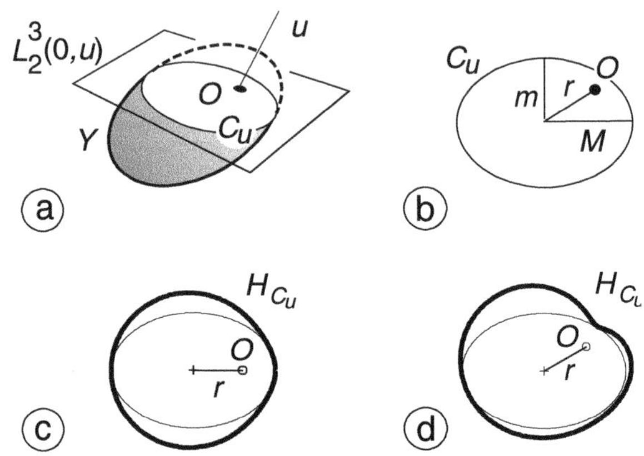

Fig. 1.8.4 (a), (b) The trace curve C_u determined by a pivotal plane in the boundary of a triaxial ellipsoid Y is an ellipse. (c), (d) For a given value of r, the area of the support set of a given ellipse is constant, see Eq. (1.8.19).

If $O \notin Y$, however, then Eq. (1.8.20) no longer holds.

To derive Eq. (1.8.19), define a rectangular frame Oxy with origin at the pivotal point $O \in K^\circ$ and parallel to the principal axes of C_u. Further, let (x_0, y_0) denote the rectangular coordinates of the pivotal point O with respect to the parallel frame with origin at the centre of C_u, whereby $x_0^2 + y_0^2 = r^2(u)$. Set $(M, u) \equiv (M(u), m(u))$, for short. The parametric equations of the ellipse boundary referred to the frame Oxy are

$$C_u : \begin{cases} X(t) = M\cos t - x_0, \\ Y(t) = m \sin t - y_0, \quad (0 \le t < 2\pi). \end{cases} \tag{1.8.21}$$

and the application of Eq. (1.8.17) yields the required result.

1.8.4 Surface area of a general object: the peak-and-valley formula

If the set $Y \subset \mathbb{R}^3$ is not necessarily convex, and the pivotal point O is not necessarily interior to Y, then an adaptation of the flower formula is possible using Eq. (1.8.1). As before, let $C_u \equiv \partial Y \cap L_2^3(0, u)$ denote a pivotal trace curve, see Fig. 1.8.5. The main task is to identify the number of intersections $I(C_u \cap L_1^2(r, \omega))$ in terms of (r, ω). For each direction $\omega \in [0, 2\pi)$ consider a line $L_1^2(r, \omega)$ sweeping the pivotal plane parallel to itself from $r = \infty$ down to $r = 0$, in search of critical points of the curve C_u in the given direction. In general, a critical point may be a local maximum, or minimum, with a tangent, or a local supremum, or infimum, without a tangent. To include all cases a local maximum, or supremum, will be called a peak, whereas a local minimum, or infimum, will be called a valley. If C_u has points above the axis $L_1^2(0,\omega)$, then we assign an index $\epsilon_k = +1$ if the kth critical point encountered is a peak, and $\epsilon_k = -1$ if it is a valley. The first critical point is necessarily a peak, and the second is also a peak if C_u is not convex. Thus, $\epsilon_1 = \epsilon_2 = 1$. Thereafter the critical point may be a peak, or a valley. Immediately after a peak is met, two new intersections appear whereas, after a valley is met, two intersections are lost. Suppose that m critical points are met successively, and let $h_1 > h_2 > \cdots > h_m > h_{m+1} \equiv 0$ denote their corresponding distances from the axis $L_1^2(0,\omega)$. Then,

$$I(C_u \cap L_1^2(r,\omega)) = \begin{cases} 0, & r > h_1, \\ 2 = 2\epsilon_1, & r \in (h_2, h_1], \\ 4 = 2(\epsilon_1 + \epsilon_2), & r \in (h_3, h_2], \\ \dots, & \dots, \\ 2\sum_{i=1}^{m} \epsilon_i, & r \in (h_{m+1}, h_m]. \end{cases} \tag{1.8.22}$$

Naturally m and $\{(\epsilon_k, h_k),\ k = 1, 2, \dots, m\}$ depend on (u, ω) in general. Now Eq. (1.8.1) yields

$$\begin{aligned} S &= \frac{1}{\pi} \int_{\mathbb{S}_+^2} \mathrm{d}u \int_0^{2\pi} \mathrm{d}\omega \sum_{k=1}^{m} \int_{h_{k+1}}^{h_k} \left(2 \sum_{i=1}^{k} \epsilon_i \right) r\,\mathrm{d}r \\ &= \frac{1}{\pi} \int_{\mathbb{S}_+^2} \mathrm{d}u \int_0^{2\pi} \mathrm{d}\omega \sum_{k=1}^{m} (h_k^2 - h_{k+1}^2) \sum_{i=1}^{k} \epsilon_i \\ &= \frac{1}{\pi} \int_{\mathbb{S}_+^2} \mathrm{d}u \int_0^{2\pi} \mathrm{d}\omega \sum_{k=1}^{m} \epsilon_k h_k^2, \end{aligned} \tag{1.8.23}$$

which is the integral-geometric representation of the peak-and-valley formula for the boundary surface area of a domain.

If $Y \subset \mathbb{R}^3$ is convex and $O \in Y$, then Eq. (1.8.23) reduces to

$$S = \frac{1}{\pi} \int_{\mathbb{S}_+^2} \mathrm{d}u \int_0^{2\pi} h^2(\omega)\,\mathrm{d}\omega, \tag{1.8.24}$$

which, bearing Eq. (1.8.9) in mind, coincides with Eq. (1.8.8), as expected.

1.8.5 Example

In Fig. 1.8.5(a), a line parallel to the axis $L_1^2(0,\omega)$ (which is taken to be horizontal for convenience) has swept a pivotal trace curve C_u from top to bottom, meeting $m = 3$ singular points which contribute $h_1^2 + h_2^2 - h_3^2$ to the summation in the rhs of

Eq. (1.8.23), because $\epsilon_1 = \epsilon_2 = 1$ correspond to peaks, whereas $\epsilon_3 = -1$ corresponds to a valley. If the sweeping line moves from bottom to top, then an additional peak is met which contributes with h_4^2. In Fig. 1.8.5(b), the position of the pivotal point O has been changed for the sake of illustration. Here the total contributions to the mentioned summation are $h_1^2 + h_2^2$ for the top to bottom sweeping, and $h_3^2 + h_4^2$ for the bottom to top one.

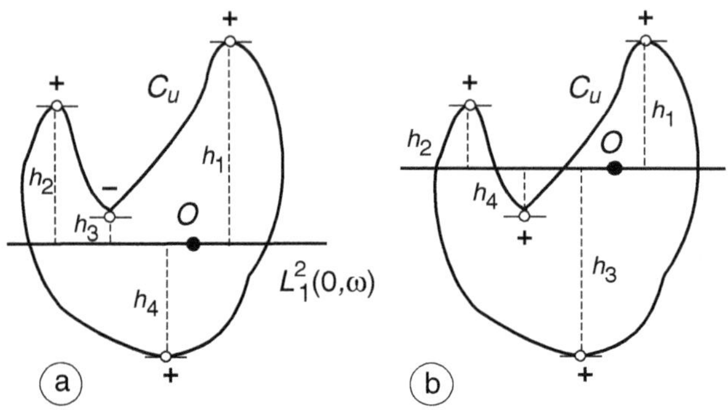

Fig. 1.8.5 Illustration of the peak-and-valley formula, see the example in Section 1.8.5.

1.8.6 For a convex set, the surfactor is equivalent to the flower formula

Let $Y \subset \mathbb{R}^3$ be a convex set with smooth boundary $\partial Y$of class C^2 and area S, and suppose that $O \in Y^\circ$. The flower formula is given by Eq. (1.8.24), namely,

$$S = \frac{1}{\pi} \int_{\mathbb{S}_+^2} \mathrm{d}u \int_0^{2\pi} h^2 \, \mathrm{d}\omega, \tag{1.8.25}$$

where $h \equiv h(\omega)$ is the support function of a pivotal section of Y. On the other hand, the surfactor Eq. (1.7.36) – with the symbol ω replaced with φ, to avoid confusion with the flower formula – may be written

$$S = \frac{1}{\pi} \int_{\mathbb{S}_+^2} du \int_0^{2\pi} \rho^2 (1 + \alpha \tan \alpha) \, \mathrm{d}\varphi, \tag{1.8.26}$$

where $\rho \equiv l_+(\varphi, u)$, the length of the radius vector of a pivotal section of Y in a direction φ within the pivotal plane, see Fig. 1.8.6. Note also that $\alpha = \varphi - \omega \in (-\pi/2, \pi/2)$, see Fig. (1.8.6). In the sequel it is important to note that, given any one of the angles $(\varphi, \omega, \alpha)$, each of the other two can be expressed as a differentiable function involving the given angle only.

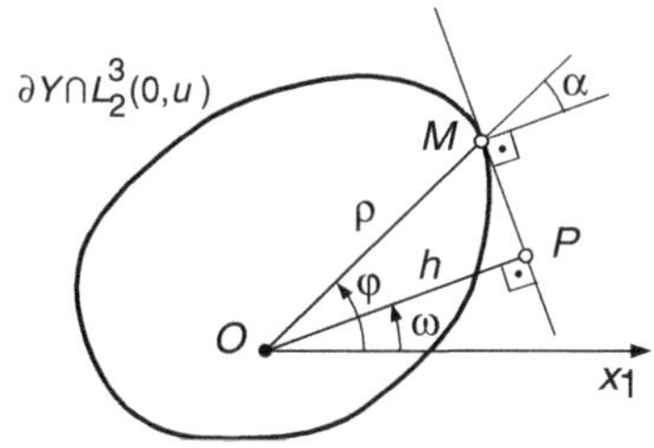

Fig. 1.8.6 Parametrization of a pivotal section of a smooth convex set, see Section 1.8.6.

We want to show that, for each orientation $u \in \mathbb{S}_+^2$ of the pivotal plane, the inner integrals in the rhs of Eqs. (1.8.25) and (1.8.26) coincide, namely,

$$\int_0^{2\pi} \rho^2(1+\alpha\tan\alpha)\,\mathrm{d}\varphi = \int_0^{2\pi} h^2\,\mathrm{d}\omega. \tag{1.8.27}$$

A pivotal section of the convex set Y is always a planar convex set K. Because $O \in K^\circ$, the area of K is

$$A(K) = \frac{1}{2}\int_0^{2\pi} \rho^2\,\mathrm{d}\varphi, \tag{1.8.28}$$

whereas $A(H_K)$ is given by Eq. (1.8.9). Consider the proper set difference $D_K = H_K \backslash K$, namely the set confined between the boundaries ∂K and ∂H_K, see Fig. 1.8.7. Then the Eq. (1.8.27) to be proved is equivalent to the following one,

$$A(D_K) = \frac{1}{2}\int_0^{2\pi} \rho^2\alpha\tan\alpha\,\mathrm{d}\varphi. \tag{1.8.29}$$

First we show that

$$A(D_K) = \frac{1}{2}\int_0^{2\pi} \rho^2\sin^2\alpha\,\mathrm{d}\omega \tag{1.8.30}$$

by a heuristic argument – for an alternative proof, see the reference in Note 4 of Section 1.8.9 below. As φ varies from 0 to 2π, the region D_K is entirely swept by a segment MP whose support axis is tangent to ∂K at M, with endpoints $M(\rho,\varphi) \in \partial K$ and $P(h,\omega) \in \partial H_K$, see Fig. 1.8.7. As the point M moves along an elementary arc of ∂K into M', the point P moves along a corresponding elementary arc of ∂H_K into P'. The element $\mathrm{d}\omega$ of angle between the vector radii OP and OP' of H_K is equal to the angle between the rays MP and $M'P'$ because, by definition, the ray MP is perpendicular to OP. Thus, the segment MP sweeps an elementary triangle of area

$$\frac{1}{2}|MP|^2\mathrm{d}\omega = \frac{1}{2}\rho^2\sin^2\alpha\,\mathrm{d}\omega, \tag{1.8.31}$$

from which Eq. (1.8.30) follows.

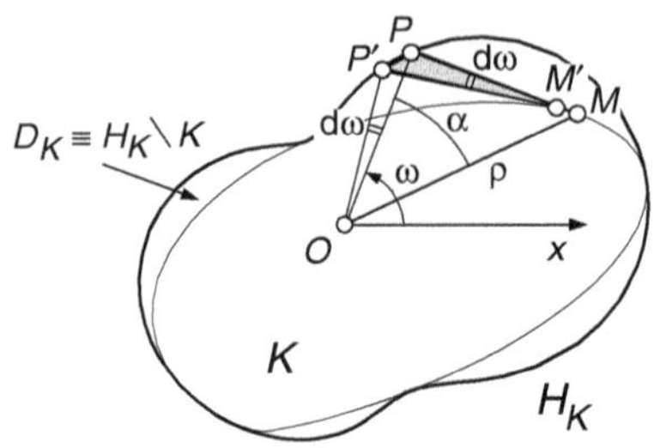

Fig. 1.8.7 Illustration of the argument used to obtain Eq. (1.8.31).

To prove the identity (1.8.30) it only remains to show that

$$\int_0^{2\pi} \rho^2 \alpha \tan\alpha \, \mathrm{d}\varphi = \int_0^{2\pi} \rho^2 \sin^2\alpha \, \mathrm{d}\omega. \tag{1.8.32}$$

From Fig. 1.8.8 we obtain the key relation

$$\tan\alpha = \frac{\mathrm{d}\rho}{\rho \, \mathrm{d}\varphi} = \frac{\rho'_\varphi}{\rho}, \tag{1.8.33}$$

whereby the lhs of Eq. (1.8.32) becomes

$$\int_0^{2\pi} \rho^2 \alpha \tan\alpha \, \mathrm{d}\varphi = \int_0^{2\pi} \rho\rho'_\varphi \alpha \, \mathrm{d}\varphi. \tag{1.8.34}$$

Integration by parts with $u = \alpha$ and $\mathrm{d}v = \rho\rho'_\varphi \, \mathrm{d}\varphi$ yields,

$$\begin{aligned}\int_0^{2\pi} \rho^2 \alpha \tan\alpha \, \mathrm{d}\varphi &= -\frac{1}{2}\int_0^{2\pi} \rho^2 \, \mathrm{d}\alpha \\ &= -\frac{1}{2}\int_0^{2\pi} \rho^2 \, \mathrm{d}\varphi + \frac{1}{2}\int_0^{2\pi} \rho^2 (\sin^2\alpha + \cos^2\alpha) \, \mathrm{d}\omega,\end{aligned} \tag{1.8.35}$$

where we have used the identity $\mathrm{d}\alpha = \mathrm{d}\varphi - \mathrm{d}\omega$. Further, $\rho\cos\alpha = h$ and therefore,

$$\begin{aligned}\int_0^{2\pi} \rho^2 \alpha \tan\alpha \, \mathrm{d}\varphi &= -A(K) + \frac{1}{2}\int_0^{2\pi} \rho^2 \sin^2\alpha \, \mathrm{d}\omega + A(H_K) \\ &= \int_0^{2\pi} \rho^2 \sin^2\alpha \, \mathrm{d}\omega,\end{aligned} \tag{1.8.36}$$

as required.

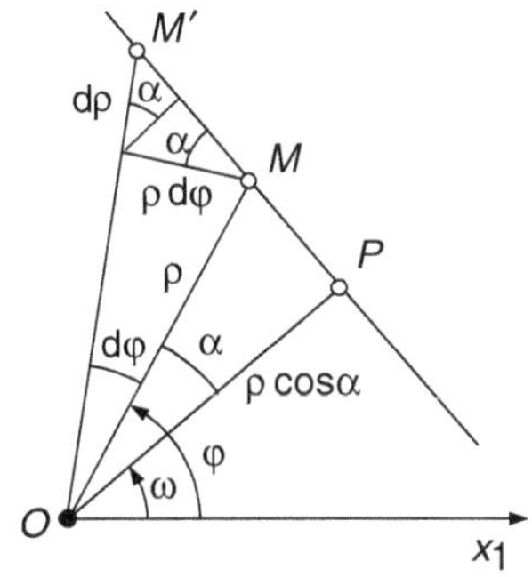

Fig. 1.8.8 Geometric detail used to obtain Eq. (1.8.33).

1.8.7 The nucleator is equivalent to the invariator representation of volume

The invariator representation of volume based on the a posteriori weighting, see Eq. (1.8.2), may be written as follows,

$$V = \frac{1}{2\pi}\int_{\mathbb{S}^2_+} \mathrm{d}u \int_0^{2\pi} \mathrm{d}\omega \int_0^{\infty} l(r,\omega,u) r \,\mathrm{d}r, \tag{1.8.37}$$

where $l(r,\omega,u) = L\{(Y \cap L_2^3(0,u)) \cap L_1^2(r,\omega;u)\}$ denotes the total linear transect length determined by the test line L_1^2 in the pivotal section. Here, the domain Y does not need to be convex, and the pivotal point O does not need to be interior to Y. First we want to show that, for each orientation $u \in \mathbb{S}^2_+$ of the pivotal plane, the functional given by the double inner integral in the rhs of the invariator Eq. (1.8.37) coincides with the corresponding functional given by the inner integral of the integrated nucleator Eq. (1.7.13), see Fig. (1.8.9), namely,

$$\int_0^{2\pi} \mathrm{d}\omega \int_0^{\infty} l(r,\omega,u) r \,\mathrm{d}r = 2\int_{Y\cap L_2^3(0,u)} \rho \,\mathrm{d}z. \tag{1.8.38}$$

Consider a point at an unsigned distance l along the line L_1^2 from an arbitrary origin on this line. Then clearly,

$$l(r,\omega,u) = \int_{(Y\cap L^3_{2[0]})\cap L_1^2} \mathrm{d}l. \tag{1.8.39}$$

The length elements $\mathrm{d}l$ and $\mathrm{d}r$ are mutually perpendicular, whereby $\mathrm{d}z = \mathrm{d}l\,\mathrm{d}r$ represents the area element at a point $z \in L^3_{2[0]}$ in the pivotal plane, see Fig. 1.8.9. Let (ρ,φ), $(\rho > 0, \varphi \in [0,2\pi))$ denote the polar coordinates of z with respect to a fixed axis Ox_1. Then $r = \rho|\cos(\omega - \varphi)|$, see Fig. 1.8.9(b), and it follows that

$$\begin{aligned}\int_0^{2\pi} d\omega \int_0^{\infty} l(r,\omega,u) r\, dr &= \int_0^{\pi} d\omega \int_{Y\cap L^3_{2[0]}} \rho|\cos(\omega-\varphi)|\, dz\\ &= \int_{Y\cap L^3_{2[0]}} \rho\, dz \int_0^{\pi} |\cos(\omega-\varphi)|\, d\omega\\ &= 2\int_{Y\cap L^2_{2[0]}} \rho\, dz, \qquad (1.8.40)\end{aligned}$$

as required. In the first identity the range of integration of ω is $(0,\pi)$ instead of $(0,2\pi)$ because in the inner integral the point z varies in the entire pivotal section.

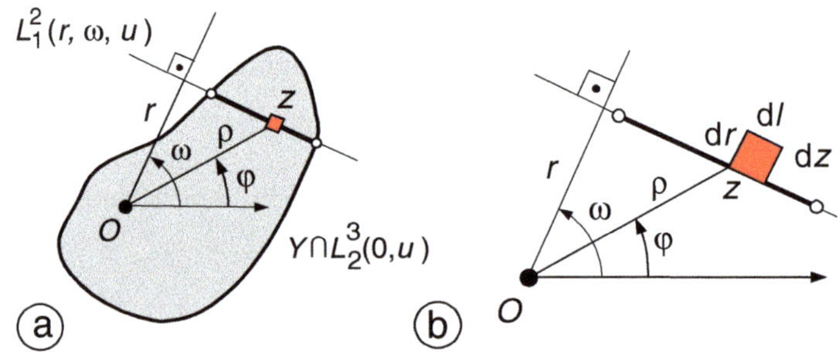

Fig. 1.8.9 Notation used in Section 1.8.7.

Further, if Y is star-shaped with respect to $O \in Y$ then, bearing in mind that $\rho\, dz = \rho^2\, d\rho d\varphi$ (see Eq. (1.7.1)), Eq. (1.8.40) becomes,

$$\int_0^{2\pi} d\omega \int_0^{\infty} l(r,\omega,u) r\, dr = \frac{2}{3}\int_0^{2\pi} l_+^3(\varphi,u)\, d\varphi, \qquad (1.8.41)$$

where $l_+(\varphi, u)$ represents the length of the radius vector of the pivotal section. Thus, for each pivotal section the invariator functional coincides with the pivotal nucleator functional defined on that section.

1.8.8 Uniqueness conjecture for the invariator

Eq. (1.8.2) may be written,

$$V = \frac{1}{2\pi}\int \alpha(Y \cap L^3_{2[0]})\, dL^3_{2[0]}, \qquad (1.8.42)$$

where

$$\alpha(Y \cap L^3_{2[0]}) = \int r \cdot L\{(Y \cap L^3_{2[0]}) \cap L^2_1\} \, \mathrm{d}L^2_1 \tag{1.8.43}$$

is a functional depending on the pivotal section $Y \cap L^3_{2[0]}$ only – and similarly for Eq. (1.8.1). For a given domain $Y \subset \mathbb{R}^3$ the preceding functional is therefore a function, $\psi(u)$ say, of $u \in \mathbb{S}^2_+$ only.

We have shown that, for any given pivotal section, the nucleator integral coincides with the invariator one and, for a convex set, the surfactor integral coincides with the flower area. It may be conjectured that, conditional on a given pivotal section, the averages of all possible functionals for V will always boil down to Eq. (1.8.43), and analogously for S. The conjecture remains open, partly because the functionals considered so far need not be the only possible ones. See Note 4 below.

1.8.9 Notes

1. Invariator weighting

In Cruz-Orive (2005), only a priori weighting, Eqs. (1.8.4)–(1.8.5), was considered. The a posteriori weighting approach based on Eqs. (1.8.1)–(1.8.2) was first developed in Cruz-Orive and Gual-Arnau (2015).

2. The flower formula for the surface area of a convex set

The identity (1.8.8) is due to Cruz-Orive (2005). Further results, notably the special case given by Eq. (1.8.20), and a closed formula due to P. Calka (personal communication) for the flower area of a convex polygon, were given in Cruz-Orive (2011). In the latter paper, the former term *support set* was replaced with *flower*, as used in another context (Calka, 2003). The concept of *wedge volume*, also developed in Cruz-Orive (2011), is not given here because it was later shown to be unnecessary (Cruz-Orive, 2012).

3. The peak-and-valley formula for the surface area of an arbitrary set

The generalization of the flower formula to arbitrarily shaped objects addressed in Section 1.8.4 is due to Thórisdóttir and Kiderlen (2014) and Thórisdóttir, Rafati, and Kiderlen (2014). The approach leading to Eq. (1.8.23), however, is basically equivalent to but simpler than the one followed by the latter authors, see Cruz-Orive and Gual-Arnau (2015). For a general treatment, see Gual-Arnau and Cruz-Orive (2016).

4. Equivalences of the invariator. The uniqueness conjecture

The material in Sections 1.8.6 and 1.8.7 is taken from Cruz-Orive (2012), which contains an alternative proof of Eq. (1.8.30), and an extension of Eq. (1.8.27) to convex polyhedra (Proposition 3 of the latter paper). For a general proof without smoothness assumptions, see Thórisdóttir and Kiderlen (2014), Proposition 4. For a general formulation of the conjecture in Section 1.8.8, see Gual-Arnau, Cruz-Orive, and Nuño-Ballesteros (2010) and Jensen and Kiderlen (2017).

1.9 Blaschke–Petkantschin Formulae for Powers of Area and Volume

1.9.1 Planar sets

Consider a planar domain $Y \subset \mathbb{R}^2$ of area A. We want to obtain Crofton formulae involving powers of linear intercept length, yielding representations of A^2. To simplify the exposition we assume that the domain Y is convex, so that every linear intercept is simply connected.

Hit the set Y with a pair $\{x_0, x_1\}$ of invariant points. Clearly,

$$A^2 = \int_Y \mathrm{d}x_0 \int_Y \mathrm{d}x_1. \tag{1.9.1}$$

Referred to the straight line $L_1^2(p, \phi)$ joining them, the two points are denoted by $\{z_0, z_1\}$, and they are translation-invariant within this line (Fig. 1.2.8). For a given line L_1^2 hitting Y, let z_-, z_+ denote the abscissas of the end points of the intercept $Y \cap L_1^2$ in the corresponding support line, so that $z_+ - z_- = L(Y \cap L_1^2)$. Application of Eq. (1.2.27) yields the following formula, first derived by M. W. Crofton,

$$\begin{aligned} A^2 &= \int \mathrm{d}L_1^2 \int_{Y\cap L_1^2} \mathrm{d}z_0 \int_{Y\cap L_1^2} |z_1 - z_0|\, \mathrm{d}z_1 \\ &= \int \mathrm{d}L_1^2 \int_{x_-}^{z_+} \mathrm{d}z_0 \left[\int_{z_-}^{z_0} (z_0 - z_1)\mathrm{d}z_1 + \int_{z_0}^{z_+} (z_1 - z_0)\, \mathrm{d}z_1\right] \\ &= \frac{1}{3}\int L^3(Y \cap L_1^2)\, \mathrm{d}L_1^2. \end{aligned} \tag{1.9.2}$$

1.9.2 Three-dimensional sets

Consider now a convex domain $Y \subset \mathbb{R}^3$ of volume V. By a procedure analogous to that used to obtain the preceding identity, Eq. (1.2.28) yields Hostinský's formula

$$V^2 = \frac{1}{6}\int L^4(Y \cap L_1^3)\,\mathrm{d}L_1^3. \tag{1.9.3}$$

Suppose that a domain $Y \subset \mathbb{R}^3$ – not necessarily convex – is hit by a pivotal plane $L_{2[0]}^3 \equiv L_2^3(0, u)$ through a pivotal point $O \in Y$. Next, hit the corresponding pivotal section with two invariant points $\{z_1, z_2\}$. Direct application of Eq. (1.2.32) yields

$$\begin{aligned} V^2 &= 2\int_{\mathbb{S}_+^2} \nabla_{2[0]}\,\mathrm{d}u \int_{Y\cap L_{2[0]}^3} \mathrm{d}z_1 \int_{Y\cap L_{2[0]}^3} \mathrm{d}z_2 \\ &= 2\int_{\mathbb{S}_+^2} \nabla_{2[0]} \cdot a_0^2\,\mathrm{d}u, \end{aligned} \tag{1.9.4}$$

where $a_0 = A(Y \cap L_{2[0]}^3)$ denotes the pivotal section area.

Finally, suppose that Y is hit directly by an invariant plane L_2^3, and then the corresponding section is hit with three invariant points $\{z_0, z_1, z_2\}$. Direct application of Eq. (1.2.34) yields

$$V^3 = 2\int_{Y\cap L_2^3 \neq \emptyset} \nabla_2 \cdot a^3\,\mathrm{d}L_2^3, \tag{1.9.5}$$

where $a = A(Y \cap L_2^3)$ denotes the section area.

1.9.3 Notes

1. Powers of planar area and volume from linear intercepts

The identity (1.9.2) was given by Crofton (1869) and it was called "Crofton's second theorem on convex figures" by Kendall and Moran (1963). The extension (1.9.3) is due to Hostinský (1925). For a complete account, see Santaló (1976). The results (1.9.4) and (1.9.5) are due to Miles (1971), Jensen and Gundersen (1983, 1985), and Jensen (1987, 1998).

1.10 The Euler–Poincaré Characteristic of a Planar Domain

1.10.1 Total curvature of a planar curve

Consider an oriented planar curve $Y \subset \mathbb{R}^2$ of finite length B, described by a position vector $\rho(b)$ joining the origin O with a point $y(b) \in Y$ at a curvilinear distance b, $0 \leq b \leq B$ from a fixed point $y(0) \in Y$ of the curve, see Fig. 1.10.1(a). Suppose first that at every point $y(b) \in Y$, the unit tangent vector $T(b) = \mathrm{d}\rho(b)/\mathrm{d}b$ exists, is unique and varies continuously, that is, the derivative $\mathrm{d}\alpha(b)/\mathrm{d}b$ is continuous, where $\alpha(b)$ represents the angle – measured counterclockwise – made by $T(b)$ with the axis of abscissas Ox_1. Actually,

$$\kappa(b) = \frac{\mathrm{d}\alpha(b)}{\mathrm{d}b} \tag{1.10.1}$$

is a scalar called the local curvature of Y at the point $y(b) \in Y$. Note that $\kappa(b)$ may be positive, or negative, according to whether $\alpha(b)$ increases, or decreases, as the point $y(b)$ moves into $y(b + \mathrm{d}b)$. The value $\kappa(b) = 0$ corresponds to an inflection point, or to a point in a straight line segment. The integral

$$C(Y) = \int_0^B \kappa(b)\,\mathrm{d}b \tag{1.10.2}$$

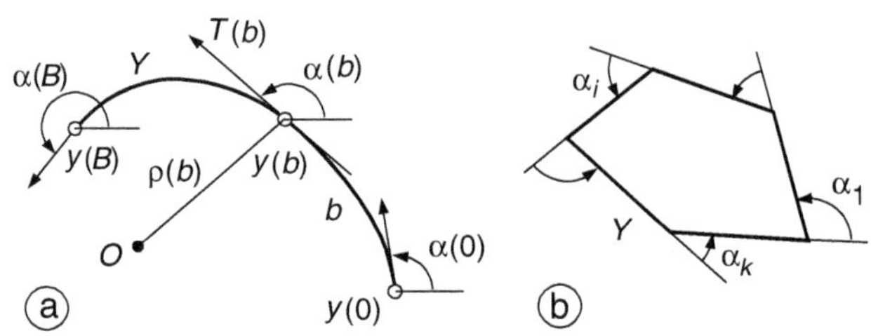

Fig. 1.10.1 Notation used in Section 1.10.1.

is a dimensionless quantity called the total curvature of the curve Y. Equivalently,

$$C(Y) = \int_0^B \mathrm{d}\alpha(b) = \alpha(B) - \alpha(0), \tag{1.10.3}$$

namely the net angle described by the tangent vector $T(b)$ as the point $y(b)$ moves along the entire curve from end to end. More generally, suppose that the curve has $k \geq 1$ corner points (excluding endpoints), and let $\{\alpha_i,\ i = 1, 2, \ldots, k\}$ represent the jumps of $\alpha(b)$ at these corners. Then,

$$C(Y) = \int_{y(b)\in Y_1} \mathrm{d}\alpha(b) + \sum_{i=1}^{k} \alpha_i, \tag{1.10.4}$$

where Y_1 represents the union of the arcs of Y with a continuous tangent. For instance, if Y is the outer boundary of a convex polygon with k vertices, then the $\{\alpha_i\}$ are the exterior angles at the vertices, see Fig. 1.10.1(b), whose sum is equal to 2π, whereas the first term in the rhs of Eq. (1.10.4) is zero. Therefore, $C(Y) = 2\pi$, which actually holds for any closed curve Y with no double points.

1.10.2 Hadwiger's representation in 2D

Consider a planar domain $Y \subset \mathbb{R}^2$ whose boundary ∂Y is a closed curve with a finite number of corner points and no double points. The curve is left-oriented, which means that a point moving counterclockwise along the curve leaves the domain on the left, see Fig. 1.10.2 (left). For such a curve we have seen that $C(\partial Y) = 2\pi$. Now suppose that Y contains a hole, see Fig. 1.10.2 (middle). The inner boundary, namely the boundary of the hole, must be right-oriented, so that the moving point also leaves the domain on the left. The total boundary ∂Y is the union of the outer boundary, which contributes 2π to $C(\partial Y)$ because it is left-oriented, and the inner boundary, which contributes -2π because it is right-oriented, whereby the total curvature of a domain containing a single hole is $C(\partial Y) = 2\pi - 2\pi = 0$.. Further, the total curvature of a domain containing two holes, see Fig. 1.10.2 (right), is $C(\partial Y) = 2\pi - 2\pi - 2\pi = -2\pi$, and so on. In fact, $C(\partial Y)$ is always an integer multiple of 2π, namely,

$$C(\partial Y) = 2\pi\chi(Y), \tag{1.10.5}$$

where

$$\chi(Y) = 1 - p \tag{1.10.6}$$

is an integer called the Euler–Poincaré characteristic of the domain Y. The non-negative integer p is called the genus, the first Betti number, or the connectivity number of Y. This number may be interpreted as the number of holes in Y, the number of extra connections or 'handles' of Y, or equivalently the maximum number of cuts one can make in Y before it falls apart into two separate pieces, see Fig. 1.10.2. If $\chi(Y) = 1$ we say that the domain Y is simply connected.

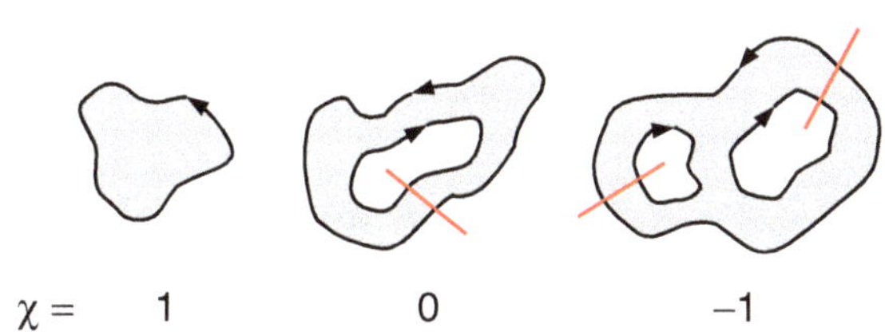

Fig. 1.10.2 Three planar domains with Euler–Poincaré characteristics 1, 0, −1, (genus 0, 1, 2), respectively, see Section 1.10.2.

The Euler–Poincaré characteristic is a topological invariant, that is, its value remains unchanged under continuous shape deformations of Y. For instance, any simply connected domain is topologically equivalent to a disk; any domain with $\chi(Y) = 0$ is topologically equivalent to an annulus, etc. Actually, $\chi(Y)$ is an additive quantity satisfying the identity $\chi(\emptyset) = 0$. If Y is a finite union of N subsets, namely if

$$Y = \bigcup_{i=1}^{N} Y_i, \tag{1.10.7}$$

then the classical inclusion-exclusion formula reads as follows,

$$\chi(Y) = \sum_{i=1}^{N} \chi(Y_i) - \sum_{1 \le i < j \le N} \chi(Y_i \cap Y_j) + \cdots + (-1)^{N-1} \chi(Y_1 \cap \cdots \cap Y_N). \tag{1.10.8}$$

Next we give an elementary derivation of H. Hadwiger's method to determine $\chi(Y)$. Consider a sequence of closed adjacent stripes $\{A_i,\ i \in \mathbb{Z}\}$ covering the plane, with arbitrary orientation relative to Y and thickness $\epsilon > 0$, namely,

$$A_i = \{(x_1, x_2) \in \mathbb{R}^2 : \ (i-1)\epsilon \le x_2 \le i\epsilon\}, \tag{1.10.9}$$

so that

$$\begin{aligned}
&\bigcup_{i \in \mathbb{Z}} A_i = \mathbb{R}^2, \\
&A_i \cap A_{i+1} \equiv L_i = \{(x_1, x_2) \in \mathbb{R}^2 : \ x_1 \in \mathbb{R},\ x_2 = i\epsilon\}, \quad i \in \mathbb{Z}, \\
&A_i \cap A_j = \emptyset, \quad |i - j| \ge 2, \\
&Y = \bigcup_{i \in \mathbb{Z}} Y \cap A_i.
\end{aligned} \tag{1.10.10}$$

Application of the inclusion-exclusion formula yields,

$$\begin{aligned}
\chi(Y) &= \sum_{i \in \mathbb{Z}} \{\chi(Y \cap A_i) - \chi(Y \cap A_i \cap A_{i+1})\} \\
&= \sum_{i \in \mathbb{Z}} \{\chi(Y \cap A_i) - \chi(Y \cap L_i)\} \\
&= \sum_{i \in \mathbb{Z}} \{\chi((Y \cap A_i) \cup L_i) - 1\},
\end{aligned} \tag{1.10.11}$$

where we have used the identity $\chi(L_i) = 1$. We can choose $\epsilon > 0$ small enough so that, whenever $Y \cap A_i \ne \emptyset$, exactly one of the following events holds, see Fig. 1.10.3(a,b).

a. $Y \cap A_i$ is simply connected and $(Y \cap A_i) \cap L_i = \emptyset$, so that $\chi((Y \cap A_i) \cup L_i) = 2$. Thus, this event contributes +1 to $\chi(Y)$, and is called an island.
b. $Y \cap A_i$ is simply connected and $(Y \cap A_i) \cap L_i \ne \emptyset$, so that $\chi((Y \cap A_i) \cup L_i) = 1$ and this event contributes 0 to $\chi(Y)$.
c. $(Y \cap A_i) \cup L_i$ contains a hole and $(Y \cap A_i) \cap L_i \ne \emptyset$, so that $\chi((Y \cap A_i) \cup L_i) = 0$. Thus, this event contributes -1 to $\chi(Y)$, and is called a bridge.

Let I, B denote the total numbers of islands and bridges, respectively, observed as a thin closed stripe sweeps the domain Y in an arbitrary direction. Then, $\chi(Y) = I - B$. To increase efficiency we may also sweep Y along the opposite direction, and redefine I, B as the respective total counts obtained in both directions. The Hadwiger representation becomes

$$\chi(Y) = \frac{1}{2}(I - B), \tag{1.10.12}$$

see Fig. 1.10.3(c,d). The subset $Y \cap A_i$ considered in cases (a)–(c) above may consist of several aligned parts – however this does not affect Eq. (1.10.12).

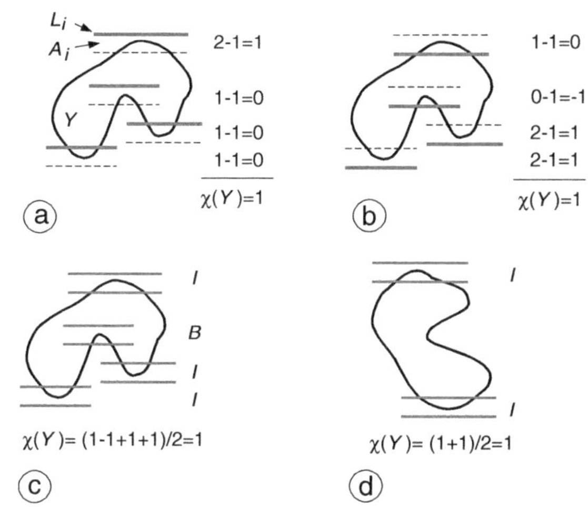

Fig. 1.10.3 (a) A stripe of infinitesimal thickness (here enlarged for clarity) sweeps a planar domain Y from top to bottom. The first event is an island – according to Eq. (1.10.11), the remaining three events contribute 0 to $\chi(Y)$. (b) The stripe sweeps Y from bottom to top and detects, in turn, two islands and a bridge. As in (a), the net contribution is $\chi(Y) = 1$, as expected. (c) Application of Eq. (1.10.12). (d) The domain orientation has been altered to illustrate the motion invariance of the rhs of Eq. (1.10.12).

The sweeping stripe may be replaced with a sweeping straight line in any arbitrary direction, whereby I and B are equivalent to the tangent counts T_+, when the local curvature at the tangent point is positive, and T_-, when such curvature is negative, respectively. Thus, recalling Eq. (1.10.5),

$$C(\partial Y) = \pi(T_+ - T_-). \tag{1.10.13}$$

In stereology, the factor $T_+ - T_-$ is called the 'net tangent count'. For instance, if Y is a disk, then $T_+ = 2$, $T_- = 0$ and $C(\partial Y) = 2\pi$, as expected. For a polygon (with possibly curved edges), the preceding identity holds with the term 'tangent count' replaced with 'contact count'.

1.10.3 Euler's formula in 2D

The representation given by Eq. (1.10.12) is convenient for practical applications. A classical representation is based on Euler's formula. Suppose that $Y \subset \mathbb{R}^2$ admits a finite partition which is effectively a graph with F simply connected subsets called faces, E simply connected edges, and V vertices or nodes, see Fig. 1.10.4. Then, Euler's formula reads as follows,

$$\chi(Y) = V - E + F. \tag{1.10.14}$$

For instance, a simply connected chord drawn through a simply connected set (e.g., a disk) generates a graph with $V = 2$, $E = 3$ and $F = 2$, whereby $\chi(Y) = 1$, as expected, see Fig. 1.10.4(a). Now suppose that Y contains a single hole. If we draw a simply connected chord lying entirely inside Y and joining the outer and the inner boundaries, then we get a tiling with $V = 2$, $E = 3$ and $F = 1$, whereby $\chi(Y) = 0$, also as expected, see Fig. 1.10.4(b).

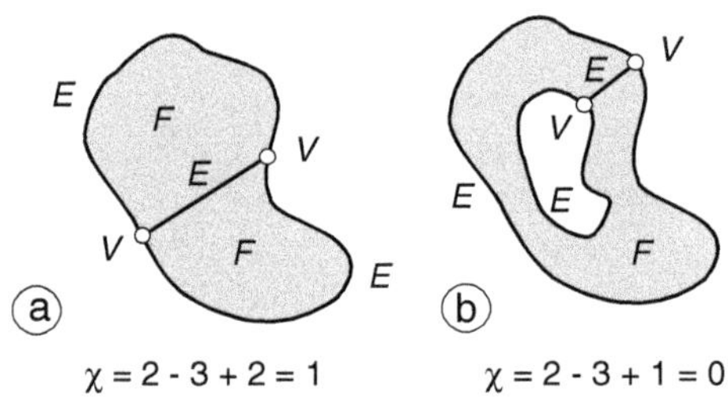

Fig. 1.10.4 Application of Euler's Eq. (1.10.14) to (a) a simply connected domain, and (b) a domain with a hole, see Section 1.10.3.

Euler's formula is a consequence of the inclusion-exclusion formula. In Eq. (1.10.8), suppose that all the $\{Y_i\}$ are simply connected faces constituting a partition of a planar domain Y. Then $Y_i \cap Y_j$ is an edge, and $Y_i \cap Y_j \cap Y_k$ is a vertex, (provided that the intersections are non-empty and the subscripts are all different). Thus,

$$\begin{aligned} \chi(Y) &= \sum_i \chi(Y_i) - \sum_{i<j} \chi(X_i \cap Y_j) + \sum_{i<j<k} \chi(Y_i \cap Y_j \cap Y_k) \\ &= F - E + V. \end{aligned} \tag{1.10.15}$$

A more general version of Euler's formula is the following,

$$\chi(Y_q) = \alpha_0 - \alpha_1 + \alpha_2 - \cdots + (-1)^q \alpha_q, \tag{1.10.16}$$

where $Y_q \subset \mathbb{R}^d$, $q \in \{0, 1, \ldots, d\}$, is a q-dimensional subset which admits a finite partitioning into α_i simply connected subsets of dimension i. For $q = 2$ we get

Eq. (1.10.14). For $q = 1$ we have

$$\chi(Y) = V - E. \tag{1.10.17}$$

For instance, if Y is a simple closed curve (e.g., a circumference), then a point in Y, ($V = 1$), determines an edge ($E = 1$), whereby $\chi(Y) = 0$.

Eq. (1.10.17) applies to a graph $Y_1 \subset \mathbb{R}^d$ consisting of V nodes and E non-intersecting branches, each connecting two nodes. Let V_i denote the number of nodes from which i branches emanate. Then,

$$\chi(Y) = V - \frac{1}{2}\sum_i iV_i, \tag{1.10.18}$$

see Fig. 1.10.5. The factor $1/2$ stems from the fact that each branch connects 2 nodes.

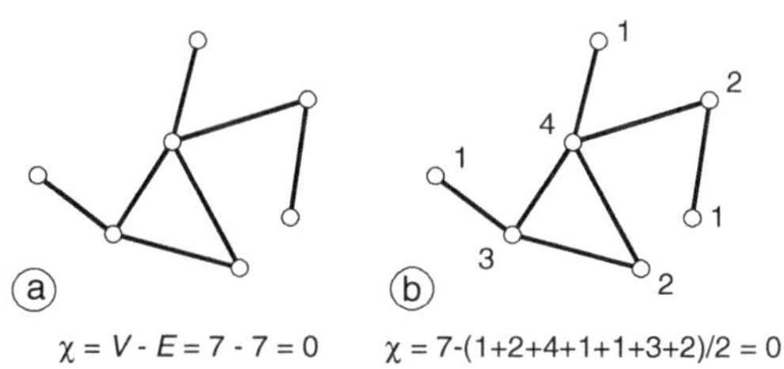

Fig. 1.10.5 (a) Illustration of Euler's Eq. (1.10.17), and (b) of Eq. (1.10.18), for a graph.

1.10.4 Notes

1. Radius of curvature

The reciprocal of the local curvature, see Eq. (1.10.1), is the radius of curvature, namely

$$R(b) = \frac{1}{\kappa(b)}. \tag{1.10.19}$$

Intuitively, as $\Delta b \to 0$, the normal axes at the points $y(b)$ and $y(b+\Delta b)$ will meet at a point called the local centre of curvature, which lies at a distance $R(b)$ from $y(b)$. The centre of curvature is the centre of a circle of radius $R(b)$ called the osculating circle, see Fig. (1.10.6).

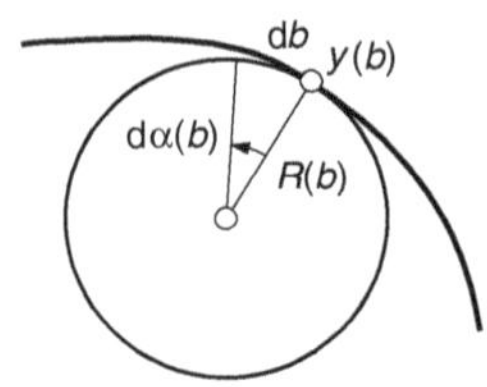

Fig. 1.10.6 The radius of curvature is the radius of the osculating circle, see Eq. (1.10.19).

2. Hadwiger's representation of the Euler–Poincaré characteristic

For a proof of the additivity and invariance properties of the Euler–Poincaré characteristic taken for granted in Section 1.10.2, see Hadwiger (1957).

3. Topological invariance of the Euler–Poincaré characteristic

Instead of Hadwiger's Eq. (1.10.12), Cahn (1967) and DeHoff (2004) proposed Eq. (1.10.13) as a representation of $C(\partial Y)$. Their derivation, however, was based on Eq. (1.10.2) and, as a consequence, their analogue to Eq. (1.10.12) was an integral over orientations, see also Weibel (1980), Serra (1982), or DeHoff (1987). On the other hand, Gundersen, Boyce, Nyengaard, and Odgaard (1993) invoked Hadwiger's result to stress the important point that Eq. (1.10.12) holds for any arbitrary orientation, see Fig. 1.10.3(c),(d). (The notations I and B, based on the sweeping stripe more than on the sweeping line concept, were also introduced in the latter paper.) Thus, different stripe orientations may yield different individual counts of I and B (or of T_+ and T_-, respectively), but their difference will always be constant for a given object at any sweeping orientation.

1.11 Crofton Formula for the Integral of Mean Curvature

Consider a surface $Y \subset \mathbb{R}^3$ of class C^2. Thus, at every point $y \in Y$ there is a unique tangent plane whose normal vector varies continuously. Our purpose is to compute the following integral,

$$\int C(Y \cap L_2^3)\, \mathrm{d}L_2^3, \tag{1.11.1}$$

namely the measure of the total curvature of the curve traces determined in Y by all the motion-invariant test planes hitting Y.

Fix an orthogonal reference trihedron $yx_1x_2x_3$ centred at a point $y \in Y$, such that the polar axis yx_3 is normal to the surface Y at y, and yx_1x_2 is the tangent plane at y, see Fig. 1.11.1. Let $L(\phi)$ represent a normal plane containing the axis yx_3, whose normal vector makes an angle $\phi \in [0, 2\pi)$ with the axis yx_1. Let $\kappa(\phi)$ denote the local curvature of the normal trace $Y \cap L(\phi)$ at the point y. It can be shown that there is an angle ϕ_1 such that $\kappa_1 \equiv \kappa(\phi_1)$ and $\kappa_2 \equiv \kappa(\phi_1 + \pi/2)$ are the minimum and the maximum values of $\kappa(\phi)$, respectively. These two curvatures are called the local principal curvatures of the surface Y at the point $y \in Y$, whereas the corresponding directions ϕ_1 and $\phi_1 + \pi/2$ are called the principal directions. For convenience we choose the axes yx_1, yx_2 along the principal directions, so that $\phi_1 = 0$. Euler's theorem states that

$$\kappa(\phi) = \kappa_1 \cos^2 \phi + \kappa_2 \sin^2 \phi. \tag{1.11.2}$$

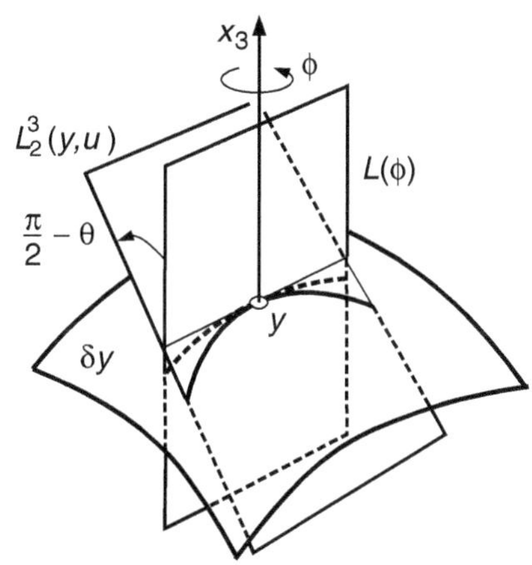

Fig. 1.11.1 Notation used to obtain the integral of mean curvature, Eq. (1.11.6), from planar curve transects.

Consider a plane $L_2^3(y, u)$ through the point $y \in Y$, with normal direction $u(\phi, \theta)$, where the origin of the colatitude angle θ is the polar axis yx_3. Let $\kappa(\phi, \theta)$ denote the local curvature of the trace $Y \cap L_2^3(y, u)$ at the point y. Meusnier's theorem states that

$$\kappa(\phi, \theta) = \frac{\kappa(\phi)}{\sin \theta}. \tag{1.11.3}$$

Let $\delta y \subset Y$ represent a local surface element of area $\mathrm{d}s$ at the point $y \in Y$. Then,

$$\int C(Y \cap L_2^3)\, \mathrm{d}L_2^3 = \int_Y \int C(\delta y \cap L_2^3)\, \mathrm{d}L_2^3. \tag{1.11.4}$$

Recall that $\mathrm{d}L_2^3 = \mathrm{d}p \sin\theta \,\mathrm{d}\phi \,\mathrm{d}\theta$. Further,

$$C(\delta y \cap L_2^3) = \kappa(\phi, \theta)\, \mathrm{d}b, \tag{1.11.5}$$

where $\mathrm{d}b = B(\delta y \cap L_2^3)$ is the trace length element. Recalling Eq. (1.5.11) and the Euler–Meusnier results we obtain

$$\begin{aligned}
&\int C(Y \cap L_2^3)\,\mathrm{d}L_2^3 \\
&\quad = \int_Y \int_{\mathbb{S}_+^2} \mathrm{d}u \int \kappa(\phi,\theta)\,\mathrm{d}b\,\mathrm{d}p \\
&\quad = \int_Y \mathrm{d}s \int_0^{2\pi} \mathrm{d}\phi \int_0^{\pi/2} \frac{1}{\sin\theta}(\kappa_1 \cos^2\phi + \kappa_2 \sin^2\phi)\sin^2\theta\,\mathrm{d}\theta \\
&\quad = 2\pi M(Y), \qquad (1.11.6)
\end{aligned}$$

where

$$M(Y) = \frac{1}{2}\int_Y (\kappa_1 + \kappa_2)\,\mathrm{d}s \qquad (1.11.7)$$

is called the integral of mean curvature of the surface Y. The integrand $\frac{1}{2}(\kappa_1 + \kappa_2)$ is called the mean local curvature at a point of the surface. Recall that $C(Y \cap L_2^3)$ may be expressed via Eq. (1.10.13).

Unlike the curvature integral of a planar curve, which is dimensionless, the integral of mean curvature $M(Y)$ of a surface has a length dimension, and it is therefore not a topological invariant under shape deformations.

1.11.1 Notes

1. Differential geometry tools

For details on Euler's and Meusnier's theorems see, for instance, Struik (1961).

2. Sectional curvature integral

The derivation in Eq. (1.11.6) is due to Miles (1981), where other sectional integral formulae for curvatures were derived.

1.12 The Euler–Poincaré Characteristic in 3D

1.12.1 Gaussian curvature

Consider a domain $Y \subset \mathbb{R}^3$ whose boundary is a closed oriented surface $\partial Y \subset \mathbb{R}^3$ of class C^2. To fix ideas it is convenient to regard Y as a connected domain (i.e., a single particle). As before, let $\mathrm{d}s$ denote the area of the surface element δy at a point $y \in \partial Y$, and κ_1, κ_2 the principal curvatures at that point. The Gaussian curvature of Y is a dimensionless quantity defined as follows,

$$G(Y) = \int_{\partial Y} \kappa_1 \kappa_2 \, \mathrm{d}s = G(\partial Y). \tag{1.12.1}$$

For instance, if Y is a ball of radius R, then

$$G(Y) = \int_{\partial Y} \frac{1}{R^2} \, \mathrm{d}s = 4\pi, \tag{1.12.2}$$

independently of R. Note that $\mathrm{d}s/R^2 = \mathrm{d}u_2$ is an element of solid angle on the unit sphere, which in this context is called the spherical image corresponding to a surface element $\delta y \subset \partial Y$. In general, if all the origins of the unit normal vectors of ∂Y are shifted to a common origin, then their end points will describe the spherical image of ∂Y on the unit sphere. Gauss' classical formula for the spherical image of a surface element reads

$$\mathrm{d}u_2 = \frac{\mathrm{d}s}{R_1 R_2}, \quad u_2 \in \mathbb{S}^2, \tag{1.12.3}$$

where R_1, R_2 are the principal radii of curvature at the point $y \in \partial Y$. Thus $G(Y)$ is the net spherical image of the surface ∂Y. (By analogy, recall that the total curvature $C(Y)$ of a planar curve was a net angle.)

1.12.2 The Euler–Poincaré characteristic: definition and Hadwiger's representation

By analogy with Eq. (1.10.5), the Gauss–Bonnet theorem states that

$$G(\partial Y) = 4\pi\chi(Y), \tag{1.12.4}$$

where

$$\chi(Y) = 1 - p \tag{1.12.5}$$

is the Euler–Poincaré characteristic of the domain Y. Now the connectivity number p may be regarded as the number of 'tunnels' or 'handles' of Y, namely the number of extra connections. The counterpart of a 'hole' contained in a two-dimensional particle is now a 'cavity', or an additional particle, with $p = 0$, in the interior of Y, and it contributes to $\chi(Y)$ as a separate particle. Thus, if Y is a finite union of separate particles, then

$$\begin{aligned}\chi = N(\text{separate particles}) - N(\text{extra connections}) \\ + N(\text{enclosed simply connected domains}),\end{aligned} \tag{1.12.6}$$

where $N(\cdot)$ denotes number.

The Hadwiger representation of $\chi(Y)$, useful in applications, is analogous to that leading to Eq. (1.10.12) in the two-dimensional case. By redefining A_i, L_i as a closed slab and a plane, respectively, Eq. (1.10.11) holds unchanged. Now the slab thickness

$\epsilon > 0$ can be chosen small enough so that, whenever $Y \cap A_i \neq \emptyset$, exactly one of the following events (illustrated in Fig. 1.12.1) holds.

a. $Y \cap A_i$ is simply connected and $(Y \cap A_i) \cap L_i = \emptyset$, so that $\chi((Y \cap A_i) \cup L_i) = 2$, and the event is called an island, which contributes +1 to $\chi(Y)$.
b. $Y \cap A_i$ is simply connected and $(Y \cap A_i) \cap L_i \neq \emptyset$, so that $\chi((Y \cap A_i) \cup L_i) = 1$, and the contribution of this event to $\chi(Y)$ is 0.
c. $(Y \cap A_i) \cup L_i$ contains a tunnel and $(Y \cap A_i) \cap L_i \neq \emptyset$, so that $\chi((Y \cap A_i) \cup L_i) = 0$, and the event is called a bridge, which contributes -1 to $\chi(Y)$.
d. $(Y \cap A_i) \cup L_i$ contains a simply connected domain and $(Y \cap A_i) \cap L_i \neq \emptyset$, so that $\chi((Y \cap A_i) \cup L_i) = 2$. This event is called a hole, and its contribution to $\chi(Y)$ is +1.

Let I, B, H denote the total numbers of islands, bridges, and holes, respectively, observed as a thin closed slab sweeps the domain Y in an arbitrary direction, and then in the opposite direction. Then, the Hadwiger representation of the Euler–Poincaré characteristic reads as follows,

$$\chi(Y) = \frac{1}{2}(I - B + H), \tag{1.12.7}$$

see Fig. 1.12.1 for an example.

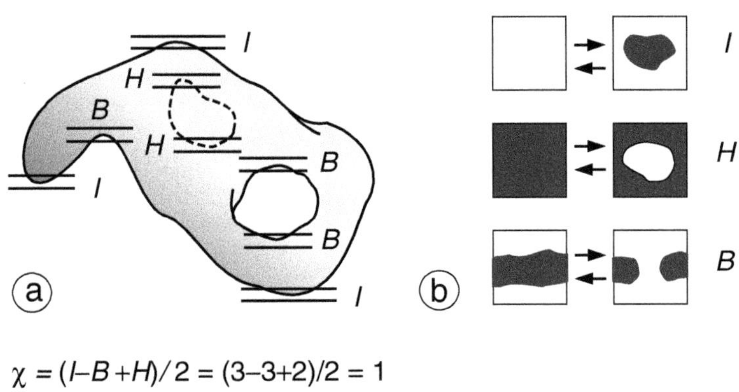

Fig. 1.12.1 Illustration of Eq. (1.12.7) for a particle with a cavity and a tunnel. If the particle is observable as a whole, then Eq. (1.12.6) yields $\chi = 1 - 1 + 1 = 1$ directly.

By analogy with Eq. (1.10.13), if a sweeping plane is used in any given direction, then, recalling Eq. (1.12.4),

$$G(Y) = G(\partial Y) = 2\pi(T_{++} - T_{+-} + T_{--}), \tag{1.12.8}$$

where T_{++}, T_{+-} and T_{--} correspond to convex, saddle and concave contacts, respectively.

1.12.3 Euler's formula in 3D

As in the planar case, an alternative representation of $\chi(Y)$ for a three-dimensional particle is provided by Euler's formula. Partition the boundary ∂Y of Y into F simply connected faces, E simply connected edges and V vertices, and apply Eq. (1.10.16) with $q = 2$, whereby, setting $\alpha_0 \equiv V$, $\alpha_1 \equiv E$, $\alpha_2 \equiv F$, we get

$$\chi(\partial Y) = V - E + F. \qquad (1.12.9)$$

For instance, if the particle is a solid tetrahedron, then we may use the natural tiling of its boundary with $F = 4$, $E = 6$, $V = 4$, whereby $\chi(\partial Y) = 2$. If we want $\chi(Y)$ for the solid tetrahedron itself, however, then we apply Eq. (1.10.16) with $q = 3$ and note that α_3 represents the solid tetrahedron, namely a simplex, hence $\alpha_3 = 1$ and

$$\chi(Y) = V - E + F - 1 = 1 = \frac{1}{2}\chi(\partial Y). \qquad (1.12.10)$$

With the practical representation given by Eq. (1.12.7), we immediately see that $\chi(Y) = 1$, because for any convex particle and any sweeping direction we will always observe $I = 2$, $B = 0$, $H = 0$.

In general, for a domain $Y \subset \mathbb{R}^3$ with boundary ∂Y, we have,

$$\chi(\partial Y) = 2\chi(Y), \qquad (1.12.11)$$

see the first paragraph of the following subsection.

1.12.4 Notes

The results in Sections 1.10 and 1.12 are special cases of the general account given in Santaló (1976), Section 13.5. Eq. (1.12.11) is a special case of Proposition 18.6.10, p. 456, from Dieck (2008).

With regard to Eq. (1.12.6), one can have a nested finite sequence: cavity $\supset$ particle $\supset$ cavity $\cdots$. If the items involved are simply connected, then each of them contributes +1 to $\chi(Y)$.

Hadwiger's representation, see Eq. (1.12.7), is equivalent to the classical representation given by Eq. (1.12.8), see DeHoff (1987, 2004). As pointed out in Section 1.10.4, however, it is not necessary to integrate over sweeping plane orientations (Gundersen, Boyce, Nyengaard, & Odgaard, 1993). For further refinements, see Ohser and Nagel (1996).

1.13 Bounded Probes: The Kinematic Density

1.13.1 Bounded probes on an axis, and in the plane

A single bounded probe on a fixed axis Ox_1 may be a test point T_0^1, or a segment of length $t > 0$, namely $T_1^1(x) = [x, x+t)$. The subscript of $T_1^1(x)$ represents the dimension of the probe, the superscript the dimension of the containing space, and $x \in \mathbb{R}$ the left endpoint of the probe segment. The associated point (AP) of a probe is a point rigidly attached to the probe which determines its position. For convenience the point x may be adopted as the AP of the probe, see Fig. 1.13.1(a). The invariant density of $T_1^1 \equiv T_1^1(x)$ is that of its AP, namely,

$$\mathrm{d}T_1^1 = \mathrm{d}x, \quad x \in \mathbb{R}. \tag{1.13.1}$$

A bounded probe in the plane may be a test point T_0^2, a bounded curve T_1^2, (e.g. a test segment), of finite length $l > 0$, or a bounded set T_2^2, generally called a quadrat, (e.g. a square), of area $a > 0$. Fix a rectangular frame Ox_1x_2 in the plane. The position and orientation of a planar probe other than a point is determined by a unit vector (x, ω) rigidly attached to the probe, emanating from a point $x \in \mathbb{R}^2$, and making an angle $\omega \in [0, 2\pi)$ with Ox_1, see Fig. 1.13.1(b). The point $x = (x_1, x_2)$ is the AP of the probe, and the vector (x, ω) is its associated vector (AV). The choice of the AV is arbitrary, but once it is chosen, it must remain rigidly attached to the probe. When $x = (0, 0)$ and $\omega = 0$, the corresponding initial position of the probe is denoted by $T_r^2(0, 0)$, $r \in \{1, 2\}$. Under a rigid motion applied to the vector (x, ω), namely a rotation ω about the origin followed by a translation x, or equivalently a translation x followed by a rotation ω about the point x, the probe $T_r^2(0, 0)$ is transformed into the probe $T_r^2(x, \omega)$. For convenience the AP may be the lower left corner of the smallest horizontal rectangle containing $T_r^2(0, 0)$. Thus, a probe does not need to contain its AP.

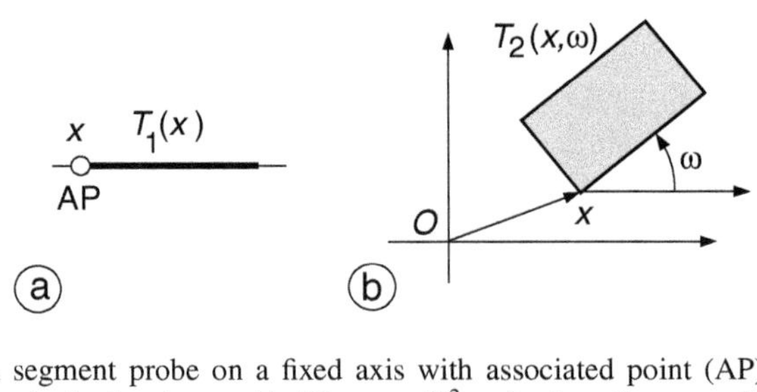

Fig. 1.13.1 (a) A segment probe on a fixed axis with associated point (AP) at $x \in \mathbb{R}$. (b) A two-dimensional probe in the plane, with AP at $x \in \mathbb{R}^2$ and orientation $\omega \in [0, 2\pi)$. See Section 1.13.1.

The motion-invariant density of $T_r^2 \equiv T_r^2(x, \omega)$ is called the kinematic density in the plane, namely

$$dT_r^2 = dx\, d\omega, \quad x \in \mathbb{R}^2, \ \omega \in \mathbb{S}^2, \ r = 1, 2, \tag{1.13.2}$$

where $dx = dx_1\, dx_2$ is the area element in the plane, and $d\omega$ the arc element of the unit circle $\mathbb{S}^1$.

1.13.2 Bounded probes in 3D

Similarly, a bounded, piecewise smooth probe $T_r^3 \subset \mathbb{R}^3$ may be a point probe T_0^3, a curve T_1^3 of finite length $l > 0$, a surface T_2^3 of area $a > 0$, or a three-dimensional set T_3^3, generally called a brick (e.g. a solid cube), of volume $v > 0$. The kinematic density of T_r^3 is the volume element of the special group of motions G_3, namely

$$dT_r^3 = dx\, du_3, \quad x \in \mathbb{R}^3, \ u_3 \in G_{3[0]}, \tag{1.13.3}$$

where $x = (x_1, x_2, x_3) \in \mathbb{R}^3$ is the AP of the probe, and dx is the corresponding volume element. On the other hand, u_3 is an element of the special group of rotations $G_{3[0]}$ about a fixed point in $\mathbb{R}^3$. Let (x, u_2) denote the AV of the probe. The rotation u_3 is decomposed into the rotation vector u_2 on the unit sphere, plus a rotation u_1 on the unit circle around u_2, namely

$$\begin{aligned} u_3 &= (u_2, u_1), \\ u_2 &\equiv u_2(\phi, \theta) \in \mathbb{S}^2, \quad \phi \in [0, 2\pi), \ \theta \in [0, \pi], \\ u_1 &\equiv \tau \in [0, 2\pi). \end{aligned} \tag{1.13.4}$$

and the rotation-invariant density becomes

$$du_3 = [du_2][d\tau] = [\sin\theta\, d\phi\, d\theta][d\tau]. \tag{1.13.5}$$

The location of the probe is determined by that of its AP, which requires three parameters, whereas the orientation of the probe requires another three parameters, namely (ϕ, θ, τ), which makes six in total. A practical expression of the kinematic density of the probe $T_r^3(x, u_2, \tau)$ is

$$dT_r^3 = dx\, du_2\, d\tau, \quad x \in \mathbb{R}^3, \ u_2 \in \mathbb{S}^2, \ \tau \in \mathbb{S}^1. \tag{1.13.6}$$

As indicated in Section 1.2.7, unlike the symbol $u \in \mathbb{S}_+^2$, used so far to represent an axial direction, now $u_2 \in \mathbb{S}^2$ represents a vector orientation.

1.13.3 Indirect generation of bounded invariant probes

Consider a bounded probe $T_r^2 \subset \mathbb{R}^2$ which can be contained in a straight line. Thus, the probe may be either a point T_0^2, or a straight line segment T_1^2.

The indirect generation of an invariant point $T_0^2 \equiv L_0^2$ within translation-invariant straight lines and planes was considered in Section 1.2.4.

In stereological applications a straight line segment $T_1^2 \equiv T_1^2(x, \omega)$ in the plane may be considered to be unoriented, in which case the associated point x is taken at the midpoint of the segment. Then, the invariant density of the segment is

$$\mathrm{d}T_1^2 = \mathrm{d}x\, \mathrm{d}\omega, \quad x \in \mathbb{R}^2, \ \omega \in [0, \pi). \tag{1.13.7}$$

Suppose that $T_1^2(x, \omega) \subset L_1^2(p, \phi)$, and let $z \equiv L_0^1$ represent the associated point x within the straight line L_1^2, see Fig. 1.13.2(a). Because $\omega = \pi/2+\phi$ we have $\mathrm{d}\omega = \mathrm{d}\phi$, and therefore,

$$\mathrm{d}T_1^2 = [\mathrm{d}x\, \mathrm{d}\omega] = [\mathrm{d}p\, \mathrm{d}z][\mathrm{d}\omega] = [\mathrm{d}p\, \mathrm{d}\phi][\mathrm{d}z] = \mathrm{d}L_1^2\, \mathrm{d}T_1^1, \tag{1.13.8}$$

which means that an unoriented segment T_1^2 with motion-invariant density in the plane is equivalent to a translation-invariant segment within a motion-invariant test line.

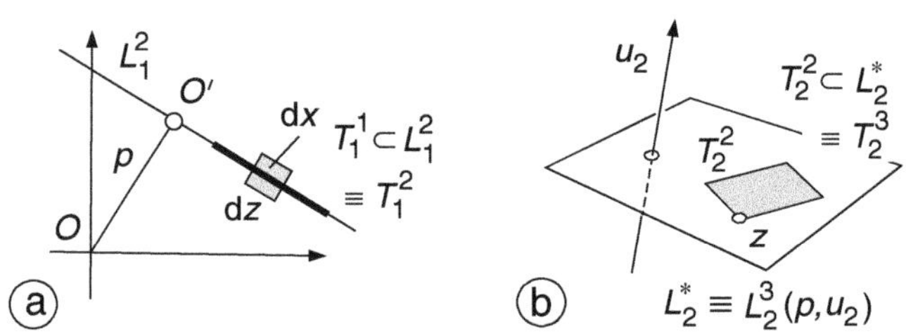

Fig. 1.13.2 (a) An invariant segment in the plane may be generated as a translation-invariant segment within an invariant line. (b) A planar invariant probe in space may be generated as an invariant probe within an invariant oriented plane.

Likewise, a motion-invariant straight line segment in space is equivalent to a translation-invariant straight line segment sliding within a motion-invariant straight line, namely

$$\mathrm{d}T_1^3 = \mathrm{d}L_1^3\, \mathrm{d}T_1^1. \tag{1.13.9}$$

Suppose that a bounded probe T_r^3, $r \in \{1, 2\}$, can be contained in an oriented plane $L_2^* \equiv L_2^3(p, u_2)$, $p \in \mathbb{R}$, $u_2 \in \mathbb{S}$, namely $T_r^3 \subset L_2^* \subset \mathbb{R}^3$, and let $z \in L_2^*$ represent the associated point x of the probe within L_2^*, see Fig. 1.13.2(b). Then, $\mathrm{d}x = \mathrm{d}p\, \mathrm{d}z$, whereby,

$$\begin{aligned} dT_r^3 &= dx\, du_2\, d\tau \\ &= [dp\, dz][du_2\, d\tau] \\ &= [dp\, du_2][dz\, d\tau] \\ &= dL_2^*\, dT_r^2, \end{aligned} \tag{1.13.10}$$

which means that a motion-invariant probe which can be contained in a plane is equivalent to a motion-invariant probe within a motion-invariant oriented plane.

1.13.4 Invariant cylinders in space

A right cylinder $Z_r^3 \subset \mathbb{R}^3$ of dimension $r \in \{2, 3\}$ is defined as the set product

$$Z_r^3 = L_{1[0]}^3 \times T_{r-1}^2, \quad r \in \{2, 3\}, \tag{1.13.11}$$

where the axis $L_{1[0]}^3$ is the generator, and the planar set T_{r-1}^2 is the (piecewise smooth) directrix of the cylinder, see Section 1.6.2 ("Curtain surface") and Fig. 1.13.3. Note that $T_{r-1}^2 = Z_r^3 \cap L_{2[0]}^3$ is the cross-section of the cylinder by a plane normal to its generator. For $r = 2$, the directrix T_1^2 is a planar closed curve without double points and Z_2^3 is a cylindrical surface. If $r = 3$, then T_2^2 is a planar domain and Z_3^3 is a solid cylinder.

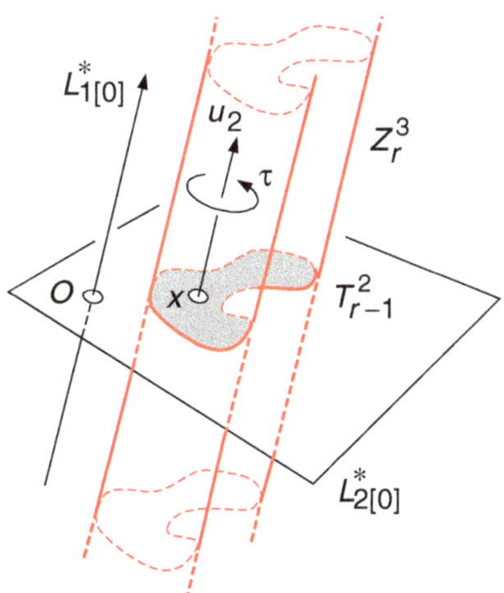

Fig. 1.13.3 Sketch of a cylinder probe whose cross section T_{r-1}^2 lies in an oriented pivotal plane $L_{2[0]}^*$. The cylinder is a straight line, a cylindrical surface, or a solid cylinder, according to whether $r = 1, 2, 3$, respectively.

Suppose that $T^2_{r-1} \equiv T^2_{r-1}(x,\tau)$, $x \in \mathbb{R}^2$, $\tau \in [0, 2\pi)$, is equipped with the kinematic density $\mathrm{d}T^2_{r-1} = \mathrm{d}x\,\mathrm{d}\tau$ within an oriented pivotal plane $L^*_{2[0]}$ with rotation-invariant density $\mathrm{d}L^*_{2[0]} = \mathrm{d}L^*_{1[0]} = \mathrm{d}u_2$, see Section 1.13.3 and Fig. 1.13.2(b). Then the motion-invariant density of Z^3_r is

$$\begin{aligned}\mathrm{d}Z^3_r &= [\mathrm{d}T^2_{r-1}][\mathrm{d}L^*_{1[0]}] \\ &= \mathrm{d}x\,\mathrm{d}u_2\,\mathrm{d}\tau, \quad x \in \mathbb{R}^2,\ u_2 \in \mathbb{S}^2,\ \tau \in \mathbb{S}^1.\end{aligned} \tag{1.13.12}$$

The axis $L^*_{1[0]}$ parallel to the generator has to be oriented, because if the orientation $u_2(\phi,\theta)$ is replaced with $u_2(\phi,\theta+\pi)$, then the resulting cylinder is in general not congruent with the original one.

1.13.5 Notes

1. Kinematic densities

The underlying theory is provided for instance by Santaló (1976), De-lin (1994), or Hadwiger (1957).

2. Invariant cylinders

The motion-invariant density for cylinders in $\mathbb{R}^3$ was given by Rey-Pastor and Santaló (1951). For a general treatment, see Santaló (1976) and Cruz-Orive and Gual-Arnau (2020). See also Note 2 in Section 1.16.5 below.

1.14 Crofton Formulae for Bounded Probes: Purpose and Preliminaries

The dimensionality relations given in Section 1.3 apply to bounded probes without change. The intersection formulae sought here involve a fixed compact set $Y \subset \mathbb{R}^d$ of q-measure $\gamma(Y)$, $q = 0, 1, \ldots, d$, hit by a bounded probe $T_r \equiv T^d_r$ of r-measure $\nu(T_r)$, $r \geq d - q$, and they are special cases of the following identity,

$$\int \alpha(Y \cap T_r)\,\mathrm{d}T_r = c_2 \cdot \nu(T_r) \cdot \gamma(Y), \tag{1.14.1}$$

where $\mathrm{d}T_r$ is the kinematic density of T_r and $c_2 \equiv c_2(q, r, d)$ is a known constant (Section 1.16.4). The case $r = 0$, when T_0 is a test point and $\nu(T_0) = 1$, was treated in Section 1.4.1. Note that in Eq. (1.14.1) the roles of Y and T_r may be exchanged: the probe T_r may be fixed and the set Y mobile and equipped with the kinematic density dY. Thus, equivalently,

$$\int \alpha(T_r \cap Y)\, dY = c_2 \cdot \nu(T_r) \cdot \gamma(Y). \tag{1.14.2}$$

1.15 Crofton Formulae for Bounded Probes of a Fixed Orientation

1.15.1 Lebesgue measure from lower-dimensional probes of a fixed orientation

If the set $Y \subset \mathbb{R}^d$ of interest is of full dimension, namely if $\dim(Y) = d$, then its Lebesgue measure can be represented by an intersection formula using a test probe $T_r^d \subset \mathbb{R}^d$ of arbitrary dimension $r \in \{0, 1, \ldots, d\}$ with a fixed orientation and translation-invariant density $dT_r^d = dx$, $x \in \mathbb{R}^d$.

Planar area from test curves

Let $Y \subset \mathbb{R}^2$ represent a planar domain of area A, and consider a test curve $T_1^2(x, \omega)$ of length l with an arbitrarily fixed orientation $\omega \in [0, 2\pi)$, which is free to move parallel to itself with translation-invariant density $dT_1^2(x, \omega) = dx$, $x \in \mathbb{R}^2$. The intercept $Y \cap T_1^2(x, \omega)$ is a finite set of curves, and we want to compute the total length measure of such intercepts as the point x moves on the entire plane. Let $dl(y)$ denote the length of an essentially linear curve element δy at a point $y \in Y \cap T_1^2(x, \omega)$. Then, for any orientation $\omega \in [0, 2\pi)$,

$$\begin{aligned}
\int_{\mathbb{R}^2} L(Y \cap T_1^2(x, \omega))\, dx &= \int_{\mathbb{R}^2} dx \int_{T_1^2(x,\omega)} 1_Y(y)\, dl(y) \\
&= \int_{\mathbb{R}^2} dx \int_{T_1^2(0,\omega)} 1_Y(y + x)\, dl(y) \\
&= \int_{T_1^2(0,\omega)} dl(y) \int_{\mathbb{R}^2} 1_Y(y + x)\, dx \\
&= lA.
\end{aligned} \tag{1.15.1}$$

Volume from bounded test curves and surfaces

Likewise, for a three-dimensional domain $Y \subset \mathbb{R}^3$ of volume V, hit by a test curve $T_1^3(x, u_2, \tau)$ of length l, with arbitrarily fixed orientation parameters $u_2 \in \mathbb{S}^2$ and $\tau \in [0, 2\pi)$, we obtain

$$\int_{\mathbb{R}^3} L(Y \cap T_1^3(x, u_2, \tau))\,\mathrm{d}x = lV. \tag{1.15.2}$$

If the probe is a test surface $T_2^3(x, u_2, \tau)$ of area a, with arbitrarily fixed orientation parameters, then,

$$\int_{\mathbb{R}^3} S(Y \cap T_2^3(x, u_2, \tau))\,\mathrm{d}x = aV. \tag{1.15.3}$$

1.15.2 Measures of arbitrary dimension from full-dimensional probes of a fixed orientation

If the test probe is of full dimension, namely $T_d^d \subset \mathbb{R}^d$, then in order to write the relevant intersection formula it suffices that T_d^d has a fixed orientation, with translation-invariant density $\mathrm{d}T_d^d = \mathrm{d}x$, $x \in \mathbb{R}^d$, whatever the dimension $q \in \{0, 1, \ldots, d\}$ of the set $Y \subset \mathbb{R}^d$ of interest.

Points, curves, and domains in the plane intersected by a quadrat

If $Y = \cup_{i=1}^N y_i$ is a fixed set of N separate point particles in the plane, then for a test quadrat $T_2^2(x, \omega)$ of area a with an arbitrarily fixed orientation $\omega \in [0, 2\pi)$, and free to move parallel to itself with translation-invariant density $\mathrm{d}T_2^2(x, \omega) = \mathrm{d}x$, we have

$$\begin{aligned}
\int_{\mathbb{R}^2} N(Y \cap T_2^2(x, \omega))\,\mathrm{d}x &= \int_{\mathbb{R}^2} \mathrm{d}x \sum_{i=1}^N 1_{T_2^2(x,\omega)}(y_i) \\
&= \int_{\mathbb{R}^2} \mathrm{d}x \sum_{i=1}^N 1_{T_2^2(0,\omega)}(y_i - x) \\
&= \sum_{i=1}^N \int_{\mathbb{R}^2} 1_{T_2^2(0,\omega)}(y_i - x)\,\mathrm{d}x \\
&= Na.
\end{aligned} \tag{1.15.4}$$

Now consider a bounded finite union of curves $T \subset \mathbb{R}^2$ of total length B. Let $\mathrm{d}b(y)$ denote the length of an essentially linear curve element δy at a point $y \in Y$. Then, for any orientation $\omega \in [0, 2\pi)$,

$$
\begin{aligned}
\int_{\mathbb{R}^2} L(Y \cap T_2^2(x,\omega))\, \mathrm{d}x &= \int_{\mathbb{R}^2} \mathrm{d}x \int_Y 1_{T_2^2(x,\omega)}(y)\, \mathrm{d}b(y) \\
&= \int_{\mathbb{R}^2} \mathrm{d}x \int_Y 1_{T_2^2(0,\omega)}(y-x)\, \mathrm{d}b(y) \\
&= \int_Y \mathrm{d}b(y) \int_{\mathbb{R}^2} 1_{T_2^2(0,\omega)}(y-x)\, \mathrm{d}x \\
&= Ba. \qquad (1.15.5)
\end{aligned}
$$

Finally let $Y \subset \mathbb{R}^2$ represent a bounded finite union of planar domains of total area A. Then,

$$
\begin{aligned}
\int_{\mathbb{R}^2} A(Y \cap T_2^2(x,\omega))\, Dx &= \int_{\mathbb{R}^2} 1_{Y \cap T_2^2(x,\omega)}(y)\, \mathrm{d}y \\
&= \int_{\mathbb{R}^2} \mathrm{d}x \int_{\mathbb{R}^2} 1_Y(y) 1_{T_2^2(x,\omega)}(y)\, \mathrm{d}y \\
&= \int_{\mathbb{R}^2} 1_Y(y)\, \mathrm{d}y \int_{\mathbb{R}^2} 1_{T_2^2(0,\omega)}(y-x)\, \mathrm{d}x \\
&= Aa. \qquad (1.15.6)
\end{aligned}
$$

Three-dimensional case

Suppose that the set $Y \subset \mathbb{R}^3$ consists of either N point particles, or finite unions of curves of total length L, surfaces of total area S, or domains of total volume V. In a similar manner it can readily be verified that if the set Y is hit by a full-dimensional test probe $T_3^3(x, u_2, \tau)$ of volume υ with a fixed orientation, then

$$
\int_{\mathbb{R}^3} \gamma(Y \cap T_3^3(x, u_2, \tau)\, \mathrm{d}x = \upsilon\gamma(Y), \qquad (1.15.7)
$$

where γ stands for either N, L, S, or V.

1.15.3 Notes

1. Probes free to rotate

If instead of having a fixed orientation the bounded probes are equipped with their full kinematic density, then the rhs of the foregoing formulae have to be multiplied by the factors

$$\int_0^{2\pi} \mathrm{d}\omega = 2\pi, \qquad (1.15.8)$$

$$\int_0^{2\pi} \mathrm{d}\phi \int_0^{\pi} \sin\theta \int_0^{2\pi} \mathrm{d}\tau = 8\pi^2, \qquad (1.15.9)$$

in the planar and spatial cases, respectively.

2. Balanzat's theorem

Eq. (1.15.6) is a special case of a general result proved by Balanzat (1940). An intuitive derivation – often used by L. A. Santaló to explain similar results in his lectures – runs as follows. If for each point $y \in Y$ we translate the probe $T(x)$ with the condition $T(x) \ni y$ (in order to ensure that $y \in Y \cap T(x)$), then the associated point x of the probe will describe the probe area a. If we repeat this procedure for each $y \in Y$, then the combined measure of the translations of y, and of x for each y, will be equal to A times a.

1.16 Crofton Formulae for Bounded Invariant Probes

Suppose that the set $Y \subset \mathbb{R}^d$ has dimension $q \in \{0, 1, \ldots, d-1\}$ and the test probe $T_r^d \subset \mathbb{R}^d$ has dimension $r \in \{1, \ldots, d-1\}$ with the condition $q + r \geq d$. Then, in order to write the relevant intersection formulae it is necessary that the probe is equipped with the kinematic density.

The following definition is useful in the present context.

1.16.1 Minkowski addition. The catching set

The Minkowski addition (or sum) of two sets $Y, T \subset \mathbb{R}^d$ is defined as follows,

$$Y \oplus T = T \oplus Y = \{y + t : \; y \in Y, \; t \in T\}. \qquad (1.16.1)$$

Suppose that the set $Y \subset \mathbb{R}^d$ is fixed whereas the set $T(x) \subset \mathbb{R}^d$ is mobile with a fixed orientation and associated point $x \in \mathbb{R}^d$. Then the geometric locus of all positions of x such that $Y \cap T(x) \neq \emptyset$, namely the positions of x such that $T(x)$ hits or 'catches' a target Y, is called the 'catching set', namely,

$$Y \oplus \check{T}(0) = \{x : \; Y \cap T(x) \neq \emptyset\}, \qquad (1.16.2)$$

see Fig. 1.16.1(b), where the set $\check{T}(0) = -T(0) = \{-t : \; t \in T(0)\}$ is the reflection of the set $T(0)$ about the origin, see Fig. 1.16.1(a). An alternative representation of the catching set is

$$Y \oplus \check{T}(0) = \bigcup_{x \in Y} \check{T}(x), \tag{1.16.3}$$

see Fig. 1.16.1(c).

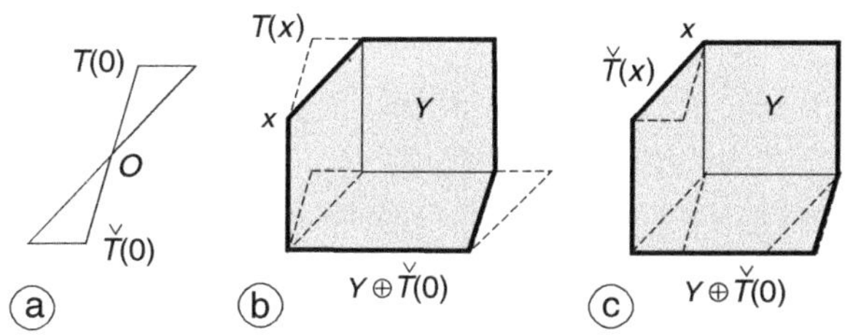

Fig. 1.16.1 (a) A bounded probe $T(0)$ and its reflection $\check{T}(0)$. (b) Example of Minkowski addition according to Eq. (1.16.2). (c) Idem according to Eq. (1.16.3).

1.16.2 Curves intersecting in the plane: Poincaré's formula

Consider a fixed planar curve $Y \subset \mathbb{R}^2$ of total length B hit by a planar mobile curve $T_1^2 \equiv T_1^2(x, \omega)$ of length l, equipped with the kinematic density $\mathrm{d}T_1^2 = \mathrm{d}x\,\mathrm{d}\omega$. We seek the total measure of the number of intersections between the fixed curve and the mobile one. Let δy be an essentially linear arc element of length $\mathrm{d}b$ at a point $y \in Y$, making an angle $\psi \equiv \psi(y) \in [0, \pi)$ with the axis Ox_1. Similarly, let $\delta t(0, \varphi)$ be an essentially linear arc element of length $\mathrm{d}l$ at a point $t \in T_1^2(0, 0)$, making an angle $\varphi \equiv [0, \pi)$ with Ox_1, see Fig. 1.16.2(a). Thus, $\delta t(x, \varphi + \omega)$ represents the arc element at a point $t \in T_1^2(x, \omega)$. By virtue of Eq. (1.16.2), the conditions $I(\delta y \cap \delta t(x, \varphi + \omega)) = 1$ and $x \in \delta y \oplus \check{\delta t}(0, \varphi + \omega)$ are equivalent. The latter Minkowski sum is a parallelogram of base length $\mathrm{d}b$ and height $\mathrm{d}l|\sin(\psi - (\varphi + \omega))|$, see Fig. 1.16.2(b), and therefore,

$$\mathrm{d}T_1^2 = |\sin(\psi - (\varphi + \omega))|\,\mathrm{d}b\,\mathrm{d}l\,\mathrm{d}\omega. \tag{1.16.4}$$

Consequently,

$$\begin{aligned} \int I(Y \cap T_1^2)\,\mathrm{d}T_1^2 &= \int_Y \mathrm{d}b \int_{T_1^2(0,0)} \mathrm{d}l \int_0^{2\pi} |\sin(\psi - (\varphi + \omega))|\,\mathrm{d}\omega \\ &= 4Bl, \end{aligned} \tag{1.16.5}$$

which is Poincaré's formula.

Similarly as in the unbounded probe case (Section 1.5.1), given Y and $T_1^2(0, 0)$, the corresponding sets of local orientation parameters $\{\psi(y),\ y \in Y\}$ and $\{\varphi(t),\ t \in T_1^2(0, 0)\}$ are fixed, and therefore they do not influence the relevant

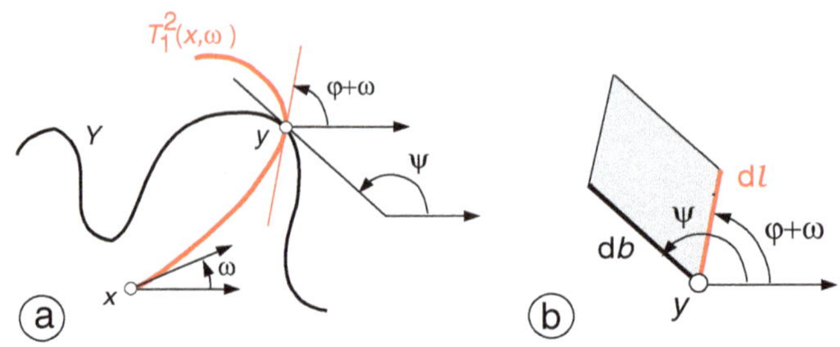

Fig. 1.16.2 (a) Notation used to derive Poincaré's formula, Eq. (1.16.5). (b) The catching set for a probe element (in red) hitting a curve element, used to obtain Eq. (1.16.4).

integral because the kinematic density is motion-invariant. Thus, henceforth local orientation parameters will not be considered in this context either.

1.16.3 Intersection formulae involving bounded curves and surfaces in space

Surface intersected by a curve

Let $Y \subset \mathbb{R}^3$ be a fixed surface of area S and $T_1^3 \equiv T_1^3(x, u_2, \tau)$ a mobile curve of length l equipped with the kinematic density $\mathrm{d}T_1^3 = \mathrm{d}x\,\mathrm{d}u_2\,\mathrm{d}\tau$.

First we seek the total measure of the number of intersections between the fixed surface and the mobile curve. Let δy represent an essentially planar surface element of area $\mathrm{d}s$ at a point $y \in Y$ and δt an essentially linear curve element of length $\mathrm{d}l$ at a point $t \in T_1^3$. Because the density $\mathrm{d}T_1^3$ is motion-invariant, the normal to the surface element δy may be taken along the polar axis Ox_3. Now, ignoring the local orientation of δt for $t \in T_1^3(0,0,0)$, the Minkowski sum $\delta y \oplus \check{\delta t}(0, u_2, \tau)$ is effectively an oblique prism of base area $\mathrm{d}s$ and height $\mathrm{d}l|\cos\theta|$, see Fig. 1.16.3(a), whereby,

$$\mathrm{d}T_1^3 = |\cos\theta|\,\mathrm{d}s\,\mathrm{d}l\,\mathrm{d}u_2\,\mathrm{d}\tau, \tag{1.16.6}$$

and we obtain,

$$\begin{aligned}
&\int I(Y \cap T_1^3)\,\mathrm{d}T_1^3 \\
&\quad = \int_Y \mathrm{d}s \int_{T_1^3(0,0,0)} \mathrm{d}l \int_0^{2\pi} \mathrm{d}\tau \int_0^{2\pi} \mathrm{d}\phi \int_0^{\pi} |\cos\theta| \sin\theta\,\mathrm{d}\theta \\
&\quad = 4\pi^2 S l.
\end{aligned} \tag{1.16.7}$$

The dual result arises when a fixed curve $Y \subset \mathbb{R}^3$ of length L is hit by a mobile test surface $T_2^3 \equiv T_2^3(x, u_2, \tau)$ of area a equipped with the kinematic density. With the pertinent change of notation of I, S, l with Q, a, L, respectively, the preceding argument shows that the total measure of the number of intersections between the fixed curve and the mobile test surface is

$$\int Q(Y \cap T_2^3)\, \mathrm{d}T_2^3 = 4\pi^2 La. \tag{1.16.8}$$

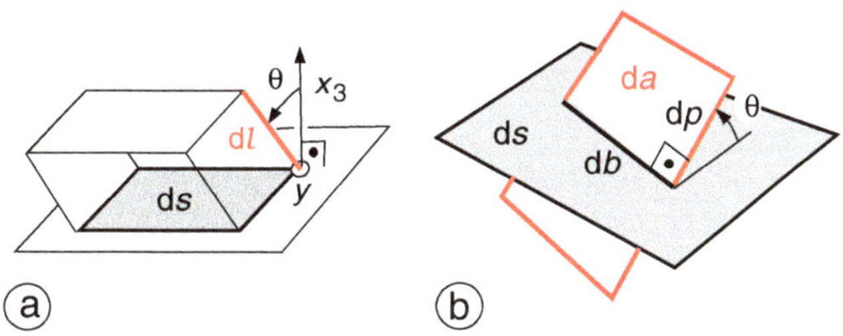

Fig. 1.16.3 (a) Catching set used to derive Eq. (1.16.6). (b) Notation used in Eq. (1.16.9).

Surface intersected by another surface

Consider a fixed surface $Y \subset \mathbb{R}^3$ of area S, and a mobile test surface $T_2^3(x, u_2, \tau)$ of area a equipped with the kinematic density $\mathrm{d}T_2^3 = \mathrm{d}x\, \mathrm{d}u_2\, \mathrm{d}\tau$. Our purpose is to compute the total measure of the trace length determined in the fixed surface by the mobile one. Let δy, δt denote essentially planar surface elements of areas $\mathrm{d}s$, $\mathrm{d}a$ at the points $y \in Y$, $t \in T_2^3(x, u_2, \tau)$, respectively. As before, the normal to the element δy is taken along the Ox_3 axis. Whenever $\delta y \cap \delta t \neq \emptyset$, set $B(\delta y \cap \delta t) = \mathrm{d}b$, the length of the corresponding trace element. The area element $\mathrm{d}a$ may now be expressed as $\mathrm{d}b\, \mathrm{d}p$, where $\mathrm{d}p$ is a length element contained in δt and perpendicular to the trace element, see Fig. 1.16.3(b). Now the volume element $\mathrm{d}x$ may be represented by an oblique prism of base area $\mathrm{d}s$ and height $\mathrm{d}p \sin\theta$, whereby $\mathrm{d}T_2^3 = [\sin\theta\, \mathrm{d}s\, \mathrm{d}p][\mathrm{d}u_2\, \mathrm{d}\tau]$ and,

$$\begin{aligned} \mathrm{d}T_2^3\, \mathrm{d}b &= [\sin\theta\, \mathrm{d}s\, \mathrm{d}b\, \mathrm{d}p][\mathrm{d}u_2\, \mathrm{d}\tau] \\ &= [\sin\theta\, \mathrm{d}s\, \mathrm{d}a][\sin\theta\, \mathrm{d}\phi\, \mathrm{d}\theta\, \mathrm{d}\tau]. \end{aligned} \tag{1.16.9}$$

Therefore, the required measure becomes

$$\begin{aligned}\int B(Y\cap T_2^3)\,\mathrm{d}T_2^3 &= \int \mathrm{d}T_2^3 \int_{Y\cap T_2^3} \mathrm{d}b \\ &= \int_Y \mathrm{d}s \int_{T_2^3(0,0,0)} \mathrm{d}a \int_0^{2\pi} \mathrm{d}\phi \int_0^{\pi} \sin^2\theta\,\mathrm{d}\theta \int_0^{2\pi} \mathrm{d}\tau \\ &= 2\pi^3 Sa. \end{aligned} \tag{1.16.10}$$

1.16.4 General Crofton formula for bounded probes

Analogously as in Section 1.5.5, the foregoing Crofton formulae are special cases of Eq. (1.14.1), or of Eq. (1.14.2), with

$$c_2(q,r,d) = c_{20}(d)\cdot\frac{O_d O_{q+r-d}}{O_q O_r}, \tag{1.16.11}$$

where

$$c_{20}(d) \equiv \int_{G_{d[0]}} \mathrm{d}u_d = O_{d-1}O_{d-2}\cdots O_1. \tag{1.16.12}$$

Here u_d, already introduced in Section 1.13.2 for $d = 3$, is an element of the special group of rotations $G_{d[0]}$, isomorphic to $\mathrm{SO}(d)$, about a fixed point in $\mathbb{R}^d$, and O_k is given by Eq. (1.5.16). For instance, the coefficients in Eq. (1.15.8) and Eq. (1.15.9) are

$$c_{20}(2) = 2\pi, \quad c_{20}(3) = 8\pi^2. \tag{1.16.13}$$

The coefficients obtained in Sections 1.16.2 and 1.16.3 are,

$$\begin{aligned} &c_2(1,1,2) = 4, \\ &c_2(2,1,3) = c_2(1,2,3) = 4\pi^2, \\ &c_2(2,2,3) = 2\pi^3. \end{aligned} \tag{1.16.14}$$

1.16.5 Notes

1. Crofton formulae for bounded probes

The formulae derived in the present section have essentially been known for a long time, e.g. Barbier (1860) – for pertinent comments see Hykšová et al. (2012). The coefficient in Eq. (1.16.11) is that of Eq. (15.20) from Santaló (1976), or of Eq. (6.8.16) from De-lin (1994).

2. Crofton formulae for cylinders

A general Crofton intersection formula for cylinders in $\mathbb{R}^d$ was given by Gual-Arnau and Cruz-Orive (2020), who included historical notes on the problem. A derivation for $\mathbb{R}^3$ is deferred to Section 2.27.3 because it uses test systems as a prerequisite. Schneider and Weil (2008) consider cylinders with a convex directrix.

3. Poincaré's formula

As pointed out by Rey-Pastor and Santaló (1951) and Santaló (1976), Eq. (1.16.5) was called Poincaré's formula by Blaschke (1936–1937, p. 24), although Poincaré (1912, p. 143) only hints the analogous result for intersecting curves on a sphere. Barbier (1860) gives the result for planar curves without proof. An early proof was given by Santaló (1936a).

1.17 Surface Area From Vertical Sections and Cycloids

1.17.1 Integral formula for a bounded curve intersected by a cycloid with a fixed orientation in the plane

The vertical section identity (1.6.1) is based on a straight line $L^2_{1\cdot v}$ contained in a vertical plane and with density proportional to $\sin\theta$, where θ is the angle of the line with the vertical axis Ox_3, see Eq. (1.2.16).

Fix a rectangular frame Ox_1x_2 in the plane and consider a bounded planar test curve $T^2_1(z)$ of finite length l, with associated point $z \in \mathbb{R}^2$, and with translation-invariant density $\mathrm{d}T^2_1(z) = \mathrm{d}z$, which means that $T^2_1(z)$ has a fixed orientation in the plane. Instead of a sine-weighted test line, we seek a test curve $T^2_1(0)$ of fixed orientation with the property that the length $\mathrm{d}l(\theta)$ of the curve element $\delta t \equiv \delta t(\theta)$ at a point $t \in T^2_1(0)$ is proportional to $\sin\theta$, namely, $\mathrm{d}l(\theta) = c|\sin\theta|\,\mathrm{d}\theta$, where c is a constant and $\theta \in [0, 2\pi)$ is the angle of the oriented tangent to $T^2_1(0)$ at the point t, with the half axis Ox_2, see Fig. 1.17.1(b). The parametric equations of $T^2_1(0)$ become

$$
\begin{aligned}
x_1(p) &= \int_0^p \sin\theta\,\mathrm{d}l(\theta) = (c/2)(p - \sin p\cos p), \quad p \in [0, 2\pi), \\
x_2(p) &= \begin{cases} \int_0^p \cos\theta\,\mathrm{d}l(\theta) = (c/2)\sin^2 p, & p \in [0, \pi), \\ -(c/2)\sin^2 p, & p \in [\pi, 2\pi). \end{cases}
\end{aligned}
\tag{1.17.1}
$$

Thus $T^2_1(0)$ is a cycloid curve consisting of the union of an upper arc (with $x_2 \geq 0$), and a lower one ($x_2 < 0$), which is the reflection of the upper arc with respect to the point $(x_1(\pi), 0)$, see Fig. 1.17.1(a). The complete length of the curve is,

$$l = \int_{T_1^2(0)} \mathrm{d}l = c \int_0^{2\pi} |\sin\theta|\,\mathrm{d}\theta = 4c, \tag{1.17.2}$$

whereby $c = l/4$. The minor and major principal axes of each arc have lengths $x_2(\pi/2) = l/8$ and $x_1(\pi) = \pi l/8$ respectively, and the length of each arc is equal to $l/2$.

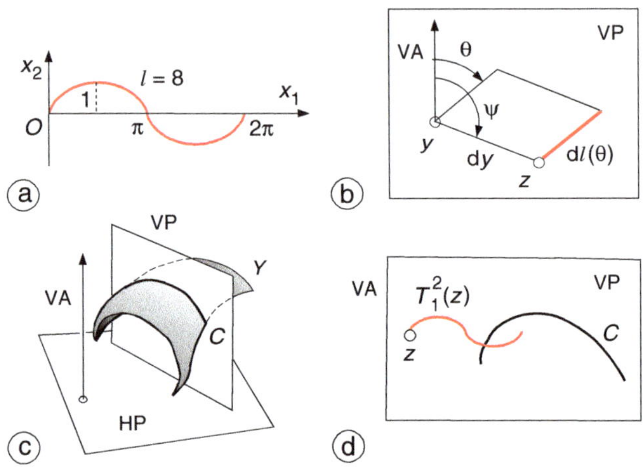

Fig. 1.17.1 (a) Cycloid test probe, Eq. (1.17.1). (b) Notation used in Eq. (1.17.3). (c) Vertical trace curve C of a surface Y. (d) Notation used in Eq. (1.17.4).

Consider now a fixed bounded curve $C \subset \mathbb{R}^2$. We seek the total measure of the number of intersections between C and the preceding cycloid test curve $T_1^2(z)$ with translation-invariant density $\mathrm{d}T_1^2(z) = \mathrm{d}z$ in the plane, and with the minor principal axes parallel to the vertical axis Ox_2, see Fig. 1.17.1(d). Let δy represent an essentially linear element of length $\mathrm{d}y$ at a point $y \in C$, and $\psi \equiv \psi(y) \in [0, \pi)$ the angle of δy with the vertical axis Ox_3. Now the angle ψ has to be taken into account because the probe density is not rotation-invariant. The conditions $I(\delta y \cap \delta(\theta)) = 1$ and $z \in \delta y \oplus \breve{\delta} t(\theta)$ are equivalent. The latter Minkowski sum is a parallelogram of base $\mathrm{d}y$ and height $|\sin(\psi - \theta)|\,\mathrm{d}l(\theta)$, see Fig. 1.17.1(b), whereby,

$$\mathrm{d}z = |\sin(\psi - \theta)|\,\mathrm{d}y\,\mathrm{d}l(\theta). \tag{1.17.3}$$

Thus,

$$
\begin{aligned}
\int_{\mathbb{R}^2} I(C \cap T_1^2(z))\, \mathrm{d}z &= \int_C \int_{T_1^2(0)} |\sin(\psi - \theta)|\, \mathrm{d}y\, \mathrm{d}l(\theta) \\
&= \frac{l}{4} \int_C \mathrm{d}y \int_0^{2\pi} |\sin(\psi - \theta)||\sin\theta|\, \mathrm{d}\theta \\
&= \frac{l}{2} \int_C \mathrm{d}y \int_0^{\pi} |\sin(\psi - \theta)| \sin\theta\, \mathrm{d}\theta \\
&= \frac{l}{2} W(C), \qquad (1.17.4)
\end{aligned}
$$

where

$$
W(C) = \int_{y \in C} (\sin\psi + (\frac{\pi}{2} - \psi)\cos\psi)\, \mathrm{d}y(\psi), \qquad (1.17.5)
$$

as in Eq. (1.6.4).

1.17.2 Surface area using a cycloid test curve with a fixed orientation on a vertical plane

Consider a surface $Y \subset \mathbb{R}^3$ of area S and the vertical design described in Section 1.6, see Fig. 1.17.1(c). Combining Eq. (1.6.3) and Eq. (1.17.4) with $C \equiv Y \cap L^3_{2\cdot v}$ yields the following identity,

$$
\begin{aligned}
S &= \frac{1}{\pi} \int W(Y \cap L^3_{2\cdot v})\, \mathrm{d}L^3_{2\cdot v} \\
&= \frac{2}{\pi l} \int \mathrm{d}L^3_{2\cdot v} \int_{L^3_{2\cdot v}} I\{(Y \cap L^3_{2\cdot v}) \cap T_1^2(z)\}\, \mathrm{d}z, \qquad (1.17.6)
\end{aligned}
$$

which shows that using a sine-weighted test line on a vertical plane, see Eq. (1.6.1), is equivalent to using a translation-invariant cycloid test curve $T_1^2(z)$ on the vertical plane, with the minor principal axes parallel to the vertical axis.

1.17.3 Notes

1. Vertical sections and cycloids

The representation (1.17.6) is due to Baddeley (1985), see also Baddeley and Jensen (2005) and Cruz-Orive, Gelšvartas, and Roberts (2014, Appendix A).

2. On the cycloid

The cycloid is a remarkable curve with a long history. A cycloid arc, see Eq. (1.17.1) with $p \in [0, \pi)$, is described by a fixed point (initially at the origin) of a circle, as the circle rolls an entire turn along the Ox_1 axis. Its properties were the object of study among mathematicians and physicists during the 16th and the 17th centuries. Well known are the brachystochrone property, found by Jakob Bernoulli, and the tautochrone and the isochrone properties, discovered by Christiaan Huygens. The latter property – together with the fact that the evolute of a half cycloid arc is a congruent half cycloid arc – is the basis of Huygens' isochronous cycloidal pendulum. Lockwood (1961) is an excellent treatise.

1.18 The Euler–Poincaré Characteristic From Bounded Probes

The identities (1.10.12) and (1.12.7) are based on thin sweeping stripes and slabs, respectively, and therefore they implicitly assume that the domains of interest are observable in their entirety. The identities given below use bounded translation-invariant probes, and they involve information contained solely in the probes.

1.18.1 The shell formula for a planar domain

Let $Y \subset \mathbb{R}^2$ represent a planar domain whose boundary ∂Y is piecewise smooth with no double points. Consider a series of adjacent horizontal stripes $\{A_i,\ i \in \mathbb{Z}\}$ of thickness $a > 0$, namely,

$$A_i = \{(x_1, x_2) \in \mathbb{R}^2 :\ x_1 \in \mathbb{R},\ (i-1)a \leq x_2 \leq ia\}. \tag{1.18.1}$$

By the first Eq. (1.10.11), the Euler–Poincaré characteristic of Y may be expressed as follows,

$$\chi(Y) = \sum_{i \in \mathbb{Z}} \{\chi(Y \cap A_i) - \chi(Y \cap A_i \cap A_{i+1})\}. \tag{1.18.2}$$

Now consider a perpendicular series of adjacent vertical stripes $\{B_j,\ j \in \mathbb{Z}\}$ of thickness $b > 0$, namely,

$$B_j = \{(x_1, x_2) \in \mathbb{R}^2 :\ (j-1)b \leq x_1 \leq jb,\ x_2 \in \mathbb{R}\}. \tag{1.18.3}$$

Application of Eq. (1.18.2) to $\chi(Y \cap A_i)$ and to $\chi(Y \cap A_i \cap A_{i+1})$ with respect to the series of vertical stripes, yields

$$\chi(Y \cap A_i) - \chi(Y \cap A_i \cap A_{i+1}) = \sum_{j \in \mathbb{Z}} \{\chi(Y \cap T_{ij}) - \chi(Y \cap S_{ij})\}, \tag{1.18.4}$$

where

$$T_{ij} = A_i \cap B_j \tag{1.18.5}$$

is the ijth rectangular quadrat from the grid determined by the two stripe series, whereas

$$S_{ij} = T_{ij} \cap (A_{i+1} \cup B_{j+1}), \tag{1.18.6}$$

is called the 'shell' of the quadrat T_{ij}, namely the union of upper and the rhs edges of T_{ij}, see Fig. 1.18.1(a). Substitution into the rhs of Eq. (1.18.4) yields the shell formula, namely,

$$\begin{aligned}\chi(Y) &= \sum_{i,j\in\mathbb{Z}} \{\chi(Y \cap T_{ij}) - \chi(Y \cap S_{ij})\} \\ &= \sum_{i,j\in\mathbb{Z}} \{\chi((Y \cap T_{ij}) \cup S_{ij}) - 1\},\end{aligned} \tag{1.18.7}$$

which means that the Euler–Poincaré characteristic of the contents of each quadrat is first determined as if the shell S_{ij} was a part of the set Y, and then the contribution of the shell (which is equal to 1) is subtracted, see Fig. 1.18.1(b).

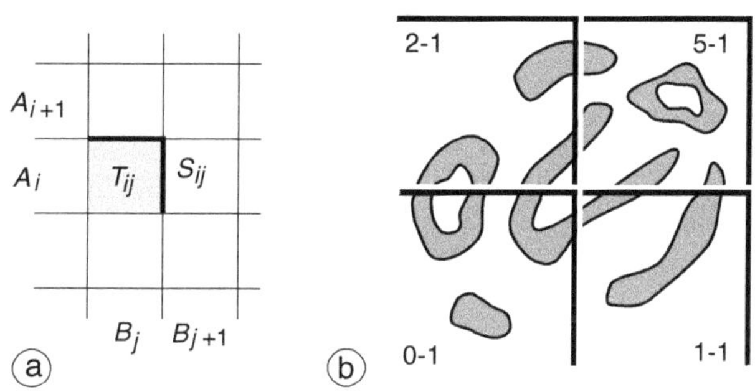

Fig. 1.18.1 (a) Definitions used to derive the shell formula, Eq. (1.18.7). (b) In the upper left quadrat, the Euler–Poincaré characteristic of the union of the two edges (thicker segments) of the shell, plus the object transect, is equal to 2. Now the Euler–Poincaré characteristic of the union of the two edges, namely 1, has to be subtracted, whereby the contribution of the first quadrat to the total $\chi(Y)$ is $2 - 1 = 1$, and analogously for the remaining quadrats. The global sum is $\chi(Y) = 4$, as expected.

Let $T(x) \equiv T_{00} + x$ represent a translate of the quadrat T_{00} by the vector $x \in \mathbb{R}^2$, and let $S(x)$ denote the corresponding shell. Then,

$$\begin{aligned}\int_{T_{00}} \chi(Y)\,\mathrm{d}x &= \int_{T_{00}} \sum_{i,j\in\mathbb{Z}} \{\chi(Y \cap T(x + \tau_{ij})) - \chi(Y \cap S(x + \tau_{ij}))\}\,\mathrm{d}x \\ &= \int_{\mathbb{R}^2} \{\chi(Y \cap T(x)) - \chi(Y \cap S(x))\}\,\mathrm{d}x,\end{aligned} \tag{1.18.8}$$

where $\{\tau_{ij} = (ia, jb),\ i, j \in \mathbb{Z}\}$ is the family of translations which take the rectangular grid of quadrats into itself. Because $\chi(Y)$ does not depend on x, we have the following integral formula,

$$\chi(Y) = \frac{1}{A(T(0))} \int_{\mathbb{R}^2} \{\chi((Y \cap T(x)) \cup S(x)) - 1\}\,\mathrm{d}x, \tag{1.18.9}$$

which will be used in Section 2.17. Note that the preceding identity does not depend on quadrat orientation.

1.18.2 The shell formula for a three-dimensional domain

The shell integral formula given by Eq. (1.18.9) holds in higher dimensions by redefining the pertinent symbols. For instance, if $Y \subset \mathbb{R}^3$ is a closed three-dimensional domain with piecewise smooth boundary, then $T(0)$ may be a rectangular box, in which case $S(0)$ is the union of three planar faces of $T(0)$ meeting at a given vertex, and the area $A(T(0))$ should be replaced with the volume $V(T(0))$ of the box.

1.18.3 Notes

1. The shell formulae

The identities (1.18.7) and (1.18.9) are due to Bhanu Prasad, Lantuejoul, Jernot, and Chermant (1989). The three-dimensional case is considered by Bhanu Prasad and Jernot (1991).

1.19 Hitting Measures and Projection Formulae

1.19.1 Introduction

Consider a fixed compact set $Y \subset \mathbb{R}^d$, $(d = 1, 2, \ldots)$, hit by a probe T (either unbounded, or bounded) equipped with the motion-invariant density $\mathrm{d}T$. The measure of all the probes hitting the set is the corresponding hitting measure, which we denote occasionally by $h(Y)$, namely,

$$h(Y) = \mathrm{meas}\{T : Y \cap T \neq \emptyset\} = \int_{Y \cap T \neq \emptyset} \mathrm{d}T. \tag{1.19.1}$$

A hitting measure is the measure of the sample space of a hitting experiment, and it is therefore necessary to construct a probability measure for the experiment (Section 2.1). Indeed, the ensuing formulae exhibit their true meaning only after incorporating probability and mean values.

The simplest hitting formula corresponds to a point probe hitting a domain, in which case the relevant integral is the Lebesgue measure of the domain, see Eq. (1.4.1).

Another simple example is the measure of the invariant test lines hitting a disk of radius r in the plane. For each orientation, a sweeping test line hitting the disk will travel a distance $2r$ parallel to itself. Integrating over orientations, the required hitting measure is equal to $2\pi r$, namely the boundary length of the disk. Less obviously, this result extends to convex sets, as shown below.

In the preceding example, the distance travelled by a sweeping test line hitting a disk at a given orientation is the length of the orthogonal linear projection of the disk onto an axis normal to the line. Thus, hitting and projection formulae – generally called Cauchy's formulae – are often related.

With the exception of the unidimensional case, see next, this section is concerned with unbounded probes.

1.19.2 Hitting and projection formulae on an axis, and in the plane

Unidimensional case

Let $Y = [0, H] \subset \mathbb{R}$ be a connected interval of length $H \geq 0$ on a fixed axis, and $T_1^1(x)$, $x \in \mathbb{R}$, a mobile segment probe of length $t \geq 0$ on the same axis, equipped with the translation-invariant density $\mathrm{d}T_1^1 = \mathrm{d}x$. Then the measure of all segment probes hitting Y is

$$h(Y) = \text{meas}\{x \in \mathbb{R} : \; Y \cap T_1^1(x) \neq \emptyset\} = \int_{-t}^{H} \mathrm{d}x = H + t. \tag{1.19.2}$$

Test lines hitting a planar domain

The measure of all the invariant test lines hitting a domain $Y \subset \mathbb{R}^2$ is

$$\begin{aligned} \int_{Y \cap L_1^2 \neq \emptyset} \mathrm{d}L_1^2 &= \int_0^{\pi} \mathrm{d}\phi \int_{Y \cap L_1^2 \neq \emptyset} \mathrm{d}p \\ &= \int_0^{\pi} H(\phi)\, \mathrm{d}\phi, \end{aligned} \tag{1.19.3}$$

where $H(\phi)$ denotes the total length of the orthogonal projection $Y'(\phi)$ of Y onto an axis of direction ϕ, which may consist of several separate segments if Y is not connected. If Y is connected, then $H(\phi)$ is called the caliper length, the breadth, or the Feret diameter of Y along the direction ϕ. In this case $H(\phi)$ is the thickness of the thinnest stripe $L^2_{1,t}(p, \phi)$ that can contain Y, see Fig. 1.19.1(a).

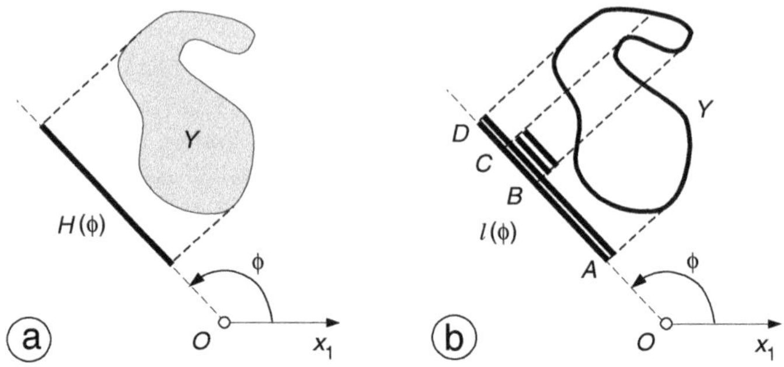

Fig. 1.19.1 (a) Orthogonal projected length, or caliper length, $H(\phi)$, of a connected set Y along a direction ϕ. (b) Total orthogonal projected length – in the example, $l(\phi) = 2|AB|+4|BC|+2|CD|$.

Case of a planar convex set

Suppose that the domain $Y \subset \mathbb{R}^2$ is convex, with boundary ∂Y of length $B > 0$. By convexity, $I(\partial Y \cap L^2_1) = 2$ for almost all test lines hitting Y, whereby the Crofton Eq. (1.5.8) yields the following hitting formula

$$\int_{Y \cap L^2_1 \neq \emptyset} \mathrm{d}L^2_1 = B. \tag{1.19.4}$$

The combination of Eqs. (1.19.3) and (1.19.4) yields the corresponding Cauchy projection formula, namely,

$$\int_0^{\pi} H(\phi)\, \mathrm{d}\phi = B. \tag{1.19.5}$$

For a disk of radius r one has $H(\phi) = 2r$ for all ϕ, and Eq. (1.19.5) yields $2\pi r = B$, as expected.

Cauchy's projection formula for a planar curve

Consider a piecewise smooth curve $Y \subset \mathbb{R}^2$, not necessarily closed, of length $B > 0$. Let $l(\phi)$, $\phi \in [0, \pi)$, denote the total orthogonal projected length of Y onto an axis making an angle $\phi \in [0, \pi)$ with the fixed axis Ox_1, with all points counted in their multiplicity, see Fig. 1.19.1(b). With the notation introduced in Section 1.5.2 we have

$$l(\phi) = \int_Y |\cos(\psi(y) - \phi)| \, \mathrm{d}b(y), \tag{1.19.6}$$

from which we obtain the corresponding Cauchy projection formula, namely

$$\begin{aligned} \int_0^\pi l(\phi)\, \mathrm{d}\phi &= \int_Y \mathrm{d}b(y) \int_0^\pi |\cos(\psi(y) - \phi)| \, \mathrm{d}\phi \\ &= 2B, \end{aligned} \tag{1.19.7}$$

independently of $\psi(y)$, as anticipated in Section 1.5.2. The aforementioned multiplicity at each point of the projection axis is effectively the number of intersections determined in the curve Y by a straight line L_1^2 perpendicular to the axis at that point. This argument shows the basic equivalence between Cauchy's Eq. (1.19.7) and Crofton's Eq. (1.5.8).

In all cases, it is equivalent – and often more convenient in practice – to fix the projection axis (for instance along Ox_1) and rotate the set $Y \equiv Y(\phi)$ with invariant density $\mathrm{d}\phi$, where ϕ now denotes the orientation of an AV previously attached to Y.

1.19.3 Hitting and projection formulae in 3D

Test lines hitting a set in 3D

Consider a surface $Y \subset \mathbb{R}^3$ of area S. From Crofton's Eq. (1.5.10) we have

$$\begin{aligned} \pi S = \int I(Y \cap L_1^3)\, \mathrm{d}L_1^3 &= \int_{\mathbb{S}_+^2} \mathrm{d}u \int I(Y \cap L_1^3)\, \mathrm{d}z \\ &= \int_{\mathbb{S}_+^2} a(u)\, \mathrm{d}u, \end{aligned} \tag{1.19.8}$$

where $a(u)$ denotes the area of the total orthogonal projection of Y onto a plane $L_2^3(0, u)$, namely the area of the orthogonal projection with all points counted in their multiplicity.

The measure of all the invariant test lines hitting a domain $Y \subset \mathbb{R}^3$, not necessarily convex, is,

$$\begin{aligned} \int_{Y \cap L_1^3 \neq \emptyset} \mathrm{d}L_1^3 &= \int_{\mathbb{S}_+^2} \mathrm{d}u \int_{Y \cap L_1^3 \neq \emptyset} \mathrm{d}z \\ &= \int_{\mathbb{S}_+^2} A(u)\, \mathrm{d}u, \end{aligned} \tag{1.19.9}$$

where $A(u)$ denotes the area of the orthogonal projection or 'shadow' $Y'(u)$ of the domain Y onto a projection plane $L_2^3(0, u)$ normal to the direction $u \in \mathbb{S}_+^2$, see Fig. 1.19.2(a).

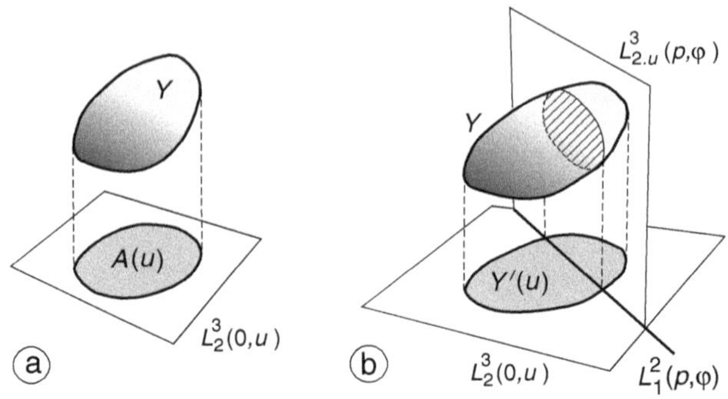

Fig. 1.19.2 (a) Illustration of Cauchy's projection formula Eq. (1.19.10). (b) Notation used to derive Eq. (1.19.14).

Test lines hitting a convex set in 3D

If the domain $Y \subset \mathbb{R}^3$ is convex with boundary ∂Y of area S, then for each direction $u \in \mathbb{S}_+^2$, the total projected area $a(u)$ is twice the shadow area $A(u)$, whereby Eq. (1.19.8) yields another Cauchy projection formula, namely,

$$\int_{\mathbb{S}_+^2} A(u)\,\mathrm{d}u = \frac{\pi}{2} S. \tag{1.19.10}$$

If $Y$is a ball of radius R, then $A(u) = \pi R^2$, and Eq. (1.19.10) yields $2\pi^2 R^2 = (\pi/2)S$, as expected.

Test planes hitting a set

The measure of all the invariant test planes hitting a domain $Y \subset \mathbb{R}^3$ is

$$\begin{aligned}\int_{Y\cap L_2^3\neq\emptyset} \mathrm{d}L_2^3 &= \int_{\mathbb{S}_+^2} \mathrm{d}u \int_{Y\cap L_2^3\neq\emptyset} \mathrm{d}p \\ &= \int_{\mathbb{S}_+^2} H(u)\,\mathrm{d}u, \end{aligned} \tag{1.19.11}$$

where $H(u)$ denotes the total length of the orthogonal linear projection $Y'(u)$ of Y onto an axis of direction $u \in \mathbb{S}_+^2$. If the set Y is connected, then $H(u)$ is the caliper length of Y along the mentioned axis – equivalently the thickness of the thinnest slab $L_{2,t}^3(p,u)$ that can contain Y.

Test planes hitting a convex set. Minkowski's theorem

Progress on the preceding formula can be made if the domain $Y \subset \mathbb{R}^3$ is convex. In this case we can take advantage of the fact that every planar section is convex, whereby $C(Y \cap L_2^3) = 2\pi$ for almost all planes hitting Y. Application of Eq. (1.11.6) yields the following hitting formula for convex sets,

$$\int_{Y \cap L_2^3 \neq \emptyset} \mathrm{d}L_2^3 = M, \tag{1.19.12}$$

where M denotes the integral of mean curvature of the convex boundary surface ∂Y. The combination of Eq. (1.19.11) and Eq. (1.19.12) yields a theorem due to H. Minkowski (see Section 1.19.7, Note 2), namely,

$$\int_{\mathbb{S}_+^2} H(u)\,\mathrm{d}u = M. \tag{1.19.13}$$

By analogy with Eq. (1.19.5), the preceding formula may also be regarded as a Cauchy projection formula.

A corollary of Minkowski's theorem is a Cauchy projection formula which may be regarded as a companion, or dual, of Eq. (1.19.10), namely,

$$\int_{\mathbb{S}_+^2} B(u)\,\mathrm{d}u = \pi M, \tag{1.19.14}$$

where $B(u)$ denotes the boundary length of the orthogonal projection $Y'(u)$ of a convex set $Y \subset \mathbb{R}^3$ onto a pivotal projection plane, see Fig. 1.19.2(b). To show the preceding identity consider a pivotal projection plane $L_{2[0]}^3 \equiv L_2^3(0, u)$ with rotation-invariant density $\mathrm{d}u$. For each direction $u \in \mathbb{S}_+^2$ consider a plane $L_{2\cdot u}^3(p, \varphi)$ perpendicular to $L_{2[0]}^3$, such that the trace $L_{2[0]}^3 \cap L_{2\cdot u}^3$ is a straight line $L_1^2(p, \varphi)$ with invariant density $\mathrm{d}L_1^2 = \mathrm{d}p\,\mathrm{d}\varphi$ within $L_{2[0]}^3$. Then,

$$[\mathrm{d}L_{2[0]}^3][\mathrm{d}L_{2\cdot u}^3] = [\mathrm{d}u][\mathrm{d}p\,\mathrm{d}\varphi] = [\mathrm{d}\varphi][\mathrm{d}p\,\mathrm{d}u] = [\mathrm{d}\varphi][\mathrm{d}L_2^3]. \tag{1.19.15}$$

Now, making use of Eq. (1.19.12) and Eq. (1.19.4),

$$\begin{aligned} \pi M &= \int_0^\pi \mathrm{d}\varphi \int_{Y \cap L_2^3 \neq \emptyset} \mathrm{d}L_2^3 = \int_{\mathbb{S}_+^2} \mathrm{d}u \int_{Y \cap L_{2\cdot u}^3 \neq \emptyset} \mathrm{d}L_{2\cdot u}^3 \\ &= \int_{\mathbb{S}_+^2} \mathrm{d}u \int_{Y'(u) \cap L_1^2 \neq \emptyset} \mathrm{d}L_1^2 = \int_{\mathbb{S}_+^2} B(u)\,\mathrm{d}u, \end{aligned} \tag{1.19.16}$$

as we wanted to show.

Special cases of Minkowski's theorem

If Y is a ball of diameter H, then the measure of all the invariant test planes hitting it is equal to $M = 2\pi H$, in accordance with Eq. (1.19.13). Explicit formulae are available to compute M for a variety of model objects. As a special case of interest, suppose that Y is a convex polyhedron. Let $l_i > 0$ and $\alpha_i \in (0, \pi)$ denote respectively the length of the ith edge and the corresponding dihedral angle, $i = 1, 2, \ldots, n$. The exterior parallel set, or dilated set, Y_ϵ of Y at distance $\epsilon \geq 0$ is the Minkowski sum

$$Y_\epsilon = Y \oplus D_3(0, \epsilon) = \bigcup_{x \in Y} D_3(x, \epsilon), \tag{1.19.17}$$

where $D_3(x, \epsilon)$ is a ball of radius ϵ with centre at $x \in \mathbb{R}^3$. Thus, the boundary of Y_ϵ is the union of flat faces with $\kappa_1 = \kappa_2 = 0$, plus n right circular cylinder sectors of angles $\{\pi - \alpha_i\}$, radius ϵ and lengths $\{l_i\}$, see Fig. 1.19.3(a), plus fragments constituting a partition of a sphere of radius ϵ, see Fig. 1.19.3(b). Therefore,

$$M(\partial Y_\epsilon) = \frac{1}{2}\left(\frac{1}{\epsilon} \cdot \sum_{i=1}^{n} (\pi - \alpha_i) \cdot \epsilon \cdot l_i + \frac{2}{\epsilon} \cdot 4\pi\epsilon^2\right), \tag{1.19.18}$$

and the integral of mean curvature of the convex polyhedron is given by the limit of $M(\partial Y_\epsilon)$ as $\epsilon \longrightarrow 0$, namely,

$$M = \frac{1}{2} \sum_{i=1}^{n} l_i(\pi - \alpha_i). \tag{1.19.19}$$

As another special case consider a convex platelet, namely a bounded planar subset $Y \subset \mathbb{R}^3$ whose boundary ∂Y is a planar convex curve of length $B > 0$. Here ∂Y_ϵ is the union of two opposite flat faces plus a cylindrical surface whose directrix is ∂Y and its generator is a semicircle of radius ϵ, so that

$$M(Y_\epsilon) = \frac{1}{2} \cdot \frac{1}{\epsilon} \cdot \pi \cdot \epsilon \cdot B = \frac{\pi}{2} B = M, \tag{1.19.20}$$

and by Minkowski's theorem the measure of all planes hitting a convex platelet is,

$$\int_{Y \cap L_2^3} \mathrm{d}L_2^3 = \frac{\pi}{2} B. \tag{1.19.21}$$

The same result is obtained by noting that $Q(\partial Y \cap L_2^3) = 2$ for almost all planes hitting Y, in which case the Crofton Eq. (1.5.14) with $L \equiv B$ yields Eq. (1.19.21), as expected.

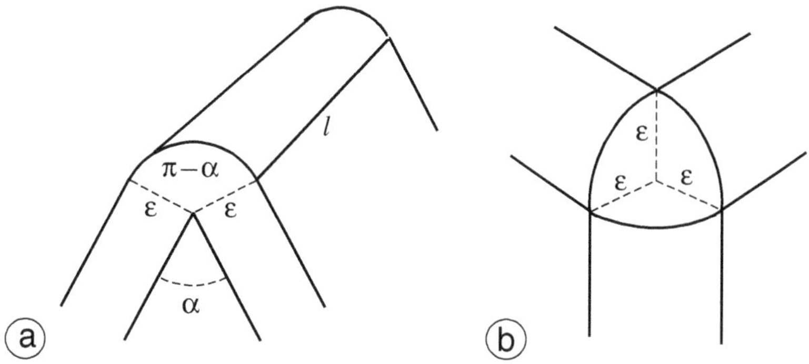

Fig. 1.19.3 Geometry involved in the derivation of Eq. (1.19.19) for the integral of mean curvature of a convex polyhedron.

Cauchy's projection formula for a spatial curve

Let $Y \subset \mathbb{R}^3$ represent a bounded curve of finite length $L > 0$, and let $l(u)$, $u \equiv u(\phi, \theta) \in \mathbb{S}^2_+$, denote the total orthogonal projected length of Y onto a plane $L^3_2(0, u)$ endowed with rotation-invariant density $\mathrm{d}u = \sin\theta \, \mathrm{d}\phi \, \mathrm{d}\theta$. Let δy represent an essentially linear curve element of length $\mathrm{d}l$ at a point $y \in Y$. By rotation invariance, the integral with respect to $\mathrm{d}u$ of the length $\mathrm{d}l(y)$ of the orthogonal projection $\delta y'$ of δy onto $L^3_2(0, u)$ does not depend on the initial orientation of δy relative to a fixed axis, (see Section 1.5.2). Thus we may take $\mathrm{d}l(y') = \mathrm{d}l(y) \sin\theta$, and the Cauchy formula for the orthogonal projection of a spatial curve onto a plane becomes

$$\begin{aligned}\int_{\mathbb{S}^2_+} l(u)\,\mathrm{d}u &= \int_{\mathbb{S}^2_+} \mathrm{d}u \int_Y \mathrm{d}l(y') \\ &= \int_Y \mathrm{d}l(y) \int_0^{2\pi} \mathrm{d}\phi \int_0^{\pi/2} \sin^2\theta\,\mathrm{d}\theta \\ &= \frac{\pi^2}{2} L. \end{aligned} \tag{1.19.22}$$

Cauchy's formula for the orthogonal projection of a motion-invariant slab hitting a spatial curve

Hit the curve $Y \subset \mathbb{R}^3$ with a motion-invariant slab $L_t(p, u)$ of thickness $t > 0$. Let $Y'(u)$ denote the orthogonal projection of Y onto a plane parallel to the slab, and $(Y \cap L_t(p, u))'$ the corresponding projection of the portion of curve contained in the slab. With a similar notation as above, for a given orientation $u \in \mathbb{S}^2_+$ we have $\mathrm{d}l(y') = \mathrm{d}l(y) \sin(\theta(y))$, where $\theta(y)$ is the angle of δy with u. Then, analogously as in Eq. (1.4.10) we obtain

$$\begin{aligned}\int_{\mathbb{R}} L\{(Y\cap L_t(p,u))'\}\,\mathrm{d}p &= \int_{\mathbb{R}} \mathrm{d}p \int_{y\in Y} 1_{L_t(p,u)}(y)\,\mathrm{d}l(y)\sin(\theta(y))\\ &= \int_{Y'(u)} \mathrm{d}l(y') \int_{\mathbb{R}} 1_{L_t(0,u)}(y-p)\,\mathrm{d}p\\ &= t\cdot L(Y'(u)).\end{aligned} \tag{1.19.23}$$

By Eq. (1.19.22), integration over directions yields the required identity,

$$\begin{aligned}\int_{\mathbb{S}_+^2} \mathrm{d}u \int_{\mathbb{R}} L\{(Y\cap L_t(p,u))'\}\,\mathrm{d}p &= t\int_{\mathbb{S}_+^2} L(Y'(u))\,\mathrm{d}u\\ &= \frac{\pi^2}{2}\cdot tL.\end{aligned} \tag{1.19.24}$$

1.19.4 Curve length in 3D from total vertical projections

As an alternative to the construction in Section 1.5.4, consider a vertical projection plane $L^3_{2[1]}(\phi)$, see Fig. 1.19.4(a), containing the vertical polar axis Ox_3, whose trace Ox_1' with the horizontal plane Ox_1x_2 is an axis making an angle $\phi \in [0,\pi)$ with the Ox_1 axis. Thus, $L^3_{2[1]}(\phi)$ is free to rotate by an angle ϕ around the polar axis with invariant density $\mathrm{d}L^3_{2[1]} = \mathrm{d}\phi$. A plane $L^3_2 \equiv L^3_2(p, u(\phi,\theta))$, $\phi \in [0,\pi)$, $\theta \in [0,\pi)$, with invariant density $\mathrm{d}L^3_2 = \mathrm{d}p\,\mathrm{d}u$, is perpendicular to $L^3_{2[1]}(\phi)$, and their intersection is just the orthogonal projection of L^3_2 onto $L^3_{2[1]}$, namely a straight line $L^2_{1\cdot v} \equiv L^2_{1\cdot v}(p, \pi/2-\theta)$ referred to the orthogonal frame $Ox_1'x_3$ within the vertical projection plane. Thus,

$$\mathrm{d}L^3_2 = [\mathrm{d}p][\sin\theta\,\mathrm{d}\phi\,\mathrm{d}\theta] = [\mathrm{d}\phi][\sin\theta\,\mathrm{d}p\,\mathrm{d}\theta] = [\mathrm{d}L^3_{2[1]}][\mathrm{d}L^2_{1\cdot v}]. \tag{1.19.25}$$

The preceding decomposition implies that an invariant test plane L^3_2 hitting a curve Y in space is equivalent to a test line $L^2_{1\cdot v}$ hitting the orthogonal projection $Y'(\phi)$ of the curve onto a vertical projection plane normal to L^3_2, see Fig. 1.19.4(a), with density $\mathrm{d}L^2_{1\cdot v} = \sin\theta\,\mathrm{d}p\,\mathrm{d}\theta$. Moreover, $Q(Y\cap L^3_2) = I(Y'(\phi)\cap L^2_{1\cdot v})$ for almost all test planes hitting the curve. Consequently,

$$\begin{aligned}\pi L &= \int Q(Y\cap L^3_2)\,\mathrm{d}L^3_2\\ &= \int_0^\pi \mathrm{d}\phi \int I(Y'(\phi)\cap L^2_{1\cdot v})\,\mathrm{d}L^2_{1\cdot v}.\end{aligned} \tag{1.19.26}$$

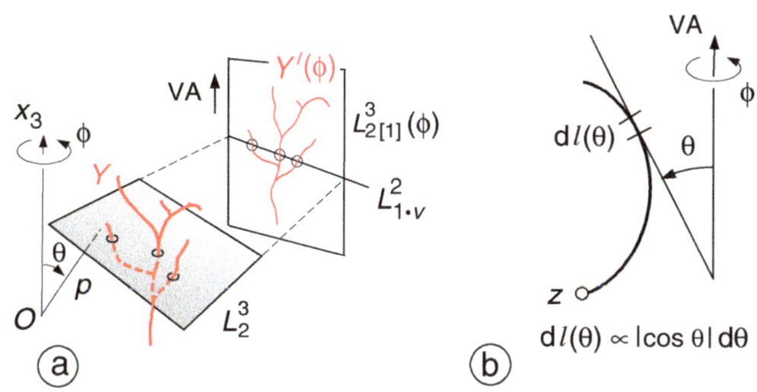

Fig. 1.19.4 (a) Geometry of the total projection $Y'(\phi)$ of a curve Y onto a vertical plane $L^3_{2[1]}(\phi)$, used to derive Eq. (1.19.26). (b) Cycloid test arc $T^2_1(z)$ involved in Eq. (1.19.30).

In order to progress we proceed similarly as in the vertical sections case (Section 1.6). Thus, let δy be an arc element of length $\mathrm{d}y$ at a point $y \in Y'(\phi)$ of the curve projection, and $\psi \equiv \psi(y) \in [0, \pi)$ the angle of δy with the vertical axis Ox_3. Then,

$$\mathrm{d}p = |\cos(\psi - \theta)|\,\mathrm{d}y, \tag{1.19.27}$$

and therefore,

$$\begin{aligned}\pi L &= \int_0^\pi \mathrm{d}\phi \int_{y\in Y'(\phi)} \mathrm{d}y \int_0^\pi |\cos(\psi-\theta)|\sin\theta\,\mathrm{d}\theta \\ &= \int_0^\pi Z(Y'(\phi))\,\mathrm{d}\phi,\end{aligned} \tag{1.19.28}$$

where

$$Z(Y'(\phi)) = \int_{y\in Y'(\phi)} (\cos\psi + \psi\sin\psi)\,\mathrm{d}y(\psi). \tag{1.19.29}$$

Finally, we may replace a test line $L^2_{1\cdot v}(p, \pi/2 - \theta)$ whose density is weighted by $\sin\theta$, with a bounded test curve $T^2_1(z)$, $z \in \mathbb{R}$, of finite length l, in the vertical projection plane, whose arc length element satisfies $\mathrm{d}l(\theta) = c|\cos\theta|\,\mathrm{d}\theta$, where $\theta \in [0, 2\pi)$ is the angle of the arc element with the vertical axis. The test curve is a cycloid consisting of two arcs whose major principal axes are parallel to the vertical axis – thus, here T^2_1 is the result or rotating the cycloid test curve arising in the vertical sections case (Section 1.17) by an angle of $\pi/2$, see Fig. 1.19.4(b). The relevant identity is the dual of Eq. (1.17.6), namely,

$$L = \frac{2}{\pi l}\int_0^\pi \mathrm{d}\phi \int_{\mathbb{R}^2} I(Y'(\phi) \cap T^2_1(z))\,\mathrm{d}z. \tag{1.19.30}$$

1.19.5 Curve length in 3D from vertical slab projections

Hit the curve $Y \subset \mathbb{R}^2$ with a vertical slab $L_t(p, \phi)$ of thickness $t > 0$, namely a slab whose faces are parallel vertical planes a distance t apart. Thus the slab is equipped with the motion-invariant density of the linear trace determined by one of its faces on the HP, namely $\mathrm{d}L_t = \mathrm{d}p\,\mathrm{d}\phi$, $p \in (-\infty, \infty)$, $\phi \in [0, \pi)$. The orthogonal projection of the portion of curve contained in the vertical slab onto a plane parallel to the slab is a curve denoted by $(Y \cap L_t(p, \phi))'$. By a similar argument to that used in the IUR case (Eq. (1.19.24)), the vertical slab version of Eq. (1.19.28) is

$$\int_0^\pi \mathrm{d}\phi \int_{\mathbb{R}} Z\{(Y \in L_t(p, \phi))'\}\,\mathrm{d}p = \pi t L. \tag{1.19.31}$$

Further, if we intersect the projection of the slice with a translation-invariant two-arc cycloid test curve $T_1^2(z)$ of length l, with the major principal axes parallel to the vertical axis, see Fig. 1.19.5, then the vertical slab version of Eq. (1.19.30) is

$$L = \frac{2}{\pi t l} \int_0^\pi \mathrm{d}\phi \int_{\mathbb{R}} \mathrm{d}p \int_{\mathbb{R}^2} I\{(Y \cap L_t(p, \phi))' \cap T_1^2(z)\}\,\mathrm{d}z. \tag{1.19.32}$$

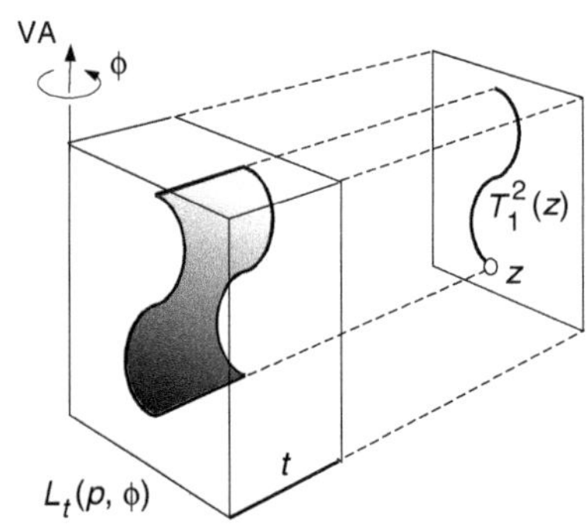

Fig. 1.19.5 A vertical slab L_t contains a portion of a cylinder whose directrix is a vertical cycloid arc T_1^2. Let $Y \subset \mathbb{R}^3$ be a target curve (not included in the figure). Eq. (1.19.32) is valid for the slab contents $Y \cap L_t$ intersected by the cylinder, or for the corresponding vertical curve projection intersected by T_1^2.

1.19.6 Integral of mean curvature of a convex domain from vertical projections

The integral of mean curvature of a convex domain $Y \subset \mathbb{R}^3$ coincides with the measure of invariant planes hitting the domain, see Eq. (1.19.12). As an alternative, consider the same construction as in Section 1.19.4. Thus, the invariant plane $L_2^3 \equiv L_2^3(p, u(\phi, \theta))$ is normal to a vertical projection plane $L_{2[1]}^3 \equiv L_{2[1]}^3(\phi)$ and the corresponding trace is the straight line $L_{1\cdot v}^2(p, \pi/2 - \theta)$. Let $Y'(\phi)$ denote the boundary of the vertical projection of Y onto the plane $L_{2[1]}^3(\phi)$, namely a closed convex curve. In the same manner as in Eq. (1.19.28) we obtain

$$\begin{aligned} M &= \int_0^\pi \mathrm{d}\phi \int_{Y'(\phi)\cap L_{1\cdot v}^2 \neq \emptyset} \sin\theta \, \mathrm{d}p \, \mathrm{d}\theta \\ &= \frac{1}{2} \int_0^\pi Z(Y'(\phi)) \, \mathrm{d}\phi, \end{aligned} \tag{1.19.33}$$

where the functional $Z(Y'(\phi))$ is given by Eq. (1.19.29). The factor $1/2$ is due to the fact that the rhs of the first Eq. (1.19.33) is equal to the rhs of Eq. (1.19.26) with $I(Y'(\phi) \cap L_{1\cdot v}^2) = 2$, because now $Y'(\phi)$ is a closed convex curve.

Further, if we use the same cycloid test curve $T_1^2(z)$ considered above, with the major principal axes parallel to the vertical axis, then,

$$M = \frac{1}{l} \int_0^\pi \mathrm{d}\phi \int_{\mathbb{R}^2} I(Y'(\phi) \cap T_1^2(z)) \, \mathrm{d}z. \tag{1.19.34}$$

Example

To illustrate Eq. (1.19.34), suppose that Y is a ball of radius R so that, by Eq. (1.19.13), $M = 4\pi R$. The boundary of the vertical projection $Y'(\phi)$ is a circle of radius R and arc element $\mathrm{d}y(\psi) = R\,\mathrm{d}\psi$, whereby – similarly as in Eq. (1.6.5) – Eq. (1.19.29) yields

$$Z(Y'(\phi)) = 4R \int_0^{\pi/2} (\cos\psi + \psi \sin\psi) \, \mathrm{d}\psi = 8R, \tag{1.19.35}$$

and Eq. (1.19.33) gives $M = 4\pi R$, as expected.

To illustrate Eq. (1.19.34),

$$\int_{\mathbb{R}^2} I(Y'(\phi) \cap T_1^2(z)) \, \mathrm{d}z = \frac{4 \cdot 2\pi R \cdot l}{2\pi}, \tag{1.19.36}$$

namely the rhs of Poincaré's formula, Eq. (1.16.5), divided by 2π, because $T_1^2(z)$ has a fixed orientation, and therefore integration over orientations can be omitted in Eq. (1.16.5). Inserting the preceding result into the rhs of Eq. (1.19.34) we obtain $M = 4\pi R$, as expected.

1.19.7 Notes

1. Hitting and projection formulae

Eq. (1.19.7) and Eq. (1.19.8) were derived by Cauchy (1832) – for a review see Hykšová et al. (2012). For general versions, and later contributions, see Santaló (1976).

2. Minkowski's theorem

The simple proof of Eq. (1.19.13), is due to Miles (1981). Compare with the classical proof, e.g. in Kendall and Moran (1963), who refer to the result as 'Minkowski's theorem'.

3. Vertical projections

The vertical slab representation given by Eq. (1.19.32) is due to Gokhale (1990). Later, Cruz-Orive and Howard (1991) extended the result for total vertical projections, Eq. (1.19.30). On the other hand, the representation (1.19.34) for the integral of mean curvature of a convex body from total vertical projections is due to Gokhale and Beneš (1998).

1.20 Hitting Measures for Bounded Probes. Kinematic Formulae

1.20.1 Introduction

Consider a fixed domain $Y \subset \mathbb{R}^d$, $(d = 1, 2, \ldots)$ with piecewise smooth boundary, hit by a mobile compact set $T \equiv T_r^d(x, u_d) \subset \mathbb{R}^d$ which plays the role of a bounded test probe equipped with the kinematic density $\mathrm{d}T = \mathrm{d}x\,\mathrm{d}u_d$. Here x denotes the associated point of T, whereas u_d represents the vector of the orientation parameters from the special group $G_{d[0]}$ of rotations about the origin. For instance, $u_1 \equiv \omega$, whereas $u_3 \equiv (u_2, \tau)$, etc., see Section 1.13 and Section 1.16.4. By virtue of Eq. (1.16.2), the relevant hitting measure reads as follows,

$$\begin{aligned} h(Y) &= \text{meas}\{T : Y \cap T \neq \emptyset\} \\ &= \int_{Y \cap T \neq \emptyset} \mathrm{d}T \\ &= \int_{G_{d[0]}} \mathrm{d}u_d \int_{Y \oplus \check{T} \neq \emptyset} \mathrm{d}x \\ &= \int_{G_{d[0]}} V(Y \oplus \check{T}(0, u_d))\, \mathrm{d}u_d, \end{aligned} \tag{1.20.1}$$

namely the Lebesgue measure of the Minkowski sum $Y \oplus \check{T}(0, u_d)$, integrated over orientations. If the sets Y and T have arbitrary shapes, then $h(Y)$ may be difficult to compute. However, if $Y \cap T$ is always a simply connected subset, namely if the corresponding Euler–Poincaré characteristic satisfies $\chi(Y \cap T) = 1$ for almost all positions and orientations of T hitting Y, then $h(Y)$ is accessible in terms of the size parameters of Y and T by means of the kinematic formula of Blaschke–Santaló, which is concerned with the following integral,

$$\int_{Y \cap T \neq \emptyset} \chi(Y \cap T)\, \mathrm{d}T. \tag{1.20.2}$$

Crofton's intersection formulae for bounded probes involving dimensionless integrands such as $N(Y \cap T)$, or $I(Y \cap T)$, emerge as special cases. In the special case $\chi(Y \cap T) = 1$ (for which it is sufficient, but not necessary, that both Y and T are convex), the kinematic formula yields the corresponding hitting measure $h(Y)$ directly. In particular, if the set Y is convex and T is a ball of radius $r > 0$ then $Y \oplus \check{T}$ is the exterior parallel set of Y at distance r, and the kinematic formula yields Steiner's formula for the Lebesgue measure of a parallel set.

The aforementioned formulae hold unchanged if the roles of Y and T are exchanged, namely if Y is mobile and T is fixed.

1.20.2 Kinematic formulae on an axis

Let $Y \subset \mathbb{R}$ represent a bounded finite union of $\chi(Y)$ separate connected segments of total length $L(Y)$, and let $T \equiv T_1^1(x)$, $x \in \mathbb{R}$, be a bounded probe consisting of $\chi(T)$ separate test segments of total length $L(T)$ on the same axis, equipped with the translation-invariant density $\mathrm{d}T = \mathrm{d}x$. In this case the kinematic formula reads

$$\int_{\mathbb{R}} \chi(Y \cap T)\, \mathrm{d}T = L(Y) \cdot \chi(T) + L(T) \cdot \chi(Y). \tag{1.20.3}$$

Special unidimensional cases

1. Suppose that T is a segment of length $t > 0$. If Y is a segment of length H, then Eq. (1.20.3) with $\chi(Y) = \chi(T) = 1$ becomes the hitting formula (1.19.2).
2. If T is as above, and Y is a bounded finite union of N separate points, then the rhs of Eq. (1.20.3) with $L(Y) = 0$, $\chi(Y) = N$ yields tN, the unidimensional version of the Crofton Eq. (1.15.7), see also Eq. (1.4.8).

1.20.3 Steiner's formulae

The planar case

The exterior parallel set of a planar convex set $K \in \mathbb{R}^2$ at distance $r > 0$ is also a convex set defined as follows,

$$K_r = K \oplus D_2(0, r) = \bigcup_{x \in K} D_2(x, r), \tag{1.20.4}$$

where $D_2(x, r)$ is a disk of radius r with centre at $x \in \mathbb{R}^2$, see Fig. 1.20.1(a). Steiner's formulae express the area $A(K_r)$ and the boundary length $B(\partial K_r)$ of the exterior parallel set in terms of $A(K)$, $B(\partial K)$ and r.

Let $\mathrm{d}b$ and $\mathrm{d}b_r$ denote the arc length elements of ∂K and ∂K_r, respectively. By Eq. (1.10.1) and Eq. (1.10.19), setting $R \equiv R(b)$ and $\alpha \equiv \alpha(b)$ we have,

$$\begin{aligned} \mathrm{d}b &= R\,\mathrm{d}\alpha, \\ \mathrm{d}b_r &= (R + r)\,\mathrm{d}\alpha, \end{aligned} \tag{1.20.5}$$

whereby,

$$\mathrm{d}b_r = \frac{R + r}{R} \cdot \mathrm{d}b. \tag{1.20.6}$$

Therefore,

$$\begin{aligned} B(\partial K_r) &= \int_{\partial K} \left(1 + \frac{r}{R}\right) \mathrm{d}b \\ &= \int_{\partial K} \mathrm{d}b + r \int_{\partial K} \mathrm{d}\alpha(b) \\ &= B(\partial K) + 2\pi r, \end{aligned} \tag{1.20.7}$$

which is the Steiner formula for the boundary length of the parallel set at distance r of a planar convex set.

The corresponding Steiner formula for the area of the parallel set is obtained by integrating the ring area element $B(\partial K_\rho)\,\mathrm{d}\rho$ from $\rho = 0$ to $\rho = r$, and adding the result to $A(K)$, namely,

$$\begin{aligned} A(K_r) &= A(K) + \int_0^r (B(\partial K) + 2\pi\rho)\,\mathrm{d}\rho \\ &= A(K) + B(\partial K)r + \pi r^2. \end{aligned} \tag{1.20.8}$$

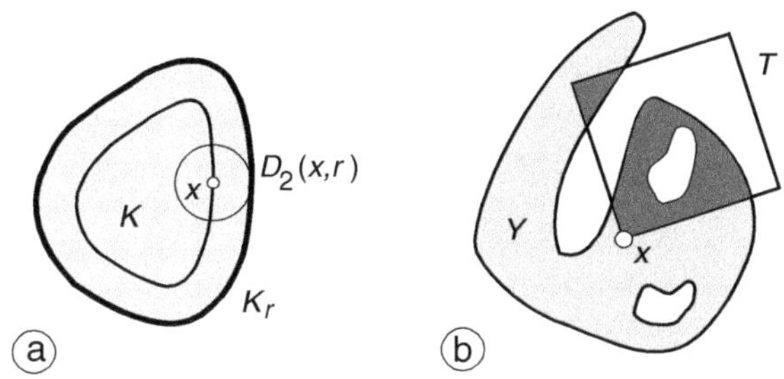

Fig. 1.20.1 (a) Parallel set K_t of a convex set K, see Eq. (1.20.4). (b) A motion-invariant quadrat T in the plane hits a domain Y, yielding $\chi(Y \cap T) = 1$. The kinematic formula, Eq. (1.20.12), gives the corresponding integral.

Steiner's formulae in 3D

The exterior parallel set K_r of a convex set $K \in \mathbb{R}^3$ at distance $r > 0$ is defined similarly as in the planar case by replacing the disk $D_2(x,r)$ with a ball $D_3(x,r)$ in Eq. (1.20.4).

Let $\mathrm{d}s$ and $\mathrm{d}s_r$ denote the surface area elements of ∂K and ∂K_r, respectively. By Gauss' formula Eq. (1.12.3),

$$\mathrm{d}s_r = \frac{(R_1 + r)(R_2 + r)}{R_1 R_2} \cdot \mathrm{d}s, \tag{1.20.9}$$

whereby,

$$\begin{aligned} S(\partial K_r) &= \int_{\partial K} (1 + r/R_1)(1 + r/R_2)\,\mathrm{d}s \\ &= S(\partial K) + 2M(\partial K)r + 4\pi r^2, \end{aligned} \tag{1.20.10}$$

which is the Steiner formula for the surface area of the parallel set at distance r of a three-dimensional convex set.

The corresponding Steiner formula for the volume of the parallel set is obtained analogously as in the planar case, namely

$$V(K_r) = V(K) + \int_0^r (S(\partial K) + 2M(\partial K)\rho + 4\pi\rho^2)\,\mathrm{d}\rho$$
$$= V(K) + S(\partial K)r + M(\partial K)r^2 + \frac{4}{3}\pi r^3. \qquad (1.20.11)$$

1.20.4 Kinematic formulae in 2D and 3D

The planar case

Consider a planar domain $Y \subset \mathbb{R}^2$ with boundary ∂Y, hit by a mobile one T, see Fig. 1.20.1(b), with boundary ∂T, equipped with the kinematic density $\mathrm{d}T = \mathrm{d}x\,\mathrm{d}\omega$, see Eq. (1.13.2). The kinematic formula in the plane reads

$$\int \chi(Y \cap T)\,\mathrm{d}T = 2\pi[A(Y)\cdot\chi(T) + A(T)\cdot\chi(Y)]$$
$$+ B(\partial Y)\cdot B(\partial T). \qquad (1.20.12)$$

Special cases in the plane

1. If $\chi(Y \cap T) = 1$, then Eq. (1.20.1) yields the hitting measure. Further, if $\chi(Y) = \chi(T) = 1$ then,

$$\int_{Y\cap T\neq\emptyset} \mathrm{d}T = 2\pi[A(Y) + A(T)] + B(\partial Y)B(\partial T). \qquad (1.20.13)$$

   If T is a disk of radius $r > 0$, then Eq. (1.20.13) divided by 2π formally coincides with the Steiner formula (1.20.8).
2. If Y and T are planar curves of lengths B and l, respectively, then $A(Y) = A(T) = 0$ whereas $B(\partial Y) = 2B$ and $B(\partial T) = 2l$ because Y and T have to be regarded as degenerate sets of zero area. Further, $\chi(Y \cap T) = 2\pi I(Y \cap T)$, and the kinematic formula (1.20.12) yields

$$\int_{Y\cap T\neq\emptyset} I(Y \cap T)\,\mathrm{d}T = 4Bl, \qquad (1.20.14)$$

   which is Poincaré's formula, Eq. (1.16.5).

Kinematic formulae in 3D

If $Y, T \subset \mathbb{R}^3$ are full-dimensional and T is equipped with the kinematic density $\mathrm{d}T = \mathrm{d}x\,\mathrm{d}u_2\,\mathrm{d}\tau$, (recall Eq. (1.13.6)), then the kinematic formula of Blaschke–Santaló reads

$$\int_{Y\cap T\neq\emptyset} \chi(Y\cap T)\,\mathrm{d}T = 8\pi^2[V(Y)\cdot\chi(T) + V(T)\cdot\chi(Y)] + 2\pi\cdot S(\partial Y)\cdot M(\partial T) + 2\pi\cdot S(\partial T)\cdot M(\partial Y). \tag{1.20.15}$$

Special cases in 3D

If $\chi(Y\cap T) = 1$, then Eq. (1.20.15) yields the hitting measure. Further, if $\chi(Y) = \chi(T) = 1$, then

$$\int_{Y\cap T\neq\emptyset} \mathrm{d}T = 8\pi^2[V(Y) + V(T)] + 2\pi\cdot S(\partial Y)\cdot M(\partial T) + 2\pi\cdot S(\partial T)\cdot M(\partial Y). \tag{1.20.16}$$

If T is a ball of radius $r > 0$, then the rhs of Eq. (1.20.16) divided by $8\pi^2$ formally coincides with the Steiner formula (1.20.11).

1.20.5 Notes

1. Steiner formulae

These are due to Steiner (1840) – for further references and notes, see Hadwiger (1957). The exterior parallel set of a convex body is a special case of the Minkowski addition of a set by a ball, namely the dilatation of the set by a ball, which plays an important role in mathematical morphology, see for instance Serra (1982). For a general treatment of Steiner formulae, see Santaló (1976) and Schneider and Weil (2008).

2. Kinematic formulae

Early versions in the plane and in space are due to Santaló (1936a, 1936c) and Blaschke (1936–1937), see also Santaló (1953). For general versions in Euclidean and non-Euclidean spaces, see Santaló (1976).

1.21 Test Systems

1.21.1 Purpose and definitions

For estimation purposes, single probes are generally unsuitable, first because the sampling design is inconvenient, and second because the hitting measures depend on unknown geometric parameters defined on the set Y of interest, as revealed by the kinematic formulae. These difficulties are removed by replacing a single test probe with a test system of probes. The key tool is Santaló's fundamental formula for test systems. As given in Eq. (1.21.7) below, however, this formula is purely integral-geometric – its stereological relevance will better emerge in the light of geometrical probability and estimation concepts, see Section 2.24.

Consider a partition of $\mathbb{R}^d$ into congruent regions or tiles generated by a fundamental tile J_0, namely,

$$\mathbb{R}^d = \bigcup_{i \in \mathbb{Z}} J_i, \quad J_i \cap J_j = \emptyset, \quad i \neq j, \tag{1.21.1}$$

which implies that every point $x \in \mathbb{R}^d$ belongs to exactly one tile J_i. Further, every J_i can be brought to coincide with J_0 by a motion g_i that takes the whole partition into itself.

The set of motions $\{g_i\}$ such that $g_i J_i = J_0$ is a discrete subgroup of the group of motions. In practice, g_i is usually a translation.

Consider now a bounded probe $T(0,0) \equiv T_r^d(0,0)$ (e.g. a domain or quadrat, a finite set of rectifiable curves, a finite set of test points, etc.), called the fundamental probe, contained in the fundamental tile J_0 with an arbitrarily fixed position and orientation $(x, u_d) \equiv (0,0)$, say, namely $T(0,0) \subset J_0$. A fixed test system, see Fig. 1.21.1(b), is defined by

$$\Lambda_{0,0} = \bigcup_{i \in \mathbb{Z}} g_i^{-1} T(0,0), \tag{1.21.2}$$

namely the result of copying the probe $T(0,0)$ into each of the tiles of the partition, preserving the same relative position and orientation within each tile.

1.21.2 Santaló's fundamental formula for test systems

The purpose of Santaló's formula is to express the integral in the lhs of Eq. (1.14.1), namely,

$$\int_{Y \cap T \neq \emptyset} \alpha(Y \cap T)\, \mathrm{d}T, \tag{1.21.3}$$

in a form suitable for estimation purposes. The set of interest $Y \subset \mathbb{R}^d$ and the probe $T \subset \mathbb{R}^d$ are bounded and of finite measure. As indicated in Section 1.14, by the invariance of the kinematic density the roles of Y and T may be exchanged without altering the value of the integral. To simplify the following derivations it is convenient to regard the probe T as fixed, and the set $Y \equiv Y(x, u_d)$, $x \in \mathbb{R}^d$, $u_d \in G_{d[0]}$ as mobile with the kinematic density $\mathrm{d}Y = \mathrm{d}x\,\mathrm{d}u_d$, as in Eq. (1.14.2) – see also Section 1.20.1 and Fig. 1.21.1(a). Now, making use of the fact that the tiles constitute a partition of $\mathbb{R}^d$, we have

$$\int_{Y \cap T \neq \emptyset} \alpha(Y \cap T)\,\mathrm{d}Y = \sum_{i \in \mathbb{Z}} \int_{J_i} \alpha(Y \cap T)\,\mathrm{d}Y, \tag{1.21.4}$$

where $T \equiv T(0, 0)$ and, for each i, the last integral is extended over $x \in J_i$ and over the relevant rotation $u_d \in G_{d[0]}$. The summation and the integral signs may be interchanged because all the integrals involved are finite. Thus,

$$\int_{Y \cap T \neq \emptyset} \alpha(Y \cap T)\,\mathrm{d}Y = \int_{J_0} \sum_{i \in \mathbb{Z}} \alpha(g_i Y \cap T)\,\mathrm{d}Y, \tag{1.21.5}$$

where we have used the motion invariance property $\mathrm{d}(g_i Y) = \mathrm{d}Y$. Moreover the intersection $g_i Y \cap T$ is congruent with $Y \cap g_i^{-1} T$, whereby

$$\begin{aligned} \int_{Y \cap T \neq \emptyset} \alpha(Y \cap T)\,\mathrm{d}Y &= \int_{J_0} \sum_{i \in \mathbb{Z}} \alpha(Y \cap g_i^{-1} T)\,\mathrm{d}Y \\ &= \int_{J_0} \alpha(Y \cap \Lambda_{0,0})\,\mathrm{d}Y, \end{aligned} \tag{1.21.6}$$

which is Santaló's fundamental formula for test systems. It means that the integral of $\alpha(Y \cap T)$ with $x \in Y \oplus \check{T}$ may be replaced with the integral of $\alpha(Y \cap \Lambda_{0,0})$ with $x \in J_0$, which constitutes a decisive simplification of the design. Combination with Eq. (1.14.2) yields

$$c_2 \cdot \gamma(Y) \cdot \nu(T) = \int_{G_{d[0]}} \mathrm{d}u_d \int_{J_0} \alpha(Y(x, u_d) \cap \Lambda_{0,0})\,\mathrm{d}x. \tag{1.21.7}$$

To keep the test system $\Lambda_{0,0}$ fixed and the set Y mobile with the kinematic density $\mathrm{d}Y$, is equivalent to fixing Y and equipping the fundamental probe $T \subset J_0$, and thereby the entire test system, with the kinematic density $\mathrm{d}T = \mathrm{d}x\,\mathrm{d}u_d$. To achieve this, the test system $\Lambda_{0,0}$ is first translated into $\Lambda_{x,0}$, $x \in J_0$, and then the entire $\Lambda_{x,0}$ is rotated by $u_d \in G_{d[0]}$ about the point x, to obtain Λ_{x,u_d} as a result. Combination with Eq. (1.14.1) yields the alternative to Eq. (1.21.7), namely

$$c_2 \cdot \gamma(Y) \cdot \nu(T) = \int_{G_{d[0]}} \mathrm{d}u_d \int_{J_0} \alpha(Y \cap \Lambda_{x,u_d})\,\mathrm{d}x. \tag{1.21.8}$$

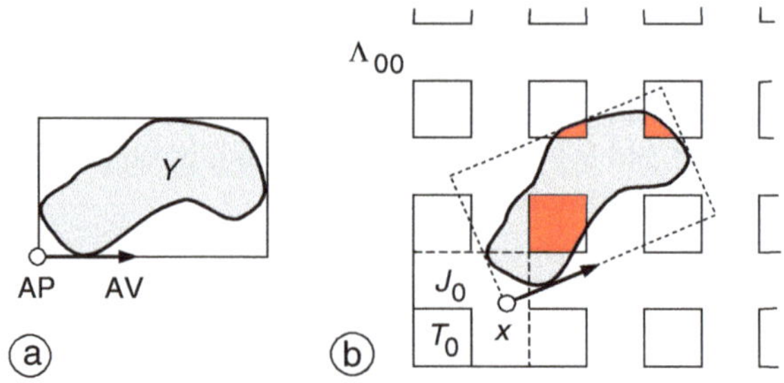

Fig. 1.21.1 (a) A bounded target set Y in the plane, with its associated point AP, and vector AV. (b) Superimposition of the set Y, assumed to be mobile with the kinematic density, onto a fixed test system Λ_{00} of quadrats, whereby Santaló's formula Eq. (1.21.7) holds. J_0 denotes the fundamental tile of the test system, and $T_0 \equiv T(0,0) \subset J_0$ the fundamental probe (quadrat here). It is equivalent to superimpose a motion-invariant test system onto a fixed Y, see Eq. (1.21.8).

As an example, suppose that $Y \subset \mathbb{R}^2$ is a bounded planar curve of length B, and adopt a test system $\Lambda_{0,0}$ with a fundamental tile J_0, the fundamental probe being a test curve $T \subset J_0$ of length $l > 0$. Application of Poincaré's formula (Eq. (1.16.5)) to the lhs of Eq. (1.21.8) yields

$$4Bl = \int_0^{2\pi} \mathrm{d}\omega \int_{J_0} I(Y \cap \Lambda_{x,\omega})\, \mathrm{d}x. \tag{1.21.9}$$

1.21.3 Santaló's fundamental formula for test systems of unbounded probes

In Eq. (1.21.3) the probe may be unbounded, in which case we have Eq. (1.3.4). As above, consider a curve $Y \subset \mathbb{R}^2$ of length B, and adopt a test system $\Lambda_{0,0}$ whose fundamental tile is a square, namely $J_0 = [0,h)^2$, whereas the fundamental probe is the straight line segment $T = [0,h)$. Then clearly $\Lambda_{0,0}$ becomes a test system of parallel straight lines a constant distance $h > 0$ apart, namely,

$$\Lambda_{0,0} = \{L_1^2(kh, 0),\ k \in \mathbb{Z}\}. \tag{1.21.10}$$

Thus, in general Eq. (1.21.7), (1.21.8) can be readily adapted to unbounded probes by a proper choice of J_0 and T.

More directly, if $T \equiv L_1^2(p,\phi)$ is a straight line with the motion-invariant density $\mathrm{d}L_1^2 = \mathrm{d}p\,\mathrm{d}\phi$ in the plane, then Eq. (1.3.4) becomes Eq. (1.5.8), and we may write

$$\begin{aligned} 2B &= \int_0^{\pi} \mathrm{d}\phi \int_{-\infty}^{\infty} I(Y \cap L_1^2)\,\mathrm{d}p \\ &= \int_0^{\pi} \mathrm{d}\phi \sum_{k\in\mathbb{Z}} \int_{(k-1)h}^{kh} I(Y \cap L_1^2)\,\mathrm{d}p \\ &= \int_0^{\pi} \mathrm{d}\phi \int_0^{h} I(Y \cap \Lambda_{z,\phi})\,\mathrm{d}z, \end{aligned} \tag{1.21.11}$$

where

$$\Lambda_{z,\phi} = \{L_1^2(z+kh,\phi),\ k \in \mathbb{Z}\}, \quad z \in [0,h),\ \phi \in [0,\pi), \tag{1.21.12}$$

is the translated and rotated version of $\Lambda_{0,0}$, see Fig. 1.21.2.

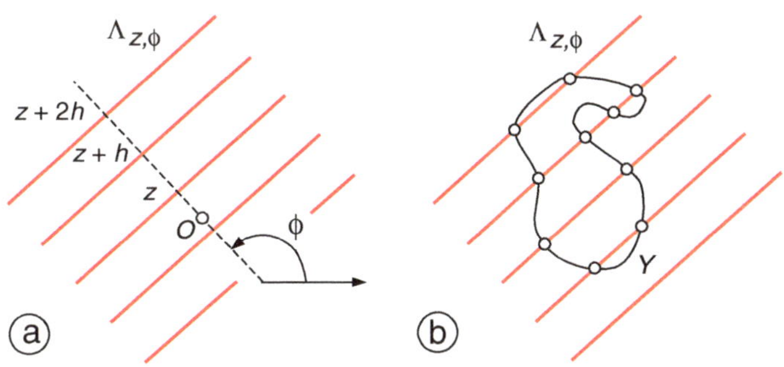

Fig. 1.21.2 (a) A test system of parallel lines, see Eq. (1.21.12). (b) The test system determines $I = 10$ intersections with a curve Y.

Consider now an r-plane $T \equiv L_r^d(z,u)$, $z \in L_{d-r}^d(0,u)$, $u \in G_{r,d-r}$, where $G_{r,d-r}$ is the Grassmannian, see Section 1.5.5. For each direction u we first construct a UR test system of points $\{z + t_k\}$, with fundamental tile $J_{0,u}$, within the orthogonal complement $L_{d-r}^d(0,u)$ of $L_r^d(z,u)$. Here $z \in J_{0,u}$, and $\{t_k\}$ is a set of translations such that each $-t_k$, $k \in \mathbb{Z}$, brings the entire test system of points into itself. Finally we construct the corresponding test system of r-planes, namely

$$\Lambda_{z,u} = \{L_r^d(z+t_k,u),\ k \in \mathbb{Z}\}, \quad z \in J_{0,u},\ u \in G_{r,d-r}. \tag{1.21.13}$$

Thus, instead of Eq. (1.21.8), we have

$$c_1 \cdot \gamma(Y) = \int_{G_{r,d-r}} \mathrm{d}u \int_{J_{0,u}} \alpha(Y \cap \Lambda_{z,u})\,\mathrm{d}z, \tag{1.21.14}$$

where the lhs is the rhs of Eq. (1.3.4), and c_1 is given by Eq. (1.5.15). In the special case of Eq. (1.21.11), for each direction $\phi \in [0, \pi)$ we generate a test system of points $\{z \in kh,\ k \in \mathbb{Z}\}$ along the axis $L_1^2(0, \phi)$, namely along the orthogonal complement of a test line, and through such points we construct the test system $\Lambda_{z,\phi}$ of parallel test lines given by Eq. (1.21.12), see Fig. 1.21.2(a).

1.21.4 Note

To our knowledge, test systems were first proposed and studied by Santaló (1939) under the name '*figuras ilimitadas*' (unlimited figures), and elaborated further in Santaló (1953, 1956, 1976) as 'lattices of figures', see Fig. 1.21.3. Mean values relevant to stereology are given in Sections 2.24–2.26.

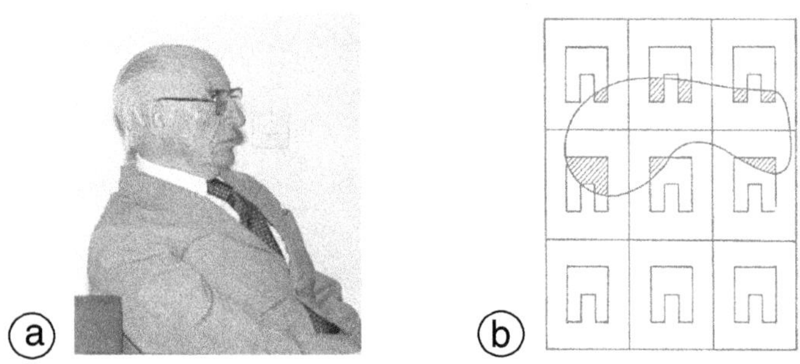

Fig. 1.21.3 (a) Luis Antonio Santaló Sors (1911–2001) at the 4th International Conference in Stereology and Stochastic Geometry, CH-Bern, 1987. (b) Original sketch of a portion of a test system from Santaló (1976).

Chapter 2
Basic Ideas of Geometric Sampling

2.1 Background and Purposes

2.1.1 Main purposes

Most of the material in Chapter 1 is concerned with integral-geometric formulae for motion-invariant measures of the intersection between a compact set and a probe (unbounded, or bounded). No clue is given there, however, of how to implement a sampling design which respects motion invariance and opens the way to the estimation of geometric properties. In this chapter, probability concepts are incorporated to describe sampling strategies, and to obtain mean values and ratio identities for the relevant parameters, known as fundamental equations of stereology. The basic concepts and results are derived first for a single test probe hitting a bounded target set. The corresponding results involving test systems, more common in practical stereology (Chapter 4) are deferred to Sections 2.24–2.31. The approach is entirely design-based: the target set is fixed and non-random, whereas the probe is equipped with a probability measure based on the motion-invariant density – principles and methods of model-based stereology, in which the target is modelled by a stationary random set whereas the location of the probe may be fixed, are described in Chapter 3.

2.1.2 Motion-invariant probability elements: concept

Pertinent prerequisites and notation are described in Section A.1.4.

In design-based geometric sampling the parameters which determine the position and orientation of a probe T (also called a 'test probe' in this context) are random variables. Suppose that T is a test point $x \in \mathbb{R}$ with translation-invariant density $\mathrm{d}x$, see Eq. (1.2.1). If (i) the probability element $\mathbb{P}(\mathrm{d}x)$ is proportional to $\mathrm{d}x$, and (ii) $x \in D \subset \mathbb{R}$, where D is an interval of a fixed finite length $H > 0$, then $\mathbb{P}(\mathrm{d}x)$ is

L. M. Cruz-Orive, *Stereology*, Interdisciplinary Applied Mathematics 59,
https://doi.org/10.1007/978-3-031-52451-6_2

given by Eq. (2.2.1) below, which means that x is a uniform random (UR) variable, see Section A.1.5. The extension to higher dimensions is straightforward, see Eq. (2.2.2) below.

In order to define a convenient joint probability element $\mathbb{P}(\mathrm{d}t)$ for the vector t of the parameters describing the position and orientation of a general geometric test probe T, we impose the following two conditions:

1. $\mathbb{P}(\mathrm{d}t)$ is proportional to the motion-invariant density $\mathrm{d}T$ of T.
2. The vector t is defined in a bounded domain of variation determined by the hitting condition $D \cap T \neq \emptyset$, where D is a fixed domain called the reference domain.

In order to construct $\mathbb{P}(\mathrm{d}t)$ we use the property given by Eq. (A.1.31). As defined in Section 1.20, the relevant hitting measure is

$$h(D) = \mathrm{meas}\{T : \ D \cap T \neq \emptyset\} = \int_{D\cap T\neq\emptyset} \mathrm{d}T, \tag{2.1.1}$$

whereby, conditional on the hitting event $D \cap T \neq \emptyset$, the adopted probability element is

$$\mathbb{P}(\mathrm{d}t) = \frac{\mathrm{d}T}{h(D)}. \tag{2.1.2}$$

For instance, if T is a motion-invariant test line in the plane, then $t := (p, \phi)$ and $\mathrm{d}T$ is given by the rhs Eq. (1.2.5). In order to define $\mathbb{P}(\mathrm{d}t)$ we adopt a disk D of finite diameter $H > 0$ as the reference domain, whereby $\mathbb{P}(\mathrm{d}t)$ is given by the rhs of Eq. (2.4.2).

In the sequel we consider relevant special cases. For coherence with the notation used in Chapter 1, random variables will often be denoted by lower case symbols.

2.1.3 Outline of this chapter

Consider a set $Y \subset \mathbb{R}^d$ of interest, and let $\gamma(Y)$ denote a target parameter, namely a relevant geometric measure defined on Y. The comprehension of what follows is not affected by d – the reader may think of $d = 3$. Indeed, the results in this chapter are derived for $d \leq 3$. Consider also an r-plane probe $L_r^d \equiv L_r$, for short, with motion-invariant density $\mathrm{d}L_r$. Define a reference ball D of a fixed and known diameter containing Y, that is, $Y \subset D \subset \mathbb{R}^d$. By Eq. (2.1.2), the motion-invariant probability element for L_r hitting D is $\mathrm{d}L_r/h_1(D)$, where

$$h_1(D) = \int_{D\cap L_r\neq\emptyset} \mathrm{d}L_r \tag{2.1.3}$$

is the relevant hitting measure. Now, recalling Eq. (1.3.4), we obtain

$$\mathbb{E}\{\alpha(Y \cap L_r)\} = \int_{D\cap L_r \neq \emptyset} \alpha(Y \cap L_r) \cdot \frac{\mathrm{d}L_r}{h_1(D)}$$
$$= \frac{c_1}{h_1(D)} \cdot \gamma(Y), \tag{2.1.4}$$

the expectation being with respect to the aforementioned motion-invariant probability element. Note that this expectation includes the zeros arising whenever the probe misses the target set, namely whenever $Y \cap L_r = \emptyset$. Thus, the target parameter may be expressed as follows,

$$\gamma(Y) = \frac{h_1(D)}{c_1} \cdot \mathbb{E}\{\alpha(Y \cap L_r)\}. \tag{2.1.5}$$

All the quantities in the rhs of the preceding identity are either known, i.e. c_1, $h_1(D)$, or observable from a proper sampling experiment, i.e. $\mathbb{E}\{\alpha(Y \cap L_r)\}$. Examples of $\gamma(\cdot)$ and $\alpha(\cdot)$ are given in the paragraphs following Eq. (1.3.4).

Similarly, if $Y \subset D \subset \mathbb{R}^d$, and D is hit by a bounded probe $T_r^d \equiv T_r$, then Eq. (1.14.1) yields,

$$\gamma(Y) = \frac{h_2(D)}{c_2 \cdot v(T_r)} \cdot \mathbb{E}\{\alpha(Y \cap T_r)\}, \tag{2.1.6}$$

where

$$h_2(D) = \int_{D\cap T_r \neq \emptyset} \mathrm{d}T_r \tag{2.1.7}$$

is the corresponding hitting measure.

Special cases of the preceding sampling designs for $d = 1, 2, 3$ are worked out in Sections 2.3–2.10 for a test point, a line, a plane, or a slab, and in Sections 2.11–2.14 for a bounded probe; the resulting identities constitute the classical equations of stereology. Apart from their didactic purpose, the pertinent results may assist in the correct design of computer simulations involving single probes, or independent replications of them. In stereological practice, however, the standard probes are test systems (Section 1.21), which circumvent the need to define a reference set $D \supset Y$; the corresponding mean value identities are developed in Sections 2.24–2.31. Local probes are treated in Sections 2.18–2.21, and in Section 2.32. Further, Sections 2.15–2.17 are devoted to particle number and connectivity, and Sections 2.22–2.23 to the stereology of projections.

As mentioned above, with the mean value identities developed in this chapter – especially those involving test systems – the construction of practical, ready-to-use estimators of the target parameter $\gamma(Y)$ is straightforward (Chapter 4).

2.2 A Single Uniform Random Test Point

2.2.1 Definition

Consider a point x in a bounded reference interval $D = [0, H)$ of the real axis. The point x is said to be uniform random (UR) in D if its probability element is proportional to its translation-invariant density $\mathrm{d}x$, see Eq. (1.2.2), that is,

$$\mathbb{P}(\mathrm{d}x) = \frac{\mathrm{d}x}{H}, \quad x \in [0, H), \tag{2.2.1}$$

which corresponds to a UR variable, and we write $x \sim \mathrm{UR}[0, H)$, or $x \sim \mathrm{UR}(D)$, see Section A.1.5. The condition $\mathbb{P}(\mathrm{d}x) = 0$, $x \notin [0, H)$, or its analogues in higher dimensions, will always be understood.

If we consider a d-cube $D = [0, H)^d \subset \mathrm{R}^d$, then Eq. (1.2.2) and the extension of Eq. (A.1.36) yield

$$\mathbb{P}(\mathrm{d}x) = \frac{\mathrm{d}x_1\,\mathrm{d}x_2\,\cdots\,\mathrm{d}x_d}{H^d}, \quad x \in [0, H)^d, \tag{2.2.2}$$

which means that the probability that a UR point in D hits an elementary cube of volume $\mathrm{d}x$ in D, is proportional to $\mathrm{d}x$, irrespective of the location of the elementary cube in D.

2.2.2 Generation of a UR test point in a cube

In Eq. (2.2.2), the Cartesian coordinates $(x_1, x_2, \ldots, x_d)$ of x are UR in the interval $[0, H)$, and independent. To generate a realization of $x \sim \mathrm{UR}(D)$, we take $x_i = H \cdot U_i$, $i = 1, 2, \ldots, d$, where the $\{U_i\}$ are $\mathrm{UR}[0, 1)$, and independent.

2.2.3 Mean values, ratios, and probabilities for a test point

Consider a domain $Y \subset D = [0, H)^d$ of volume V. For a single realization of a test point $x \sim \mathrm{UR}(D)$, let $P(Y \cap x)$ denote the number (either 0, or 1) of test points hitting Y or, equivalently, the indicator random variable $1_Y(x)$. Then, using the probability element given by Eq. (2.2.2),

$$\mathbb{E}\{P(Y \cap x)\} = \int_{\mathbb{R}^d} 1_Y(x)\mathbb{P}(\mathrm{d}x) = \frac{V}{H^d}. \tag{2.2.3}$$

For two domains $Y_1, Y_2 \subset D$, of volumes V_1, V_2, respectively, the preceding result yields,

$$\frac{\mathbb{E}(P_2)}{\mathbb{E}(P_1)} = \frac{V_2/H^d}{V_1/H^d} = \frac{V_2}{V_1}, \tag{2.2.4}$$

where $P_i \equiv P(Y_i \cap x)$, $i = 1, 2$, for short. In the classical stereological notation,

$$P_P = V_V, \tag{2.2.5}$$

which is a special case of a fundamental equation of stereology.

The probability that a test point $x \sim \mathrm{UR}(D)$ hits the set $Y \subset D$ is precisely $\mathbb{E}\{P(Y \cap x)\}$, namely,

$$\mathbb{P}(Y \cap x \neq \emptyset) \equiv \mathbb{P}(x \uparrow Y) = \int_Y \mathbb{P}(\mathrm{d}x) = \frac{1}{H^d}\int_Y \mathrm{d}x = \frac{V}{H^d}, \tag{2.2.6}$$

where the symbol '↑' reads 'hits'.

In stereology, a familiar setup is $Y_2 \subset Y_1 \subset D$ with $V_2 < V_1$. In this case, the probability that a test point $x \sim \mathrm{UR}(D)$ hits Y_2 given that it hits Y_1 is

$$\mathbb{P}(x \uparrow Y_2 | x \uparrow Y_1) = \frac{\mathbb{P}(x \uparrow Y_2)}{\mathbb{P}(x \uparrow Y_1)} = \frac{V_2/H^d}{V_1/H^d} = \frac{V_2}{V_1}, \tag{2.2.7}$$

which does not depend on the size of the reference domain D.

2.2.4 Generation of a UR test point hitting a domain by a direct rejection method

With the preceding setup $Y \subset D = [0, H)^d$, generate a sequence of independent realizations of $x \sim \mathrm{UR}(D)$, and retain the first realization satisfying $x \cap Y \neq \emptyset$, see Fig. 2.2.1(b). Then $x \sim \mathrm{UR}(Y)$.

In fact, recalling Eq. (2.2.6),

$$\mathbb{P}(\mathrm{d}x | x \uparrow Y) = \frac{\mathbb{P}(\mathrm{d}x)}{\mathbb{P}(x \uparrow Y)} = \frac{\mathrm{d}x/H^d}{V/H^d} = \frac{\mathrm{d}x}{V}, \tag{2.2.8}$$

which means that, if a test point $x \sim \mathrm{UR}(D)$ hits a domain $Y \subset D$, then $x \sim \mathrm{UR}(Y)$.

The number of rejections will be reduced if D is the smallest cube containing the set Y.

2.2.5 Properties of a UR test point hitting a planar domain

For a planar domain $Y \subset \mathbb{R}^2$ of area $A > 0$ with an arbitrary, fixed orientation relative to an orthogonal reference frame Ox_1x_2, Eq. (2.2.8) may be written as follows,

$$\mathbb{P}(\mathrm{d}x_1, \mathrm{d}x_2 | \uparrow) = \frac{\mathrm{d}x_1\,\mathrm{d}x_2}{A}, \quad x_1 \in Y', \; (x_1, x_2) \in Y \cap L_1^2(x_1, 0), \tag{2.2.9}$$

where '$\uparrow$' is shorthand for the hitting event '$x \uparrow Y$'. Further, Y' represents the orthogonal linear projection of the set Y onto the horizontal axis Ox_1, and $Y \cap L_1^2(x_1, 0)$ is the vertical intercept (of total length $l(x_1)$, say) determined in Y by the vertical straight line $L_1^2(x_1, 0)$, see Fig. 2.2.1(c).

Unless Y is a rectangle with the sides parallel to the coordinate axes, the random variables x_1 and x_2 are not independent in general. For instance, the marginal probability element of x_1 becomes

$$\begin{aligned} \mathbb{P}(\mathrm{d}x_1 | \uparrow) &= \frac{\mathrm{d}x_1}{A} \int_{Y \cap L_1^2(x_1, 0)} \mathrm{d}x_2 \\ &= \frac{l(x_1)\,\mathrm{d}x_1}{A}, \quad x_1 \in Y', \end{aligned} \tag{2.2.10}$$

and similarly for $\mathbb{P}(\mathrm{d}x_2 | \uparrow)$, whereby

$$\mathbb{P}(\mathrm{d}x_1 | \uparrow) \cdot \mathbb{P}(\mathrm{d}x_2 | \uparrow) \neq \mathbb{P}(\mathrm{d}x_1, \mathrm{d}x_2 | \uparrow). \tag{2.2.11}$$

Eq. (2.2.10) means that, conditional on the hitting event, the marginal probability element of x_1 is proportional to $l(x_1)$ – for a simple interpretation see Fig. 2.2.1(c,d) and the next section. Moreover, conditional on x_1 and on the hitting event, the ordinate x_2 is UR within that intercept, that is,

$$\mathbb{P}(\mathrm{d}x_2 | x_1, \uparrow) = \frac{\mathrm{d}x_2}{l(x_1)}, \quad x_2 \in Y \cap L_1^2(x_1, 0), \tag{2.2.12}$$

which can also be directly verified from the identity

$$\mathbb{P}(\mathrm{d}x_1, \mathrm{d}x_2 | \uparrow) = \mathbb{P}(\mathrm{d}x_2 | x_1, \uparrow) \cdot \mathbb{P}(\mathrm{d}x_1 | \uparrow). \tag{2.2.13}$$

2.2.6 Sampling a UR point in a domain by a two-stage rejection method

The motivation of the following sampling methods is twofold: (i) The direct method (Section 2.2.4) assumes that sampling can be implemented directly in $\mathbb{R}^d$. In practice, point sampling in 3D, for instance, is usually implemented on planar UR sections, see Fig. 2.2.2; and (ii), the two-stage rejection method helps to understand the concept of point sampling, and of weighted sampling in general.

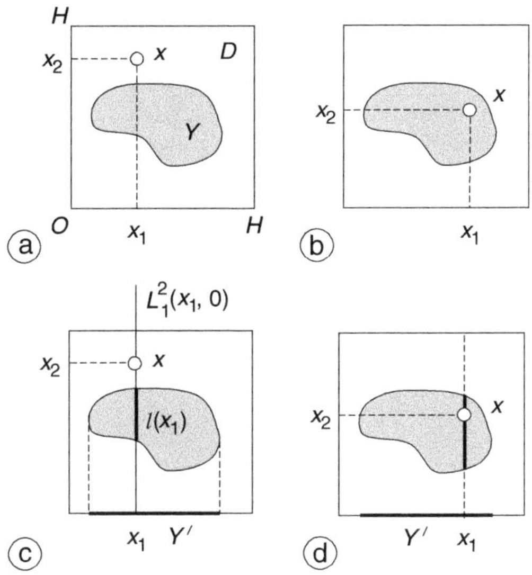

Fig. 2.2.1 (a), (b) Generation of a UR test point x hitting a planar set Y by the rejection method. The trial in (a) fails, the one in (b) succeeds. (c), (d) If $x \sim \mathrm{UR}(Y)$, then the marginal probability element of x_1 is proportional to $l(x_1)$, because x_1 'survives' only if $x \in Y \cap L_1^2(x_1, 0)$, see Eq. (2.2.10).

With reference to Eq. (2.2.10) and Fig. 2.2.1(c,d), the fact that the marginal probability element of x_1 is proportional to $l(x_1)$ is easy to interpret if we notice that the rejection principle with $Y \subset D = [0, H]^2$ may be implemented in the following two stages.

1. Generate a sequence of independent UR abscissas in the interval $[0, H)$, and retain the first abscissa, x_1 say, satisfying the condition $Y \cap L_1^2(x_1, 0) \neq \emptyset$, see Fig. 2.2.1(c).
2. Generate a UR ordinate $x_2 \sim \mathrm{UR}[0, H)$. If the point $x = (x_1, x_2)$ hits the vertical linear intercept $Y \cap L_1^2(x_1, 0)$, then $x \sim \mathrm{UR}(Y)$, otherwise return to stage 1, see Fig. 2.2.1(d).

The second stage illustrates the fact that, conditional on the hitting event, the marginal probability element of x_1 is weighted by the intercept length $l(x_1)$.

The method extends naturally to higher dimensions. For instance, in order to generate $x = (x_1, x_2, x_3) \sim \mathrm{UR}(Y)$, $Y \subset D \equiv [0, H)^3$ we may use the following two steps.

1. Generate a sequence of independent UR ordinates in $[0, H)$, and retain the first one, x_3 say, satisfying the condition $Y \cap L_2^3(x_3, 0) \neq \emptyset$, see Fig. 2.2.2(a).
2. Generate a UR point (x_1, x_2) in the horizontal square $D \cap L_2^3(x_3, 0)$. If the point (x_1, x_2, x_3) hits the horizontal planar transect $Y \cap L_2^3(x_3, 0)$, then $x \sim \mathrm{UR}(Y)$, see Fig. 2.2.2(c). Otherwise, return to stage 1.

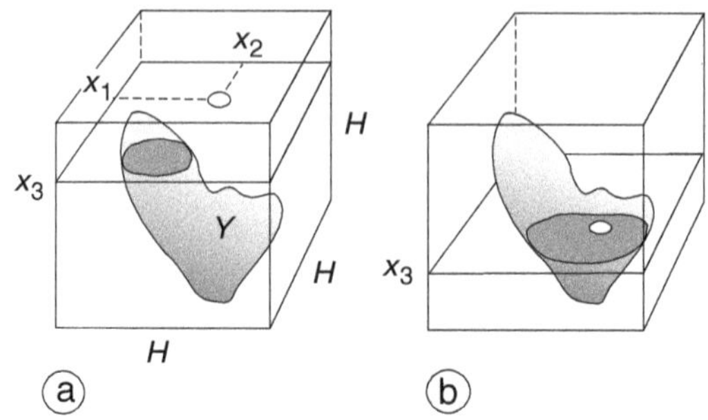

Fig. 2.2.2 Generation of a UR point in a set Y by the two-stage rejection method. The second stage fails in (a) and succeeds in (b), see Section 2.2.6.

2.3 Weighted Sampling

2.3.1 Concept

The preceding results motivate the concept of weighted sampling. Let $\mathbb{P}(\mathrm{d}x)$ be the probability element of a random variable X. If X is observed conditional on an event E, then Bayes' theorem yields

$$\mathbb{P}(\mathrm{d}x|E) \propto \mathbb{P}(E|x) \cdot \mathbb{P}(\mathrm{d}x). \tag{2.3.1}$$

Further, if $\mathbb{P}(E|x) = w(x)$, namely a known function of x, then the probability element of the conditional random variable $X|E$ is

$$\mathbb{P}(\mathrm{d}x|E) = \frac{w(x) \cdot \mathbb{P}(\mathrm{d}x)}{\int w(x) \cdot \mathbb{P}(\mathrm{d}x)} = \frac{w(x) \cdot \mathbb{P}(\mathrm{d}x)}{\mathbb{E}\{w(x)\}}, \tag{2.3.2}$$

and the distribution of $X|E$ is said to be w-weighted.

2.3.2 Example: length-weighted intercept

With the setup $Y \subset D = [0, H)^2$ considered in Section 2.2.6, generate a UR abscissa x_1 in the orthogonal projection Y' of Y, see Fig. 2.2.1(c), so that the corresponding probability element is

$$\mathbb{P}(\mathrm{d}x_1) = \frac{\mathrm{d}x_1}{L(Y')}, \quad x_1 \in Y'. \tag{2.3.3}$$

Note that

$$A = \int_{Y'} l(x_1)\,\mathrm{d}x_1 = L(Y') \cdot \mathbb{E}\{l(x_1)\}, \tag{2.3.4}$$

and therefore Eq. (2.2.10) may be written as follows,

$$\mathbb{P}(\mathrm{d}x_1 | \uparrow) = \frac{l(x_1) \cdot \mathbb{P}(\mathrm{d}x_1)}{\mathbb{E}\{l(x_1)\}}, \quad x_1 \in Y', \tag{2.3.5}$$

which means that the distribution of x_1 conditional on the hitting event is length-weighted of the form given by Eq. (2.3.2).

Let $\mathbb{P}(\mathrm{d}l)$ denote the probability element of the unweighted vertical intercept length $l \equiv l(x_1)$, $x_1 \in Y'$. Since each value of x_1 determines a value of l we have that $\mathbb{P}(\mathrm{d}x_1) = \mathbb{P}(\mathrm{d}l)$, and therefore

$$\mathbb{P}(\mathrm{d}l | \uparrow) = \frac{l \cdot \mathbb{P}(\mathrm{d}l)}{\mathbb{E}(l)}, \tag{2.3.6}$$

which means that, conditional on the hitting event $x \uparrow Y$, the distribution of l is l-weighted. We say that the corresponding vertical intercept is 'point-sampled' because it is selected by the UR test point $x \in Y$.

2.3.3 A priori and a posteriori weighting

In the preceding example, the unconditional second moment $\mathbb{E}(l^2)$ of the intercept length may be evaluated in two alternative ways.

A priori weighting

Generate a UR point $x = (x_1, x_2)$ in the set Y, and record the corresponding intercept length $l = l(x_1)$, $x_1 \in Y'$. Because this length is conditional on the hitting event $x \uparrow Y$, its ordinary mean is in fact its expected value with respect to the probability element given by Eq. (2.3.6), namely,

$$\mathbb{E}(l | \uparrow) = \frac{1}{\mathbb{E}(l)} \int l^2 \mathbb{P}(\mathrm{d}l) = \frac{\mathbb{E}(l^2)}{\mathbb{E}(l)}. \tag{2.3.7}$$

Thus, under this design the observed first moment of l is equal to the ratio of the unweighted second moment to the unweighted first moment of l, because the l-weighting is implicit in the design.

It follows that, if both $\mathbb{E}(l | \uparrow)$ and $\mathbb{E}(l) = A/L(Y')$ are known, then the square coefficient of variation of the intercept length l of a set Y is accessible without knowing the distribution of l, namely,

$$\mathrm{CV}^2(l) = \frac{\mathbb{E}(l^2)}{[\mathbb{E}(l)]^2} - 1 = \frac{\mathbb{E}(l | \uparrow)}{\mathbb{E}(l)} - 1. \tag{2.3.8}$$

A posteriori weighting

Generate a UR abscissa $x_1 \sim \mathrm{UR}(Y')$, and record the intercept length $l = l(x_1)$. By definition, the ordinary mean of the square $l^2(x_1)$ of this length with respect to the probability element given by Eq. (2.3.3) is precisely $\mathbb{E}(l^2)$, namely,

$$\mathbb{E}(l^2) = \int_{Y'} l^2(x_1)\mathbb{P}(\mathrm{d}x_1). \tag{2.3.9}$$

Thus, here the sampling is unweighted, and therefore $\mathbb{E}(l^2)$ has to be evaluated by weighting the observed measurement $l(x_1)$ a posteriori with a numerical weight equal to $l(x_1)$ itself, and taking the ordinary mean of the resulting power $l^2(x_1)$.

2.3.4 Weighted point sampling

Consider the setup $Y_2 \subset Y_1 \subset D = [0, H)^d$, and $x \sim \mathrm{UR}(D)$, as in Section 2.2.3. If we ignore a realization of the test point $x \sim \mathrm{UR}(D)$ whenever it misses the set Y_1, namely if we force that $P_1 = 1$, then $x \sim \mathrm{UR}(Y_1)$, see Section 2.2.4, and trivially,

$$\mathbb{E}\left(\frac{P_2}{P_1}\middle| x \uparrow Y_1\right) = \mathbb{E}(P_2 | x \uparrow Y_1) = \int_{Y_1} 1_{Y_2}(x) \cdot \frac{\mathrm{d}x}{V_1} = \frac{V_2}{V_1}, \tag{2.3.10}$$

or, in a more convenient notation,

$$\mathbb{E}_P\left(\frac{P_2}{P_1}\right) = \frac{V_2}{V_1}, \tag{2.3.11}$$

the subscript 'P' meaning that the expectation is over the P-weighted probability element of x.

2.3.5 Notes

1. Early history

As pointed out by Hykšová et al. (2012), toward 1664–1666 Isaac Newton (1643–1727) apparently formulated the principle that a 'random' point hitting a domain of area $A > 0$ will hit a subdomain of area $a \leq A$ with a probability equal to a/A – a special case of Eq. (2.2.7). This may not be surprising inasmuch as probability was developing at Newton's time, notably with Blaise Pascal (1623–1662), and with the early members of the Bernoulli family. Nonetheless, Newton's leap from a discrete sample space (pertinent to dice and card games) to a continuous one cannot be underestimated.

2. The rejection method

Consider a random variable X with $\mathbb{P}(\mathrm{d}x) = f(x)\,\mathrm{d}x$, with the property that the entire graph of the probability density function $f(x)$ can be enclosed in a bounded rectangle $D = [0, H_1) \times [0, H_2)$. Generate a UR point (x_1, x_2) in D, and retain the point if it hits the region under the graph of $f(x)$. By virtue of Eq. (2.2.10) (with $A = 1$), we have

$$\mathbb{P}(\mathrm{d}x_1 | \uparrow) = f(x_1)\,\mathrm{d}x_1, \qquad (2.3.12)$$

and therefore x_1 is a realization of X. Von Neumann (1951) exploited this idea to generate realizations of a random variable.

3. Weighted sampling

The basic ideas of conditioning and weighted sampling, as well as generalized versions of Eq. (2.3.11), were introduced in a stereological context by Miles and Davy (1976) and Miles (1978a).

4. Fundamental equations of stereology

In a model-based setting with applications in geology, Eq. (2.2.5) is due to Glagolev (1933), see Cruz-Orive (2017). The full meaning of these and analogous results will be better appreciated when dealing with test systems (Sections 2.24–2.31). The abbreviated notation P_P, A_A, etc., was proposed after the foundation of the ISS in the 1960s, see Underwood (1970). For details on the foundation of the ISS (now ISSIA) see Elias (1962, 1963), Bach (1963) and the history chapter of Miles (1987).

5. A remark on ratios

As in Section 2.2.3, often in this book two domains Y_1, Y_2 are considered which do not need to satisfy the condition $Y_2 \subset Y_1$ (familiar in stereology). Thus, the ratios P_P, A_A, V_V may be greater than 1.

2.4 FUR and IUR Test Lines in 2D

2.4.1 Definitions

For an axis $L^2_{1[0]} \equiv L^2_1(0, \phi)$ of direction $\phi \in [0, \pi)$ in the plane we adopt a probability element proportional to its rotation-invariant density $\mathrm{d}L^2_{1[0]} = \mathrm{d}\phi$, see Eq. (1.2.4), namely,

$$\mathbb{P}(\mathrm{d}\phi) = \frac{\mathrm{d}\phi}{\pi}, \quad \phi \in [0, \pi), \tag{2.4.1}$$

so that $\phi \sim \mathrm{UR}[0, \pi)$, and we say that the axis is isotropic random (IR).

For a straight line $L_1^2 \equiv L_1^2(p, \phi)$ in the plane we adopt a probability element proportional to its motion-invariant density $\mathrm{d}L_1^2 = \mathrm{d}p\,\mathrm{d}\phi$, see Eq. (1.2.5). Because the range of p is infinite, however, we impose the condition that L_1^2 hits a bounded reference set $D \subset \mathbb{R}^2$. A convenient choice for D is a disk of diameter $H > 0$, say, centred at the origin, see Fig. 2.4.1(a). The hitting measure is the boundary length πH of D, see Eq. (1.19.4), whereby the motion-invariant joint probability element is

$$\mathbb{P}(\mathrm{d}p, \mathrm{d}\phi) = \frac{\mathrm{d}p}{H} \cdot \frac{\mathrm{d}\phi}{\pi}, \quad p \in [-H/2, H/2],\ \phi \in [0, \pi). \tag{2.4.2}$$

Thus, $p \sim \mathrm{UR}[-H/2, H/2]$ and $\phi \sim \mathrm{UR}[0, \pi)$ are independent random variables. This means that L_1^2 has isotropic orientation, and for each orientation it has a UR position – hence we say that L_1^2 is isotropic uniform random (IUR) hitting D, and we write $L_1^2 \sim \mathrm{IUR}(D)$. The hitting condition with respect to a bounded set is important – an unrestricted motion-invariant line in the plane cannot be said to be IUR, because the definition of a probability element requires a bounded reference space.

If the direction ϕ is fixed, while p is UR, then the test line $L_1^2(p, \phi)$ is said to be UR with a fixed orientation (FUR) hitting D – that is, for any $\phi \in [0, \pi)$,

$$\mathbb{P}(\mathrm{d}p) = \frac{\mathrm{d}p}{H}, \quad p \in [-H/2, H/2]. \tag{2.4.3}$$

2.4.2 Generation of an IUR test line hitting a disk

Take two independent UR numbers $U_i \sim \mathrm{UR}[0, 1)$, $i = 1, 2$ and set the direction $\phi = U_1 \cdot \pi$ and the signed distance $p = (2U_2 - 1) \cdot R$ along that direction, where R is the radius of the disk D. Then, $L_1^2(p, \phi) \sim \mathrm{IUR}(D)$.

Indirect generation of an IR direction

Consider the setup $Y \subset D = [-R, R]^2$ of Section 2.2.4, where Y is a disk of radius R centred at the origin, and sample a point $x \sim \mathrm{UR}(Y)$ by the rejection method. Then the orientation of the ray Ox is $\mathrm{UR}[0, 2\pi)$, and the direction of its support axis is $\mathrm{UR}[0, \pi)$.

To justify this, let (r, ω), $r \in [0, R]$, $\omega \in [0, 2\pi)$ denote the polar coordinates of the point x. By Eq. (2.2.9),

$$\mathbb{P}(\mathrm{d}x) = \frac{\mathrm{d}x}{\pi R^2} = \frac{2r\,\mathrm{d}r}{R^2} \cdot \frac{\mathrm{d}\omega}{2\pi}, \tag{2.4.4}$$

which shows that $\omega \sim \mathrm{UR}[0, 2\pi)$, independently of r. Note that the probability element of r is r-weighted.

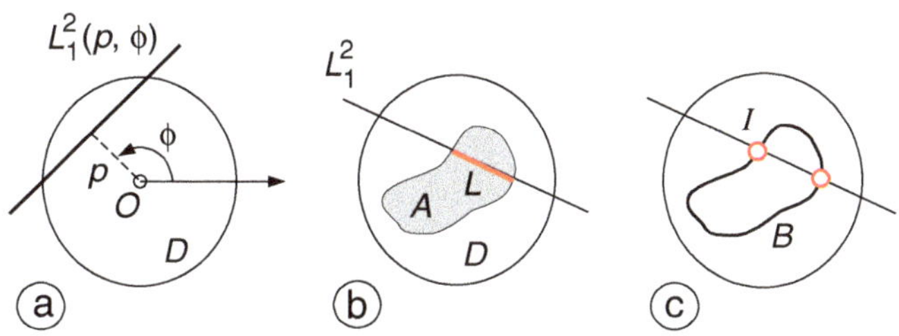

Fig. 2.4.1 (a) A test line hitting a reference disk D. (b) The mean value of the transect length L depends on whether the test line is FUR hitting the reference disk D, see Eq. (2.4.5), or conditional (FUR, or IUR), hitting a set, see Eq. (2.4.24) and Eq. (2.4.25), respectively. (c) Illustration of Eq. (2.4.6).

2.4.3 Mean values for a test line in 2D

Planar domain

Consider a planar domain $Y \subset D \subset \mathbb{R}^2$ of area $A > 0$ enclosed in a reference disk D of diameter H centred at the origin. Because $\dim(Y) = 2$, a suitable probe is a test line $L_1^2 \sim \mathrm{FUR}(D)$ with associated probability element given by Eq. (2.4.3). Set $L \equiv L(Y \cap L_1^2)$. By Eq. (1.4.4),

$$\mathbb{E}(L) = \frac{1}{H} \int_{-H/2}^{H/2} L \, \mathrm{d}p = \frac{A}{H}, \tag{2.4.5}$$

for any given orientation of the test line, see Fig. 2.4.1(b). Thus, the result is the same if $L_1^2 \sim \mathrm{IUR}(D)$.

Remark

The preceding mean value is computed including the zeros of L arising when the test line misses the set Y. This applies to all the analogous mean values given in the sequel whenever $Y \subset D \subset \mathbb{R}^2$, where D is a reference disk, or a ball in higher dimensions.

Planar curve

Here $Y \subset D \subset \mathbb{R}^2$ and Y represents a piecewise smooth curve of finite length $B > 0$. A suitable probe is $L_1^2 \sim \mathrm{IUR}(D)$, with associated probability element given by Eq. (2.4.2). Set $I \equiv I(Y \cap L_1^2)$. Then, using Eq. (1.5.8), the mean number of intersections is

$$\mathbb{E}(I) = \frac{1}{\pi H} \int_{L_1^2 \uparrow Y} I \, \mathrm{d}L_1^2 = \frac{2B}{\pi H}, \tag{2.4.6}$$

see Fig. 2.4.1(c).

2.4.4 Classical ratios for test lines in 2D

Consider two planar sets $Y_1, Y_2 \subset D \subset \mathbb{R}^2$ of positive areas A_1, A_2 respectively, and a FUR test line hitting D. Application of Eq. (2.4.5) yields the classical stereological equation

$$\frac{\mathbb{E}(L_2)}{\mathbb{E}(L_1)} = \frac{A_2/H}{A_1/H} = \frac{A_2}{A_1}, \tag{2.4.7}$$

namely,

$$A_A = L_L. \tag{2.4.8}$$

Suppose that the set Y_2 is a curve of length B_2, and the test line is IUR hitting D. Application of Eq. (2.4.5) and Eq. (2.4.6) yields another classical stereological equation, namely,

$$\frac{\mathbb{E}(I_2)}{\mathbb{E}(L_1)} = \frac{2B_2/(\pi H)}{A_1/H} = \frac{2}{\pi} \cdot \frac{B_2}{A_1}, \tag{2.4.9}$$

namely,

$$B_A = \frac{\pi}{2} \cdot I_L. \tag{2.4.10}$$

If Y_1 is also curve, and its length is B_1, then

$$\frac{\mathbb{E}(I_2)}{\mathbb{E}(I_1)} = \frac{2B_2/(\pi H)}{2B_1/(\pi H)} = \frac{B_2}{B_1}, \tag{2.4.11}$$

namely,

$$B_B = I_I. \tag{2.4.12}$$

Mean intercept length for a planar domain

Consider a planar domain Y with boundary ∂Y. The intersection $\partial Y \cap L_1^2$ will typically consist of an even number $I = 2N$ of intersection points (see Section 1.3.1), where N represents the number of separate linear intercepts (see Section 1.4.2) constituting the linear IUR transect $Y \cap L_1^2$ of total length L. Let l denote the length of a linear intercept. Then, by Eq. (2.4.9),

$$\mathbb{E}(l) = \frac{\mathbb{E}(L)}{\mathbb{E}(N)} = \frac{2\mathbb{E}(L)}{\mathbb{E}(I)} = \frac{\pi A}{B}. \tag{2.4.13}$$

Equivalently,

$$\mathbb{E}(l) = \pi / B_A = 2/I_L. \tag{2.4.14}$$

2.4.5 Sampling an IUR test line hitting a planar set by the rejection method

Consider the setup $\subset D \subset \mathbb{R}^2$, where Y may denote a planar domain, or a curve. To sample an IUR test line hitting the set Y by a rejection method, generate a sequence of independent realizations of a test line $L_1^2 \sim \mathrm{IUR}(D)$, see Section 2.4.2, and retain the first one satisfying $L_1^2 \uparrow Y$, see Fig. 2.4.2(b). Then $L_1^2 \sim \mathrm{IUR}(Y)$.

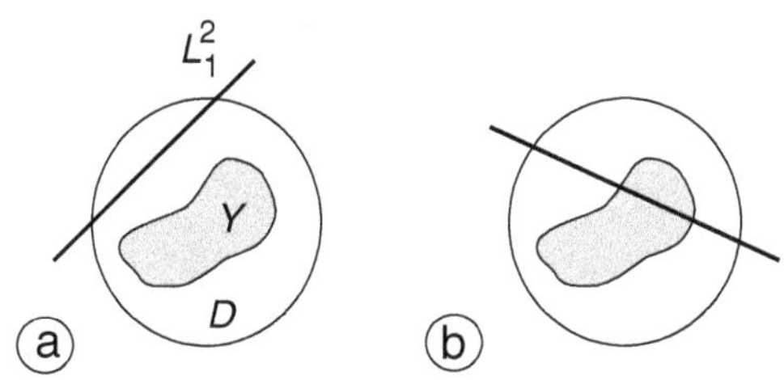

Fig. 2.4.2 Sampling an IUR test line hitting a planar set Y by the rejection method (Section 2.4.5). The experiment fails in (a), and succeeds in (b).

To justify the method, first note that

$$\mathbb{P}(L_1^2 \uparrow Y) = \int_{L_1^2 \uparrow Y} \mathbb{P}(\mathrm{d}p, \mathrm{d}\phi) = \frac{h(Y)}{\pi H}, \tag{2.4.15}$$

where $h(Y)$ is the measure of all IUR test lines hitting Y, see Eq. (1.19.1). Now, recalling Eq. (2.4.2),

$$\begin{aligned} \mathbb{P}(\mathrm{d}p, \mathrm{d}\phi | L_1^2 \uparrow Y) &= \frac{\mathbb{P}(\mathrm{d}p, \mathrm{d}\phi)}{\mathbb{P}(L_1^2 \uparrow Y)} \\ &= \frac{\mathrm{d}p\, \mathrm{d}\phi}{h(Y)}, \quad p \in Y'(\phi),\ \phi \in [0, \pi), \end{aligned} \tag{2.4.16}$$

which is the probability element of $L_1^2 \sim \mathrm{IUR}(Y)$. Here, $Y'(\phi)$ denotes the orthogonal linear projection of Y onto the axis $L_1^2(0, \phi)$ normal to the test line. If Y is not connected, then $Y'(\phi)$ may consist of several separate segments of total length $H(\phi)$. By Eq. (1.19.3),

$$h(Y) = \pi \mathbb{E}\{H(\phi)\}. \tag{2.4.17}$$

2.4.6 Conditional IUR and FUR test lines hitting a planar set

So far we have mainly used the unconditional probability element given by Eq. (2.4.2). Here we examine the properties of the conditional probability element given by Eq. (2.4.16), namely

$$\mathbb{P}(\mathrm{d}p, \mathrm{d}\phi| \uparrow) = \frac{\mathrm{d}p\,\mathrm{d}\phi}{\pi \mathbb{E}\{H(\phi)\}}. \tag{2.4.18}$$

Unless the projected length $H(\phi)$ of Y is constant, the random variables p and ϕ are no longer independent. The marginal conditional probability element of ϕ is

$$\begin{aligned}\mathbb{P}(\mathrm{d}\phi| \uparrow) &= \frac{\mathrm{d}\phi}{\pi \mathbb{E}\{H(\phi)\}} \int_{Y'(\phi)} \mathrm{d}p \\ &= \frac{H(\phi)\mathbb{P}(\mathrm{d}\phi)}{\mathbb{E}\{H(\phi)\}}, \qquad \phi \in [0, \pi).\end{aligned} \tag{2.4.19}$$

On the other hand, the expression of the marginal $\mathbb{P}(\mathrm{d}p| \uparrow)$ is usually complicated, generally containing several branches. Thus, in general,

$$\mathbb{P}(\mathrm{d}p| \uparrow) \cdot \mathbb{P}(\mathrm{d}\phi| \uparrow) \neq \mathbb{P}(\mathrm{d}p, \mathrm{d}\phi| \uparrow). \tag{2.4.20}$$

Eq. (2.4.19) shows that the marginal conditional distribution of ϕ is $H(\phi)$-weighted. This agrees with the fact that, by the rejection method, for each value of ϕ the test line $L_1^2(p, \phi)$ is retained with probability $H(\phi)/H$. Furthermore, for each value of ϕ the parameter is UR, namely

$$\mathbb{P}(\mathrm{d}p|\phi, \uparrow) = \frac{\mathbb{P}(\mathrm{d}p, \mathrm{d}\phi| \uparrow)}{\mathbb{P}(\mathrm{d}\phi| \uparrow)} = \frac{\mathrm{d}p}{H(\phi)}, \qquad p \in Y'(\phi). \tag{2.4.21}$$

If Y is a domain of area A then, conditional on the hitting event, the mean intercept length $L \equiv L(Y \cap L_1^2)$ is no longer that given by Eq. (2.4.5). If $L_1^2 \sim \mathrm{FUR}(Y)$, then, using Eq. (2.4.21),

$$\mathbb{E}(L|\phi, \uparrow) = \int L\mathbb{P}(\mathrm{d}p|\phi, \uparrow) = \frac{A}{H(\phi)}. \tag{2.4.22}$$

However, if $L_1^2 \sim \mathrm{IUR}(Y)$, then, using Eq. (2.4.19) and the preceding result,

$$\mathbb{E}(L| \uparrow) = \int \mathbb{E}(L|\phi, \uparrow)\mathbb{P}(\mathrm{d}\phi| \uparrow) = \frac{A}{\mathbb{E}\{H(\phi)\}}. \tag{2.4.23}$$

Thus, we have the following two different representations of A,

$$A = H(\phi) \cdot \mathbb{E}(L|\phi, \uparrow), \tag{2.4.24}$$

$$A = \mathbb{E}\{H(\phi)\} \cdot \mathbb{E}(L|\uparrow), \tag{2.4.25}$$

according to whether the test line is FUR, or IUR, hitting the set Y, respectively.

Explicit results are easy to obtain under shape assumptions. For instance, suppose that the set Y is connected, and let Y_0 denote its convex hull, namely the smallest convex set containing Y. If $L_1^2 \sim \text{IUR}(D)$, then the events $L_1^2 \uparrow Y$ and $L_1^2 \uparrow Y_0$ are equivalent. By Eq. (1.19.4), the hitting measure of either event is the boundary length B_0 of Y_0, whereby, using Eq. (2.4.2),

$$\mathbb{P}(L_1^2 \uparrow Y) = \frac{1}{\pi H} \int_{L_1^2 \uparrow Y} \mathrm{d}p\, \mathrm{d}\phi = \frac{B_0}{\pi H}. \tag{2.4.26}$$

As a consequence, if $Y_2 \subset Y_1 \subset D \subset \mathbb{R}^2$ are connected sets, then

$$\mathbb{P}(L_1^2 \uparrow Y_2 | L_1^2 \uparrow Y_1) = \frac{\mathbb{P}(L_1^2 \uparrow Y_2)}{\mathbb{P}(L_1^2 \uparrow Y_1)} = \frac{B_{02}}{B_{01}}, \tag{2.4.27}$$

where B_{0i} denotes the boundary length of the convex hull of Y_i, $i = 1, 2$.

Suppose that the set $Y \subset \mathbb{R}^2$ is moreover convex with area A and boundary length B. Then, by Eq. (1.19.5) and Eq. (2.4.1),

$$\mathbb{E}\{H(\phi)\} = \int_0^\pi H(\phi)\mathbb{P}(\mathrm{d}\phi) = \frac{1}{\pi} \int_0^\pi H(\phi)\mathrm{d}\phi = \frac{B}{\pi}, \tag{2.4.28}$$

which shows that the mean caliper length of Y over IR directions is directly accessible if B is known. In this case Eq. (2.4.18) becomes

$$\mathbb{P}(\mathrm{d}p, \mathrm{d}\phi | \uparrow) = \frac{\mathrm{d}p\, \mathrm{d}\phi}{B}. \tag{2.4.29}$$

Remark

The ratio identities derived in Section 2.4.4 hold unchanged under the hitting condition because the normalizing factor in the rhs of Eq. (2.4.18) cancels out in the ratios. This extends to any dimension.

2.4.7 How to make an IUR cut of a steak

Place the steak in a flat circular dish of diameter H, say, on a table. (i) Rotate the dish isotropically at random between 0° and 180°. (ii) Independently, aim with a knife at a UR position along a diameter of the dish. If the blade misses the steak, see Fig. 2.4.2(a), then reject the cutting attempt and return to step (i). Otherwise, the corresponding cut is IUR hitting the steak (Fig. 2.4.2(b)).

In step (ii), most people would probably place the blade at random within the steak. This would result in cutting the steak with probability 1 for any orientation, which we know is incorrect. In fact, from Eq. (2.4.2), $\mathbb{P}(\mathrm{d}p, \mathrm{d}\phi|\phi) = \mathbb{P}(\mathrm{d}p|\phi) = \mathrm{d}p/H$, whereby

$$\mathbb{P}(\uparrow|\phi) = \int_{L_1^2(p,\phi)\uparrow Y} \mathbb{P}(\mathrm{d}p|\phi) = \frac{H(\phi)}{H}. \tag{2.4.30}$$

Thus, for a given orientation ϕ the cutting probability has to be proportional to the caliper length $H(\phi)$ of the steak perpendicular to the blade, hence rejections must be allowed.

2.4.8 Notes

1. IUR test lines in 2D

As indicated in Note 1 of Section 1.5.5, Crofton (1868) was the first to propose the density $\mathrm{d}L_1^2 = \mathrm{d}p\,\mathrm{d}\phi$ for a straight line in the plane, with which he derived the early Crofton formulae. The theory of motion-invariant measures emerged with integral geometry in the early 20th century (Section 1.2.10, Note 1). At the time, probabilities were usually computed as ratios of such measures. The incorporation of motion-invariant probability elements (e.g. Eq. (2.4.2)) facilitated the mathematical treatment and widened the scope of applications, see for instance Miles and Davy (1976), or Miles (1978a). Randomness concepts such as isotropic (IR), uniform (UR), uniform with a fixed orientation (FUR) and isotropic uniform random (IUR) were introduced in stereology by the latter authors.

2. The Buffon needle problem – a preliminary view

Buffon's needle problem, proposed and solved by the French naturalist George Louis Leclerc, Comte de Buffon (1707–1788), see Fig. 2.4.3(a), encapsulates the art and spirit of stereology. It also pioneers a definite transition from discrete to continuous probability spaces. Its original version (Buffon (1777), see Miles and Serra (1978)), reads as follows.

> "A player throws a rod of length l at random over a board bearing an array of parallel lines a distance $h \geq l$ apart. If the rod hits a line, the player obtains a money prize m, say. How much should he/she pay for playing?"

Buffon argues that the mean gain will be m times the probability of winning. To find this probability he anticipates the basic ideas underlying the kinematic density for the rod, Fig. 2.4.3(b), and the use of a test system of parallel test lines, Fig. 2.4.3(c), which will be revisited in Section 2.26.2. To simplify the computation, we reformulate the problem into an equivalent one, as suggested by Robert Deltheil (1890–1972), (Deltheil, 1926, p. 61), namely:

> "A needle of length l is arbitrarily fixed in a disk of diameter $h > l$ in the plane. The disk is hit by an IUR test line. Calculate the probability that the straight line hits the needle", see Fig. 2.4.3(d).

Regard the needle as a flattened convex body of perimeter length $B = 2l$. By Eq. (2.4.26) the required probability is equal to $2l/(\pi h)$, which is Buffon's result.

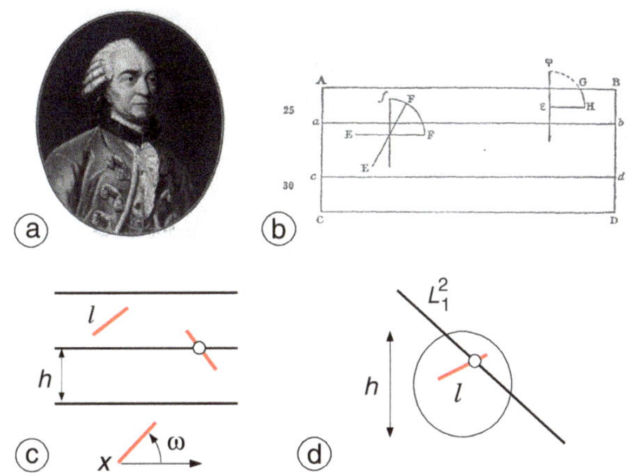

Fig. 2.4.3 (a) Portrait of Buffon by P. M. Alix (Carnavalet Museum, Paris). (b) Original illustration of his needle problem. (c) Alternative illustration. (d) Deltheil's simplification. See Section 2.4.8, Note 2.

3. Bertrand's paradoxes

In spite of their elegance, the results of M. W. Crofton and other contemporary authors aroused controversy at the time because the concept of motion invariance was not yet fully established. Different choices for the density of a geometric object gave rise to different solutions of the same problem, thus giving rise to 'paradoxes'. As an example, Joseph Louis François Bertrand (1822–1900) considered the following problem (Bertrand, 1889, pp. 4–5):

> "Take a chord of a circle at random. What is the probability that this chord is longer than the side of an inscribed equilateral triangle?"

The problem is equivalent to finding the probability that a random chord of the unit circle centred at the origin O, is longer than $\sqrt{3}$. Bertrand considered the following three possibilities.

1. The polar angles $0 \le \alpha, \beta \le 2\pi$ of the end points of a chord are assumed to be UR and independent, see Fig. 2.4.4(a). For any $\alpha \in [0, 2\pi)$, the required probability is
$$\mathbb{P}\{\beta \in [\alpha + 2\pi/3, \alpha + 4\pi/3)\} = (2\pi/3)/(2\pi) = 1/3. \tag{2.4.31}$$
2. For any orientation of the chord, its distance p from O is UR in $[0, 1]$, see Fig. 2.4.4(b). Then the required probability is $\mathbb{P}(p < 1/2) = 1/2$.
3. For any orientation of the chord, its midpoint M is UR within the unit disk, see Fig. 2.4.4(c). By Eq. (2.4.4), the distribution function of the distance $p = |OM|$ is $F(p) = p^2$, whereby the required probability is $\mathbb{P}(p < 1/2) = F(1/2) = 1/4$.

What is the right answer? Bertrand concludes that the problem is ill-posed, because the term "random chord" is not qualified. Following Kendall and Moran (1963), we next verify that the probability elements involved in paradoxes 1 and 3 are not motion-invariant, thus depend on the choice of the origin of coordinates.

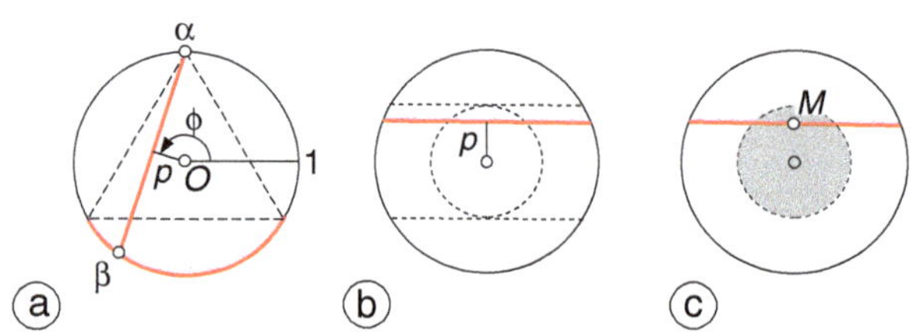

Fig. 2.4.4 Illustration of Bertrand's paradoxes, see Section 2.4.8, Note 3.

1. Let (p, ϕ) denote the normal coordinates of the support line of the chord. In paradox 1, the joint probability element of (α, β) is $\mathbb{P}(\mathrm{d}\alpha, \mathrm{d}\beta) = \mathrm{d}\alpha\,\mathrm{d}\beta/(4\pi^2)$, and $\alpha, \beta = \phi \pm \cos^{-1}(p)$. Therefore,
$$\begin{aligned}\mathbb{P}(\mathrm{d}p, \mathrm{d}\phi) &= \frac{2}{4\pi^2}\left|\frac{\partial(\alpha, \beta)}{\partial(p, \phi)}\right| \mathrm{d}p\,\mathrm{d}\phi \\ &= \frac{2}{\pi}\frac{\mathrm{d}p}{\sqrt{1-p^2}} \cdot \frac{\mathrm{d}\phi}{2\pi}, \quad p \in [0, 1],\ \phi \in [0, 2\pi),\end{aligned} \tag{2.4.32}$$
which depends on the distance p of the chord from the origin. If the latter is chosen at the centre of the circle, then the required probability is $\mathbb{P}(p < 1/2) = 1/3$, but not otherwise. Note that p and ϕ are independent and ϕ is IR, but p is not UR.
2. Here the pertinent probability element is the motion-invariant one given by Eq. (2.4.2) with $H = 2$. Clearly $\mathbb{P}(|p| < 1/2) = 1/2$, and the same value is obtained for any choice of origin.
3. By Eq. (2.4.4), $\mathbb{P}(\mathrm{d}p, \mathrm{d}\phi) \propto p$, which does therefore depend on the choice of origin.

2.5 FUR and IUR Test Planes, and Lines, in 3D

2.5.1 Definitions

The orientation of an axis $L^3_{1[0]} \equiv L^3_1(0, u)$ in space is determined by a point $u \equiv u(\phi, \theta)$ on the unit hemisphere, see Section 1.2.3. The adopted probability element of u is the normalized version of the rotation-invariant density $\mathrm{d}u$, see Eq. (1.2.6), namely

$$\mathbb{P}(\mathrm{d}u) = \frac{\mathrm{d}u}{2\pi}, \quad u \in \mathbb{S}^2_+, \tag{2.5.1}$$

in which case u is said to be isotropic random, namely $u \sim \mathrm{UR}(\mathbb{S}^2_+)$. Because $\mathrm{d}u = \sin\theta\,\mathrm{d}\phi\,\mathrm{d}\theta$, the joint probability element of the spherical polar coordinates of u is

$$\mathbb{P}(\mathrm{d}\phi, \mathrm{d}\theta) = \frac{\mathrm{d}\phi}{2\pi} \cdot \sin\theta\,\mathrm{d}\theta, \quad \phi \in [0, 2\pi),\ \theta \in [0, \pi/2), \tag{2.5.2}$$

which implies that ϕ and θ are independent and $\phi \sim \mathrm{UR}[0, 2\pi)$, whereas

$$\mathbb{P}(\mathrm{d}\theta) = \sin\theta\,\mathrm{d}\theta, \quad \theta \in [0, \pi/2). \tag{2.5.3}$$

Thus, the colatitude angle θ is not UR – this is a mere consequence of the choice of spherical polar coordinates. Uniform randomness of a random variable X requires that any interval of length $\mathrm{d}x$ has a probability proportional to $\mathrm{d}x$, irrespective of its location x. The probability of a colatitude interval of length $\mathrm{d}\theta$ is proportional to the area of the spherical zone determined by two parallel planes a distance $\cos(\theta + \mathrm{d}\theta) - \cos\theta$ apart, namely $2\pi\sin\theta\,\mathrm{d}\theta$, which depends on θ – hence cannot be UR. However, the corresponding area between two planes a distance $\mathrm{d}(\cos\theta)$ apart is $2\pi\mathrm{d}(\cos\theta)$ – thus $\cos\theta$ is UR$[0, 1)$. An isotropic realization of $u(\phi, \theta)$ may be generated with $\phi = 2\pi U_1$, $\theta = \cos^{-1}(U_2)$, $U_i \sim \mathrm{UR}[0, 1)$, $i = 1, 2$, and independent.

For a test plane $L^3_2 \equiv L^3_2(p, u)$ with density $\mathrm{d}L^3_2 = \mathrm{d}p\,\mathrm{d}u$ hitting a reference ball D of diameter H centred at O, the adopted probability element is

$$\mathbb{P}(\mathrm{d}p, \mathrm{d}u) = \frac{\mathrm{d}p}{H} \cdot \frac{\mathrm{d}u}{2\pi}, \quad p \in [-H/2, H/2],\ u \in \mathbb{S}^2_+. \tag{2.5.4}$$

Thus, $p \sim \mathrm{UR}[-H/2, H/2]$ and $u \sim \mathrm{UR}(\mathbb{S}^2_+)$ are independent, and we say that L^3_2 is IUR hitting D, namely $L^3_2 \sim \mathrm{IUR}(D)$.

If the direction $u \in \mathbb{S}^2_+$ is fixed, then the plane is said to be FUR hitting D, and the relevant probability element is the first factor in the rhs of Eq. (2.5.4).

Likewise, for a test line $L^3_1 \equiv L^3_1(z, u)$ with density $\mathrm{d}L^3_1 = \mathrm{d}z\,\mathrm{d}u$ hitting D, we adopt

$$\mathbb{P}(\mathrm{d}z, \mathrm{d}u) = \frac{\mathrm{d}z}{a} \cdot \frac{\mathrm{d}u}{2\pi}, \quad z \in D_2(u),\ u \in \mathbb{S}^2_+, \tag{2.5.5}$$

where dz is the area element in the equatorial disk $D_2(u)$ normal to u, and $a = \pi H^2/4$ is the area of $D_2(u)$. Thus, $z \sim \mathrm{UR}(D_2(u))$ and $u \sim \mathrm{UR}(\mathbb{S}^2_+)$ are independent, and we write $L^3_1 \sim \mathrm{IUR}(D)$.

If $u \in \mathbb{S}^2_+$ is fixed, then the line is FUR hitting D, and the relevant probability element is the first factor in the rhs of Eq. (2.5.5).

2.5.2 Generation IUR test planes, and lines, hitting a ball

Test plane hitting a ball

Take three independent UR numbers $\{U_i \sim \mathrm{UR}[0,1),\ i = 1,2,3\}$ and generate an IR direction $u \equiv u(\phi,\theta) \in \mathbb{S}^2_+$ using the probability element given by Eq. (2.5.2), namely $\phi = 2\pi U_1$, and $\theta = \cos^{-1}(U_2)$. Next generate a signed UR distance $p = (2U_1 - 1) \cdot H/2$ along that direction. Then, according to Eq. (2.5.4), $L^3_2(p,u) \sim \mathrm{IUR}(D)$.

Here, and in the sequel, for FUR test planes and lines all the directions involved may be chosen as convenient.

Test line hitting a ball by a direct method

Take four independent UR numbers $\{U_i \sim \mathrm{UR}[0,1),\ i = 1,\ldots,4\}$.

1. With $\{U_1, U_2\}$, generate an IR direction $u \in \mathbb{S}^2_+$ as above.
2. With $\{U_3, U_4\}$, generate an independent UR point z in the equatorial disk $D_2(u)$ by inscribing the latter in a square of side length H, and using the rejection method described in Section 2.2.4. Then, according to Eq. (2.5.5), $L^3_1(z,u) \sim \mathrm{IUR}(D)$.

Test line hitting a ball by a two-stage rejection method

The motivation of this method is analogous to that in Section 2.2.6.

1. Sample an IUR test plane $L^3_2 \equiv L^3_2(p,u)$ hitting the ball D as described above. Then, $D \cap L^3_2$ is a disk of radius $r(p) = (R^2 - p^2)^{1/2}$, where $R = H/2$ is the radius of D.
2. Within the plane $L^3_2(p,u)$, and with origin at the point of spherical polar coordinates (p,u), generate an independent, signed UR distance $q \sim \mathrm{UR}[-R,R]$. If $|q| > r(p)$, as in Fig. 2.5.1(a), then return to step 1. If $|q| \le r(p)$, then generate $\varphi \sim \mathrm{UR}[0,\pi)$, and a realization of $L^2_1(q,\varphi) \sim \mathrm{IUR}(D)$ is completed.

The preceding step is justified in Section 1.2.4, see Fig. 1.2.4(c). Note also that q must be initially sampled in the interval $[-R,R]$, and not in $[-r(p), r(p)]$, see Section 2.4.7 with $Y = D \cap L^3_2$ in the place of the steak.

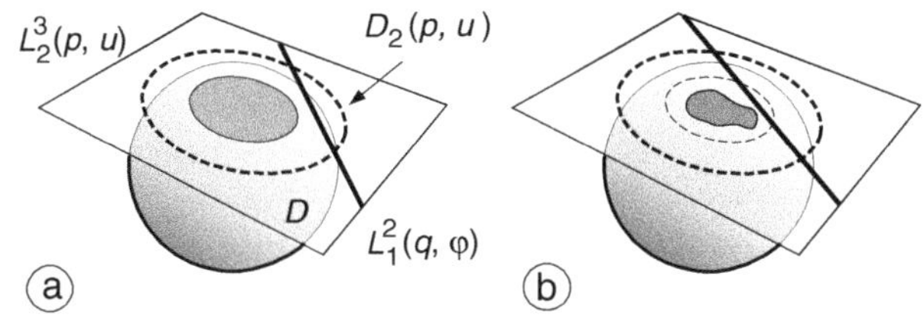

Fig. 2.5.1 (a) Two stage rejection method to generate an IUR test line hitting a reference ball D, see Section 2.5.2. Here the test line misses the ball. (b) Idem to generate an IUR test line hitting a target set Y (only its transect is represented) within a reference ball D. While the test line hits the ball, it misses the planar transect (in dark grey) of the target set, see Section 2.8.3.

Indirect generation of an IR direction on the sphere

Analogously as in Section 2.4.2, let Y be a ball of radius R centred at the origin, and $x \sim \mathrm{UR}(Y)$. Let (r, u_2) denote the spherical polar coordinates of the point x, where $u_2 \equiv (\phi, \theta)$. Then, $\mathrm{d}x = r^2\,\mathrm{d}r\,\mathrm{d}u_2$, and

$$(\mathrm{d}x) = \frac{3r^2\,\mathrm{d}r}{R^3} \cdot \frac{\mathrm{d}u_2}{4\pi}, \qquad r \in [0, R],\ u_2 \in \mathbb{S}^2, \tag{2.5.6}$$

which shows that r and u_2 are independent and $u_2 \sim \mathrm{UR}(\mathbb{S}^2)$. The direction of the support axis Ox is $\mathrm{UR}(\mathbb{S}_+^2)$.

2.6 Mean Values and Ratios for a Test Plane

2.6.1 Mean values

The path is analogous to that in Section 2.4.3. The background notation and results are in Sections 1.5 and 1.11. In the sequel we adopt the setup $Y \subset D \subset \mathbb{R}^3$, where Y is a target set of dimension $q \in \{1, 2, 3\}$, and D is a reference ball of diameter H.

Domain Y of positive volume

Let $L_2^3 \equiv L_2^3(p, u)$ denote a FUR test plane hitting D, and set $A \equiv A(Y \cap L_2^3)$, the corresponding transect area, with $A(\emptyset) = 0$. The pertinent probability element is $\mathbb{P}(\mathrm{d}p) = \mathrm{d}p/H$, namely the marginal of p from Eq. (2.5.4), and therefore,

$$\mathbb{E}(A) = \frac{1}{H}\int_{-H/2}^{H/2} A\,\mathrm{d}p = \frac{V}{H}, \tag{2.6.1}$$

irrespective of the choice of $u \in \mathbb{S}^2_+$, where V is the volume of the set Y. The result is the same if the test plane is IUR hitting D.

Bounded one-sided surface Y

Let $L^3_2 \sim \text{IUR}(D)$, and set $B \equiv B(Y \cap L^3_2)$ for transect (or 'trace') length, and $C \equiv C(Y \cap L^3_2)$ for total trace curvature. Then, recalling the probability element given by Eq. (2.5.4),

$$\mathbb{E}(B) = \frac{1}{2\pi H} \int_{L^3_2 \uparrow D} B \, \mathrm{d}L^3_2 = \frac{(\pi^2/2)S}{2\pi H} = \frac{\pi S}{4H}, \tag{2.6.2}$$

$$\mathbb{E}(C) = \frac{1}{2\pi H} \int_{L^3_2 \uparrow D} C \, \mathrm{d}L^3_2 = \frac{2\pi M}{2\pi H} = \frac{M}{H}, \tag{2.6.3}$$

where S and M denote the surface area and the integral of mean curvature of the surface Y, respectively.

Bounded curve Y

Again, hit D with an IUR test plane L^3_2, and set $Q \equiv Q(Y \cap L^3_2)$, the number of point transects. Then,

$$\mathbb{E}(Q) = \frac{1}{2\pi H} \int_{L^3_2 \uparrow D} Q \, \mathrm{d}L^3_2 = \frac{\pi L}{2\pi H} = \frac{L}{2H}. \tag{2.6.4}$$

2.6.2 Ratios: Classical equations of stereology

Consider two sets $Y_1, Y_2 \subset D \subset \mathbb{R}^3$ of volumes V_1, V_2, respectively, and a FUR test plane hitting D. Application of Eq. (2.6.1) yields the classical stereological equation,

$$\frac{\mathbb{E}(A_2)}{\mathbb{E}(A_1)} = \frac{V_2}{V_1}, \quad \text{namely } V_V = A_A, \tag{2.6.5}$$

known as the Delesse principle.

Likewise, if the set Y_2 is a surface of area S_2 and integral of mean curvature M_2, and the test plane is IUR hitting D, then,

$$\frac{\mathbb{E}(B_2)}{\mathbb{E}(A_1)} = \frac{\pi}{4} \cdot \frac{S_2}{V_1}, \quad \text{namely } S_V = \frac{4}{\pi} \cdot B_A, \tag{2.6.6}$$

$$\frac{\mathbb{E}(C_2)}{\mathbb{E}(A_1)} = \frac{M_2}{V_1}, \quad \text{namely } M_V = C_A = \pi T_A, \tag{2.6.7}$$

the last identity stemming from Eq. (1.10.13).

If Y_2 is a curve of length L_2, then

$$\frac{\mathbb{E}(Q_2)}{\mathbb{E}(A_1)} = \frac{1}{2} \cdot \frac{L_2}{V_1}, \quad \text{namely } L_V = 2Q_A, \tag{2.6.8}$$

Further, if Y_1 is also a surface, or a curve, then further equations, analogous to Eq. (2.4.12), are easily available. For instance,

$$\begin{aligned} S_S &= B_B \\ M_M &= C_C = T_T, \\ L_L &= Q_Q. \end{aligned} \tag{2.6.9}$$

2.7 Mean Values and Ratios for a Test Line in 3D

2.7.1 Mean values

Domain Y of positive volume

If $Y \subset D \subset \mathbb{R}^3$ and we hit the reference ball D with a FUR test line $L_1^3 \equiv L_1^3(z, u)$, then the relevant probability element is $\mathbb{P}(\mathrm{d}z) = \mathrm{d}z/a$, where a denotes the equatorial area of D. Set $L \equiv L(Y \cap L_1^3)$. Then,

$$\mathbb{E}(L) = \frac{1}{a} \int_{D_2} L \, \mathrm{d}z = \frac{V}{a}. \tag{2.7.1}$$

Bounded one-sided surface Y

Here $L_1^3 \sim \mathrm{IUR}(D)$. Let $I \equiv I(Y \cap L_1^3)$ denote the corresponding number of intersections with Y. Then, recalling Eq. (2.5.5),

$$\mathbb{E}(I) = \frac{1}{2\pi a} \int_{L_1^3 \uparrow D} I \, \mathrm{d}L_1^3 = \frac{\pi S}{2\pi a} = \frac{S}{2a}. \tag{2.7.2}$$

2.7.2 Ratios: Further equations of stereology

If we consider two sets $Y_1, Y_2 \subset D \subset \mathbb{R}^3$ of volumes V_1, V_2 respectively, and a FUR test line hitting D, then using Eq. (2.7.1) we obtain the following stereological equation,

$$\frac{\mathbb{E}(L_2)}{\mathbb{E}(L_1)} = \frac{V_2}{V_1}, \quad \text{namely } V_V = L_L. \tag{2.7.3}$$

If the set Y_2 is a surface of area S_2, and the test line is IUR hitting D, then by Eq. (2.7.1) and Eq. (2.7.2),

$$\frac{\mathbb{E}(I_2)}{\mathbb{E}(L_1)} = \frac{1}{2} \cdot \frac{S_2}{V_1}, \quad \text{namely } S_V = 2I_L. \tag{2.7.4}$$

If Y_1 is also a surface, then

$$\frac{\mathbb{E}(I_2)}{\mathbb{E}(I_1)} = \frac{1}{2} \cdot \frac{S_2}{S_1}, \quad \text{namely } S_S = I_I. \tag{2.7.5}$$

Mean intercept length for a three-dimensional domain

Similarly as in the planar case, see Eq. (2.4.13), consider a domain $Y \subset \mathbb{R}^3$ of volume V with a boundary ∂Y of area S. The mean individual length of the linear intercepts constituting the IUR transect $Y \cap L_1^3$ is

$$\mathbb{E}(l) = \frac{\mathbb{E}(L)}{\mathbb{E}(N)} = \frac{2\mathbb{E}(L)}{\mathbb{E}(I)} = \frac{4V}{S}. \tag{2.7.6}$$

Mean free path

Let $Y \subset D \subset \mathbb{R}^3$, where Y represents an aggregate of arbitrary particles contained in a convex reference set D, and let $L_1^3 \sim \text{IUR}(D)$. A free path is a connected linear intercept from a linear transect $L_1^3 \cap Y^c$, where the set difference $Y^c = D \backslash Y$ is the compartment external to the particles. Note that the end points of an intercept may lie in the boundary of a non-convex particle (convexity is the usual assumption in this context). By Eq. (2.7.6), the mean free path is

$$\mathbb{E}(l) = \frac{4V(Y^c)}{S(\partial Y^c)} = 4 \cdot \frac{V(D) - V(Y)}{S(\partial D) + S(\partial Y)}. \tag{2.7.7}$$

Thus, the mean free path is a simple combination of the total volume and surface area of the embedding phase Y^c, and can therefore not be used as a reliable shape descriptor in general.

2.8 Test Planes and Lines Conditional on Hitting a Set

2.8.1 Sampling an IUR test plane hitting a set

Let $Y \subset D \subset \mathbb{R}^3$, where Y is a target set of dimension $q \in \{1,2,3\}$, and D is a reference ball. An IUR realization of a test plane L_2^3 hitting the set Y may be generated as follows.

1. Generate $L_2^3 \sim \text{IUR}(D)$ as in Section 2.5.2.
2. If $L_2^3 \uparrow Y$, then $L_2^3 \sim \text{IUR}(Y)$ – otherwise go to step 1.

In fact, by Eq. (2.5.4) the hitting probability is

$$\mathbb{P}(L_2^3 \uparrow Y) = \int_{L_2^3 \uparrow Y} \mathbb{P}(\mathrm{d}p, \mathrm{d}u) = \frac{h(Y)}{2\pi H}, \tag{2.8.1}$$

where $h(Y)$ is the motion-invariant measure of all IUR test planes hitting the set Y. Let $Y_1'(u)$ denote the orthogonal linear projection of Y onto the axis $L_1^3(0,u)$ normal to the test plane. Then, setting $\uparrow \equiv L_2^3 \uparrow Y$,

$$\mathbb{P}(\mathrm{d}p, \mathrm{d}u | \uparrow) = \frac{\mathbb{P}(\mathrm{d}p, \mathrm{d}u)}{\mathbb{P}(\uparrow)} = \frac{\mathrm{d}p \, \mathrm{d}u}{h(Y)}, \quad p \in Y_1'(u), \; u \in \mathbb{S}_+^2, \tag{2.8.2}$$

which by virtue of Eq. (2.1.2) is the probability element of $L_2^3 \sim \text{IUR}(Y)$.

2.8.2 Conditional distributions, mean values and probabilities for a test plane hitting a set

By Eq. (1.19.11),

$$h(Y) = 2\pi \mathbb{E}\{H(u)\} \tag{2.8.3}$$

the expectation being with respect to the isotropic probability element given by Eq. (2.5.1). Recall that $H(u)$ is the total length of $Y_1'(u)$. Thus, Eq. (2.8.2) may be written as follows,

$$\mathbb{P}(\mathrm{d}p, \mathrm{d}u | \uparrow) = \frac{\mathrm{d}p \, \mathrm{d}u}{2\pi \mathbb{E}\{H(u)\}}, \quad p \in Y_1'(u), \; u \in \mathbb{S}_+^2. \tag{2.8.4}$$

The marginal probability element of the direction u is $H(u)$-weighted. In fact,

$$\begin{aligned} \mathbb{P}(\mathrm{d}u | \uparrow) &= \frac{\mathrm{d}u}{2\pi \mathbb{E}\{H(u)\}} \int_{Y_1'(u)} \mathrm{d}p \\ &= \frac{H(u)\mathbb{P}(\mathrm{d}u)}{\mathbb{E}\{H(u)\}}. \end{aligned} \tag{2.8.5}$$

In general, if $L_2^3 \sim \text{IUR}(Y)$ then p and u are not independent. In the usual way it is easy to show that for each direction u the component p is UR, namely,

$$\mathbb{P}(\mathrm{d}p|u,\uparrow) = \frac{\mathrm{d}p}{H(u)}, \quad p \in Y_1'(u), \tag{2.8.6}$$

which corresponds to a FUR test plane.

Eq. (2.4.24) and Eq. (2.4.25) extend easily. If $L_2^3 \sim \text{FUR}(Y)$, then, using Eq. (2.8.6),

$$V = H(u) \cdot \mathbb{E}(A|u,\uparrow), \tag{2.8.7}$$

where $A \equiv A(Y \cap L_2^3)$. However, if $L_2^3 \sim \text{IUR}(Y)$ then, averaging $\mathbb{E}(A|u,\uparrow)$ with respect to $\mathbb{P}(\mathrm{d}u|\uparrow)$, see Eq. (2.8.5), we obtain

$$V = \mathbb{E}\{H(u)\} \cdot \mathbb{E}(A|\uparrow). \tag{2.8.8}$$

If the sets $Y_2 \subset Y_1 \subset D \subset \mathbb{R}^3$ are connected, then, using Eq. (2.8.1) and Eq. (2.8.3),

$$\mathbb{P}(L_2^3 \uparrow Y_2 | L_2^3 \uparrow Y_1) = \frac{\mathbb{P}(L_2^3 \uparrow Y_2)}{\mathbb{P}(L_2^3 \uparrow Y_1)} = \frac{\mathbb{E}\{H_2(u)\}}{\mathbb{E}\{H_1(u)\}}. \tag{2.8.9}$$

Further, if Y_1 and Y_2 are convex with integrals of mean curvature M_1 and M_2, respectively, then, by virtue of Minkowski's theorem, Eq. (1.19.13), we obtain

$$\mathbb{P}(L_2^3 \uparrow Y_2 | L_2^3 \uparrow Y_1) = \frac{M_2}{M_1}. \tag{2.8.10}$$

2.8.3 Sampling an IUR test line hitting a set

The setup is $Y \subset D \subset \mathbb{R}^3$, where Y is a bounded set of dimension $q \in \{2,3\}$. To generate an IUR realization of a test line L_1^3 hitting Y,

1. Generate $L_1^3 \sim \text{IUR}(D)$ by either of the two methods described in Section 2.5.2.
2. If $L_1^3 \uparrow Y$, then $L_1^3 \sim \text{IUR}(Y)$ – otherwise (Fig. 2.5.1(b)) go to step 1.

The method may be justified similarly as in the case of an IUR test plane. By Eq. (2.5.5) and Eq. (1.19.9),

$$\mathbb{P}(L_1^3 \uparrow Y) = \int_{L_1^3 \uparrow Y} \mathbb{P}(\mathrm{d}z, \mathrm{d}u) = \frac{\mathbb{E}\{A(u)\}}{a}, \tag{2.8.11}$$

where $A(u)$ denotes the area of the orthogonal projection $Y_2'(u)$ of Y onto the plane $L_2^3(0,u)$ normal to the test line. Thus, setting $\uparrow \equiv L_1^3 \uparrow Y$,

$$\mathbb{P}(\mathrm{d}z, \mathrm{d}u | \uparrow) = \frac{\mathrm{d}z\,\mathrm{d}u}{2\pi\mathbb{E}\{A(u)\}}, \quad z \in Y_2'(u)\ u \in \mathbb{S}_+^2, \tag{2.8.12}$$

which by Eq. (2.1.2) is the probability element of $L_1^3 \sim \mathrm{IUR}(Y)$.

2.8.4 Conditional distributions, mean values and probabilities for a test line hitting a set

From Eq. (2.8.12), the marginal probability element of the direction u is $A(u)$-weighted, namely,

$$\mathbb{P}(\mathrm{d}u | \uparrow) = \frac{A(u)\mathbb{P}(\mathrm{d}u)}{\mathbb{E}\{A(u)\}}, \tag{2.8.13}$$

whereas, conditional on a given direction u, the point z is UR, namely,

$$\mathbb{P}(\mathrm{d}z | u, \uparrow) = \frac{\mathrm{d}z}{A(u)}, \quad z \in Y_2'(u). \tag{2.8.14}$$

Similarly as in the preceding section,

$$V = A(u) \cdot \mathbb{E}(L | u, \uparrow), \tag{2.8.15}$$

$$V = \mathbb{E}\{A(u)\} \cdot \mathbb{E}(L | \uparrow), \tag{2.8.16}$$

according to whether $L_1^3 \sim \mathrm{FUR}(Y)$, or $L_1^3 \sim \mathrm{IUR}(Y)$, respectively.

If the set $Y \subset \mathbb{R}^3$ is convex with surface area S then, recalling Cauchy's formula Eq. (1.19.10), and Eq. (2.5.1), we obtain,

$$\mathbb{E}\{A(u)\} = S/4, \tag{2.8.17}$$

and in this case the probability element Eq. (2.8.12) becomes

$$\mathbb{P}(\mathrm{d}z, \mathrm{d}u | \uparrow) = \frac{\mathrm{d}z\,\mathrm{d}u}{(\pi/2)S}, \quad z \in Y_2'(u),\ u \in \mathbb{S}_+^2. \tag{2.8.18}$$

Consider two connected domains Y_1, Y_2 such that $Y_2 \subset Y_1 \subset D$. If $L_1^3 \sim \mathrm{IUR}(D)$ then, applying Eq. (2.8.11), we get

$$\mathbb{P}(L_1^3 \uparrow Y_2 | L_1^3 \uparrow Y_1) = \frac{\mathbb{E}\{A_2(u)\}}{\mathbb{E}\{A_1(u)\}}. \tag{2.8.19}$$

Let S_{0i} denote the surface area of the convex hull of Y_i, $i = 1, 2$, namely the smallest convex set enclosing Y_i. Then, by Eq. (2.8.17) the preceding equation yields

$$\mathbb{P}(L_1^3 \uparrow Y_2 | L_1^3 \uparrow Y_1) = \frac{S_{02}}{S_{01}}. \tag{2.8.20}$$

2.8.5 Notes

1. Stereological equations with FUR and IUR planes and test lines

Hostinský (1925) and Czuber (1884) obtained probabilities, and mean values such as Eq. (2.7.6), under the assumption of convexity. The Delesse principle, based on FUR test planes, see Eq. (2.6.5), is due to Delesse (1847). Delesse proposed the cut-and-weight method, apparently prevailing at the time, to measure section areas. Rosiwal (1898) proposed a much more efficient method based on FUR test lines, see Eq. (2.4.8). Thus, Eq. (2.7.3) emerged, with no shape assumptions, as a combination of the areal method of Delesse with the lineal analysis method of Rosiwal. Saltykov (1946) and Smith and Guttman (1953) derived Eq. (2.7.4) and Eq. (2.6.8), which may be regarded as dual of each other by interchanging the roles of object and probe, see the last paragraph of Section 1.5.4. Mean curvature equations were derived by DeHoff (2004). A coherent account of stereological equations appeared in Underwood (1970), who introduced the shortcut notation. Miles (1972), Miles and Davy (1976), and Davy and Miles (1977) set the mathematical foundations. Independently, Mayhew and Cruz-Orive (1974) advanced that, if A_1 is a random variable, then Eq. (2.6.5) holds, but $\mathbb{E}(A_2/A_1) \neq V_2/V_1$ in general. For further historical notes, see Cruz-Orive (2017). Also, for a concise generalization of the classical stereological equations, see Appendix 1.1 in Voss and Cruz-Orive (2009).

2. Mean free path

Santaló (1943) considered mean free paths for convex particles in the plane and in space; he also considered curvilinear paths.

2.9 Sampling With a Vertical Plane

2.9.1 Preliminaries

Consider the setup $Y_2 \subset Y_1 \subset D \subset \mathbb{R}^3$ of Section 2.7.2, where Y_2 is a surface of area S contained in a domain Y_1 of volume V, whereas D is a reference ball of diameter H centred at the origin. We want to express S in terms of the information contained in a vertical plane $L^3_{2\cdot v} \equiv L^3_{2\cdot v}(p, \phi)$, see Section 1.6, hitting D and normal to the horizontal plane Ox_1x_2.

The joint probability element of (p, ϕ) is given by Eq. (2.4.2), because it corresponds to the IUR linear trace determined by $L^3_{2\cdot v}$ in the horizontal equatorial disk D_2 of D in the horizontal plane. If a $(\sin\theta)$-weighted test line $L^2_{1\cdot v} \equiv L^2_{1\cdot v}(q, \pi-\theta) \subset L^3_{2\cdot v}$ hits the reference ball D, then $L^2_{1\cdot v}$ is effectively $L^3_1 \sim \text{IUR}(D)$.

2.9.2 Vertical plane containing a sine-weighted test line

Set $A \equiv A(Y_1 \cap L^3_{2\cdot v})$, namely the area of a vertical transect through the domain Y, with $A(\emptyset) = 0$, as usual. Then,

$$\mathbb{E}(A) = \frac{1}{\pi H} \int_0^{\pi} \mathrm{d}\phi \int_{-H/2}^{H/2} A \, \mathrm{d}p = \frac{V}{H}. \tag{2.9.1}$$

On the other hand, set $W \equiv W(Y_2 \cap L^3_{2\cdot v})$, namely the functional given by Eq. (1.6.4). By Eq. (1.6.3) we have,

$$\mathbb{E}(W) = \frac{1}{\pi H} \int_0^{\pi} \mathrm{d}\phi \int_{-H/2}^{H/2} W \, \mathrm{d}p = \frac{S}{H}. \tag{2.9.2}$$

From the preceding two identities we obtain the following stereological equation,

$$\frac{\mathbb{E}(W)}{\mathbb{E}(A)} = \frac{S}{V}, \quad \text{namely } S_V = W_A. \tag{2.9.3}$$

If we generate a sine-weighted test line within a vertical plane, hitting the reference ball D, then Eqs. (2.7.4)–(2.7.6) apply directly by virtue of Eq. (1.6.1).

2.9.3 Note

Eq. (2.9.3) was first obtained by Baddeley (1985).

2.10 Sampling With a FUR Slab

2.10.1 Definitions

The motion-invariant density of a slab of thickness t is the same as that of its reference face plane, see Eq. (1.2.14). Because a slab is a full-dimensional probe, the corresponding intersection formulae are orientation-independent (Section 1.4.3). Consequently, for our present purposes it suffices to consider a FUR slab, namely a slab $L_t(p) \subset \mathbb{R}^d$ of a fixed orientation – e.g. normal to a sampling axis Ox_1 – with translation-invariant density $\mathrm{d}L_t(p) = \mathrm{d}p$, $p \in \mathbb{R}$, along that axis. The slab $L_t(0)$ determines the segment $T_t(0) = [0, t)$ in the sampling axis. Thus, there is a one-to-one correspondence between the test slab $L_t(p)$ and the test segment $T_t(p) = [p, p + t)$, see Fig. 2.10.1(a).

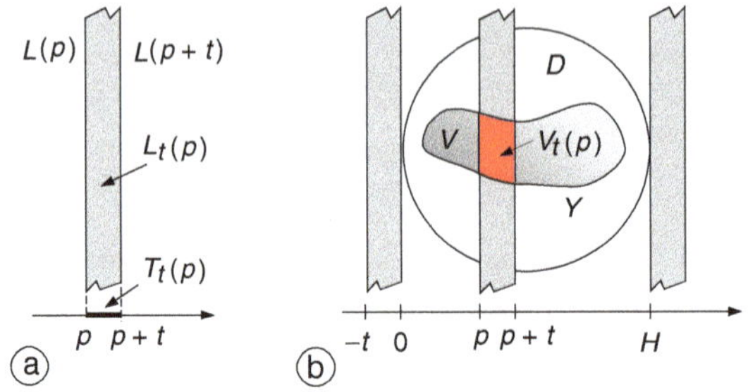

Fig. 2.10.1 (a) Orthogonal projection, onto a perpendicular plane, of a portion of a slab of thickness t normal to a sampling axis. (b) Range of the location parameter p of a FUR slab hitting a reference ball D.

If the slab $L_t(p)$ hits a reference ball D whose diameter is the segment $[0, H)$, see Fig. 2.10.1(b), then the adopted probability element is

$$\mathbb{P}(\mathrm{d}p) = \frac{\mathrm{d}p}{H+t}, \quad p \in [-t, H), \tag{2.10.1}$$

because the events $L_t(p) \uparrow D$, $T_t(p) \uparrow [0, H)$ and $p \in [-t, H)$ are equivalent. Note that the pertinent hitting measure is equal to $H+t$, see Eq. (1.19.2). Thus the test slab, or the test segment, are UR hitting the reference ball D, or the reference segment $[0, H)$, respectively.

2.10.2 FUR test slab hitting a compact set

Consider a set $Y \subset D \subset \mathbb{R}^d$ of dimension $q \in \{0, 1, \dots, d\}$ with measure $\gamma(Y)$, and hit the reference ball D with a FUR slab $L_t(p)$. Define

$$f_t(p) = \frac{1}{t} \cdot \gamma(Y \cap L_t(p)), \quad p \in [-t, H). \tag{2.10.2}$$

Then, by Eq. (2.10.1) and Eq. (1.4.12), we have

$$\gamma(Y) = (H+t) \cdot \mathbb{E}\{f_t(p)\}. \tag{2.10.3}$$

The main purpose of slab sampling is usually to work with global equations like the preceding one, rather than ratios. To fix ideas, henceforth we concentrate on the case $q = d = 3$ with $\gamma(Y) \equiv V > 0$. Define the area function

$$A(p) = A(Y \cap L(p)), \tag{2.10.4}$$

namely the section area determined in the set Y by the reference plane $L(p) \equiv L_0(p)$ of the slab $L_t(p)$. The volume of the slice $Y \cap L_t(p)$ is

$$V_t(p) = V(Y \cap L_t(p)) = \int_p^{p+t} A(x)\,\mathrm{d}x, \qquad (2.10.5)$$

see Fig. 2.10.2(a). In this case the function $f_t(p)$ is

$$A_t(p) = \frac{1}{t} \cdot V_t(p), \qquad (2.10.6)$$

see Fig. 2.10.2(b), which may be interpreted as the mean area of a planar section within the slice, namely the mean of $A(x)$ with respect to $x \sim \mathrm{UR}[p, p+t)$ for each value of p. Thus, if $L_t(p) \sim \mathrm{UR}(D)$, or equivalently if $p \sim \mathrm{UR}[-t, H)$, then Eq. (2.10.3) becomes

$$V = (H + t) \cdot \mathbb{E}\{A_t(p)\}. \qquad (2.10.7)$$

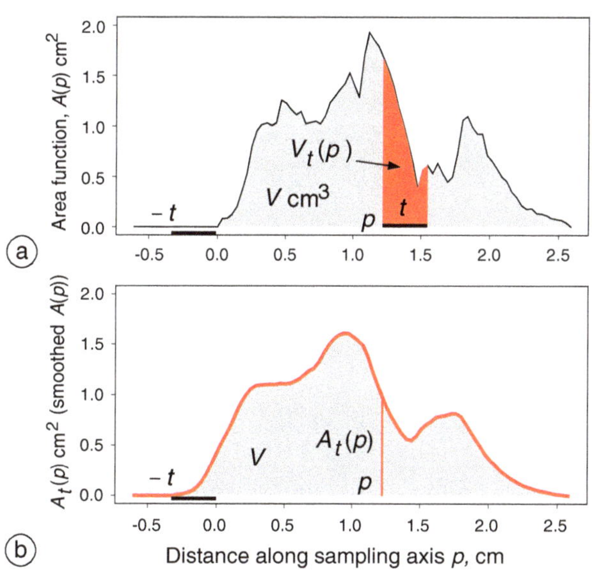

Fig. 2.10.2 (a) Approximate section area function for the union of the six lung lobes of an albino rat, see Section 2.10.4, Note 1. The area in red is the volume of a slice of thickness t. The mean section area within the slice is the smoothed, or filtered, area function by a segment of length t. (b) Graph of the smoothed area function. See Section 2.10.2.

The preceding expectation is computed including the zeros of A_t arising when the slab misses the set Y. Thus,

$$V = \frac{H+t}{t} \cdot \mathbb{E}\{V_t(p)\}, \tag{2.10.8}$$

where $\mathbb{E}\{V_t(p)\}$ represents the mean slice volume.

If the planar section area $A(p)$ defined by Eq. (2.10.4) is directly available, then

$$V = (H+t) \cdot \mathbb{E}\{A(p)\}, \tag{2.10.9}$$

where the expectation includes the zeros of $A(p)$ arising when the slab face $L(p)$ misses the target set Y.

The probability that the slab $L_t \sim \mathrm{UR}(D)$ with a given orientation $u \in \mathbb{S}_+^2$ hits a connected set $Y \subset D \subset \mathbb{R}^d$ is

$$\mathbb{P}(L_t \uparrow Y) = \int_{L_t(p)\uparrow Y} \frac{\mathrm{d}p}{H+t} = \frac{H(u)+t}{H+t}, \tag{2.10.10}$$

where $H(u)$ denotes the caliper length of Y along the direction u normal to the slab.

2.10.3 Moving average, convolution, filtering

The area function $A_t(p)$ defined above is a smoothed version of $A(p)$. In fact,

$$\begin{aligned} V_t(p) &= \int_{\mathbb{R}} 1_{[p,p+t)}(x) A(x)\,\mathrm{d}x \\ &= \int_{\mathbb{R}} 1_{[0,t)}(x-p) A(x)\,\mathrm{d}x \\ &= \int_{\mathbb{R}} 1_{[0,t)}(x) A(p+x)\,\mathrm{d}x \end{aligned} \tag{2.10.11}$$

is the regularization or filtering of the function $A(x)$ by the weight function $1_{[0,t)}(x)$.

In the discrete case, the analogous procedure to smoothen a sequence $\{f_i,\ i \in \mathbb{Z}\}$ of numerical data is to adopt a set of constant weights $\{w_0, w_1, \ldots, w_r\}$ adding up to 1. Then the element f_i may be replaced with the forward moving average

$$f_{w,i} = \sum_{k=0}^{r} w_k f_{i+k}. \tag{2.10.12}$$

For instance, if $\{f_i\} = \{5, 1, 8, 3\}$ and we adopt $w_0 = w_1 = 1/2$, then $\{f_{w,i}\} = \{3, 4.5, 5.5, 1.5\}$. In this way the jumps between consecutive data tend to be reduced, that is, the data set $\{f_{w,i},\ i \in \mathbb{Z}\}$ tends to be smoother than the original one. The degree of smoothness increases with the number of weights.

In the continuous case, the underlying concept is that of convolution. Given two integrable functions $f, w : \mathbb{R} \to \mathbb{R}_+$, the convolution of f and w is

$$(f * w)(x) = \int_{\mathbb{R}} f(x-y)w(y)\,\mathrm{d}y = \int_{\mathbb{R}} f(y)w(x-y)\,\mathrm{d}y = (w * f)(x). \quad (2.10.13)$$

Define $\check{1}_{[0,t]}(x) = 1_{[0,t]}(-x)$, the reflection of the indicator function about the origin. Then Eq. (2.10.11) may be written

$$V_t(p) = (A * \check{1}_{[0,t)})(p). \quad (2.10.14)$$

2.10.4 Notes

1. Slab sampling

Representations such as Eq. (2.10.14) can be found in Matheron (1965, pp. 26–27), see also Gual-Arnau and Cruz-Orive (1998). The data corresponding to Fig.2.10.2 was used in the latter paper. The six lung lobes of an albino rat were processed and embedded in paraffin, see Burri and Weibel (1971). Each lobe was exhaustively cut into 4 μm thick serial sections at 360 μm steps, and section area was estimated with a square grid (see Section 4.6) of 550 μm between nearest test points. The lobe section data were then arranged approximately as if the union of the six lobes had been cut simultaneously into 71 parallel extensive sections 360 μm apart. The result is plotted in Fig. 2.10.2(a).

2.11 FUR, and IUR, Bounded Test Probes

2.11.1 Definitions

The background material is given in Section 1.13. A bounded test probe of dimension $r \in \{1, 2, \ldots, d\}$ in $\mathbb{R}^d$ is denoted by $T_r^d(x, u_d)$, where $x \in \mathbb{R}^d$ is the associated point (AP), and $u_d \in G_{d[0]}$ the vector of orientation parameters, see Section 1.16.4 and Fig. 2.11.1(a). Here we adopt the following simplified version of the notation introduced in Section 1.16,

$$\begin{aligned} D_{\oplus}(u_d) &\equiv D \oplus \check{T}_r^d(0, u_d), \\ Y_{\oplus}(u_d) &\equiv Y \oplus \check{T}_r^d(0, u_d), \end{aligned} \quad (2.11.1)$$

where $D \subset \mathbb{R}^d$ represents a d-ball, and $Y \subset \mathbb{R}^d$ a set of dimension $q \geq d - r$. Thus, $D_{\oplus}(u_d)$, $Y_{\oplus}(u_d)$ represent the geometric loci of the associated point x as the FUR probe $T_r^d(x, u_d)$ hits D, Y, respectively, with a given orientation u_d, see

Fig. 2.11.1(b),(c). If D is a ball, then the volume of $D_{\oplus}(u_d)$ does not depend on u_d, that is, $V(D_{\oplus}(u_d)) = V(D_{\oplus})$, say, where $D_{\oplus} \equiv D_{\oplus}(0)$.

On a sampling axis ($d = 1$) the translation-invariant density of a test segment $T_1^1 \equiv T_1^1(p) = [p, p + t)$ is $\mathrm{d}T_1^1 = \mathrm{d}p$. If this test segment hits a fixed reference segment $D = [0, H)$, then by Eq. (2.1.2) the associated probability element is

$$\mathbb{P}(\mathrm{d}p) = \frac{\mathrm{d}p}{L(D_{\oplus})} = \frac{\mathrm{d}p}{H+t}, \quad p \in D_{\oplus} \equiv [-t, H), \tag{2.11.2}$$

as in Eq. (2.10.1).

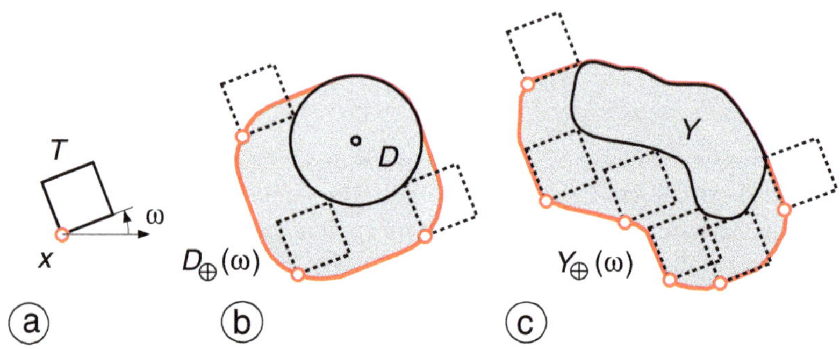

Fig. 2.11.1 (a) A bounded test probe with AP at x and orientation ω. (b) Minkowski addition of a reference disk D and the probe. (c) Minkowski addition of a set Y and the probe, see Eq. (2.11.1).

The probability element adopted for the orientation ω of a bounded IUR test probe $T_r^2 \equiv T_r^2(x, \omega)$, $(r = 1, 2)$, in the plane is proportional to its invariant density $\mathrm{d}\omega$, namely,

$$\mathbb{P}(\mathrm{d}\omega) = \frac{\mathrm{d}\omega}{2\pi}, \quad \omega \in \mathbb{S}^1 = [0, 2\pi), \tag{2.11.3}$$

that is, $\omega \sim \mathrm{UR}[0, 2\pi)$. To define a probability element for an IUR test probe in the plane it is convenient to impose the condition that T_r^2 hits a reference disk D because then $A(D_{\oplus}(\omega)) = A(D_{\oplus})$, say, does not depend on ω. Thus, by Eq. (2.1.2), Eq. (1.16.2), and Eq. (1.20.1),

$$\mathbb{P}(\mathrm{d}x, \mathrm{d}\omega) = \frac{\mathrm{d}x\,\mathrm{d}\omega}{\int_{T_r^2 \uparrow D} \mathrm{d}x\,\mathrm{d}\omega} = \frac{\mathrm{d}x}{A(D_{\oplus})} \cdot \frac{\mathrm{d}\omega}{2\pi}, \tag{2.11.4}$$

which shows that, in this case, x is UR in $D_{\oplus}(\omega)$, whereas ω is UR in $[0, 2\pi)$, and independent. We say that $T_r^2 \sim \mathrm{IUR}(D)$.

For the extension to T_r^d we may adopt a d-ball as a reference set. For instance, based on Eq. (1.13.6) we say that the probe $T_r^3 \equiv T_r^3(x, u_2, \tau) \subset \mathbb{R}^3$, $(r = 1, 2, 3)$, is IUR hitting a ball $D \subset \mathbb{R}^3$ if

$$\mathbb{P}(\mathrm{d}x, \mathrm{d}u_2, \mathrm{d}\tau) = \frac{\mathrm{d}x}{V(D_\oplus)} \cdot \frac{\mathrm{d}u_2}{4\pi} \cdot \frac{\mathrm{d}\tau}{2\pi}, \tag{2.11.5}$$

which means that the unit vector u_2 is UR in $\mathbb{S}^2$, the unit vector τ is UR in $\mathbb{S}^1$ and, for each $u_3 \equiv (u_2, \tau)$, the AP is UR in $D_\oplus(u_3)$.

2.11.2 Generation of an IUR bounded test probe hitting a ball

General case

An IUR realization of a bounded test probe $T_r^d(x, u_d)$ hitting a ball D may be generated as follows.

1. Generate an IR realization of the orientation vector u_d.
2. Generate a UR realization of the AP x in $D_\oplus(u_d)$.

In the FUR case, the orientation vector u_d is chosen as convenient.

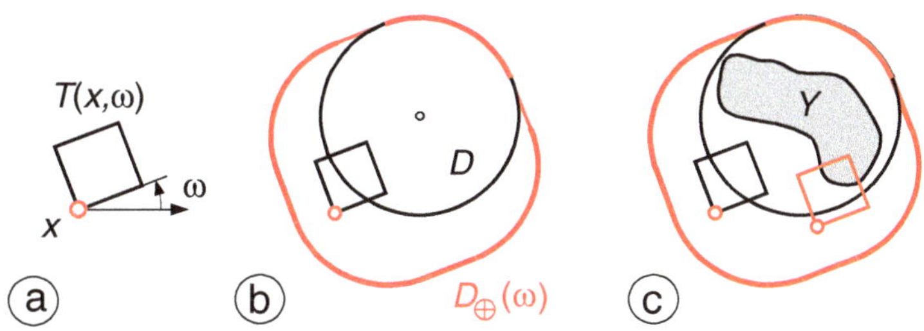

Fig. 2.11.2 (a) Test probe. (b) generation of a FUR realization of the probe hitting a disk D, see Section 2.11.2. (c) Idem to generate a FUR realization of the probe hitting a planar set $T \subset D$ by the rejection method, see Section 2.13.1.

2.12 Mean Values and Ratios for a Bounded Test Probe

The next results follow from the probability element given by Eq. (2.11.5), and from the Crofton formulae given in Sections 1.15 and 1.16. We adopt the setup $T \subset D \subset \mathbb{R}^d$, $d = 2, 3$, where Y is a target set of dimension $q \in \{1, 2, 3\}$, and D is a reference ball of diameter H.

2.12.1 Two-dimensional case

Planar domain of positive area

Because $\dim(Y) = 2$, the area A of the domain Y can be identified by means of a FUR probe $T_r^2(x)$, $r = 0, 1, 2$, hitting D. From Eq. (2.11.4),

$$\mathbb{P}(\mathrm{d}x) = \frac{\mathrm{d}x}{A(D_\oplus)}, \quad x \in D_\oplus. \tag{2.12.1}$$

Then, for a UR test point $T_0^2 \equiv x$, a FUR test curve T_1^2 of length l, or a FUR test quadrat of area a, hitting D, we have,

$$\mathbb{E}\{P(Y \cap T_0^2)\} = \frac{A}{A(D_\oplus)}, \tag{2.12.2}$$

$$\mathbb{E}\{L(Y \cap T_1^2)\} = \frac{lA}{A(D_\oplus)}, \tag{2.12.3}$$

$$\mathbb{E}\{A(Y \cap T_2^2)\} = \frac{aA}{A(D_\oplus)}, \tag{2.12.4}$$

which use Eq. (1.4.1), Eq. (1.15.1) and Eq. (1.15.6), respectively, see Fig. 2.12.1(a), (c). As usual, the expectations include the zeros arising when the test probe misses the set Y. As a consequence,

$$\mathbb{E}\{L(Y \cap T_1^2(x))\} = l \cdot \mathbb{E}\{P(Y \cap x)\}, \tag{2.12.5}$$

$$\mathbb{E}\{A(Y \cap T_2^2(x))\} = a \cdot \mathbb{E}\{P(Y \cap x)\}, \tag{2.12.6}$$

which show that mean intercept length and intersection area do not require direct measurements: it suffices to score whether the associated point x of the test probe hits Y, $(P = 1)$, or not, $(P = 0)$, and multiply the corresponding mean score by the measure of the corresponding probe, see Fig. 2.12.1(b),(d).

Planar curve

Because $\dim(Y) = 1$, the length B of Y can be identified by means of either a FUR test quadrat $T_2^2(x)$ hitting D, or by an IUR test curve $T_1^2(x, \omega)$ hitting D. Application of Eq. (1.15.5) and Eq. (1.16.5) yields, respectively,

$$\mathbb{E}\{B(Y \cap T_2^2)\} = \frac{aB}{A(D_\oplus)}, \tag{2.12.7}$$

$$\mathbb{E}\{I(Y \cap T_1^2)\} = \frac{4lB}{2\pi A(D_\oplus)} = \frac{2}{\pi} \cdot \frac{lB}{A(D_\oplus)}. \tag{2.12.8}$$

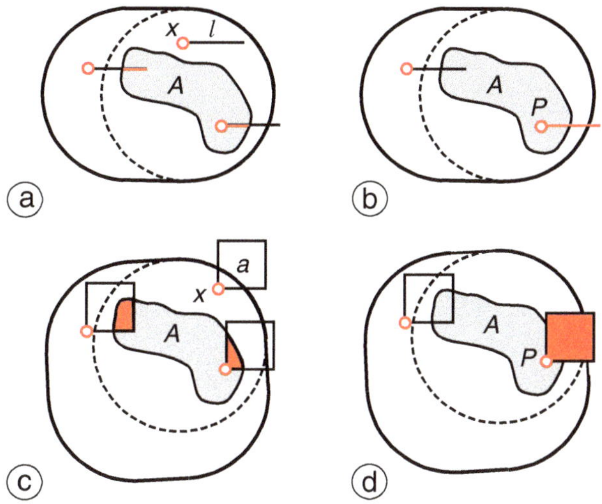

Fig. 2.12.1 The mean intercept length (in red), see (a) and Eq. (2.12.3), is equivalent to the mean number of APs of the FUR segment probe hitting the target set, times the probe length, see (b) and Eq. (2.12.5). Likewise (c), (d) illustrate Eq. (2.12.4) and Eq. (2.12.6), respectively.

Ratios

For two domains $Y_1, Y_2 \subset D \subset \mathbb{R}^2$ of areas A_1, A_2 respectively, it suffices to use a FUR test probe $T_r^2, r = 0, 1, 2$ hitting D. Of interest are mainly the lower-dimensional probes. For instance, the application of Eq. (2.12.2) and Eq. (2.12.3) yields, with the customary abbreviations, the following,

$$\frac{\mathbb{E}(P_2)}{\mathbb{E}(P_1)} = \frac{\mathbb{E}(L_2)}{\mathbb{E}(L_1)} = \frac{A_2}{A_1}, \quad \text{namely } P_P = L_L = A_A, \tag{2.12.9}$$

which have arisen before for a UR test point and a FUR test line, respectively. Note that the test curve length l is not involved. Also, recall Section 2.3.5, Note 5.

If Y_2 is a curve of length B_2, then the combination of Eq. (2.12.8) and Eq. (2.12.3) yields

$$\frac{\mathbb{E}(I_2)}{\mathbb{E}(L_1)} = \frac{2}{\pi} \cdot \frac{lB_2}{lA_1} = \frac{2}{\pi} \cdot \frac{B_2}{A_1}, \quad \text{namely } B_A = \frac{\pi}{2} \cdot I_L, \tag{2.12.10}$$

which is formally the same as Eq. (2.4.10) for a test line. However, Eq. (2.12.8) and Eq. (2.12.2) yield

$$\frac{\mathbb{E}(I_2)}{\mathbb{E}(P_1)} = \frac{2}{\pi} \cdot \frac{lB_2}{A_1}, \quad \text{namely } B_A = \frac{\pi}{2} \cdot \frac{1}{l} \cdot I_P, \tag{2.12.11}$$

which holds for a finite IUR test curve of length l with an associated test point – it can also be obtained applying Eq. (2.12.5) to Eq. (2.12.10).

If Y_1, Y_2 are curves of lengths B_1, B_2, respectively, and the probe is an IUR test curve hitting D, then

$$\frac{\mathbb{E}(I_2)}{\mathbb{E}(I_1)} = \frac{B_2}{B_1}, \quad \text{namely } B_B = I_I. \tag{2.12.12}$$

2.12.2 Three-dimensional case

Domain of volume $V > 0$

Here $Y \subset D \subset \mathbb{R}^3$, where D is a reference ball. The probability element of a FUR probe $T_r^3(x)$ is given by Eq. (2.12.1) with $V(D_\oplus)$ in the place of $A(D_\oplus)$. The relevant mean values analogous to Eqs. (2.12.2)–(2.12.4) may be written in compact form as follows,

$$\mathbb{E}\{\alpha(Y \cap T_r^3)\} = \frac{\nu(T_r^3) \cdot V}{V(D_\oplus)}, \tag{2.12.13}$$

where, for $r = 0, 1, 2, 3$, we have $\alpha \equiv P, L, S, V$ and $\nu(T_r^3) = 1, l, a, v$, respectively. Identities analogous to Eq. (2.12.5) and Eq. (2.12.6) are

$$\mathbb{E}\{\alpha(Y \cap T_r^3(x))\} = \nu(T_r^3) \cdot \mathbb{E}\{P(Y \cap x)\}, \quad r = 1, 2, 3. \tag{2.12.14}$$

Surface of area $S > 0$

In this case the probe $T_r^3 \equiv T_r^3(x, u)$, $r = 1, 2$, is IUR hitting the reference ball D, with probability element given by Eq. (2.11.5). For a bounded surface probe T_2^3 of area a, set $B = B(Y \cap T_2^3)$ (trace length); also, for a bounded curve probe T_1^3 of length l, set $I = I(Y \cap T_1^3)$ (the number of intersections), with $B = I = 0$ whenever the probe misses the surface Y. Recalling the relevant coefficients in Eq. (1.16.14), we have

$$\mathbb{E}(B) = \frac{2\pi^3 aS}{8\pi^2 V(D_\oplus)} = \frac{\pi}{4} \cdot \frac{aS}{V(D_\oplus)}, \tag{2.12.15}$$

$$\mathbb{E}(I) = \frac{4\pi^2 lS}{8\pi^2 V(D_\oplus)} = \frac{1}{2} \cdot \frac{lS}{V(D_\oplus)}. \tag{2.12.16}$$

Spatial curve of length $L > 0$

For a bounded IUR surface probe T_2^3 of area a hitting D we obtain

$$\mathbb{E}(Q) = \frac{4\pi^2 aL}{8\pi^2 V(D_\oplus)} = \frac{1}{2} \cdot \frac{aL}{V(D_\oplus)}, \tag{2.12.17}$$

where $Q \equiv Q(Y \cap T_2^3)$, the number of point transects. The preceding identity is the dual of Eq. (2.12.16), because the roles of test probe and target set are interchangeable.

Ratios

Consider two domains $Y_1, Y_2 \subset D \subset \mathbb{R}^3$ of volumes V_1, V_2 respectively, and hit the reference ball D with a FUR test probe T_r^3. Then, similarly as in Section 2.12.1, suitable combinations of the preceding mean values yield the following stereological equations,

$$V_V = A_A = L_L = P_P \tag{2.12.18}$$

for surface, curve, and point probes, namely for $r = 2, 1, 0$, respectively.

If Y_2 is a surface of area S_2, then the probe is IUR hitting D. The corresponding stereological equations are

$$S_V = \frac{4}{\pi} B_A = 2I_L = \frac{2}{l} I_P, \tag{2.12.19}$$

for surface and curve probes for the first two identities, respectively, and for a curve probe of length l, with its AP hitting the domain Y_1 for the last identity.

If both Y_1, Y_2 are surfaces of areas S_1, S_2, respectively, then the analogue to Eq. (2.12.12) is

$$S_S = B_B = I_I, \tag{2.12.20}$$

for surface and curve probes, respectively.

Finally, if Y_2 is a curve of length L and Y_1 is a domain of volume V, then

$$L_V = 2Q_A = \frac{2}{a} Q_P, \tag{2.12.21}$$

for a surface probe. In the last identity, the AP of a surface probe of area a is used to hit the domain Y_1.

2.12.3 Note

A coherent account on quadrat sampling was first provided by Miles and Davy (1977). For a condensed, general treatment, see Appendix 1.2 in Voss and Cruz-Orive (2009).

2.13 Bounded Test Probes Conditional on Hitting a Set

2.13.1 Generation of a bounded test probe hitting a set by the rejection method

General case

The purpose is to generate a realization of a FUR, or an IUR, bounded test probe $T_r^d(x, u_d)$ hitting a target set Y. We adopt the usual setup $Y \subset D \subset \mathbb{R}^d$, where D is a reference ball. The number of rejections will be reduced if D is the smallest ball containing Y. For the IUR case,

1. Generate a realization $T_r^d(x, u_d) \sim \mathrm{IUR}(D)$ as described in Section 2.11.2.
2. If the probe $T_r^d(x, u_d)$ does not hit the set Y (Fig. 2.11.2(c), black square), then return to step 1. Otherwise (Fig. 2.11.2(c), red square), the probe is IUR hitting Y.

The preceding method may be used to generate Monte Carlo realizations of the random variable (x, u_d) conditional on the hitting event without the need to compute normalizing constants such as $A(Y_\oplus(u_d))$, $\mathbb{E}\{A(Y_\oplus(u_d))\}$.

To justify the method for $d = 2$, divide the rhs of Eq. (2.11.4) by the hitting probability:

$$\mathbb{P}(T_r^2 \uparrow Y) = \int_{T_r^2 \uparrow Y} \mathbb{P}(\mathrm{d}x, \mathrm{d}\omega) = \frac{\mathbb{E}\{A(Y_\oplus(\omega))\}}{A(D_\oplus)}, \tag{2.13.1}$$

to obtain,

$$\mathbb{P}(\mathrm{d}x, \mathrm{d}\omega |\uparrow) = \frac{\mathrm{d}x\, \mathrm{d}\omega}{2\pi\mathbb{E}\{A(Y_\oplus(\omega))\}}, \quad x \in Y_\oplus(\omega),\ \omega \in [0, 2\pi), \tag{2.13.2}$$

which is the probability element associated with $T_r^2 \sim \mathrm{IUR}(Y)$.

Two-stage rejection method to generate an IUR planar probe hitting a set in 3D

Consider the case $Y \subset D \subset \mathbb{R}^3$, where D is a reference ball of radius R centred at the origin. A realization of an IUR test probe $T_r^3(x, u_2, \tau)$ hitting the target set Y may be generated in the following two steps. We assume that the probe is a bounded planar subset of dimension $r \in \{1, 2\}$. Recall Fig. 1.13.2.

1. Generate a sequence of independent, oriented planes $L_2^* \equiv L_2^3(p, u_2)$, $p \sim \mathrm{UR}[0, R]$, $u_2 \sim \mathrm{UR}(\mathbb{S}^2)$, and retain the first one satisfying the hitting event $Y \cap L_2^* \neq \emptyset$.
2. Let $D_2(p, u_2)$ denote the orthogonal projection of the ball D onto the retained plane L_2^*, see the analogous disk in Fig. 2.5.1(b). Within L_2^*, the planar probe $T_r^3(x, u_2, \tau)$ may be reparametrized as $T_r^2(z, \tau)$, see Fig. 1.13.2(b). Generate a realization of $\tau \sim \mathrm{UR}[0, 2\pi)$, and proceed as in Fig. 2.11.2(c) with $D_\oplus(\omega)$ replaced with $D_{2\oplus} \equiv D_2(p, u_2) \oplus \check{T}_r^2(0, \tau)$. If $Y \cap L_2^* \cap T_r^2(z, \tau) \neq \emptyset$, then the realization $T_r^2(z, \tau)$ is IUR hitting the target set Y – otherwise, return to step 1.

2.13.2 Conditional distributions for a bounded test probe hitting a set

Consider the case $Y \subset \mathbb{R}^2$. Integration over x in the rhs of Eq. (2.13.2) shows that the marginal probability element of the direction ω is weighted by the relevant Minkowski area, namely,

$$\mathbb{P}(\mathrm{d}\omega | \uparrow) = \frac{A(Y_\oplus(\omega))\mathbb{P}(\mathrm{d}\omega)}{\mathbb{E}\{A(Y_\oplus(\omega))\}}, \qquad \omega \in [0, 2\pi). \tag{2.13.3}$$

Unless Y is a disk, the random variables x and ω are not independent. For each orientation, the associated point x is conditionally UR, namely,

$$\mathbb{P}(\mathrm{d}x | \omega, \uparrow) = \frac{\mathrm{d}x}{A(Y_\oplus(\omega))}, \qquad x \in Y_\oplus(\omega), \tag{2.13.4}$$

which corresponds to a bounded FUR probe hitting the set Y with a fixed orientation $\omega \in [0, 2\pi)$.

Again, the extension of the preceding results to $\mathbb{R}^d$ is straightforward.

Probabilities

Suppose that $Y_2 \subset Y_1 \subset D \subset \mathbb{R}^2$ with $\dim(Y_1) = \dim(Y_2) = 2$. To illustrate the hitting probabilities it suffices to consider a test quadrat T_2^2. If T_2^2 is FUR hitting D, then Eq. (2.13.4) yields

$$\mathbb{P}(T_2^2 \uparrow Y_2 | T_2^2 \uparrow Y_1, \omega) = \frac{A(Y_{2\oplus}(\omega))}{A(Y_{1\oplus}(\omega))}, \tag{2.13.5}$$

whereas, if T_2^2 is IUR hitting the reference disk D, then Eq. (2.13.1) yields

$$\mathbb{P}(T_2^2 \uparrow Y_2 | T_2^2 \uparrow Y_1) = \frac{\mathbb{E}\{A(Y_{2\oplus}(\omega))\}}{\mathbb{E}\{A(Y_{1\oplus}(\omega))\}}. \tag{2.13.6}$$

In the latter case, if $\chi(Y_2 \cap T_2^2) = \chi(Y_1 \cap T_2^2) = 1$ for any location and orientation of T_2^2, which hold in particular if Y_1, Y_2 and T_2^2 are all convex, then the kinematic formula, Eq. (1.20.13), yields

$$\mathbb{P}(T_2^2 \uparrow Y_2 | T_2^2 \uparrow Y_1) = \frac{2\pi(a + A_2) + bB_2}{2\pi(a + A_1) + bB_1}, \tag{2.13.7}$$

where a, b denote the area and the boundary length of T_2^2, respectively. However, if T_2^2 is FUR, then no explicit results exist in general because the rhs of Eq. (2.13.5) is shape-dependent.

2.14 Sampling With a Cycloid Test Curve in a Vertical Plane

Consider the setup described in Section 2.9.1, and let

$$T_1^2 \equiv T_1^2(x, p, \phi) \subset L_{2\cdot v}^3(p, \phi) \tag{2.14.1}$$

denote a translation-invariant cycloid test curve of length l in the vertical plane $L_{2\cdot v}^3$, hitting the reference ball D with its minor axis along the vertical direction, see Fig. 1.17.1(d). A UR realization of T_1^2 hitting D may be generated in two stages as described in Section 2.11.2 for a planar IUR probe by replacing L_2^* with $L_{2\cdot v}^3(p, \phi)$, and denoting the translated equatorial disk $D_2(p, u_2)$ by $D_2(p, \phi)$. Set

$$D_{2\oplus}(p, \phi) \equiv D_2(p, \phi) \oplus \check{T}_1^2(0, p, \phi). \tag{2.14.2}$$

For $x \in D_{2\oplus}(p, \phi)$, $p \in [-H/2, H/2]$, $\phi \in [0, \pi)$, the probability element associated with T_1^2 is

$$\mathbb{P}(\mathrm{d}x, \mathrm{d}p, \mathrm{d}\phi) = \frac{\mathrm{d}x}{A(D_{2\oplus})} \cdot \frac{\mathrm{d}p}{H} \cdot \frac{\mathrm{d}\phi}{\pi}, \tag{2.14.3}$$

where $A(D_{2\oplus}) \equiv A(D_{2\oplus}(p, \phi))$ does not depend on the parameters (p, ϕ). Set $I \equiv I((Y_2 \cap L_{2\cdot v}^3) \cap T_1^2)$. By Eq. (1.17.6), we have

$$\begin{aligned} \mathbb{E}(I) &= \frac{1}{\pi H A(D_{2\oplus})} \int_0^\pi \mathrm{d}\phi \int_{-H/2}^{H/2} \mathrm{d}p \int_{D_{2\oplus}(p,\phi)} I \,\mathrm{d}x \\ &= \frac{lS}{2HA(D_{2\oplus})}. \end{aligned} \tag{2.14.4}$$

Further, set $P \equiv P(x \cap Y_1)$. Then,

$$\begin{aligned}\mathbb{E}(P) &= \frac{1}{\pi H A(D_{2\oplus})}\int_0^{\pi} \mathrm{d}\phi \int_{-H/2}^{H/2} \mathrm{d}p \int_{D_{2\oplus}(p,\phi)} P\,\mathrm{d}x \\ &= \frac{V}{HA(D_{2\oplus})},\end{aligned} \tag{2.14.5}$$

which combined with Eq. (2.14.4) gives us

$$\frac{\mathbb{E}(I)}{\mathbb{E}(P)} = \frac{l}{2}\frac{S}{V}, \text{ namely } S_V = \frac{2}{l} I_P. \tag{2.14.6}$$

As expected, the last identity coincides formally with Eq. (2.12.19), which corresponds to an IUR test curve of length l hitting D directly in $\mathbb{R}^3$.

2.15 Particle Number

2.15.1 Introduction

Consider a bounded and finite set $Y = \{y_1, y_2, \dots, y_N\} \subset D \subset \mathbb{R}^d$ of N distinct point particles, where y_i denotes the ith point particle, enclosed in a reference d-ball of diameter H. Let $L_t \equiv L_t(p)$ denote a FUR slab of thickness t hitting D with probability element given by Eq. (2.10.1). Recalling Eq. (1.4.8), and Section 2.10.2,

$$N = \frac{H+t}{t} \cdot \mathbb{E}\{Q(p)\}, \tag{2.15.1}$$

where $Q(p)$ denotes the random number of point particles captured by the slab.

Likewise, let $T_2^2(x)$ denote a FUR test quadrat of area a hitting a reference disk $D \supset Y$ in the plane. Then, recalling Eq. (1.15.4), and Eq. (2.12.1),

$$N = \frac{A(D_{\oplus})}{a} \cdot \mathbb{E}\{Q(x)\}, \tag{2.15.2}$$

where $Q(x)$ now represents the random number of point particles captured by the quadrat. The extension to $\mathbb{R}^d$ is straightforward.

The concept of 'particles captured by a slab' may not be that obvious if each element of Y is not just a point particle but a general particle, namely a bounded and connected subset Y_i of $\mathbb{R}^d$, of arbitrary shape and dimension $q \in \{1, 2, \dots, d\}$, separated from other particles. In fact, if a particle hits a face of a slab, then it is necessary to establish a rule to decide whether the particle is 'captured' by the slab, or not, and the same problem arises for a quadrat, or indeed for any bounded d-probe in $\mathbb{R}^d$. The underlying artifact is called the edge effect.

2.15.2 Counting rules for arbitrary particles

Number is a quantity of dimension 0. Therefore, the purpose of a counting rule for arbitrary particles is to remove the size and shape properties of the particles, making them equivalent to point particles for the sake of counting with a full-dimensional probe.

The associated point rule

To each particle Y_i of the set $Y = \{Y_1, Y_2, \ldots, Y_N\}$, an associated point (AP) y_i is assigned by a rule fixed a priori for all particles. The AP is defined in Section 1.13, for a bounded test probe. Because an AP is equivalent to a point particle, the number of particles captured by a full-dimensional probe is defined as the number of APs captured by the probe, see Fig. 2.15.1, whereby Eq. (2.15.1) and Eq. (2.15.2) hold without any further qualification.

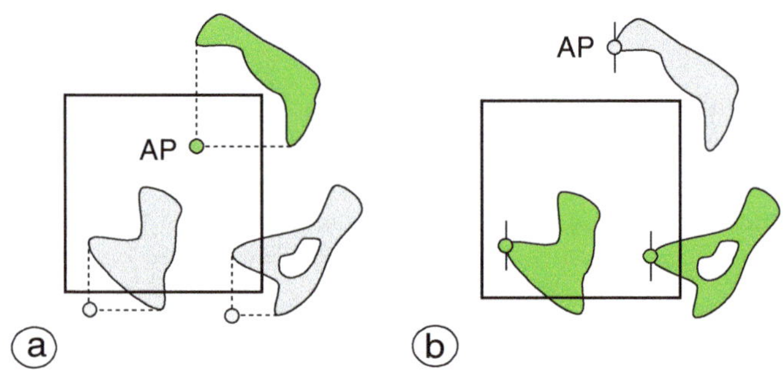

Fig. 2.15.1 Green particles are counted in, or selected by the quadrat, grey ones are not. The counting event depends on the definition of the AP, but the mean count does not if the quadrat is FUR hitting a reference set containing the particles.

Remarks

1. Because the probe is mobile with uniform randomness, the event that an AP lies in the probe boundary has probability 0 and may be ignored.
2. Depending on its definition, the AP of a particle may lie outside the particle – thus a particle may be selected by a probe without being hit by the probe, see Fig. 2.15.1(a).
3. For a given set of particles, different definitions of the AP may lead to different counts for a given position of the probe, but the aforementioned mean value identities will always hold.

The scanning principle and the slab disector

To fix ideas, consider a set $Y \subset D \subset \mathbb{R}^2$ of N arbitrary particles in the plane, where D is a reference disk whose diameter is the segment $[0, H)$, $H > 0$. As a vertical straight line $L(p)$ scans D from $p = 0$ to $p = H$, it will meet each particle exactly once for the first time, and the total number of such encounters will always be equal to N, irrespective of particle location and orientation relative to the line. This is the scanning principle, and it applies to particles in $\mathbb{R}^d$ by regarding $L(p)$ as a scanning hyperplane. In the sequel, $L(p)$ is referred to as a 'plane'. In microscopy, a planar optical section scanning a specimen in search of cells, organelles, or grains, is an example of the scanning principle in $\mathbb{R}^3$. For any distance $t > 0$ travelled by $L(p)$, this plane sweeps a portion of the containing space, and therefore scan sampling is a variant of sampling with a full-dimensional probe.

For a given orientation, the scanning plane $L(p)$ will generally meet a particle for the first time at a unique contact point which may be defined as a unique AP of the particle, see Fig. 2.15.1(b) and Fig. 2.15.2(a). (If the contact subset is a planar face of the particle, then a unique AP can be defined in that face.) Therefore, the scanning method may be regarded as a one-to-one mapping of a set of N particles into a corresponding set of N APs or point particles. Let $Q(p)$ denote the number of such APs encountered as the plane moves continuously from $L(p)$ to $L(p + t)$. This number is identical to the number of APs in the slab $L_t(p)$, see Fig. 2.15.2(b). Equivalently, $Q(p)$ is also identical to the number $Q^-(p)$ of particles captured by the slab disector $L_t^-(p)$, which is a geometric probe defined indirectly by the following rule: A particle is captured by $L_t^-(p)$ if it has at least one point in common with the slab $L_t(p)$, but it does not hit the face $L(p)$ (called the look-up face) of the slab. This exclusion face motivates the upper index "–" for disector counts of arbitrary particles.

The disector is a useful practical tool because it avoids the direct tagging and scoring of virtual APs in a slab probe. It is equivalent to a plane that sweeps space from $L(p)$ to $L(p + t)$, scoring particles encountered for the first time along the way. In fact, the latter method is often implemented by optical scanning under a microscope, and it is called the optical (slab) disector in stereology.

Counting with a bounded probe

The slab disector rule can be extended to count particles in a bounded quadrat by replacing continuous with discrete scanning. Consider a partition of space into a series of adjacent, non-overlapping slabs $\{L_t(kt),\ k \in \mathbb{Z}\}$ ordered from left to right in sequential order, see Fig. 2.15.3(a). For each particle Y_i, $i \in \{1, 2, \ldots, N\}$, as k increases we will always find a unique index k_i such that $L_t(kt) \cap Y_i = \emptyset$ for all $k < k_i$, but $L_t(k_i t) \cap Y_i \neq \emptyset$. In other words, there is always a unique slab $L_t(k_i t)$ that hits the particle for the first time. This amounts to saying that the particle is captured or selected by, or counted in, the slab disector $L_t^-(k_i t)$.

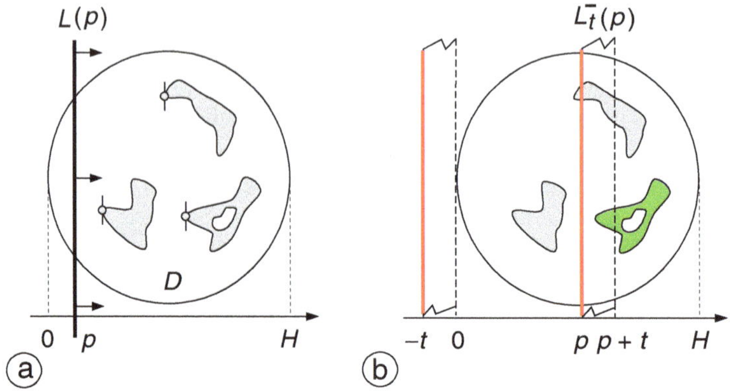

Fig. 2.15.2 Illustration of the concepts of (a) the scanning principle, and (b) the slab disector principle.

Consider now a partition of the plane into a grid of non-overlapping rectangular quadrats of side lengths t_1, t_2, namely,

$$\mathbb{R}^2 = \sum_{k \in \mathbb{Z}} \sum_{l \in \mathbb{Z}} T(k, l),$$
$$T(k, l) = [kt_1, (k + 1)t_1) \times [lt_2, (l + 1)t_2). \tag{2.15.3}$$

Enumerate the quadrat columns from left to right by increasing k in sequential order, and within each column enumerate the quadrats from bottom to top by increasing l in sequential order. Proceeding in this way, for each particle Y_i, we will always find a unique quadrat $T(k_i, l_i)$ that hits the particle for the first time, see the green quadrat in Fig. 2.15.3(b). This implies that neither the half space $x_1 \leq k_i t_1$ nor the vertical stripe portion $\{(x_1, x_2) : \ k_i t_1 < x_1 \leq (k_i + 1)t_1, \ x_2 \leq l_i t_2\}$ below this quadrat will have any points in common with the particle. Therefore, a particle is captured by a quadrat if it has at least one point in common with the quadrat, but the particle does not hit any of the extended edges determined by the preceding restrictions. These extended edges, in red in Fig. 2.15.3(c), constitute a 'forbidden line' for a particle – hence the counting rule, due to H. J. G. Gundersen (1943–2021), is called the forbidden line rule.

Next we show that Eq. (2.15.2) holds for arbitrary particles sampled with the forbidden line rule. Shift the origin to a point $x \in T(0, 0)$, dragging the entire partition together by the vector x, and set $T_{kl}(x) = T(k, l) + x$. Further, let $Q^-(\cdot)$ denote the number of particles captured by a quadrat according to the forbidden line rule, and set $Q^-_{kl}(x) \equiv Q^-(T_{kl}(x) \cap Y)$ and $Q^-_{kli}(x) \equiv Q^-(T_{kl}(x) \cap Y_i)$. For each $x \in T(0, 0)$,

$$\sum_{k \in \mathbb{Z}} \sum_{l \in \mathbb{Z}} Q^-_{kli}(x) = 1, \quad i = 1, 2, \ldots, N, \tag{2.15.4}$$

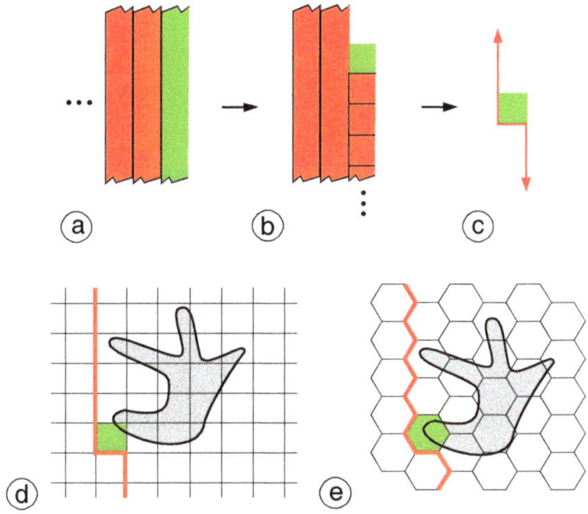

Fig. 2.15.3 (a) The green slab has met a particle for the first time. (b) The green quadrat has met the particle for the first time. (c) Gundersen's forbidden line rule to remove edge effects in particle counting. (c, d) With the forbidden line rule a bounded particle is counted in exactly one quadrat from a partition.

because the forbidden line rule warrants that each particle Y_i is necessarily captured by exactly one quadrat, see Fig. 2.15.3(d) for an example. Because the N particles are disjoint,

$$\sum_{k\in\mathbb{Z}}\sum_{l\in\mathbb{Z}} Q^{-}_{kl}(x) = \sum_{i=1}^{N}\sum_{k\in\mathbb{Z}}\sum_{l\in\mathbb{Z}} Q^{-}_{kli}(x) = N. \tag{2.15.5}$$

Let $T_2^2(x)$, $x \in \mathbb{R}^2$ denote a FUR quadrat equipped with the forbidden line rule and congruent with $T(0,0)$, and set $Q^-(x) \equiv Q^-(T_2^2(x) \cap Y)$. Then,

$$\int_{\mathbb{R}^2} Q^-(x)\,\mathrm{d}x = \int_{T(0,0)} \mathrm{d}x \sum_{k\in\mathbb{Z}}\sum_{l\in\mathbb{Z}} Q^{-}_{kl}(x) = aN, \tag{2.15.6}$$

similarly as in Eq. (1.15.4) (which was derived for point particles), where $a = t_1 t_2$ is the quadrat area. If $Y \subset D \subset \mathbb{R}^2$ and $T_2^2(x) \sim \mathrm{FUR}(D)$, then

$$\mathbb{E}\{Q^-(x)\} = \frac{a}{A(D_\oplus)} \cdot N, \tag{2.15.7}$$

which is the equivalent to Eq. (2.15.2) for general particles sampled with the forbidden line rule.

Remarks

1. The forbidden line rule is different from the AP rule. For a single particle with AP y_i, it is immediate that

$$\int_{\mathbb{R}^2} 1_{T_2^2(x)}(y_i)\,\mathrm{d}x = a \quad i \in \{1, 2, \ldots, N\}. \tag{2.15.8}$$

   For the forbidden line rule, however, the global identity Eq. (2.15.5) for the union of all particles, leading to Eq. (2.15.7), is also easy to show, but the counterpart of Eq. (2.15.8) for a single particle is not obvious. The shape of the geometric locus of all positions of x such that $T_2^2(x)$ captures an arbitrary particle Y_i by the forbidden line rule depends on the shape of Y_i. The area of this geometric locus (which may even not be connected in general) must indeed be equal to a, but a direct, general proof is to our knowledge not available.
2. The forbidden line rule applies to any quadrat shape that can partition the plane without overlapping, so that each quadrat can be brought to coincide with a given quadrat by a translation that leaves the entire partition invariant. See Fig. 2.15.3(e).
3. The extension of the forbidden line rule to $\mathbb{R}^3$ is available, and the corresponding probe is called the unbiased brick, see Fig. 2.15.4(b). Extensions to higher dimensions have apparently not been considered.

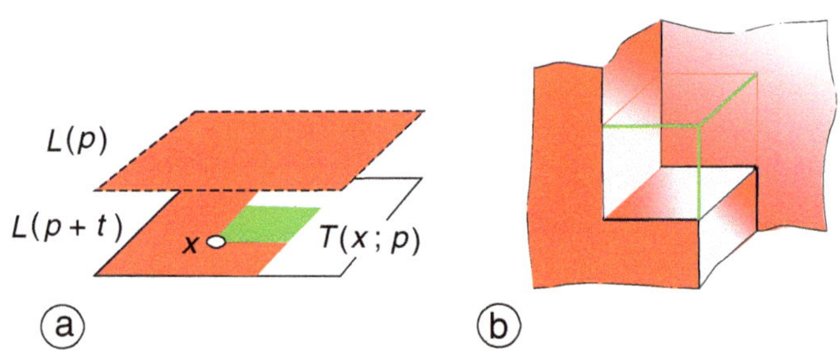

Fig. 2.15.4 (a) Sketch of the bounded disector. (b) Idem of the unbiased brick. A particle is directly counted, or selected, in 3D, if it has points in common with the sampling cube but not with any of the (red coloured) extended planar surfaces.

The bounded disector

This is a counting tool designed to subsample particles within a slab disector. Recall that a slab disector $L_t^-(p)$ is unbounded, hence, if $D \subset \mathbb{R}^3$ is the reference set containing the particle population Y, then the counting procedure requires the inspection of the entire slice $D \cap L_t(p)$, which may preclude its practical implementation. The first stage sampling identity is given by Eq. (2.15.1), namely,

$$N = \frac{H+t}{t} \cdot \mathbb{E}\{Q^{-}(p)\}, \tag{2.15.9}$$

where $Q^{-}(p)$ denotes the number of particles hitting the face $L(p+t)$ but not the look-up face $L(p)$, plus the number of particles entirely contained in the slab $L_t(p)$. For convenience we assume that D is a ball whose diameter is the segment $[0, H)$, $H > 0$. The idea of the bounded disector is to subsample a FUR bounded quadrat $T(x; p)$ hitting the reference section $D \cap L(p+t)$, see Fig. 2.15.4(a). The fact that this section is empty for $p \in [H-t, H)$ may be circumvented by choosing the origin and the diameter H in such a way that there can be no particles captured by the slab disector $L_t^{-}(H-t)$ in any direction. The total number of particles captured by the quadrat $T(x; p)$ may be expressed as follows,

$$Q^{-}(p,x) = Q_1^{-}(p,x) + Q_2^{-}(p,x), \tag{2.15.10}$$

where:

- $Q_1^{-}(p,x)$ is the number of particles satisfying: (1) the particles hit the reference section $D \cap L(p+t)$ but not the look-up section $D \cap L(p)$, and (2) the particle transects in the reference section are captured by $T(x; p)$ according to the forbidden line rule. If the particle is not convex, then a transect may consist of several separate parts, called profiles. In this case the transect is redefined for counting purposes as a connected set consisting of the union of the profiles and a set of arbitrary straight line segments connecting them.
- $Q_2^{-}(p,x)$ is the number of the individual orthogonal projections of the particles entirely contained in the slab $L_t(p)$, onto the reference section $D \cap L(p+t)$, which are captured by the quadrat $T(x; p)$ according to the forbidden line rule. Overlapping artifacts are assumed to be either identifiable, or non-existent – such artifacts can usually be minimized by a proper choice of the disector thickness t.

For the preceding design, the joint probability element of (p,x) is

$$\mathbb{P}(\mathrm{d}x, \mathrm{d}p) = \frac{\mathrm{d}x}{A(D_{2\oplus})} \cdot \frac{\mathrm{d}p}{H+t}, \quad x \in D_{2\oplus}(p), \; p \in [-t, H], \tag{2.15.11}$$

where $D_{2\oplus}(p) \equiv D_2(p) \oplus \check{T}(0, p)$ and $D_2(p)$ is the orthogonal projection of D onto the plane $L(p+t)$. The expectation of the disector count $Q^{-}(p,x)$ with respect to the preceding probability element is

$$\begin{aligned} \mathbb{E}\{Q^{-}(p,x)\} &= \frac{1}{(H+t)A(D_{2\oplus})} \int_{-t}^{H} \mathrm{d}p \int_{D_{2\oplus}(p)} Q^{-}(p,x)\,\mathrm{d}x \\ &= \frac{1}{(H+t)A(D_{2\oplus})} \int_{-t}^{H} aQ^{-}(p)\,\mathrm{d}p \\ &= \frac{atN}{(H+t)A(D_{2\oplus})}, \end{aligned} \tag{2.15.12}$$

where t, a denote disector thickness and quadrat area, respectively. The second identity follows from Eq. (1.15.4). Thus,

$$N = \frac{(H+t)A(D_{2\oplus})}{at} \cdot \mathbb{E}\{Q^{-}(p,x)\}. \tag{2.15.13}$$

Alternatively, Eq. (2.15.12) may be rearranged as follows,

$$\frac{N}{(H+t)A(D_{2\oplus})} = \frac{\mathbb{E}\{Q^{-}(p,x)\}}{at}, \text{ or } N_V = Q_V^{-}, \tag{2.15.14}$$

a stereological ratio equation which is intuitively plausible because $Q^{-}(p,x)$ is a direct particle count in space, whereas the product at may be interpreted as the volume of the bounded disector.

2.15.3 Notes

1. The associated point (AP) rule

The idea was hinted at by Miles (1974, p. 232) and precisely described in Miles (1978b). For further details and extensions see Jensen and Sundberg (1986a). Thompson (1932, p. 26) anticipated the AP rule by stating that an unbiased definition of the number of objects sampled by a slab could be the number of object centroids in the slab.

2. The scanning principle and the slab disector

Sweeping lines were proposed by DeHoff (1967) and Cahn (1967) to score tangent counts in planar sections, see Weibel (1979, p. 40). Thompson et al. (1932, p. 37) removed the need to use the AP rule and proposed to count "objects having a part in the master section but not in a given adjoining section", which is the slab disector idea – see also Miles (1972, Section 7). Bendtsen and Nyengaard (1989), however, reveal that slab disectors with no edge effects (i.e. affecting entire organs) were used by anatomists to count organelles already from the late 19th century.

3. The forbidden line rule for quadrat sampling

The rule was discovered by Gundersen (1977). The 'first hit' principle underlying the rule is described in Jensen (1998), see also Baddeley and Jensen (2005). For the three-dimensional version (Fig. 2.15.4(b)) see Howard, Reid, Baddeley, and Boyde (1985).

4. The bounded, and the optical disector

The bounded disector was first described by Sterio (1984). The early version used just two parallel physical sections a known distance apart, playing the role of the reference and the look-up sections. The optical disector was proposed by Gundersen (1986), see also Gundersen et al. (1988) – for further details see Howard and Reed (2005) and West (2012). 'D. C. Sterio' is an anagram of 'disector', adopted as a *nom de plume* by H. J. G. Gundersen.

2.16 Particle Number and Size From Slab and Plane Probes

2.16.1 Preliminaries

The target is an aggregate $Y = \{Y_1, Y_2, \ldots, Y_N\} \subset D \subset \mathbb{R}^3$ of particles enclosed in a reference ball D of diameter H. Let $H_i(u)$ and V_i denote the caliper length along a direction $u \in \mathbb{S}^2_+$, and the volume of the particle Y_i, respectively. The mean particle caliper length along u, and the mean particle volume, are

$$\mathbb{E}\{H(u)\} = \frac{1}{N}\sum_{i=1}^{N} H_i(u), \quad \mathbb{E}(V) = \frac{1}{N}\sum_{i=1}^{N} V_i, \tag{2.16.1}$$

respectively. (The notation $\mathbb{E}\{H(u)\}$ was also used in Section 2.8.2 for the mean caliper length of an individual particle over orientations.)

2.16.2 Mean particle caliper length

Let $L_t \equiv L_t(p, u)$ denote a FUR slab of thickness t hitting the reference ball D. We want to express $\mathbb{E}\{H(u)\}$ in terms of counts.

Define

$$Q(p, u) = \sum_{i=1}^{N} 1_i(p, u), \tag{2.16.2}$$

the total number of particles hit by the slab $L_t(p, u)$ (not to be confused with the number of particle fragments in the slab), where

$$1_i(p, u) = \begin{cases} 1 \text{ if } L_t(p, u) \uparrow Y_i \\ 0 \text{ otherwise.} \end{cases} \tag{2.16.3}$$

By Eq. (2.10.10), the mean over p conditional on a given orientation is

$$\mathbb{E}\{Q(p, u)|u\} = \sum_{i=1}^{N} \mathbb{E}\{1_i(p, u)|u\} = \sum_{i=1}^{N} \frac{H_i(u) + t}{H + t}. \tag{2.16.4}$$

On the other hand, let $Q^-(p,u)$ denote the total number of particles sampled by the FUR slab disector $L_t^-(p,u)$. By Eq. (2.15.9),

$$\mathbb{E}\{Q^-(p,u)\} = \frac{tN}{H+t}. \tag{2.16.5}$$

From the preceding two identities we obtain

$$\mathbb{E}\{H(u)\} = t\left\{\frac{\mathbb{E}\{Q(p,u)|u\}}{\mathbb{E}\{Q^-(p,u)|u\}} - 1\right\}, \tag{2.16.6}$$

which expresses the mean particle caliper length in a given direction in terms of mean counts only.

2.16.3 Particle number and mean volume

Particle number

Let $L(p,u)$ denote a planar face of a FUR slab $L_t(p,u)$ hitting the reference ball D. By Eq. (2.10.9),

$$\mathbb{E}\{A(D \cap L(p,u))\} = \frac{V(D)}{H+t}, \tag{2.16.7}$$

which, combined with Eq. (2.16.5) yields the following stereological equation for particle counts with a slab disector,

$$\frac{N}{V(D)} = \frac{\mathbb{E}\{Q^-(p,u)\}}{t \cdot \mathbb{E}\{A(D \cap L(p,u))\}}, \text{ or } N_V = \frac{1}{t} \cdot Q_A^-, \tag{2.16.8}$$

analogous to Eq. (2.15.14). However, if we use Eq. (2.16.4), then we obtain the classical stereological equation,

$$\begin{aligned} \frac{N}{V(D)} &= \frac{\mathbb{E}\{Q(p,u)\}/\mathbb{E}\{A(D \cap L(p,u))\}}{\mathbb{E}\{H(u)\}+t}, \\ N_V &= \frac{Q_A}{\mathbb{E}\{H(u)\}+t}, \end{aligned} \tag{2.16.9}$$

which is not quite useful because the mean particle caliper length is not accessible from slab measurements alone.

Particle size with a FUR plane probe

Next we examine particle size using only a FUR plane $L(p,u)$ hitting the reference ball D. Set $A_i \equiv A_i(p,u) = A(Y_i \cap L(p,u))$, the individual planar transect area through a particle Y_i. By Eq. (2.8.7) we have

$$\mathbb{E}\{A_i|u,\uparrow\} = V_i/H_i(u), \quad i = 1,2,\ldots,N, \tag{2.16.10}$$

where $\uparrow$ denotes the hitting event $L(p,u) \uparrow Y_i$. On the other hand, similarly as in Eq. (2.4.30),

$$\mathbb{P}\{\uparrow|H_i(u)\} = H_i(u)/H, \tag{2.16.11}$$

whereby,

$$\begin{aligned}\mathbb{P}\{H_i(u)|u,\uparrow\} &\propto \mathbb{P}\{\uparrow|H_i(u)\}\cdot\mathbb{P}\{H_i(u)\}\\ &\propto H_i(u)\cdot\mathbb{P}\{H_i(u)\}.\end{aligned} \tag{2.16.12}$$

Without loss of generality we may suppose that all the values taken by $H_i(u)$ are different, in which case $\mathbb{P}\{H_i(u)\} = 1/N$. Normalization of the rhs of Eq. (2.16.12) yields the probability function of particle caliper length conditional on the hitting event, namely,

$$\mathbb{P}\{H_i(u)|u,\uparrow\} = \frac{H_i(u)}{N\cdot\mathbb{E}\{H(u)\}}, \quad i = 1,2,\ldots,N, \tag{2.16.13}$$

where $\mathbb{E}\{H(u)\}$ is defined in Eq. (2.16.1). Therefore, the mean area A of the transect determined in a particle by a hitting FUR plane $L(p,u)$, averaged over all particles, is

$$\begin{aligned}\mathbb{E}(A|u,\uparrow) &= \sum_{i=1}^{N}\mathbb{E}(A_i|u,\uparrow)\cdot\mathbb{P}\{H_i(u)|u,\uparrow\}\\ &= \sum_{i=1}^{N}\frac{V_i}{H_i(u)}\cdot\frac{H_i(u)}{N\mathbb{E}\{H(u)\}}\\ &= \frac{\mathbb{E}(V)}{\mathbb{E}\{H(u)\}},\end{aligned} \tag{2.16.14}$$

where $\mathbb{E}(V)$ is defined in Eq. (2.16.1). Thus, the mean particle transect area is not a good descriptor of particle size because $\mathbb{E}\{H(u)\}$ is not identifiable from the information available in a section.

Access to the mean particle volume requires slab disector counting. In fact,

$$\begin{aligned}\mathbb{E}(V) &= \frac{V(Y)}{N}\\ &= \frac{V(Y)}{V(D)}\cdot\frac{V(D)}{N}\\ &= \frac{V_V}{N_V},\end{aligned} \tag{2.16.15}$$

and N_V is given by Eq. (2.16.8).

For a FUR plane $L(p,u)$ hitting D, the following identity is also worth mentioning,

$$\begin{aligned}\mathbb{E}\left(\frac{1}{H_i(u)}\bigg|u,\uparrow\right) &= \sum_{i=1}^{N} \frac{1}{H_i(u)} \cdot \frac{H_i(u)}{N\mathbb{E}\{H(u)\}} \\ &= \frac{1}{\mathbb{E}\{H(u)\}}.\end{aligned} \tag{2.16.16}$$

The combination of the preceding identity with Eq. (2.16.9) with $t = 0$ yields

$$N_V = Q_A \cdot \mathbb{E}\left(\frac{1}{H_i(u)}\bigg|u,\uparrow\right). \tag{2.16.17}$$

Further, by an argument similar to that leading to Eq. (2.3.8),

$$\mathrm{CV}^2\{H(u)\} = \frac{\mathbb{E}\{H(u)|u,\uparrow\}}{\mathbb{E}\{H(u)\}} - 1. \tag{2.16.18}$$

Particle size with an IUR plane probe

If the plane $L(p,u)$ is IUR hitting D, then recalling Eq. (2.8.8) we have, instead of Eq. (2.16.10),

$$\begin{aligned}&\mathbb{E}(A_i|\uparrow) = V_i/H_i, \\ &H_i \equiv \mathbb{E}_u\{H_i(u)\} = \int_{\mathbb{S}_+^2} H_i(u)\mathbb{P}(\mathrm{d}u).\end{aligned} \tag{2.16.19}$$

The mean versions of Eq. (2.16.11) and Eq. (2.16.12) over isotropic orientations read respectively as follows,

$$\mathbb{P}(\uparrow|H_i) = \int_{\mathbb{S}_+^2} \mathbb{P}\{\uparrow|H_i(u)\}\mathbb{P}(\mathrm{d}u) = H_i/H, \tag{2.16.20}$$

$$\mathbb{P}(H_i|\uparrow) \propto H_i\mathbb{P}(H_i), \tag{2.16.21}$$

and $\mathbb{P}(H_i) = 1/N$. Normalization of the rhs of Eq. (2.16.21) yields the probability function of the individual mean (over isotropic directions) of particle caliper length, conditional on the hitting event, namely,

$$\mathbb{P}(H_i|\uparrow) = \frac{H_i}{N\mathbb{E}(H)}, \tag{2.16.22}$$

where

$$\mathbb{E}(H) = \frac{1}{N}\sum_{i=1}^{N} \mathbb{E}\{H_i(u)\}. \tag{2.16.23}$$

The mean version of Eq. (2.16.14) over isotropic orientations reads

$$\begin{aligned}\mathbb{E}(A|\uparrow) &= \sum_{i=1}^{N} \mathbb{E}(A_i|\uparrow) \cdot \mathbb{P}(H_i|\uparrow) \\ &= \sum_{i=1}^{N} \frac{V_i}{H_i} \cdot \frac{H_i}{N\mathbb{E}(H)} \\ &= \frac{\mathbb{E}(V)}{\mathbb{E}(H)}. \end{aligned} \tag{2.16.24}$$

Similarly, the isotropic versions of Eq. (2.16.16) and Eq. (2.16.18) result on replacing $H(u)$ with $\mathbb{E}_u\{H(u)\}$ and $\mathbb{E}\{H(u)\}$ with $\mathbb{E}(H)$.

2.16.4 Notes

1. Sampling particles with a test plane

Early theory can be found in Miles (1972, Section 5). Some basic principles were already known to Thompson (1932) – for instance, Eq. (2.16.17) is equivalent to Eq. (12) from the latter paper. The latter result was rediscovered by Cruz-Orive (1980b). Analogous results are typical of renewal theory, see e.g. Cox (1962).

2. Sampling particles with a slab

Eq. (2.16.9) was proposed by Abercrombie (1946) who, to make it usable, assumed that the particles were spheres. This, and earlier attempts at estimating particle number from sections, are reviewed in Baddeley and Jensen (2005). For $t = 0$ the result is implicit in Wicksell (1925, p. 89), who pioneered the so-called unfolding methods to estimate the distribution of H from the observable distribution of circular profiles in a FUR section of spherical particles. The literature on such methods grew up to a large extent in the following decades – for an early review, see Cruz-Orive (1983). After the publication of the disector in 1984, see Eq. (2.16.8), or Eq. (2.15.14), unfolding methods decayed. Eq. (2.16.6) was derived by Gundersen (1992).

2.17 Connectivity

The background material is described in Sections 1.10, 1.12 and 1.18, which also include notes and references. Let $Y \subset D \subset \mathbb{R}^3$ represent a domain with Euler–Poincaré characteristic χ – which is the target quantity – enclosed in a reference ball D of diameter H.

Consider first a FUR slab disector $L_t^-(p)$, as defined in Section 2.15.2, hitting D. By Eq. (1.12.7), the contribution of this disector to χ is

$$\chi^-(p) = \frac{1}{2}(I(p) - B(p) + H(p)), \tag{2.17.1}$$

where the relevant terms are total counts scored when $L(p)$, and then when $L(p+t)$, are adopted in turn as look-up faces (hence the factor $1/2$). Because $\chi(Y)$ is dimensionless, the argument leading to Eq. (2.15.9) holds, and therefore,

$$\chi = \frac{H+t}{t} \cdot \mathbb{E}\{\chi^-(p)\}. \tag{2.17.2}$$

A proper method to represent χ with a bounded FUR probe $T(x)$ hitting D, consisting of a three-dimensional box, is the shell method described in Section 1.18. Then,

$$\frac{\chi}{V(D_\oplus)} = \frac{\mathbb{E}\{\chi^-(x)\}}{V(T(0))}, \text{ or } \chi_V = \chi_V^-, \tag{2.17.3}$$

where, as indicated in Section 1.18.2,

$$\chi^-(x) = \chi((Y \cap T(x)) \cup S(x)) - 1. \tag{2.17.4}$$

2.18 Sampling With a Local Probe

2.18.1 Definitions

The local probes of interest here are described in Section 1.2.7. A local probe can only rotate about a given point, or a given axis – thus the probability element adopted for a local probe is the normalized version of its invariant density, namely an isotropic probability element.

For a ray $L_{1+}^2(\phi)$ emanating from a fixed point in the plane, or for a half-plane $L_{2+}^3(\phi)$ revolving around a fixed axis in space, the pertinent probability element is

$$\mathbb{P}(\mathrm{d}\phi) = \frac{\mathrm{d}\phi}{2\pi}, \quad \phi \in [0, 2\pi). \tag{2.18.1}$$

For a ray $L_{1+}^3(u_2)$ emanating from a fixed point in space,

$$\mathbb{P}(\mathrm{d}u_2) = \frac{\mathrm{d}u_2}{4\pi}, \quad u_2 \in \mathbb{S}^2. \tag{2.18.2}$$

Such rays are said to be isotropic random (IR). The joint probability element of the spherical polar coordinates of $u_2 \equiv u_2(\phi, \theta)$ is

$$\mathbb{P}(\mathrm{d}\phi, \mathrm{d}\theta) = \frac{\mathrm{d}\phi}{2\pi} \cdot \frac{1}{2} \sin\theta \, \mathrm{d}\theta, \quad \phi \in [0, 2\pi), \ \theta \in [0, \pi), \tag{2.18.3}$$

which implies that ϕ and θ are independent and $\phi \sim \mathrm{UR}[0, 2\pi)$, whereas

$$\mathbb{P}(\mathrm{d}\theta) = \frac{1}{2} \sin\theta \, \mathrm{d}\theta, \quad \theta \in [0, \pi). \tag{2.18.4}$$

An isotropic realization of $u_2(\phi, \theta)$ may be generated by setting

$$\phi = 2\pi U_1, \quad \theta = \cos^{-1}(1 - 2U_2), \tag{2.18.5}$$

where $U_i \sim \mathrm{UR}[0, 1)$, $i = 1, 2$, and independent.

For an axis $L^2_{1[0]}$ in the plane, or for a plane $L^3_{2[1]}$ containing a fixed axis in space, the adopted probability element is given by Eq. (2.4.1). For an axis $L^3_{1[0]}$, or for a plane $L^3_{2[0]}$ through a fixed point in space, the corresponding element is given by Eq. (2.5.1).

In the sequel we consider a domain $Y \subset \mathbb{R}^3$ of volume V with piecewise smooth boundary of area S, which are the target quantities. The background results are given in Section 1.7.

The target of the nucleator is only V, the one of the surfactor is only S. The invariator can deal with V and with S simultaneously.

2.18.2 Nucleator versions

The nucleator requires the presence of a distinguishable point O with a fixed location relative to the set Y. This model typically applies to a neuron containing a single nucleolus. In general, it is not necessary that $O \in Y$.

The direct, and the vertical nucleator

An isotropic ray $L^3_{1+}(u_2)$, $u_2 \in \mathbb{S}^2$, emanating from O directly in space, and hitting Y, will generate an intersection consisting of say $m(u_2) \geq 1$ separate intercept segments. The closer and farther distances of the end points of the ith intercept segment from O are denoted by $l_{i-}(u_2)$ and $l_{i+}(u_2)$, respectively. By Eq. (1.7.8),

$$\begin{aligned} V &= 4\pi \mathbb{E}\{f(u_2)\}, \\ f(u_2) &= \frac{1}{3} \sum_{i=1}^{m(u_2)} (l^3_{i+}(u_2) - l^3_{i-}(u_2)), \end{aligned} \tag{2.18.6}$$

the expectation being with respect to $\mathbb{P}(\mathrm{d}u_2)$, see Eq. (2.18.2). If the ray misses Y, then $f(u_2) = 0$. Thus, if $O \notin Y^\circ$ then the expectation includes zeros. If $O \in Y^\circ$, then no zeros arise and $l_{1-}(u_2) = 0$ for all u_2. If the polar axis Ox_3 of the reference trihedron is adopted as a vertical axis (VA), then an IR ray may be generated in a vertical plane $L^3_{2\cdot v}(0, \phi)$ through O – it suffices to observe Eq. (2.18.5), where θ is the colatitude of the ray with origin in the VA. The corresponding version is called the vertical nucleator, which is just a convenient way to implement the direct nucleator, recall Fig. 1.7.2(b).

The pivotal nucleator

Besides doing it on a vertical plane, the nucleator may also be implemented on a pivotal plane, namely on an isotropic plane $L^3_2(0, u)$ through the fixed pivotal point O. Thus, an IR ray $L^3_{1+}(u_2)$, $u_2 \in \mathbb{S}^2$ is equivalent to an IR ray $L^2_{1+}(\omega) \subset L^3_2(0, u)$, $\omega \in [0, 2\pi)$, emanating from O within the pivotal plane, see Fig. 1.7.2(d). The corresponding joint probability element is the normalized version of the product density $\mathrm{d}L^3_{2[0]}\, \mathrm{d}L^2_{1+}$, namely,

$$\mathbb{P}(\mathrm{d}u, \mathrm{d}\omega) = \frac{\mathrm{d}u}{2\pi} \cdot \frac{\mathrm{d}\omega}{2\pi}, \qquad u \in \mathbb{S}^2_+,\ \omega \in [0, 2\pi). \tag{2.18.7}$$

The extension of Eq. (1.7.14) to an arbitrary domain Y yields an identity analogous to Eq. (2.18.6), namely,

$$\begin{aligned} V &= 4\pi \mathbb{E}\{f(\omega, u)\}, \\ f(\omega, u) &= \frac{1}{3} \sum_{i=1}^{m(\omega,u)} (l^3_{i+}(\omega, u) - l^3_{i-}(\omega, u)), \end{aligned} \tag{2.18.8}$$

the expectation being with respect to the joint probability element given by Eq. (2.18.7). Note that if $O \notin Y^\circ$, then that expectation includes the zeros arising either when the pivotal plane misses Y, or when the ray misses the pivotal section. In the latter case, a new pivotal plane has to be generated before sampling another IR ray.

The integrated nucleator

Here it is convenient to enclose the set Y in a ball D of equatorial area a, centred at the pivotal point O. Generate a pivotal plane $L^3_2(0, u)$ and then an independent UR test point z in the corresponding equatorial section. The pertinent joint probability element is the normalized version of $\mathrm{d}L^3_{2[0]}\, \mathrm{d}z$, namely,

$$\mathbb{P}(\mathrm{d}u, \mathrm{d}z) = \frac{\mathrm{d}u}{2\pi} \cdot \frac{\mathrm{d}z}{a}, \qquad u \in \mathbb{S}^2_+,\ z \in D \cap L^3_2(0, u). \tag{2.18.9}$$

Let $r(z, u)$ denote the distance $|z|$ of the test point z from O whenever z hits the pivotal section $Y \cap L_2^3(0, u)$, and set $r(z, u) = 0$ otherwise. Then, recalling Eq. (1.7.13), we have

$$V = 2a\mathbb{E}\{r(z, u)\}, \tag{2.18.10}$$

the expectation being with respect to the joint probability element given by Eq. (2.18.9). Again this expectation includes the zeros arising either when the pivotal plane misses Y, or when the test point misses the pivotal section. In the latter case, to preserve independence a new pivotal plane has to be generated before sampling another test point.

As a cross-check, suppose that Y is a ball D of radius R centred at O. Then, for any $u \in \mathbb{S}_+^2$ we have $\mathbb{E}\{r(z, u)\} = 2R/3$, namely the mean of the distance r from O of a UR point in an equatorial disk of D, because by Eq. (2.4.4) the probability density of r is $f(r) = 2r/R^2$, $0 \le r \le R$. Therefore, Eq. (2.18.10) yields $V = (4\pi/3)R^3$.

2.18.3 The surfactor

The sampling design is that of the pivotal nucleator. For simplicity we assume that Y is a smooth convex set, and also that $O \in Y$. In addition to the radius vector $l_+(\omega, u)$ measured along the ray $L_{1+}^2(\omega)$ in the pivotal plane, it is necessary to measure the angle $\alpha \equiv \alpha(\omega, u) \in (-\pi/2, \pi/2)$ between the ray and the normal to the trace $\partial Y \cap L_2^3(0, u)$ at the intersection with the ray. By Eq. (1.7.36),

$$S = 4\pi\mathbb{E}\{l_+^2(\omega, u)(1 + \alpha \tan \alpha)\}, \tag{2.18.11}$$

the expectation being with respect to the joint probability element given by Eq. (2.18.7). As a cross-check, if Y is a ball of radius R centred at O then $l_+^2(\omega, u) = R$ and $\alpha = 0$, whereby Eq. (2.18.11) yields the expected result $S = 4\pi R^2$.

2.18.4 Coaxial planes: Pappus–Guldin formulae

The purpose is to derive mean value versions of the formulae given in Section 1.7.2 for the volume V of a domain $Y \subset \mathbb{R}^3$.

The intercepts representation of volume

Here it is convenient to include the set Y in a reference circular cylinder D of radius R and height H, with the base centred at O and the axis along the reference axis Ox_3. The joint probability element adopted for the relevant subset (ϕ, x_3) of cylindrical coordinates of a point $x \subset \mathbb{R}^3$ is the normalized version of the corresponding invariant densities, namely,

$$\mathbb{P}(\mathrm{d}\phi, \mathrm{d}x_3) = \frac{\mathrm{d}\phi}{2\pi} \cdot \frac{\mathrm{d}x_3}{H}. \tag{2.18.12}$$

Thus, by Eq. (1.7.17) we have,

$$V = \pi H \cdot \mathbb{E}\left\{\sum_{i=1}^{m} (l_{i+}^2 - l_{i-}^2)\right\}, \tag{2.18.13}$$

where the expectation is with respect to the preceding probability element, and it includes the zeros arising whenever the ray $L_{1+}^2(x_3, \phi)$ misses Y. When $Y \equiv D$, then $m = 1$, $l_{1-} = 0$ and $l_{1+} = R$ for any values of (ϕ, x_3), whereby the preceding formula yields $V = \pi H R^2$, as expected.

The Pappus–Guldin formula

Here it is not necessary to include the set Y in a reference circular cylinder, because the only relevant coordinate is the angle $\phi \sim \mathrm{UR}[0, 2\pi)$. Then Eq. (1.7.19) yields

$$V = 2\pi \cdot \mathbb{E}\{l_+(\phi) \cdot A_+(\phi)\}, \tag{2.18.14}$$

and the expectation also includes the zeros arising when the half-plane L_{2+}^3 misses the set Y. If Y is the cylinder considered in the preceding section, then $Y \cap L_{2+}^3(\phi)$ is a rectangle of base R and height H for any value of ϕ, so that $l_+(\phi) = R/2$, $A = RH$, and the preceding formula again yields $V = \pi R^2 H$, as expected.

2.18.5 Local slabs for particle number. The Horvitz–Thompson twist

The purpose is to elaborate further on the results given in Section 1.7.4.

Isotropic slab around a fixed axis

The probability element of an isotropic slab around a fixed axis is

$$\mathbb{P}(\mathrm{d}\phi) = \frac{\mathrm{d}\phi}{\pi}, \quad \phi \in [0, \pi), \tag{2.18.15}$$

whereby Eq. (1.7.38), with $Q(\phi) \equiv Q(Y \cap L_t(\phi))$, for short, yields

$$\mathbb{E}\{Q(\phi)\} = \frac{1}{\pi} \sum_{i=1}^{N} \alpha_i, \tag{2.18.16}$$

which is not useful because, while $Q(\phi)$ is observable, not all the N distances $\{h_i\}$ are known.

The preceding result is also immediate if, for each point particle y_i, we define the sampling indicator function, namely

$$1_i(\phi) = \begin{cases} 1 \text{ if } y_i \in L_{t+}(\phi), \text{ with probability } \pi_i, \\ 0 \text{ otherwise}, \end{cases} \tag{2.18.17}$$

where

$$\pi_i = \frac{\alpha_i}{\pi} > 0, \quad i = 1, 2, \ldots, N, \tag{2.18.18}$$

is the positive a priori probability that y_i is sampled by the slab, and the angle α_i is given by Eq. (1.7.39). Thus,

$$\mathbb{E}\{Q(\phi)\} = \sum_{i=1}^{N} \mathbb{E}\{1_i(\phi)\} = \sum_{i=1}^{N} \pi_i, \tag{2.18.19}$$

which is Eq. (2.18.16). Instead, the Horvitz–Thompson idea is to evaluate

$$\sum_{i=1}^{N} \mathbb{E}\left(\frac{1_i(\phi)}{\pi_i}\right) = \sum_{i=1}^{N} \frac{\pi_i}{\pi_i} = N, \tag{2.18.20}$$

whereby,

$$\begin{aligned} N &= \pi \mathbb{E} \sum_{i \in \mathcal{S}} \frac{1}{\alpha_i} \\ &\approx \frac{\pi}{t} \mathbb{E} \sum_{i \in \mathcal{S}} h_i \text{ if } t \ll 2h_{(1)}, \end{aligned} \tag{2.18.21}$$

where $h_{(1)} = \min\{h_1, h_2, \ldots, h_N\}$, and

$$\mathcal{S} = \{i \in \{1, 2, \ldots, N\} : \ y_i \in L_t(\phi)\}, \tag{2.18.22}$$

is the subset indexing the sampled particles. Thus, the scenario is drastically changed, because now the $\{h_i\}$ in Eq. (2.18.21) are the distances of the sampled particles only, which can in principle be observed.

Isotropic slab about a fixed point

Here the probability element of the oriented slab $L_t^*(u_2)$ is

$$\mathbb{P}(\mathrm{d}u_2) = \frac{\mathrm{d}u_2}{4\pi}, \quad u_2 \in \mathbb{S}^2, \tag{2.18.23}$$

and the a priori probability that the point particle y_i is sampled by the slab is

$$\pi_i = \frac{\omega}{4\pi} > 0, \tag{2.18.24}$$

where the solid angle ω_i is given by Eq. (1.7.41). Application of the Horvitz–Thompson idea now yields

$$\begin{aligned} N &= 4\pi \mathbb{E} \sum_{i \in \mathcal{S}} \frac{1}{\omega_i} \\ &= \frac{2}{t} \mathbb{E} \sum_{i \in \mathcal{S}} d_i \text{ if } t < 2d_{(1)}, \end{aligned} \tag{2.18.25}$$

where $d_{(1)} = \min\{d_1, d_2, \ldots, d_N\}$, and

$$\mathcal{S} = \{i \in \{1, 2, \ldots, N\} : \; y_i \in L_t^*(u_2)\} \tag{2.18.26}$$

is the subset indexing the sampled particles.

2.18.6 Notes

1. Local stereology

The identities given in the present section are mean values of the ones given in Section 1.7 with respect to isotropic orientations of the local probe. Pertinent references are given in Section 1.7.5.

2. Local slabs for particle number

The slab design given in Section 2.18.5, with Eq. (2.18.21), are due to Evans and Gundersen (1989), whereas the isotropic slab design with Eq. (2.18.25), was proposed by Gundersen et al. (1988). The probability argument is due to Horvitz and Thompson (1952). For a general treatment of local slabs, and further, related references, see Jensen (1998).

2.19 Point-Sampled Intercepts

2.19.1 Representation of particle volume from IUR, and from point-sampled intercepts

Consider first a convex particle $Y \subset \mathbb{R}^3$ of volume V and surface area S. Generate an IUR test line $L_1^3 \equiv L_1^3(z, u)$ hitting Y – so that the pertinent probability element is given by Eq. (2.8.18) – and set $l \equiv l(z, u) \equiv L(Y \cap L_1^3)$. Conditional on the hitting event $L^3 \uparrow Y$, and by virtue of Eq. (1.9.3), we have

$$\mathbb{E}(l^4) = \int l^4(z,u)\mathbb{P}(\mathrm{d}z, \mathrm{d}u|\uparrow) = \frac{12}{\pi} \cdot \frac{V^2}{S}. \tag{2.19.1}$$

On the other hand,

$$\mathbb{E}(l) = \int l(z,u)\mathbb{P}(\mathrm{d}z, \mathrm{d}u|\uparrow) = 4 \cdot \frac{V}{S}, \tag{2.19.2}$$

which is Eq. (2.7.6). Thus,

$$V = \frac{\pi}{3} \cdot \frac{\mathbb{E}(l^1)}{\mathbb{E}(l)}, \tag{2.19.3}$$

the expectations being over IUR test lines hitting the convex set, that is, with respect to the unweighted probability element $\mathbb{P}(\mathrm{d}l)$ of the intercept length l.

Alternatively, generate a UR point x in Y, and then an IR axis $L_1^3(x,u)$ through x, so that the corresponding joint probability element is the normalized version of the invariant product density $\mathrm{d}x\,\mathrm{d}u$, namely,

$$\mathbb{P}(\mathrm{d}x, \mathrm{d}u) = \frac{\mathrm{d}x}{V} \cdot \frac{\mathrm{d}u}{2\pi}, \quad x \in Y,\ u \in \mathbb{S}_+^2, \tag{2.19.4}$$

see Fig. 2.19.1(a). The linear intercept $Y \cap L_1^3(x,u)$ is point-sampled in the sense described in Section 2.3.2, and therefore the extension of Eq. (2.3.6) to $\mathbb{R}^3$ yields

$$\mathbb{P}(\mathrm{d}l|x \uparrow Y) = \frac{l\mathbb{P}(\mathrm{d}l)}{\mathbb{E}(l)}, \tag{2.19.5}$$

where $\mathbb{P}(\mathrm{d}l)$ is the unweighted probability element of intercept length involved in the expectations of Eq. (2.19.3). Thus, analogously as in Eq. (2.3.7),

$$\mathbb{E}(l_0^3) = \frac{\mathbb{E}(l^4)}{\mathbb{E}(l)}, \tag{2.19.6}$$

which combined with Eq. (2.19.3) yields

$$V = \frac{\pi}{3} \cdot \mathbb{E}(l_0^3), \tag{2.19.7}$$

where $l_0 \equiv l(x,u) = L(Y \cap L_1^3(x,u))$ is standard notation in the literature.

If the target set Y is not necessarily convex, then

$$\begin{aligned} V &= \pi\mathbb{E}\{f_0(x,u)\}, \\ f_0(x,u) &= \frac{1}{3}\left\{l_0^3(x,u) + 2\sum_{i=1}^{m(u)} (l_{i+}^3(x,u) - l_{i-}^3(x,u))\right\}, \end{aligned} \tag{2.19.8}$$

where $m(u)$ is the total number of intercept segments which do not contain the point x, on both sides of x. The expectation is with respect to the joint probability element given by Eq. (2.19.4).

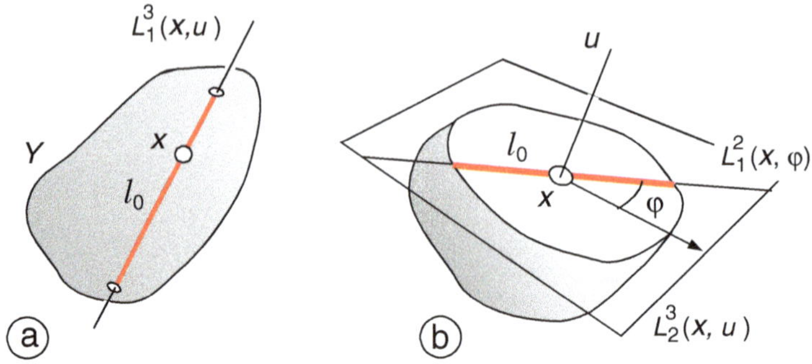

Fig. 2.19.1 (a) Point-sampled intercept satisfying Eq. (2.19.8). (b) Point sampled intercept on a pivotal plane, see Eq. (2.19.9).

Similarly as in the pivotal nucleator, an IR axis $L_1^3(x,u)$, $u \sim \mathrm{UR}(\mathbb{S}_+^2)$, is equivalent to an IR axis $L_1^2(x,\varphi)$, $\varphi \sim \mathrm{UR}[0,\pi)$ within an IR pivotal plane $L_2^3(x,u)$, $u \sim \mathrm{UR}(\mathbb{S}_+^2)$, see Fig. 2.19.1(b). In this case,

$$\begin{aligned} V &= \pi\mathbb{E}\{f_0(\varphi,u)\}, \\ f_0(\varphi,u) &= \frac{1}{3}\left\{l_0^3(\varphi,u) + 2\sum_{i=1}^{m(u)}(l_{i+}^3(\varphi,u) - l_{i-}^3(\varphi,u))\right\}. \end{aligned} \tag{2.19.9}$$

The expectation is with respect to the joint probability element of (x,u,φ), where $x \sim \mathrm{UR}(Y)$, $u \sim \mathrm{UR}(\mathbb{S}_+^2)$, $\varphi \sim \mathrm{UR}[0,\pi)$, and mutually independent.

2.19.2 Case of a population of particles

Consider a population of N particles $Y = \{Y_1, Y_2, \ldots, Y_N\} \subset \mathbb{R}^3$ which for simplicity of exposition are all assumed to be convex. Let V_i, S_i denote the volume and the surface area of the particle Y_i, respectively, and set

$$V(Y) \equiv \sum_{i=1}^{N} V_i = N \cdot \mathbb{E}(V), \quad S(Y) \equiv \sum_{i=1}^{N} S_i = N \cdot \mathbb{E}(S), \tag{2.19.10}$$

the total particle volume and surface area, respectively.

Let $L_1^3 \sim \mathrm{IUR}(Y)$ denote an IUR test line hitting the particle population Y. By an argument similar to that leading to Eq. (2.16.13), and recalling Eq. (2.8.20), we have

$$\mathbb{P}(S_i|L_1^3 \uparrow Y_i) = S_i/S(Y), \quad i = 1,2,\ldots,N. \tag{2.19.11}$$

Further let $l_i = L(Y_i \cap L_1^3)$ denote the linear intercept length corresponding to the particle Y_i. By virtue of Eq. (2.19.1) for a single particle, the population mean of l_i^4

is

$$\begin{aligned}\mathbb{E}(l^4) &= \sum_{i=1}^{N} \mathbb{E}(l_i^4 | L_1^3 \uparrow Y_i) \cdot \mathbb{P}(S_i | L_1^3 \uparrow Y_i) \\ &= \frac{12}{\pi} \sum_{i=1}^{N} \frac{V_i^2}{S_i} \cdot \frac{S_i}{S(Y)} \\ &= \frac{12}{\pi} \cdot \frac{\mathbb{E}(V^2)}{\mathbb{E}(S)}. \end{aligned} \tag{2.19.12}$$

Likewise, recalling Eq. (2.19.2) for a single particle, the population mean of the individual particle intercept length is

$$\mathbb{E}(l) = 4 \cdot \frac{\mathbb{E}(V)}{\mathbb{E}(S)}. \tag{2.19.13}$$

From the preceding two identities, the population volume weighted mean particle volume is

$$\mathbb{E}_V(V) \equiv \frac{\mathbb{E}(V^2)}{\mathbb{E}(V)} = \frac{\pi}{3} \cdot \frac{\mathbb{E}(l^1)}{\mathbb{E}(l)}. \tag{2.19.14}$$

Now let $x \sim \text{UR}(Y)$ be a UR point hitting the particle population, and let $l_{0i} = L(Y_i \cap L_1^3(x, u))$ denote the point-sampled linear intercept length corresponding to the particle Y_i. Recalling Eq. (2.2.3),

$$\mathbb{P}(V_i | x \uparrow Y_i) = V_i / V(Y), \quad i = 1, 2, \ldots, N. \tag{2.19.15}$$

By virtue of Eq. (2.19.7) for a single particle, the population mean of the cubic power of the point-sampled intercept length of a particle becomes

$$\begin{aligned}\mathbb{E}(l_0^3) &= \sum_{i=1}^{N} \mathbb{E}(l_{0i}^3 | x \uparrow Y_i) \cdot \mathbb{P}(V_i | x \uparrow Y_i) \\ &= \frac{3}{\pi} \sum_{i=1}^{N} V_i \cdot \frac{V_i}{V(Y)} \\ &= \frac{3}{\pi} \cdot \frac{\mathbb{E}(V^2)}{\mathbb{E}(V)}. \end{aligned} \tag{2.19.16}$$

Consequently, a representation of the population volume-weighted mean particle volume alternative to Eq. (2.19.13), in terms of the mean of the point-sampled particle intercept length, is

$$\mathbb{E}_V(V) = \frac{\pi}{3} \cdot \mathbb{E}(l_0^3). \tag{2.19.17}$$

Remarks

1. The rhs of Eq. (2.19.13) is based on an IUR test line hitting the particle population Y in 3D. By Eq. (1.2.11), the test line may be generated as an IUR test line through an IUR planar section of Y.
2. In contrast, Eq. (2.19.17) is based on an IR point-sampled test line in 3D, see Fig. 2.19.1(a). The test line may be generated as an IR axis within an IR pivotal plane through a UR test point in Y, see Fig. 2.19.1(b).

2.19.3 Notes

1. Point-sampled intercepts

Early results for a single particle were given by Miles (1979), who also extended them to particle populations (Miles, 1983, 1985). Simplified versions such as Eq. (2.19.8) for a single particle, the corresponding sampling design (Section 2.19.1), and their extension to a population of particles, (Section 2.19.2), were provided by Gundersen and Jensen (1983, 1985) and by Jensen and Gundersen (1985, 1989).

2.20 The Invariator

Consider a domain $Y \subset D \subset \mathbb{R}^3$ of volume V, with piecewise smooth boundary ∂Y of area S, enclosed in a reference ball D centred at a pivotal point O which does not need to be included in Y. Further, let $H = 2R$ and $a = \pi R^2$ denote the diameter and the equatorial area of D, respectively. The background results are in Section 1.8. Each of the Sections 2.20.1 and 2.20.2 applies to S and V simultaneously, whereas Sections 2.20.3 and 2.20.4 apply to S only. The a priori and a posteriori weighting concepts were introduced in Section 2.3.3.

The invariator is a two-stage design. In the first stage a pivotal plane, namely an IR plane $L^3_{2[0]} \equiv L^3_2(0, u)$ is generated through O with $u \sim \mathrm{UR}(\mathbb{S}^2_+)$. Thus, the first stage is common to local stereology – the second stage, however, is usually not.

2.20.1 A posteriori weighted test lines for volume and surface area

Generate a test line $L^2_1 \equiv L^2_1(r, \omega) \sim \mathrm{IUR}(D_2(u))$, $r > 0$, $\omega \in [0, 2\pi)$, hitting the equator $D_2(u) = D \cap L^3_1(0, u)$ determined in D by the pivotal plane. Thus, $\mathrm{d}L^2_1 = \mathrm{d}r\, \mathrm{d}\omega$. The adopted joint probability element is

$$\mathbb{P}(\mathrm{d}r, \mathrm{d}\omega, \mathrm{d}u) = \frac{\mathrm{d}r}{R} \cdot \frac{\mathrm{d}\omega}{2\pi} \cdot \frac{\mathrm{d}u}{2\pi}. \tag{2.20.1}$$

By virtue of Eq. (1.8.1) and Eq. (1.8.2), we obtain

$$S = 2\pi H \cdot \mathbb{E}\{r \cdot I((\partial Y \cap L^3_{2[0]}) \cap L^2_1)\}, \tag{2.20.2}$$

$$V = \pi H \cdot \mathbb{E}\{r \cdot L((Y \cap L^3_{2[0]}) \cap L^2_1)\}, \tag{2.20.3}$$

the expectations being with respect to the preceding probability element. As usual with single probes, the mentioned expectations include zeros arising (i) when the pivotal plane misses the set Y, in which case an independent IR pivotal plane has to be generated, and (ii) when the test line misses the pivotal section, in which case an independent IR pivotal plane has to be generated as well, etc.

2.20.2 A priori weighted test lines for volume and surface area

Generate a test point $z \equiv z(r, \omega) \sim \mathrm{UR}(D_2(u))$ of polar coordinates (r, ω), $r > 0$, $\omega \in [0, 2\pi)$, so that the area element at z is $\mathrm{d}z = r\,\mathrm{d}r\,\mathrm{d}\omega$. Now, draw a test line $L^2_{1(z)} \subset L^3_{2[0]}$ through z and perpendicular to the vector Oz. The invariant density $\mathrm{d}L^2_{1(z)}$ of this line is that of the point z, namely the area element at z, see Eq. (1.2.22). The adopted joint probability element is

$$\mathbb{P}(\mathrm{d}z, \mathrm{d}u) = \frac{\mathrm{d}z}{a} \cdot \frac{\mathrm{d}u}{2\pi}. \tag{2.20.4}$$

By virtue of Eq. (1.8.4) and Eq. (1.8.5), we obtain

$$S = 2a \cdot \mathbb{E}\{I((\partial Y \cap L^3_{2[0]}) \cap L^2_{1(z)})\}, \tag{2.20.5}$$

$$V = a \cdot \mathbb{E}\{L((Y \cap L^3_{2[0]}) \cap L^2_{1(z)})\}. \tag{2.20.6}$$

Analogous remarks as in the preceding section have to be observed concerning the inclusion of zeros in the preceding expectations.

2.20.3 The flower formula for the surface area of a convex body

Here we assume that the set $Y \subset D \subset \mathbb{R}^2$ is convex, but it does not need to contain the pivotal point O. From Eq. (1.8.8),

$$S = 4\mathbb{E}\left\{A\left(H_{Y\cap L^3_2(0,u)}\right)\right\}, \tag{2.20.7}$$

where $H_{Y\cap L^3_2(0,u)}$ denotes the flower of a pivotal section of Y. Again, the preceding expectation includes the zeros arising when the pivotal plane misses Y.

Suppose that $O \in Y^\circ$, and let $h(\omega, u)$ denote the radius vector of the flower of a pivotal section $Y \cap L_2^3(0, u)$. Then, recalling Eq. (1.8.9),

$$S = 4\pi\mathbb{E}\{h^2(\omega, u)\}, \tag{2.20.8}$$

the expectation being with respect to the joint probability element

$$\mathbb{P}(\mathrm{d}\omega, \mathrm{d}u) = \frac{\mathrm{d}\omega}{2\pi} \cdot \frac{\mathrm{d}u}{2\pi}. \tag{2.20.9}$$

As a cross-check, Eq. (2.20.8) holds when Y is a sphere centred at O.

Case of a triaxial ellipsoid

If the set Y is a triaxial ellipsoid and $O \in Y^\circ$, then the pivotal section is an ellipse whose principal semiaxes $0 < m(u) \le M(u)$, say, are random variables, as well as the distance $r(u)$ of the centre of the pivotal ellipse from O. Then, by Eq. (1.8.20),

$$S = 2\pi\mathbb{E}\{M^2(u) + m^2(u) + r^2(u)\}. \tag{2.20.10}$$

Again, note that the preceding identity holds for a sphere centred at O.

2.20.4 The peak-and-valley formula for the surface area of an arbitrary set

If the set Y is not convex then, by virtue of Eq. (1.8.23), a generalization of Eq. (2.20.8) is possible. It is moreover not necessary that the pivotal point O lies in the interior of Y.

In a pivotal plane $L_2^3(0, u)$, generate an IR axis $L_1^2(0, \omega)$, $\omega \in [0, 2\pi)$. If the pivotal trace curve $C_u \equiv \partial Y \cap L_2^3(0, u)$ has points above this axis, then sweep the curve with a mobile test line $L_1^2(r, \omega)$ from $r = H/2$ down to $r = 0$, and for each critical point encountered in the trace curve C_u, score: (i) the index $\epsilon_k = +1$ if the kth critical point is a peak, and $\epsilon_k = -1$ if it is a valley, and (ii) the corresponding distance h_k of the critical point from the axis $L_1^2(0, \omega)$. Then,

$$S = 4\pi\mathbb{E}\sum_{k=1}^{m} \epsilon_k h_k^2, \tag{2.20.11}$$

where m is the number of critical points encountered in the left half-plane, say, determined by the axis $L_1^2(0, \omega)$. The expectation is with respect to the joint probability element of (ω, u), which formally coincides with Eq. (2.20.9). As usual, if $O \notin Y^\circ$ then zeros have to be included in the expectation whenever the pivotal plane misses Y.

If the set Y is convex and $O \in Y^\circ$, then $m = 1$ and $\epsilon_1 = +1$ almost surely for all (ω, u), and Eq. (2.20.11) reduces to Eq. (2.20.8).

2.20.5 Note

The results given in the present section are mean value versions of the ones derived in Section 1.8. Pertinent notes and references are given in Section 1.8.9.

2.21 Blaschke–Petkantschin Formulae for Higher Moments of Particle Volume

2.21.1 Single particle sampled with a pivotal plane

Consider a domain $Y \subset \mathbb{R}^3$ of volume V containing a pivotal point O and a pivotal plane $L^3_{2[0]} \equiv L^3_2(0, u)$. In the pivotal transect $Y \cap L^3_2(0, u)$ (which may consist of several separate profiles if Y is not convex), generate two independent UR points (z_1, z_2), and let $\nabla_{2[0]} \equiv \nabla_{2[0]}(z_1, z_2, u)$ represent the area of the triangle of vertices (O, z_1, z_2), see Fig. 2.21.1(a). Further, let $a_0 \equiv a_0(u)$ denote the pivotal transect area. By Eq. (1.9.4),

$$V^2 = 4\pi \mathbb{E}(\nabla_{2[0]} \cdot a_0^2), \tag{2.21.1}$$

with $\mathbb{P}(\mathrm{d}z_1, \mathrm{d}z_2, \mathrm{d}u) = \mathrm{d}z_1 \, \mathrm{d}z_2 \, \mathrm{d}u/(2\pi a_0^2)$.

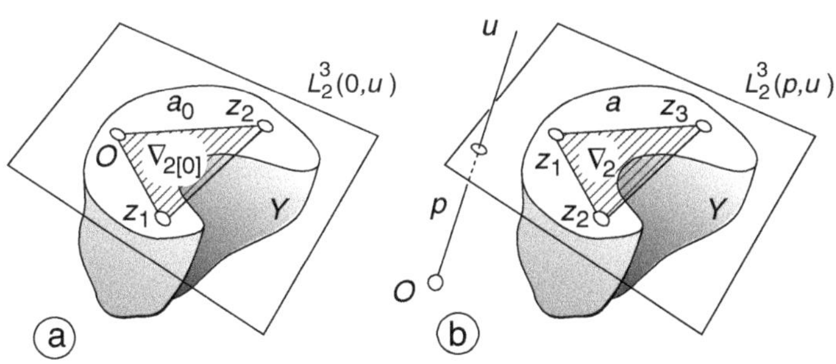

Fig. 2.21.1 (a) Geometric elements involved in Eq. (2.21.1). (b) Idem in Eq. (2.21.2).

2.21.2 Single particle sampled with an IUR plane

Instead of a pivotal plane, consider now an IUR plane $L_2^3(p,u)$ hitting the domain Y, whereby $\mathbb{P}(\mathrm{d}p, \mathrm{d}u)$ is given by Eq. (2.8.4). In the transect $Y \cap L_2^3(p,u)$, generate three independent UR points (z_1, z_2, z_3), and let $a = a(u)$, $\nabla_2 \equiv \nabla_2(z_1, z_2, z_3, u)$, denote the areas of the transect and of the triangle of vertices (z_1, z_2, z_3), respectively, see Fig. 2.21.1(b). Then, by Eq. (1.9.5),

$$V^3 = 4\pi \mathbb{E}(H(u))\mathbb{E}(\nabla_2 \cdot a^3), \tag{2.21.2}$$

with $\mathbb{P}(\mathrm{d}z_1, \mathrm{d}z_2, \mathrm{d}z_3, \mathrm{d}u) = \mathrm{d}z_1\, \mathrm{d}z_2\, \mathrm{d}z_3\, \mathrm{d}u/(2\pi a^3)$.

The expectations in Eq. (2.21.1) and Eq. (2.21.2) include no zeros, because the UR points involved are forced to hit each transect.

2.21.3 Case of a population of particles

Consider a population $Y = \{Y_1, Y_2, \ldots, Y_N\} \subset \mathbb{R}^3$ of N arbitrary particles. Let V_i, $H_i \equiv \mathbb{E}_u(H_i(u))$ denote the volume and the mean caliper length of the particle Y_i, respectively, and set

$$V(Y) \equiv \sum_{i=1}^{N} V_i = N \cdot \mathbb{E}(V), \quad H(Y) \equiv \sum_{i=1}^{N} H_i = N \cdot \mathbb{E}(H). \tag{2.21.3}$$

Consider also a test plane $L_2^3 \sim \mathrm{IUR}(Y)$. From Eq. (2.21.2) for a single particle, the population version of the expectation is

$$\begin{aligned} \mathbb{E}(\nabla_2 \cdot a^3) &= \sum_{i=1}^{N} \mathbb{E}(\nabla_{2i} \cdot a_i^3 | L_2^3 \uparrow Y_i) \cdot \mathbb{P}(H_i | L_2^3 \uparrow Y_i) \\ &= \frac{1}{4\pi} \cdot \sum_{i=1}^{N} \frac{V_i^3}{H_i} \cdot \frac{H_i}{H(Y)} \\ &= \frac{1}{4\pi} \cdot \frac{\mathbb{E}(V^3)}{\mathbb{E}(H)}. \end{aligned} \tag{2.21.4}$$

On the other hand, by Eq. (2.16.24),

$$\mathbb{E}(a) = \frac{\mathbb{E}(V)}{\mathbb{E}(H)}, \tag{2.21.5}$$

whereby the second moment of the volume-weighted particle volume from plane-sampled particles may be expressed as follows,

$$\mathbb{E}_V(V^2) \equiv \frac{\mathbb{E}(V^3)}{\mathbb{E}(V)} = 4\pi \cdot \frac{\mathbb{E}(\nabla_2 \cdot a^3)}{\mathbb{E}(a)}. \qquad (2.21.6)$$

Now suppose that $x \sim \mathrm{UR}(Y)$ is a UR point hitting the particle population Y. Similarly as in Eq. (2.19.15), if $x \uparrow Y$, then x will sample the particle Y_i with a probability proportional to V_i. Moreover, regarding the point x as the pivotal point in Fig. 2.21.1(a), Eq. (2.21.1) holds for the particle Y_i. Therefore, for the particle population,

$$\begin{aligned}\mathbb{E}(\nabla_{2[0]} \cdot a_0^2) &= \sum_{i=1}^{N} \mathbb{E}(\nabla_{2[0]i} \cdot a_{0i}^2 | x \uparrow Y_i) \cdot \mathbb{P}(V_i | x \uparrow Y_i) \\ &= \frac{1}{4\pi} \cdot \sum_{i=1}^{N} V_i^2 \cdot \frac{V_i}{V(Y)} \\ &= \frac{1}{4\pi} \cdot \frac{\mathbb{E}(V^3)}{\mathbb{E}(V)}, \end{aligned} \qquad (2.21.7)$$

whereby the alternative to Eq. (2.21.6) for point-sampled particles is

$$\mathbb{E}_V(V^2) \equiv \frac{\mathbb{E}(V^3)}{\mathbb{E}(V)} = 4\pi \cdot \mathbb{E}(\nabla_{2[0]} \cdot a_0^2). \qquad (2.21.8)$$

Remarks

1. When the basic probe is an IUR plane, the particle transects involved in the computation of the rhs of Eq. (2.21.6) are selected in the sectioning plane with equal probabilities. Then, three UR points are selected in each sampled transect.
2. However, to compute the rhs of Eq. (2.21.8) only point-sampled transects are considered. Again, the point $x \sim \mathrm{UR}(Y)$ may be selected in an IUR plane transect by the rejection method. Then, two additional UR points are selected in the corresponding point-sampled particle transect.

2.21.4 Notes

1. Higher moments of particle volume

The results in Section 2.21 are due to Jensen and Gundersen (1983, 1985), Gundersen and Jensen (1985), Miles (1985) and Jensen (1987, 1998). Further practical details are given in Gundersen (1986), Jensen and Sørensen (1991) and Sørensen (1991).

2.22 Isotropic Projections

2.22.1 Total isotropic projections onto an axis in 2D

Let $Y \subset \mathbb{R}^2$ represent a piecewise smooth curve of length $B > 0$. From Eq. (1.19.7) we obtain Cauchy's projection formula for curves, namely,

$$B = \frac{\pi}{2} \cdot \mathbb{E}\{l(\phi)\}, \tag{2.22.1}$$

where $l(\phi)$, $\phi \in [0, \pi)$, denotes the total orthogonal projected length (multiplicities included) of the curve onto an axis $L^2(0, \phi)$. The expectation is over isotropic orientations, namely with respect to $\mathbb{P}(\mathrm{d}\phi) = \mathrm{d}\phi/\pi$.

If the curve is the boundary of a convex set, then $l(\phi) = 2H(\phi)$ for all ϕ, and therefore

$$B = \pi \cdot \mathbb{E}\{H(\phi)\}, \tag{2.22.2}$$

which is Eq. (2.4.28).

2.22.2 Total isotropic projection of a curve onto a plane

Analogously, for a piecewise smooth curve $Y \subset \mathbb{R}^3$ of length $L > 0$, by Eq. (1.19.22), Cauchy's projection formula reads

$$L = \frac{4}{\pi} \cdot \mathbb{E}\{l(u)\}, \tag{2.22.3}$$

where $l(u)$, $u \equiv u(\phi, \theta) \in \mathbb{S}^2_+$ denotes the length of the total orthogonal projection $Y'(u)$ of Y onto a plane $L^3_2(0, u)$. The expectation is with respect to the UR probability element $\mathbb{P}(\mathrm{d}u) = \mathrm{d}u/(2\pi)$.

2.22.3 Orthogonal projection of an IUR slab hitting a curve, onto a plane parallel to the slab

Enclose the curve Y in a reference ball whose diameter is the segment $[0, H)$, $H > 0$. The sampling procedure consists of the following two steps.

1. Generate an IUR slab $L_t(p, u)$ of thickness $t > 0$ hitting the ball. If the slab misses Y, then score a zero and generate an independent IUR slab.
2. Project the curve portion contained in the slab orthogonally onto a plane parallel to the slab. Let $(Y \cap L_t(p, u))'$ denote the corresponding curve projection.

The joint probability element associated with the slab is

$$\mathbb{P}(\mathrm{d}p, \mathrm{d}u) = \frac{\mathrm{d}p}{H+t} \cdot \frac{\mathrm{d}u}{2\pi}, \quad p \in [-t, H), \; u \in \mathbb{S}_+^2, \tag{2.22.4}$$

which applied to Eq. (1.19.24) yields

$$L = \frac{4}{\pi} \cdot \frac{H+t}{t} \cdot \mathbb{E}\{L(Y \cap L_t(p,u))'\}. \tag{2.22.5}$$

The expectation includes the zeros arising when the slab misses the curve. Combining the preceding identity with Eq. (2.10.7) yields

$$\frac{L}{V} = \frac{4}{\pi} \cdot \frac{1}{t} \cdot \frac{\mathbb{E}\{L(Y \cap L_t(p,u))'\}}{\mathbb{E}\{A_t(p)\}}, \text{ or } L_V = \frac{4}{\pi} \cdot \frac{1}{t} \cdot L_A'. \tag{2.22.6}$$

2.22.4 Total isotropic projections of a set onto a plane

In general, projection measures do not supply information about area in the plane, nor volume in space. Here we derive some results for surface area, and for the integral of mean curvature of a convex set.

Consider a surface $Y \subset \mathbb{R}^3$ of area S, and let $a(u)$ denote the total orthogonal projected area of Y onto an isotropic projection plane $L_2^3(0, u)$. From Eq. (1.19.8) we have

$$S = 2\mathbb{E}\{a(u)\}. \tag{2.22.7}$$

If Y represents the boundary of a convex set, then

$$S = 4\mathbb{E}\{A(u)\}, \tag{2.22.8}$$

where $A(u)$ is the projected or 'shadow' area of Y. Note the formal analogy between the preceding identity and the flower formula Eq. (2.20.7).

Consider a convex set $Y \subset \mathbb{R}^3$ with surface area S and integral of mean curvature M. Let $B(u)$, $H(u)$ denote the perimeter and the caliper lengths, respectively, of the orthogonal isotropic projection of Y. Then, Minkowski's theorem (Eq. (1.19.13) and Eq. (1.19.14)) yields

$$M = 2\mathbb{E}\{B(u)\} = 2\pi\mathbb{E}\{H(u)\}. \tag{2.22.9}$$

2.22.5 Notes

1. Isotropic projections of a curve

Cauchy's Eq. (2.22.1) was used by Steinhaus (1930) and Moran (1966), see also Cruz-Orive (1989b) and Gómez, Cruz, and Cruz-Orive (2016).

As hinted in Section 1.19.2, the isotropic projection design may be regarded as the first stage of the intersection design with an IUR test line. Enclose a curve $Y \subset \mathbb{R}^2$ in a disk of diameter H centred at the origin. For each orientation $\phi \in [0, \pi)$ of the projection axis, let $I(p, \phi)$ denote the number of intersections determined in Y by a test line normal to the projection axis at an abscissa $p \in [-H/2, H/2]$. Then,

$$l(\phi) = \int_{-H/2}^{H/2} I(p, \phi)\, \mathrm{d}p = H \cdot \mathbb{E}\{I(p, \phi)|\phi\}, \tag{2.22.10}$$

the expectation being with respect to $\mathbb{P}(\mathrm{d}p) = \mathrm{d}p/H$. Substitution into the rhs of Eq. (2.22.1) recovers Eq. (2.4.6).

Similarly, enclose a curve $Y \subset \mathbb{R}^3$ in a ball of diameter H centred at the origin. For each orientation $u \in \mathbb{S}_+^2$ of the projection plane, the number Q of intersections determined in Y by a test plane normal to the projection plane at a distance $p \in [-H/2, H/2]$ from the origin, is the same as the number I of intersections determined in $Y'(u)$ by an IUR test line in the projection plane. By Eq. (2.4.6), with $B.\, I$ replaced with $l(u)$, Q, respectively,

$$l(u) = \frac{\pi}{2} \cdot H \cdot \mathbb{E}(Q|u), \tag{2.22.11}$$

which, substituted into the rhs of Eq. (2.22.3), returns Eq. (2.6.4).

2. Isotropic projections of a set

In general, set projections may provide information on the volume of the set only in special cases, see e.g. Wulfsohn, Gundersen, Jensen, and Nyengaard (2004). Cauchy's Eq. (2.22.8) for surface area, valid only for convex bodies, was studied by Moran (1944). Again, the more general Eq. (2.22.7) is closely related with Eq. (2.7.2) for an IUR test line.

2.23 Vertical Projections

2.23.1 Total vertical projections of a curve

The functional method

Consider a piecewise smooth curve $Y \subset \mathbb{R}^3$ of length $L > 0$ and a projection plane $L^3_{2[1]}(\phi)$ with isotropic orientation $\phi \sim \mathrm{UR}[0, \pi)$ around a fixed vertical axis (VA). By Eq. (1.19.28) we have,

$$L = \mathbb{E}\{Z(Y'(\phi))\}, \tag{2.23.1}$$

where $Y'(\phi)$ denotes the total vertical projection of the curve, and the functional $Z(\cdot)$ is given by Eq. (1.19.29).

The test cycloid method

The functional $Z(\cdot)$ may be replaced with the mean number of intersections between $Y'(\phi)$ and a FUR test curve $T_1^2(z)$ in the vertical projection plane, consisting of two cycloid arcs of total length l, with the major principal axis parallel to the vertical axis. To define the joint probability element of the associated point x of the test curve it is convenient to enclose the curve Y in a ball D. In this way, the vertical projection $Y'(\phi)$ will always be contained in the projection $D_2'(\phi)$ of the equatorial disk $D_2(\phi)$ parallel to the vertical projection plane. Set $D_{2\oplus}'(\phi) \equiv D_2'(\phi) \oplus \check{T}_1^2(0)$, whose area $A(D_{2\oplus}')$, say, does not depend on ϕ. Then,

$$\mathbb{P}(\mathrm{d}\phi, \mathrm{d}z) = \frac{\mathrm{d}x}{A(D_{2\oplus}')} \cdot \frac{\mathrm{d}\phi}{\pi}, \quad x \in D_{2\oplus}'(\phi), \ \phi \in [0, \pi), \tag{2.23.2}$$

and using Eq. (1.19.30),

$$L = 2 \cdot \frac{A(D_{2\oplus}')}{l} \cdot \mathbb{E}\{I(Y'(\phi) \cap T_1^2(z))\}. \tag{2.23.3}$$

The expectation includes the zeros arising when the test cycloid misses the total vertical projection of the curve, in which case a new vertical projection with $\phi \sim \mathrm{UR}[0, \pi)$ has to be generated. The preceding result is a projection counterpart of Eq. (2.14.4).

2.23.2 Vertical slice projections of a curve

Enclose the curve Y of length $L > 0$ into a ball of diameter H.

The functional method

The sampling procedure consists of the following two steps.

1. Generate a vertical slab $L_t \equiv L_t(p, \phi)$ of thickness $t > 0$ hitting the ball. If the slab misses the curve, then score a zero and generate an independent vertical slab, otherwise go to the next step.
2. Project the curve portion contained in the slab orthogonally onto a vertical plane parallel to the slab. Let $(Y \cap L_t(p, \phi))'$ denote the corresponding vertical curve projection.

The joint probability element associated with the slab is

$$\mathbb{P}(\mathrm{d}p, \mathrm{d}\phi) = \frac{\mathrm{d}p}{H+t} \cdot \frac{\mathrm{d}\phi}{\pi}, \quad p \in [-t, H),\ \phi \in [0, \pi), \tag{2.23.4}$$

which applied to Eq. (1.19.31) yields

$$L = \frac{H+t}{t} \cdot \mathbb{E}\{Z((Y \cap L_t(p, \phi))')\}. \tag{2.23.5}$$

The expectation includes the aforementioned zeros. Combining the preceding identity with Eq. (2.10.7) yields

$$\frac{L}{V} = \frac{1}{t} \cdot \frac{\mathbb{E}\{Z((Y \cap L_t(p, \phi))')\}}{\mathbb{E}\{A_t(p)\}}, \text{ or } L_V = \frac{1}{t} \cdot Z_A. \tag{2.23.6}$$

The test cycloid method

The design is analogous to that adopted in Section 2.14, with the only difference that now a vertical slice projection $(Y \cap L_t(p, \phi))'$, parallel to the slab, is used instead of a vertical plane section. The pertinent joint probability element is therefore similar to that of Eq. (2.14.3), namely,

$$\mathbb{P}(\mathrm{d}x, \mathrm{d}p, \mathrm{d}\phi) = \frac{\mathrm{d}x}{A(D'_{2\oplus})} \cdot \frac{\mathrm{d}p}{H+t} \cdot \frac{\mathrm{d}\phi}{\pi},$$
$$x \in D'_{2\oplus}(\phi),\ p \in [-t - H/2, H/2],\ \phi \in [0, \pi), \tag{2.23.7}$$

where $D'_{2\oplus}(\phi)$ is defined as in Eq. (2.23.2). The expectation of the rhs of Eq. (1.19.32) with respect to the preceding probability element yields

$$L = 2 \cdot \frac{A(D'_{2\oplus})}{l} \cdot \frac{H+t}{t} \cdot \mathbb{E}(I'), \text{ or } L_V = \frac{2}{t} \cdot I'_L, \tag{2.23.8}$$

where $I' \equiv I((Y \cap L_t(p, \phi))' \cap T_1^2(x))$.

2.23.3 Integral of mean curvature of a convex set from total vertical projections

The design is the same as that for the vertical projections of a curve (Section 2.23.1). The object of interest is a convex set $Y \subset \mathbb{R}^3$ with piecewise smooth boundary, and the target quantity is its integral of mean curvature M. Now $Y'(\phi)$ denotes a closed convex curve, namely the boundary of the orthogonal projection of the set Y onto the projection plane $L^3_{2[1]}(\phi)$.

The functional method

Eq. (1.19.33) directly yields

$$M = \frac{\pi}{2} \cdot \mathbb{E}\{Z(Y'(\phi))\}. \tag{2.23.9}$$

For instance, if Y is a ball of diameter H, then $Y'(\phi)$ is a great circle and, by Eq. (1.19.35), $Z(Y'(\phi)) = 4H$, whereby Eq. (2.23.9) yields $M = 2\pi H$, as expected.

The test cycloid method

With the same setup used for a curve (Section 2.23.1), Eq. (1.19.34) yields

$$M = \pi \cdot \frac{A(D'_{2\oplus})}{l} \cdot \mathbb{E}\{I(Y'(\phi) \cap T_1^2(x))\}. \tag{2.23.10}$$

The expectation is with respect to the probability element given by Eq. (2.23.2), and it includes the zeros arising when the test cycloid misses $Y'(\phi)$, in which case a new total vertical projection has to be generated with $\phi \sim \mathrm{UR}[0, \pi)$.

2.23.4 Note

For references pertaining to vertical projections of curves, see Note 3 in Section 1.19.7.

2.24 Test Systems: Preliminary Comments

The sampling designs hitherto developed in this chapter are based on probability elements requiring a bounded reference set D containing the target set Y, see Eq. (2.1.2). For instance, for a UR test point, or for a FUR unbounded probe, a suitable choice of D is a bounded segment on an axis, or generally a hypercube. For an IUR unbounded probe, D is a disk, or a ball. The relevant mean values are over independent realizations of a single FUR, IUR, or vertical probe hitting D. As indicated in Section 2.1.3, such designs are intended to introduce the main concepts in a general setting, which may also be of interest in computer science.

For practical stereology requiring invasive sampling (e.g. physical sectioning), however, independent sampling is clearly precluded. For bounded probes, an additional snag is that the pertinent probability elements involve the hitting measure $h_2(D)$, see Eq. (2.1.7), which in turn depends on the Minkowski sums $D_\oplus(u_d)$, or $Y_\oplus(u_d)$, see Sections 2.11–2.14.

Apart from convenience, efficiency aspects usually favour systematic over independent designs. The pertinent tool for systematic sampling is a test system, whose mathematical concept was introduced in Section 1.21. The corresponding mean values, which are prerequisites for stereological inference, are given below by incorporating probability elements which render the mathematics simple, and the designs easy to implement. The stereological equations for ratios remain formally unaffected.

Without affecting their comprehension, the introductory Sections 2.25.1 and 2.26.1 are formulated for $\mathbb{R}^d$, but the rest concentrate on $d = 1, 2, 3$.

2.25 FUR Test Systems

2.25.1 Introduction

FUR test systems are adequate if at least one among the target set $Y \subset \mathbb{R}^d$, and the test system, are of full dimension (Section 1.3). For a FUR test system with a bounded fundamental test probe $T_0 \subset J_0 \subset \mathbb{R}^d$, Santaló's formula, Eq. (1.21.8), reduces to

$$\gamma(Y)\nu(T_0) = \int_{J_0} \alpha(Y \cap \Lambda_x)\,\mathrm{d}x, \tag{2.25.1}$$

because orientation is not involved. Here $\gamma(Y)$ is the target measure, whereas $\nu(T_0)$ is the known measure of the fundamental probe. Based on the fact that the associated point x of the test system Λ_x is equipped with the translation-invariant density $\mathrm{d}x$, $x \in J_0$, we adopt a UR probability element for x, namely,

$$\mathbb{P}(\mathrm{d}x) = \frac{\mathrm{d}x}{V(J_0)}, \quad x \in J_0, \tag{2.25.2}$$

where the normalizing constant $V(J_0) > 0$ represents the known volume of the fundamental tile J_0. From the preceding two identities we obtain the basic one for FUR test systems,

$$\gamma(Y) = \frac{V(J_0)}{\nu(T_0)} \cdot \mathbb{E}\{\alpha(Y \cap \Lambda_x)\}, \tag{2.25.3}$$

where $\mathbb{E}\{\alpha(Y \cap \Lambda_x)\}$ is the mean value, with respect to $\mathbb{P}(\mathrm{d}x)$, of the contents of the intersection between the target set and the union of all the probes of the test system. Instead of $V(J_0)$, the analogous identity for a single test probe involves the hitting measure $h_2(D)$, see for instance Eq. (2.1.6) with $c_2 = 1$.

For a FUR test system of unbounded test probes, the corresponding Santaló formula, Eq. (1.21.14), becomes

$$\gamma(Y) = \int_{J_0} \alpha(Y \cap \Lambda_z)\,\mathrm{d}z. \tag{2.25.4}$$

If the fundamental probe is a FUR r-plane L_r^d, then the auxiliary test system of points – at which the r-plane translates are placed in order to construct the test system – is contained in the orthogonal complement $L_{d-r[0]}^d$ of L_r^d. We adopt a UR probability element for the associated point z of the test system in this subspace, whereby the basic identity for FUR test systems of unbounded probes reads

$$\gamma(Y) = V(J_0) \cdot \mathbb{E}\{\alpha(Y \cap \Lambda_z)\}, \tag{2.25.5}$$

where $V(J_0) > 0$ is the volume of the $(d-r)$-dimensional fundamental tile J_0. The alternative for a single unbounded test probe is Eq. (2.1.5) with $c_1 = 1$.

The quantity $\alpha(\cdot)$ is a finite random variable called the measurement function.

2.25.2 FUR test systems of points

In practice, test systems are often used on planar sections, see Fig. 2.25.1(b). For a target domain $Y \subset \mathbb{R}^2$ of area $A > 0$ and a FUR test system of points with a fundamental tile of area $a > 0$ and a fundamental probe consisting of a finite set of $p \geq 1$ test points (see Fig. 2.25.1(b) with $p = 1$), the basic Eq. (2.25.3) yields

$$A = \frac{a}{p} \cdot \mathbb{E}\{P(Y \cap \Lambda_x)\}. \tag{2.25.6}$$

Similarly, for a domain $Y \subset \mathbb{R}^3$ of volume $V > 0$, the adopted FUR test system is a spatial grid of test points with fundamental tile volume v, and the corresponding identity follows from the preceding one by replacing the symbols A, a with V, v respectively, and similarly for any spatial dimension, see Eq. (2.29.1).

For the simple case where $Y \subset \mathbb{R}$ is a bounded and finite union of segments of total length $L > 0$ on a sampling axis, the test system may consist of test points a constant distance $T > 0$ apart, namely,

$$\Lambda_z = \{z + kT,\ k \in \mathbb{Z}\}, \quad z \sim \mathrm{UR}[0, T), \tag{2.25.7}$$

and Eq. (2.25.5) yields

$$L = T \cdot \mathbb{E}\{P(Y \cap \Lambda_z)\}. \tag{2.25.8}$$

The distance T is called the sampling period.

2.25.3 FUR Cavalieri design

Cavalieri planes

Here the target is usually the volume $V > 0$ of a domain $Y \subset \mathbb{R}^3$. The fundamental probe is a plane normal to an axis, called the sampling axis, with an arbitrarily fixed orientation $u \in \mathbb{S}_+^2$, see Fig. 2.25.1(a). On this axis we define a UR test system of points with fundamental tile $J_0 = [0, T)$, see Eq. (2.25.7). The corresponding Cavalieri test system consists of a series of FUR planes a constant distance T apart, namely

$$\Lambda_z = \{L_2^3(z + kT, u),\ k \in \mathbb{Z}\}, \quad z \sim \text{UR}[0, T), \tag{2.25.9}$$

and Eq. (2.25.5) yields the Cavalieri identity

$$V = T \cdot \mathbb{E}\{A(Y \cap \Lambda_z)\}. \tag{2.25.10}$$

It is didactic to derive the preceding identity from first principles. Recall that $\mathbb{P}(\mathrm{d}z) = \mathrm{d}z/T$, $z \in [0, T)$, and set $A(p) \equiv A(Y \cap L_2^3(p, u))$ in Eq. (1.4.5). Then,

$$\begin{aligned} V &= \int_{\mathbb{R}} A(p)\,\mathrm{d}p \\ &= \sum_{k\in\mathbb{Z}} \int_{kT}^{(k+1)T} A(p)\,\mathrm{d}p \\ &= \sum_{k\in\mathbb{Z}} \int_0^T A(z + kT)\,\mathrm{d}z \\ &= T \int_0^T \sum_{k\in\mathbb{Z}} A(z + kT)\mathbb{P}(\mathrm{d}z), \end{aligned} \tag{2.25.11}$$

which is Eq. (2.25.10). For parallel test lines in the plane, or in general for parallel hyperplanes normal to a fixed sampling axis, the Cavalieri identity is formally the same with the pertinent measurement function. This sampling design is related to the Cavalieri principle, see Note 3 in Section 1.4.4, hence its name.

The number $Q(Y \cap \Lambda_z)$ of Cavalieri planes hitting the set Y is generally random. By Eq. (2.25.8),

$$\mathbb{E}\{Q(Y \cap \Lambda_z)\} = \frac{H(u)}{T}, \tag{2.25.12}$$

where $H(u)$ is the total length of the orthogonal projection of Y onto the sampling axis. Therefore, the mean individual area $\mathbb{E}(A)$ of a planar Cavalieri transect is

$$\mathbb{E}(A) = \frac{\mathbb{E}\{A(Y \cap \Lambda_z)\}}{\mathbb{E}\{Q(Y \cap \Lambda_z)\}} = \frac{V}{H(u)}, \tag{2.25.13}$$

which coincides with Eq. (2.8.7) for a single FUR plane probe hitting the domain Y.

Cavalieri planes subsampled with test points

A further representation of V, in terms of point counts, is obtained by using a FUR test system Λ_x of points with fundamental tile J_0, (of area a, containing p test points), on each Cavalieri transect $Y \cap \Lambda_z$, and applying Eq. (2.25.6) to the rhs of Eq. (2.25.10), that is,

$$V = T \cdot \frac{a}{p} \cdot \mathbb{E}\{P(Y \cap \Lambda_z \cap \Lambda_x)\}, \tag{2.25.14}$$

where $z \sim \mathrm{UR}[0, T)$ and, independently, $x \sim \mathrm{UR}(J_0)$ for each $z \in [0, T)$.

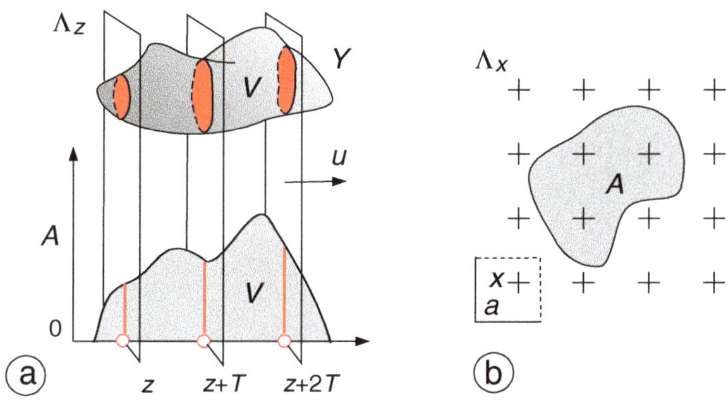

Fig. 2.25.1 (a) Test system Λ_z of Cavalieri planes, see Eq. (2.25.9), hitting a domain Y. (b) A portion of a test system Λ_x of points hitting a planar domain, e.g. a planar Cavalieri transect. A = transect area.

2.25.4 The FUR fakir probe

Again, the usual target is the volume $V > 0$ of a domain $Y \subset \mathbb{R}^3$. The fundamental probe is a FUR straight line normal to a sampling plane with a fixed orientation $u \in \mathbb{S}^2_+$. On this plane we define a FUR test system of points with a fundamental tile J_0 of area a. Through each of these points a straight line is drawn normal to the sampling plane, and the result is the FUR fakir probe, see Fig. 2.25.2(a), namely,

$$\Lambda_z = \{L_1^3(z + t_k, u),\ k \in \mathbb{Z}\}, \quad z \sim \mathrm{UR}(J_0), \tag{2.25.15}$$

where $\{t_k\}$ is the pertinent family of translations. By Eq. (2.25.5) the fakir identity reads

$$V = a \cdot \mathbb{E}\{L(Y \cap \Lambda_z)\}. \tag{2.25.16}$$

The name of the probe evokes the popular idea of a fakir bed with needles emanating from a flat board.

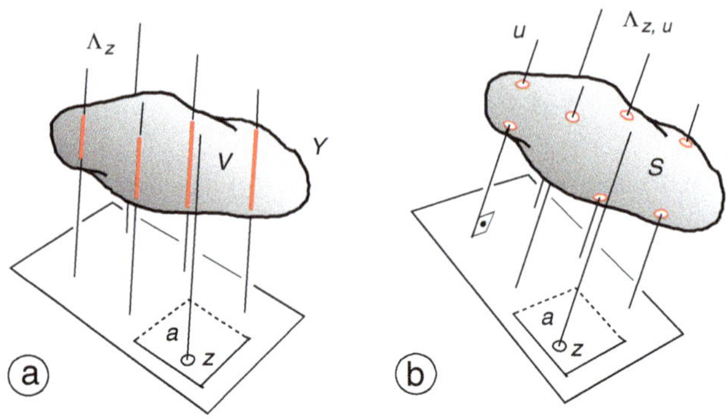

Fig. 2.25.2 (a) A FUR fakir probe for volume, see Section 2.25.4. (b) A IUR fakir probe for surface area, see Section 2.26.4.

By an argument analogous to that leading to Eq. (2.25.12), the mean of the number of fakir test lines hitting the set Y is

$$\mathbb{E}\{Q(Y \cap \Lambda_z)\} = \frac{A(u)}{a}, \tag{2.25.17}$$

where $A(u)$ is the area of the orthogonal projection of the domain Y onto the sampling plane. Likewise, the mean linear transect length per hitting test line is

$$\mathbb{E}(L) = \frac{V}{A(u)}, \tag{2.25.18}$$

which coincides formally with Eq. (2.8.15) for a single FUR line probe.

2.25.5 FUR Cavalieri slabs, and slab disectors

Stripes in $\mathbb{R}^2$, slabs in $\mathbb{R}^3$, or in general hyperslabs (also simply called slabs here) in $\mathbb{R}^d$, are all full-dimensional probes, and therefore the target may be the measure $\gamma(Y)$ of a set $Y \subset \mathbb{R}^d$ with $\dim(Y) \in \{0, 1, \ldots, d\}$. Recall that the density of a slab is that of its reference face (Section 1.2.5), whereby a FUR test system of slabs of a fixed thickness $t \in (0, T]$, where T is the fixed distance between consecutive reference faces, is one in which the reference faces are FUR Cavalieri planes with period T, see Fig. 2.25.3(b). Therefore, similarly as in Eq. (2.25.9) a test system of

FUR Cavalieri slabs may be represented as follows,

$$\Lambda_{t,z} = \{L_t(z + kT, u),\ k \in \mathbb{Z}\}, \quad z \sim \mathrm{UR}[0, T). \tag{2.25.19}$$

Application of Santaló's technique to Eq. (1.4.12) yields the basic identity for slabs, namely,

$$\gamma(Y) = \frac{T}{t} \cdot \mathbb{E}\{\gamma(Y \cap \Lambda_{t,z}\}. \tag{2.25.20}$$

For $\gamma \equiv V$, the preceding identity is analogous to Eq. (2.10.8), valid for a single FUR slab.

An interesting case arises when Y is a bounded and finite set of N particles and the target is $\gamma(Y) \equiv N(Y) = N$, whereby a proper fundamental probe is the slab disector (Section 2.15.2), and

$$N = \frac{T}{t} \cdot \mathbb{E}\{Q^-(Y \cap \Lambda_{t,z}^-)\}, \tag{2.25.21}$$

namely the sampling period T/t times the mean value of the total number of particles captured by all the Cavalieri slab disectors, see Fig. 2.25.3(a). The superscript '–' indicates that Q^- corresponds to disector counts – compare with Eq. (2.16.5) for a single slab disector. Eq. (2.25.21) holds for any dimension $d \geq 1$. A similar identity, analogous to Eq. (2.17.2) for a single FUR slab, holds for the Euler–Poincaré characteristic of a domain Y, namely,

$$\chi(Y) = \frac{T}{t} \cdot \mathbb{E}\{\chi^-(Y \cap \Lambda_{t,z}^-)\}. \tag{2.25.22}$$

If $\dim(Y) > 0$, then $\dim(Y \cap \Lambda_{t,z}) = \dim(Y)$, so that $\gamma(Y \cap \Lambda_{t,z})$ is not directly accessible by counting only. For instance, if $\gamma(Y) = L, S$, or V, then $\gamma(Y \cap \Lambda_{t,z})$ also represents L, S, or V, respectively.

Now we extend Eq. (2.16.9) to Cavalieri slabs. Consider a bounded and finite set $Y = \{Y_1, Y_2, \ldots, Y_N\} \subset \mathbb{R}^d$ of N particles hit by a FUR test system $\Lambda_{t,z}$ of Cavalieri slabs normal to a sampling axis of a fixed direction u. As defined by Eq. (2.16.2), let $Q(z, u)$ and $Q(Y \cap \Lambda_{t,z})$ denote the total number of particles hit by the slab $L_t(z, u)$ and by the test system $\Lambda_{t,z}$, respectively (not just the number of particle fragments in the slab, see Fig. 2.25.3(b)). By an argument analogous to that used in Eq. (2.25.11), we have

$$\int_{\mathbb{R}} Q(z, u)\, \mathrm{d}z = \int_0^T Q(Y \cap \Lambda_{t,z})\, \mathrm{d}z. \tag{2.25.23}$$

By Eq. (2.16.4) the lhs is equal to $N \cdot \{\mathbb{E}H(u) + t\}$, whereby the test system version of Eq. (2.16.9) is

$$N = \frac{T \cdot \mathbb{E}\{Q(Y \cap \Lambda_{t,z})\}}{\mathbb{E}\{H(u)\} + t}. \tag{2.25.24}$$

For $H_i(u) = 0$, $i = 1, 2, \ldots, N$, we get Eq. (2.25.20) for Cavalieri slab disectors sampling point particles. For $N = 1$ we get the mean number of FUR Cavalieri slabs hitting a particle, namely,

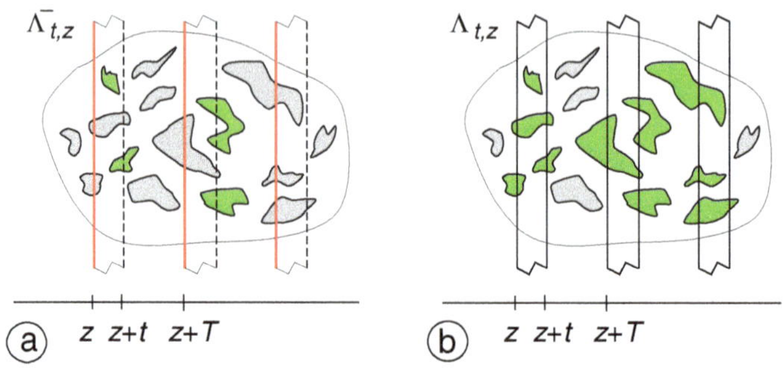

Fig. 2.25.3 (a) Cavalieri slab disectors sampling $Q^{-} = 4$ particles (in green) out of an aggregate of $N = 15$ particles. These quantities are simply and directly related by Eq. (2.25.21) (here $T/t = 3$). Note that a particle transect in the second slab consists of two fragments – hence, applying the disector rule may occasionally require additional information on particle shape outside a slab. (b) Cavalieri slabs hitting $Q = 10$ particles (in green). Unfortunately the mean of Q is not related to N only – particle size comes in, see Eq. (2.25.24).

$$\mathbb{E}\{Q(Y \cap \Lambda_{t,z})\} = \frac{H(u)+t}{T}. \tag{2.25.25}$$

For $t = 0$, the preceding identity becomes Eq. (2.25.12).

From Eq. (2.25.21) and Eq. (2.25.24) we may express the mean particle caliper length along the direction u by an identity analogous to Eq. (2.16.6).

2.25.6 FUR quadrats and disectors

The planar case

Quadrats in the plane, or generally blocks in $\mathbb{R}^d$ are also full-dimensional probes, hence the targets are as described above for slabs. In the planar case, which is of special practical interest, the fundamental probe is a FUR quadrat $T_0 \equiv T_2^2(0, \omega) \subset J_0 \subset \mathbb{R}^2$ of area $a_0 > 0$ with a fixed direction $\omega \in [0, 2\pi)$, and the FUR test system of quadrats may be represented as follows,

$$\Lambda_x = \{T_2^2(x + t_k, \omega),\ k \in \mathbb{Z}\}, \quad x \sim \mathrm{UR}(J_0). \tag{2.25.26}$$

Application of the results in Section 1.15.2 to the lhs of Eq. (2.25.1) yields

$$\gamma(Y) \cdot a_0 = \int_{J_0} \gamma(Y \cap \Lambda_x)\,\mathrm{d}x = a \cdot \mathbb{E}\{\gamma(Y \cap \Lambda_x)\}, \tag{2.25.27}$$

where $a \geq a_0$ is the tile area. Thus, the basic identity for FUR systematic quadrats reads

$$\gamma(Y) = \frac{a}{a_0} \cdot \mathbb{E}\{\gamma(Y \cap \Lambda_x)\}, \tag{2.25.28}$$

where $\gamma = N, B, A$ according to whether $\dim(Y) = 0, 1, 2$ respectively.

Number and connectivity in 3D

By an analogous argument to that leading to Eq. (2.25.14), the design corresponding to Eq. (2.25.21) may be completed by using a FUR test system Λ_x of bounded disectors of quadrat area a_0 and thickness t_0 (see Section 2.15.2) on each of the Cavalieri slices from $Y \cap \Lambda_{t,z}$, and applying Eq. (2.25.28) with $\gamma \equiv N$, namely,

$$N = \frac{T}{t_0} \cdot \frac{a}{a_0} \cdot \mathbb{E}\{Q^-(Y \cap \Lambda_{t,z} \cap \Lambda_x^-)\}, \tag{2.25.29}$$

which does not depend on the thickness $t \geq t_0$ of the Cavalieri slabs. A similar design holds for $\gamma \equiv \chi$, the Euler–Poincaré characteristic.

2.25.7 FUR test systems of circles and spheres

Lower-dimensional FUR probes are suitable to access planar area, or generally a volume. For instance, for a bounded domain $Y \subset \mathbb{R}^2$ of area $A > 0$ we have the identity (2.25.6) for a test system of points. Likewise, for a FUR test system in the plane whose fundamental probe is a bounded curve $T_0 \subset J_0 \subset \mathbb{R}^2$ of finite length $l > 0$ and arbitrary shape, direct application of Eq. (2.25.3) yields

$$A = \frac{a}{l} \cdot \mathbb{E}\{B(Y \cap \Lambda_x)\}, \tag{2.25.30}$$

namely the sampling period a/l times the mean of the total curve length determined in the target domain Y by all the curves of the test system. The generalization to $\mathbb{R}^d$ is straightforward. However, if the fundamental probe incorporates the isotropy condition in its geometry, then the corresponding FUR test system can also access curve length, or surface area, as we show next.

Consider a planar curve $Y \subset \mathbb{R}^2$ of length $B > 0$ intersected by a FUR test system $\Lambda_x \subset \mathbb{R}^2$ whose fundamental probe $T_1^2(x) \subset J_0 \subset \mathbb{R}^2$ is a circle of perimeter length $l = 2\pi r > 0$. In Eq. (1.16.5), ω may be arbitrarily fixed, whereas $\mathrm{d}l = r\,\mathrm{d}\varphi$. The innermost integral (now with respect to $\mathrm{d}\varphi$) is equal to 4. Thus,

$$\int_{\mathbb{R}^2} I(Y \cap T_1^2(x))\,\mathrm{d}x = 4rB = \frac{2}{\pi} \cdot Bl, \tag{2.25.31}$$

and Eq. (1.21.8) yields

$$B = \frac{\pi}{2} \cdot \frac{a}{l} \cdot \mathbb{E}\{I(Y \cap \Lambda_x)\}, \tag{2.25.32}$$

the expectation being with respect to $\mathbb{P}(\mathrm{d}x) = \mathrm{d}x/a$, where a is the area of J_0.

Likewise, let $Y \subset \mathbb{R}^3$ be a curve of length $L > 0$ intersected by a FUR test system $\Lambda_x \subset \mathbb{R}^3$ whose fundamental probe $T_2^3(x) \subset J_0 \subset \mathbb{R}^3$ is a sphere of area $a = 4\pi r^2 > 0$. The orientation parameters of the sphere probe may be arbitrarily fixed, whereas its surface area element is $\mathrm{d}a = r^2\,\mathrm{d}u_2$, so that the analogue of Eq. (1.16.6)) is $\mathrm{d}T_2^3 = |\cos\theta| r^2\,\mathrm{d}u_2\,\mathrm{d}l$, and the analogue of Eq. (1.16.8) becomes

$$\int_{\mathbb{R}^3} Q(Y \cap T_2^3(x))\,\mathrm{d}x = 2\pi r^2 L = \frac{1}{2}\cdot La. \tag{2.25.33}$$

Now, applying Eq. (1.21.8),

$$L = 2\cdot\frac{v}{a}\cdot\mathbb{E}\{Q(Y\cap\Lambda_x)\}, \tag{2.25.34}$$

where v is the volume of J_0.

For completeness, consider the same FUR test system as above hitting a surface $Y \subset \mathbb{R}^3$ of area $S > 0$. By a similar argument, Eq. (1.16.10) becomes

$$\int_{\mathbb{R}^3} B(Y \cap T_2^3(x))\,\mathrm{d}x = \pi^2 r^2 S = \frac{\pi}{4}\cdot Sa, \tag{2.25.35}$$

hence,

$$S = \frac{4}{\pi}\cdot\frac{v}{a}\cdot\mathbb{E}\{B(Y\cap\Lambda_x)\}. \tag{2.25.36}$$

2.25.8 Notes

1. General test systems

General mean values obtained from Santaló's fundamental formulae for test systems (Santaló, 1976) can be seen in Cruz-Orive (2002) and in Voss and Cruz-Orive (2009). See also Jensen and Gundersen (1982) and Cruz-Orive (1982). The historical notes given below borrow in part from the survey of Cruz-Orive (2017).

2. Test systems of points, Section 2.25.2

Decades after Delesse and Rosiwal formulated their principles, see Note 1 of Section 2.8.5, Glagolev (1933) proposed to estimate section area using a square grid of test points. Based on Henri Lebesgue's definition, the area of a planar set is commonly conceived as the limit of the sum of the areas of squares covering the set, as the size of each square tends to zero. This might have inspired the 'areal method' of Thomson (1930). In contrast, A. A. Glagolev's definition of area is essentially statistical, namely the area of the set is equal to the mean number of points of a UR test system hitting the set, times the tile area, see Eq. (2.25.6). Incidentally, Glagolev socialized with Andréi Kolmogorov, founder of modern probability.

3. FUR Cavalieri design

The first rigorous version of the Cavalieri identity, see Eq. (2.25.10), appears in Moran (1950) in the context of Monte Carlo integration. In fact, many stereological methods may be regarded as Monte Carlo integration methods. The essential condition that the offset z has to be UR$[0, T)$ was also given by Matheron (1971, p. 21), Cruz-Orive and Weibel (1981), Thioulouse, Mathy, and Ploye (1985) and Gundersen and Jensen (1987).

4. FUR fakir probe

This is a special spatial grid with a fixed orientation, designed for volume – it was proposed by Cruz-Orive (1993).

5. FUR Cavalieri slabs, and slab disectors

The use of Cavalieri slabs to estimate the number of organelles in histology goes back to the late 19th century (Bendtsen & Nyengaard, 1989). The corresponding design was studied by Gual-Arnau and Cruz-Orive (1998), Cruz-Orive (1999, 2004, 2006), McNulty, Cruz-Orive, Roberts, Holmes, and Gual-Arnau (2000) and Maletti and Wulfsohn (2006).

6. FUR quadrats, and bounded disectors

The use of FUR systematic quadrats is old – for early references see, for instance, Ripley (1981). Miles (1978b) treats test systems of quadrats in a stereological context. Test systems of bounded disectors, see Eq. (2.25.29), are extensively used in stereology (for references, see Section 2.15.3). For an account of bounded and unbounded disector designs, see Cruz-Orive and Geiser (2004).

7. FUR test systems of circles and spheres

The Merz grid, see e.g. Howard and Reed (2005, Fig. 12.5), is composed of wavy chains of half circles. The fundamental tile is a rectangle $J_0 = [0, 2h) \times [0, h)$, and the fundamental probe is the union of two half circles of diameter h. For this grid, Eq. (2.25.32) holds with $a/l = (2/\pi)h$. On the other hand, Mouton, Gokhale, Ward, and West (2002) proposed spherical probes for curve length in space, see Eq. (2.25.34) and West (2012).

2.26 IUR Test Systems

2.26.1 Introduction

Bounded fundamental probes

An IUR test system $\Lambda_{x,u_d} \subset \mathbb{R}^d$ with a bounded fundamental probe $T_0 \equiv T_r^d(0,0) \subset J_0 \subset \mathbb{R}^d$ yields mean value identities for a target set $Y \subset \mathbb{R}^d$ of arbitrary dimension $q \leq d$, provided that $q + r \geq d$ (Section 1.3). Then, Santaló's formula, Eq. (1.21.8), reads

$$c_2\gamma(Y)v(T_0) = \int_{G_{d[0]}} \mathrm{d}u_d \int_{J_0} \alpha(Y \cap \Lambda_{x,u_d})\,\mathrm{d}x, \tag{2.26.1}$$

where the coefficient c_2 is given by Eq. (1.16.11). The joint probability element adopted for the orientation and the location parameters is proportional to their respective motion-invariant densities, namely

$$\mathbb{P}(\mathrm{d}x, \mathrm{d}u_d) = \frac{\mathrm{d}x}{V(J_0)} \cdot \frac{\mathrm{d}u_d}{c_{20}}, \quad x \in J_0,\ u_d \in G_{d[0]}, \tag{2.26.2}$$

where the normalizing constant c_{20} is given by Eq. (1.16.12). Thus, location and orientation are respectively UR and IR, and independent. Now, using Eq. (2.26.1), the basic identity for an IUR test system of bounded probes hitting the target set reads

$$\gamma(Y) = \frac{c_{20}}{c_2} \cdot \frac{V(J_0)}{v(T_0)} \cdot \mathbb{E}\{\alpha(Y \cap \Lambda_{x,u_d})\}, \tag{2.26.3}$$

where the expectation is with respect to the joint probability element given by Eq. (2.26.2).

Unbounded probes

Similarly, for an IUR test system of unbounded r-planes, Santaló's formula, Eq. (1.21.14), reads

$$c_1 \cdot \gamma(Y) = \int_{G_{r,d-r}} \mathrm{d}u \int_{J_0} \alpha(Y \cap \Lambda_{z,u})\,\mathrm{d}z, \tag{2.26.4}$$

where the coefficient c_1 is given by Eq. (1.5.15). We adopt the following joint probability element,

$$\mathbb{P}(\mathrm{d}z, \mathrm{d}u) = \frac{\mathrm{d}z}{V(J_0)} \cdot \frac{\mathrm{d}u}{c_{10}}, \; z \in J_0, \; u \in G_{r,d-r}, \tag{2.26.5}$$

where the normalizing constant c_{10} is given by Eq. (1.5.17). Here the fundamental tile J_0 has dimension $d - r$ because it lies in the orthogonal complement of the fundamental r-plane. The corresponding identity reads

$$\gamma(Y) = \frac{c_{10}}{c_1} \cdot V(J_0) \cdot \mathbb{E}\{\alpha(Y \cap \Lambda_{z,u})\}. \tag{2.26.6}$$

Remark

By Eq. (1.5.15) and Eq. (1.16.11) it follows that

$$\frac{c_{20}}{c_2} = \frac{c_{10}}{c_1} = \frac{O_q O_r}{O_d O_{q+r-d}}. \tag{2.26.7}$$

Consequently, the numerical factors involved in the preceding expressions of $\gamma(Y)$ are the same for either bounded or unbounded probes, provided that the parameters $\{q, r, d\}$ are the same in either case.

2.26.2 Buffon–Steinhaus test system

Consider a curve $Y \subset \mathbb{R}^2$ of finite length $B > 0$. The Buffon test system, see Eq. (1.21.12) and Fig. 1.21.2(a), consists of an IUR series of parallel test lines a constant distance $T > 0$ apart, namely,

$$\Lambda_{z,\phi} = \{L_1^2(z + kT, \phi), \; k \in \mathbb{Z}\}, \quad z \sim \mathrm{UR}[0, T), \; \phi \sim \mathrm{UR}[0, \pi). \tag{2.26.8}$$

Thus, the fundamental probe is a straight line $L_1^2(0, \phi)$, and the fundamental tile for the auxiliary test system of points on an axis normal to the test lines is the segment $J_0 = [0, T)$. The adopted joint probability element is

$$\mathbb{P}(\mathrm{d}z, \mathrm{d}\phi) = \frac{\mathrm{d}z}{T} \cdot \frac{\mathrm{d}\phi}{\pi}, \quad z \in [0, T), \; \phi \in [0, \pi), \tag{2.26.9}$$

and the basic identity is a special case of Eq. (2.26.6) with $c_1 = 2$ and $c_{10} = \pi$, namely,

$$B = \frac{\pi}{2} \cdot T \cdot \mathbb{E}\{I(Y \cap \Lambda_{z,\phi})\}. \tag{2.26.10}$$

More generally, consider an IUR test system whose fundamental probe $T_0 \equiv T_1^2(0, 0) \subset J_0 \subset \mathbb{R}^2$ is a test curve of finite length $l > 0$ contained in a fundamental tile J_0 of area a. With the coefficients $c_2 = 4$ and $c_{20} = 2\pi$, the basic Eq. (2.26.3) yields

$$B = \frac{\pi}{2} \cdot \frac{a}{l} \cdot \mathbb{E}\{I(Y \cap \Lambda_{x,\omega})\}, \quad x \sim \mathrm{UR}(J_0),\ \omega \sim \mathrm{UR}[0, 2\pi), \tag{2.26.11}$$

which is equivalent to Eq. (2.25.32). If J_0 is a square of side length $T > 0$ and T_0 is a side of J_0, then the test system becomes a Buffon test system, see Eq. (2.26.8), and the preceding identity returns Eq. (2.26.10) back because $a/l = T^2/T = T$.

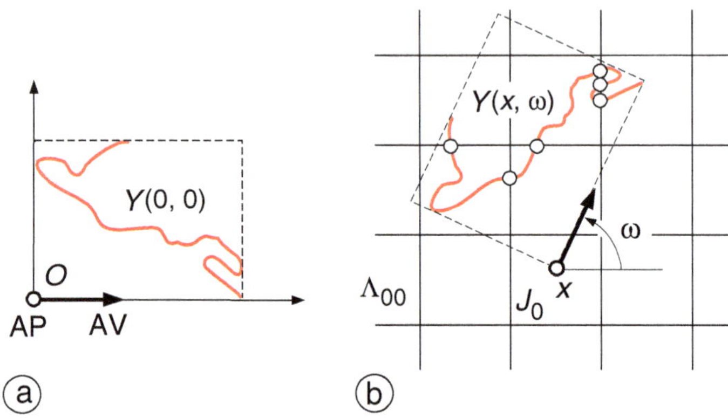

Fig. 2.26.1 (a) A target curve $Y(0, 0)$ with its AP at the origin and its AV being the vector $(1, 0)$. (b) An IUR superimposition $Y(x, \omega)$ of the curve on a fixed Buffon–Steinhaus (or square grid) test system. Here $I = 6$ intersections. See Eq. (2.26.12).

The Buffon–Steinhaus test system, also called the square grid, is the union of two mutually perpendicular Buffon test systems, see Fig. 2.26.1(b). Thus, the test length per unit area is doubled, and the factor T in Eq. (2.26.9) should be replaced with $T/2$. Formally, for the above curve test system let T_0 be the union of two perpendicular sides of the square tile J_0. Then we have a square grid with $a/l = T^2/(2T) = T/2$, whereby

$$B = \frac{\pi}{4} \cdot T \cdot \mathbb{E}\{I(Y \cap \Lambda_{x,\omega})\}. \tag{2.26.12}$$

Alternatively the test system may be kept fixed, see Fig. 2.26.1, and the target set $Y \equiv Y(x, \omega)$ mobile – recall Eq. (1.21.7) – whereby

$$B = \frac{\pi}{4} \cdot T \cdot \mathbb{E}\{I(Y(x, \omega) \cap \Lambda_{0,0})\}. \tag{2.26.13}$$

2.26.3 The Isotropic Cavalieri (ICav) design

A test system of ICav planes of period $T > 0$ is one in which the sampling axis is isotropic random. Thus

$$\Lambda_{z,u} = \{L_2^3(z + kT, u),\ k \in \mathbb{Z}\}, \quad z \sim \mathrm{UR}[0, T),\ u \sim \mathrm{UR}(\mathbb{S}_+^2), \tag{2.26.14}$$

which is Eq. (2.25.9) with the additional property that the direction u normal to the test planes is IR on the unit hemisphere, and independent from the UR offset z of the planes along that direction, see Fig. 2.26.2(a). Thus the test system is IUR, and therefore the target set $Y \subset \mathbb{R}^3$ may be not only a domain of volume $V > 0$, but a bounded surface of area $S > 0$ and integral of mean curvature M, or a bounded curve of length $L > 0$. From Eq. (1.5.12), Eq. (1.11.6), and Eq. (1.5.14), the basic Eq. (2.26.6) yields respectively the following ICav identities,

$$S = \frac{4}{\pi} \cdot T \cdot \mathbb{E}\{B(Y \cap \Lambda_{z,u})\}, \tag{2.26.15}$$

$$M = T \cdot \mathbb{E}\{C(Y \cap \Lambda_{z,u})\}, \tag{2.26.16}$$

$$L = 2T \cdot \mathbb{E}\{Q(Y \cap \Lambda_{z,u})\}, \tag{2.26.17}$$

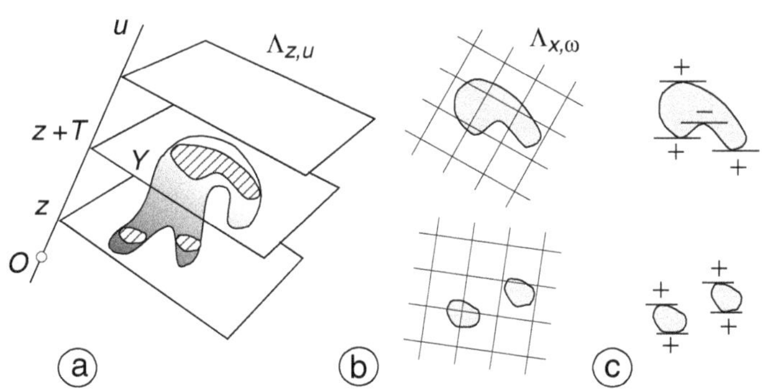

Fig. 2.26.2 (a) ICav planes hitting a target domain Y. (b) Each of the two transects is submitted to a square grid which is IUR relative to the transects, in order to access the volume of Y by point counting via Eq. (2.25.14), or the surface area of the boundary of Y by intersection counting via Eq. (2.26.18). (c) Illustration of the net tangent count, see Eq. (1.10.13), to access the integral of mean curvature of the boundary of Y via Eq. (2.26.16). The total curvature is $C = 2\pi$ for the upper trace, and $C = 4\pi$ for the lower one.

An alternative to Eq. (2.26.15), based on intersection counts, is obtained by using an IUR test line grid $\Lambda_{x,\omega}$ on each ICav section, see Fig. 2.26.2(b), and applying Eq. (2.26.11), namely,

$$S = 2T \cdot \frac{a}{l} \cdot \mathbb{E}\{I(Y \cap \Lambda_{z,u} \cap \Lambda_{x,\omega})\}, \tag{2.26.18}$$

where the random pairs (z, u) and (x, ω) are distributed independently as indicated in Eq. (2.26.14) and Eq. (2.26.11), respectively. Further, recall that, for each planar section, the total curvature of a curve trace entering in Eq. (2.26.16) may be expressed in terms of the net tangent count by means of a sweeping line of an arbitrary direction using Eq. (1.10.13), see Fig. 2.26.2(c).

2.26.4 The isotropic fakir probe

This is an isotropic version of the FUR fakir probe described in Section 2.25.4, namely,

$$\Lambda_{z,u} = \{L_1^3(z + t_k, u),\ k \in \mathbb{Z}\}, \quad z \sim \mathrm{UR}(J_0),\ u \sim \mathrm{UR}(\mathbb{S}_+^2). \tag{2.26.19}$$

The test system is IUR, and therefore the target parameter may be not only a volume but a surface area, see Fig. 2.25.2(b). In the basic Eq. (2.26.6) we have $c_{10} = 2\pi$ and $c_1 = \pi$, whereby the basic identity for the area $S > 0$ of a bounded surface $Y \subset \mathbb{R}^3$ hit by a IUR fakir probe reads

$$S = 2a \cdot \mathbb{E}\{I(Y \cap \Lambda_{z,u})\}. \tag{2.26.20}$$

2.26.5 Notes

1. Buffon–Steinhaus test system

Based on Buffon's result, see Note 2 in Section 2.4.8, Steinhaus (1930) proposed to estimate the length of a planar curve using an IUR square grid of test lines via Eq. (2.26.12). The method, and its precision, were revisited by Gómez et al. (2016).

2. The Isotropic Cavalieri (ICav) design

This design was proposed by Cruz-Orive, Ramos-Herrera, and Artacho-Pérula (2010), see also González-Villa, Cruz, and Cruz-Orive (2018).

3. The isotropic fakir probe

This probe was implemented by Kubínová and Janáček (2001) and Kubínová, Mao, Janáček, and Archembeau (2003).

2.27 Test Systems of Cylinders

2.27.1 Construction

On an oriented plane $L_2^* \equiv L_2^*(0, u_2), u_2 \in \mathbb{S}^2$, construct a test system with a bounded fundamental probe $T_{r-1}^2 \subset J_0^2 \subset L_2^*$, $r \in \{2, 3\}$, where J_0^2 is a fundamental tile of area $a > 0$. The probe T_1^2 is a curve of length b, whereas T_2^2 is a domain of area a_0. Fig. 2.27.1(a) is a sketch of an IUR test system $\Lambda_Z \subset \mathbb{R}^3$ of cylinders congruent with Z_r^3, see Eq. (1.13.11) and Fig. 1.13.3, namely,

$$\Lambda_Z = \{Z_r^3(z + t_k, u_2, \tau),\ k \in \mathbb{Z}\}, \tag{2.27.1}$$

$$\mathbb{P}(\mathrm{d}z, \mathrm{d}u_2, \mathrm{d}\tau) = \frac{\mathrm{d}z}{a} \cdot \frac{\mathrm{d}u_2}{4\pi} \cdot \frac{\mathrm{d}\tau}{2\pi}, \quad z \in J_0^2,\ u_2 \in \mathbb{S}^2,\ \tau \in \mathbb{S}^1, \tag{2.27.2}$$

where $\{t_k\}$ is a suitable system of translations in the plane. The test system consists of cylindrical surfaces, or solid cylinders, according to whether $r = 2$, or $r = 3$, respectively. For $r = 1$ the test system is an IUR fakir probe, see Fig. 2.25.2(b).

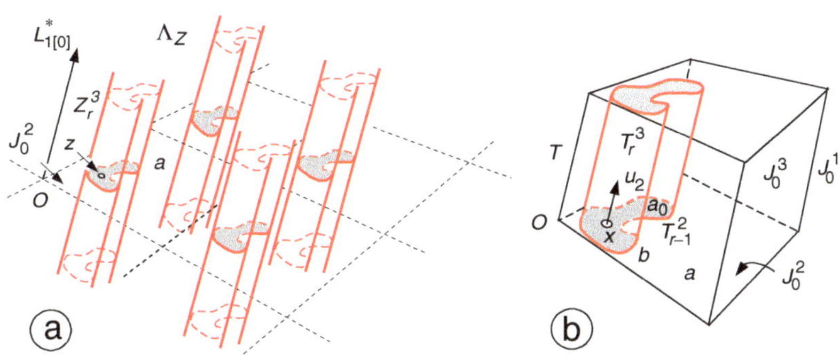

Fig. 2.27.1 (a) A portion of a test system of cylinders. (b) Fundamental tile in $\mathbb{R}^3$ containing a bounded cylinder as fundamental probe. By construction, the corresponding test system coincides with the one in (a). See Section 2.27.1.

Consider now an IUR test system $\Lambda_T \subset \mathbb{R}^3$ whose fundamental tile is a box $J_0^3 = J_0^2 \times J_0^1$ of volume $v = aT$. The fundamental probe is a bounded cylinder $T_r^3 = T_{r-1}^2 \times J_0^1 \subset J_0^3$, $r = 2, 3$, see Fig. 2.27.1(b), whereby

$$\Lambda_T = \{T_r^3(x + t_k, u_2, \tau),\ k \in \mathbb{Z}\}, \tag{2.27.3}$$

$$\mathbb{P}(\mathrm{d}x, \mathrm{d}u_2, \mathrm{d}\tau) = \frac{\mathrm{d}x}{aT} \cdot \frac{\mathrm{d}u_2}{4\pi} \cdot \frac{\mathrm{d}\tau}{2\pi}, \quad z \in J_0^3,\ u_2 \in \mathbb{S}^2,\ \tau \in \mathbb{S}^1, \tag{2.27.4}$$

where $\{t_k\}$ is now a suitable system of translations in space. Any stack of bounded cylinders from Λ_T is in fact a cylinder Z_r^3, and this implies that the test systems Λ_T and Λ_Z coincide.

Let $Y \subset \mathbb{R}^3$ denote a target set of dimension $q \in \{0, 1, 2, 3)$ and measure $\gamma(Y)$. Provided that $q + r - 3 \geq 0$, it follows that

$$\mathbb{E}\{\alpha(Y \cap \Lambda_Z)\} = \mathbb{E}\{\alpha(Y \cap \Lambda_T)\}, \tag{2.27.5}$$

the expectations being with respect to the corresponding probability elements given by Eq. (2.27.2) and Eq. (2.27.4), respectively. Thus, the expectation for Λ_Z can be evaluated indirectly via the one for Λ_T by applying the basic identities for test systems of bounded probes, namely Eq. (2.25.3) in the FUR case, or Eq. (2.26.3) in the IUR case.

2.27.2 FUR cylinders

If $\dim(Y) = q \in \{0, 1, 2, 3\}$, as above, then we may use a FUR test system Λ_Z of solid cylinders ($r = 3$). Eq. (2.25.3) applies to Λ_T with $r = 3$, namely with $V(J_0^3) = aT$ and $V(T_3^3) = a_0 T$. Further, bearing Eq. (2.27.5) in mind we obtain

$$\gamma(Y) = \frac{a}{a_0} \cdot \mathbb{E}\{\gamma(Y \cap \Lambda_Z)\}, \tag{2.27.6}$$

where $\gamma = N, L, S, V$ according to whether $q = 0, 1, 2, 3$, respectively.

Likewise, let $Y \subset \mathbb{R}^3$ be a domain of volume V, and Λ_Z a FUR test system of cylindrical surfaces. Then,

$$V = \frac{a}{b} \cdot \mathbb{E}\{S(Y \cap \Lambda_Z)\}. \tag{2.27.7}$$

2.27.3 IUR cylinders

Now the dimension q of Y and the dimension r of Λ_Z may be both less than 3 provided that $q + r - 3 \geq 0$.

For $q = r = 2$, namely for a target surface of area S hit by a test system Λ_Z of cylindrical surfaces we have

$$S = \frac{4}{\pi} \cdot \frac{a}{b} \cdot \mathbb{E}\{B(Y \cap \Lambda_Z)\}. \tag{2.27.8}$$

We have first applied Eq. (2.26.3) to Λ_T with $c_{20} = 8\pi^2$ and $c_2 = 2\pi^3$, see Section 1.16.4. Further, $V(J_0^3) = aT$ and $S(T_2^3) = bT$. Finally, recalling Eq. (2.27.5), Eq. (2.27.8) is obtained.

Likewise, if Y is a target curve of length L we may use a test system Λ_Z of cylindrical surfaces. Now $c_{20} = 8\pi^3$ and $c_2 = 4\pi^2$, whereby

$$L = 2 \cdot \frac{a}{b} \cdot \mathbb{E}\{Q(Y \cap \Lambda_Z)\}. \tag{2.27.9}$$

2.27.4 Direct derivation of Crofton formulae for a cylinder probe

Let $Y_q \subset \mathbb{R}^3$ represent a target set of dimension $q = 1, 2, 3$, that is $\gamma(Y_q) = L, S, V$, respectively, intersected by a cylindrical surface $Z_2 \equiv Z_2^3$ endowed with the kinematic density $\mathrm{d}Z_2$, see Eq. (1.13.12). Next we apply the foregoing results to obtain the corresponding Crofton intersection formulae. This is achieved by using the reverse of Eq. (2.26.1) in each case. Note also that the normalization constant in Eq. (2.27.2) is equal to $8\pi^2 a$.

From Eq. (2.27.9),

$$L = 2 \cdot \frac{a}{b} \cdot \frac{1}{8\pi^2 a} \int_{\mathbb{R}^2 \times \mathbb{S}^2 \times \mathbb{S}^1} Q(Y_1 \cap Z_2)\, \mathrm{d}Z_2, \tag{2.27.10}$$

whereby

$$\int Q(Y_1 \cap Z_2)\, \mathrm{d}Z_2 = 4\pi^2 bL. \tag{2.27.11}$$

Likewise, Eq. (2.27.8) leads to

$$\int B(Y_2 \cap Z_2)\, \mathrm{d}Z_2 = 2\pi^3 bS. \tag{2.27.12}$$

Finally, Eq. (2.27.7) also holds if Λ_Z is IUR, whereby

$$V = \frac{a}{b} \cdot \frac{1}{8\pi^2 a} \int_{\mathbb{R}^2 \times \mathbb{S}^2 \times \mathbb{S}^1} S(Y_3 \cap Z_2)\, \mathrm{d}Z_2, \tag{2.27.13}$$

that is,

$$\int S(Y_3 \cap Z_2)\, \mathrm{d}Z_2 = 8\pi^2 bV. \tag{2.27.14}$$

2.27.5 Note

The results given in Section 2.27 are particular cases of the general treatment of Cruz-Orive and Gual-Arnau (2020).

2.28 Vertical Designs

2.28.1 Vertical Cavalieri (VCav) planes

The fundamental probe of a test system $\Lambda_{z,\phi}$ of VCav planes is a vertical plane $L^3_{2\cdot v}(0,0)$, see Section 1.6. Thus,

$$\Lambda_{z,\phi} = \{L^3_{2\cdot v}(z + kT, \phi),\ k \in \mathbb{Z}\}. \quad z \sim \mathrm{UR}[0,T),\ \phi \sim \mathrm{UR}[0,\pi), \tag{2.28.1}$$

see Fig. 2.28.1(a). For a surface $Y \subset \mathbb{R}^3$ of area S, application of Eq. (2.26.6) to Eq. (1.6.3) with $c_{10} = c_1 = \pi$ yields

$$S = T \cdot \mathbb{E}\{W(Y \cap \Lambda_{z,\phi})\}, \tag{2.28.2}$$

where, for each vertical section, the functional $W(\cdot)$ is given by Eq. (1.6.4).

The Cavalieri identity (2.25.10) for volume also holds for VCav planes with an arbitrary orientation $\phi \in [0,\pi)$.

2.28.2 VCav planes bearing FUR cycloid test systems

A convenient alternative to Eq. (2.28.2) is obtained by using a FUR test system Λ_x of cycloids on each vertical plane section, see Fig. 2.28.1(b). For each $z \in [0,T)$, the fundamental probe $T^2_1(0) \subset J_0 \subset \mathbb{R}^2$ of Λ_x may be the union of two FUR cycloid arcs of total length $l > 0$, which must have their minor principal axes parallel to the vertical axis. The minor and major principal axes of each arc have lengths $l/8$ and $\pi l/8$ respectively, see Section 1.17.1. As fundamental tile we may adopt the rectangle $J_0 = [0, \pi l/4) \times [0, l/4)$. For $z \sim \mathrm{UR}[0,T)$, $\phi \sim \mathrm{UR}[0,\pi)$, and $x \sim \mathrm{UR}(J_0)$, application of Eq. (2.26.1) to Eq. (1.17.6) yields

$$\frac{\pi l S}{2} = T\pi a \cdot \mathbb{E}\{I(Y \cap \Lambda_{z,\phi} \cap \Lambda_x)\}, \tag{2.28.3}$$

where a is the area of J_0. Thus we obtain the identity

$$S = 2T \cdot \frac{a}{l} \cdot \mathbb{E}\{I(Y \cap \Lambda_{z,\phi} \cap \Lambda_x)\}, \tag{2.28.4}$$

which is the vertical design version of Eq. (2.26.18) – this could be anticipated, because the vertical design is only a convenient resource to generate IUR test curve elements in space.

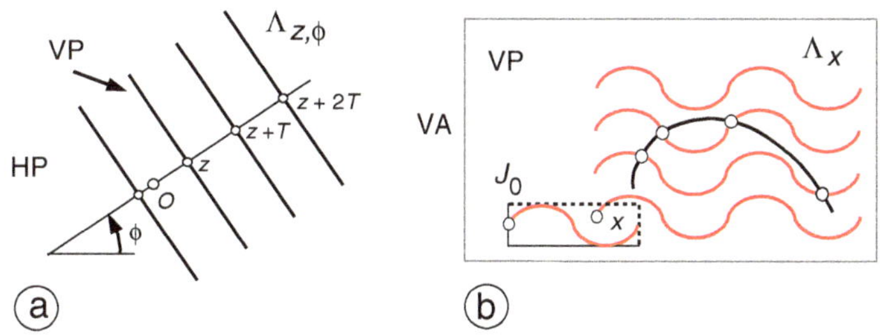

Fig. 2.28.1 (a) Projection of a portion of a test system of VCav planes (VP) onto the horizontal plane (HP). (b) A portion of a UR test system of cycloids superimposed onto a vertical trace curve determined in a surface by a vertical plane. See Section 2.28.2.

Remark

The fundamental probe $T_1^2(0)$ may consist of one or more cycloid arc fragments, provided that the union of all such fragments is equivalent – up to translations – to the double cycloid arc just described. There is also freedom to choose the fundamental tile $J_0 \supset T_1^2(0)$.

2.28.3 Total vertical projections of a curve

The target $Y \subset \mathbb{R}^3$ is a smooth curve of length $L > 0$, which is projected orthogonally onto a plane $L_{2[1]}^3(\phi)$ containing a fixed vertical axis, with isotropic orientation $\phi \sim \mathrm{UR}[0, \pi)$ around that axis. The total curve projection is denoted by $Y'(\phi)$, see Section 2.23.1 and Fig. 2.28.2(b).

For the functional method, Eq. (2.23.1) remains unchanged, namely,

$$L = \mathbb{E}\{Z(Y'(\phi))\}, \tag{2.28.5}$$

$$Z(Y'(\phi)) = \int_{y \in Y'(\phi)} [\cos(\psi(y)) + \psi(y)\sin(\psi(y))]\, \mathrm{d}y, \tag{2.28.6}$$

where $\psi(y) \in [-\pi/2, \pi/2)$ is the angle between the vertical axis and an essentially linear arc element δy of length $\mathrm{d}y$, at a point $y \in Y'(\phi)$.

For the cycloid test system method, the cycloid arcs constituting the fundamental probe must have their major principal axes parallel to the vertical axis, that is, the probe may be the one considered in the preceding section rotated by 90°, with the corresponding fundamental tile $J_0 = [0, l/4) \times [0, \pi l/4)$. For $\phi \sim \mathrm{UR}[0, \pi)$, $x \sim \mathrm{UR}(J_0)$, Eq. (1.19.30) yields

$$\pi l L = 2\pi a \cdot \mathbb{E}\{I(Y'(\phi) \cap \Lambda_x)\}, \tag{2.28.7}$$

whereby

$$L = 2 \cdot \frac{a}{l} \cdot \mathbb{E}\{I(Y'(\phi) \cap \Lambda_x)\}, \tag{2.28.8}$$

(compare with Eq. (2.23.3) for a single cycloid test probe).

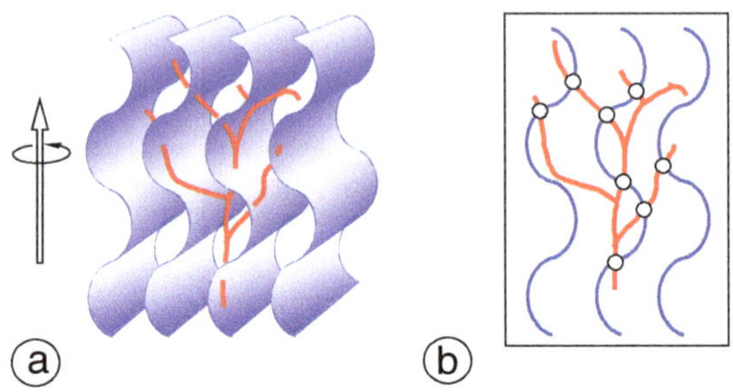

Fig. 2.28.2 (a) A portion of a test system of vertical cylindrical surfaces of cycloidal directrix, whose fundamental probe is sketched in Fig. 1.19.5, hitting a bounded curve of length L (in red). Here $L = 2(v/a)\mathbb{E}(Q)$, see also Eq. (2.26.17) and Eq. (2.29.5), where a and v denote the surface area of the fundamental probe and the volume of the fundamental tile, respectively. (b) Orthogonal projection onto a vertical plane, leading to a more convenient design, see Eq. (2.28.8).

2.28.4 Vertical slice projections of a curve

The target curve $Y \subset \mathbb{R}^3$ is first hit by a test system of VCav slabs of period $T > 0$ and thickness $0 < t \leq T$, namely,

$$\Lambda_{t,z,\phi} = \{L_t(z + kT, \phi),\ k \in \mathbb{Z}\}, \quad z \sim \mathrm{UR}[0, T),\ \phi \sim \mathrm{UR}[0, \pi). \tag{2.28.9}$$

Then, for each orientation $\phi \in [0, \pi)$, the contents $Y \cap L_t$ of each slice is projected orthogonally onto a plane parallel to the slabs, to yield a set $(Y \cap L_t)'$ of projected curve fragments. With the aid of Eq. (1.19.31) we obtain the functional method identity

$$\pi t L = \pi T \cdot \mathbb{E}\{Z((Y \cap \Lambda_{t,z,\phi})')\}, \tag{2.28.10}$$

that is,

$$L = \frac{T}{t} \cdot \mathbb{E}\{Z((Y \cap \Lambda_{t,z,\phi})')\}, \tag{2.28.11}$$

where the functional $Z(\cdot)$ is given by Eq. (2.28.6).

For each orientation $\phi \in [0, \pi)$, the FUR test system Λ_x of cycloids described in the preceding section may be superimposed onto each slice projection. By virtue of Eq. (1.19.32), and with the pertinent probability elements, in the usual manner we get

$$L = 2 \cdot \frac{T}{t} \cdot \frac{a}{l} \cdot \mathbb{E}\{I((Y \cap \Lambda_{t,z,\phi})' \cap \Lambda_x)\}, \tag{2.28.12}$$

(compare with Eq. (2.23.8)).

2.28.5 Total vertical projections of a convex set

With the setup of Section 2.23.3, the version of Eq. (2.23.10) when the FUR test system Λ_x of cycloids just described is used on the vertical projection plane reads

$$M = \pi \cdot \frac{a}{l} \cdot \mathbb{E}\{I(Y'(\phi) \cap \Lambda_x)\}. \tag{2.28.13}$$

2.28.6 Notes

1. VCav planes

Relevant references are Baddeley (1985), Baddeley, Gundersen, and Cruz-Orive (1986), Michel and Cruz-Orive (1988), and Cruz-Orive et al. (2014). For early applications see Cruz-Orive and Hunziker (1986) and Vesterby, Kragstrup, Gundersen, and Melsen (1987). For historical details, see Cruz-Orive (2017).

2. Vertical projections

For pertinent references see Note 3 in Section 1.19.7. The slice projection Eq. (2.28.12) is due to Gokhale (1990), see also Gokhale (1992). This design was applied by Batra, König, and Cruz-Orive (1995). For early applications of the total vertical projections design for curve length (Cruz-Orive & Howard, 1991), see Roberts, Howard, Cruz-Orive, and Edwards (1991), Howard, Cruz-Orive, and Yaegashi (1992), and Wulfsohn, Nyengaard, Gundersen, Cutler, and Squires (1999).

2.29 Spatial Grids

2.29.1 Introduction

A spatial grid is a test system in $\mathbb{R}^d$, $d \geq 3$. Thus, the results given in Sections 2.25.1 and 2.26.1 hold directly for FUR and IUR spatial grids, respectively. Examples in $\mathbb{R}^3$ are given next. For convenience, it may be assumed that the fundamental tile is a cube $J_0 = [0, T)^3 \subset \mathbb{R}^3$ of volume $v = T^3 > 0$. The target parameter is the measure $\gamma(Y)$ of a set $Y \subset \mathbb{R}^3$, where $\gamma = N, \chi, L, S, M, V$ according to whether $\dim(Y) = 0, 0, 1, 2, 2, 3$, respectively.

2.29.2 FUR spatial grids. Examples

FUR test systems described in Section 2.25, such as grids of points in $\mathbb{R}^3$, Cavalieri planes and slabs, the fakir probe, or test systems of blocks, spheres, and cylinders (Section 2.27.1), constitute examples of FUR spatial grids.

Fig. 2.29.1(a,d) illustrates a FUR cubic grid Λ_x of test points, for which the three-dimensional extension of Eq. (2.25.6) reads

$$V = \frac{v}{p} \cdot \mathbb{E}\{P(Y \cap \Lambda_x)\}. \tag{2.29.1}$$

A special FUR spatial grid Λ_x of test lines is the union of three mutually perpendicular, and intersecting, fakir probes, see Fig. 2.29.1(b,e). To construct it, we may adopt a bounded test probe consisting of the union of the three edges of a cubic fundamental tile J_0 of side length T, which meet at the origin, see Fig. 2.29.1(b). Thus, $l = 3T$, and

$$V = \frac{v}{l} \cdot \mathbb{E}\{L(Y \cap \Lambda_x)\}, \tag{2.29.2}$$

where $v/l = T^3/(3T) = T^2/3$ in this case.

Note that the vertices of the preceding grid constitute a cubic grid of test points. Thus we say that, in a cubic grid of test points, the latter are "perfectly registered", that is, they are perfectly aligned along the three main directions of their support lines. In contrast, test points generated by superimposing a UR test system of points independently on each plane of a Cavalieri series will not be perfectly registered in general. The corresponding design, leading to Eq. (2.25.14), is therefore different from that leading to Eq. (2.29.1).

2.29.3 IUR spatial grids. Examples

If the FUR spatial grid of test lines just described is endowed with the joint probability element given by Eq. (2.26.2) with $d = 3$, namely,

$$\mathbb{P}(\mathrm{d}x, \mathrm{d}u_2, \mathrm{d}\tau) = \frac{\mathrm{d}x}{v} \cdot \frac{\mathrm{d}u_2}{4\pi} \cdot \frac{\mathrm{d}\tau}{2\pi}, \quad x \in J_0,\ u_2 \in \mathbb{S}^2,\ \tau \in [0, 2\pi), \tag{2.29.3}$$

then the grid is IUR, and the area of a target surface $Y \subset \mathbb{R}^3$ can be accessed. It may be convenient to keep the grid fixed and to apply the IUR motion to Y, whereby

$$S = 2 \cdot \frac{v}{l} \cdot \mathbb{E}\{I(Y(x, u_2, \tau) \cap \Lambda_{0,0,0})\}, \tag{2.29.4}$$

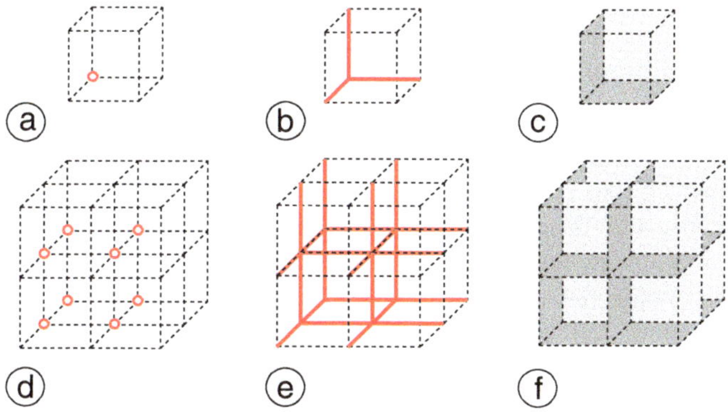

Fig. 2.29.1 (a,b,c) Cubic fundamental tiles containing a fundamental probe. (d,e,f) Corresponding spatial grids of test points, lines and planes, respectively.

which holds in general for a fundamental curve probe of arbitrary shape, of length $l > 0$, within a fundamental tile $J_0 \in \mathbb{R}^3$ of volume v.

The union of three mutually perpendicular series of ICav planes, see Fig. 2.29.1(f), is a IUR test grid of planes. For a cubic fundamental tile of side length T, the fundamental probe may be the union of the three square faces of the tile which meet at the origin, see Fig. 2.29.1(c). In addition to V, S, an IUR grid of planes can access integral of mean curvature M, and curve length L. For instance, for the latter,

$$L = 2 \cdot \frac{v}{a} \cdot \mathbb{E}\{Q(Y(x, u_2, \tau) \cap \Lambda_{0,0,0})\}, \tag{2.29.5}$$

where $v/a = T^3/(3T^2) = T/3$ for the preceding construction. Again, the preceding identity holds for an arbitrarily shaped fundamental surface probe of area $a > 0$. For the integral of mean curvature, however, the fundamental probe must consist of planar domains.

2.29.4 Note

The spatial grid of test lines illustrated in Fig. 2.29.1(b,e) was proposed by Sandau (1987), see also Sandau and Hahn (1994). Spatial grids of test points, lines, and planes, were further described by Cruz-Orive (1997). For further implementations, see Pache, Roberts, Zimmerman, Vock, and Cruz-Orive (1993), Kubínová and Janáček (2001), Kubínová et al. (2003) and Howard and Reed (2005). A vertical spatial grid of cycloid arcs was proposed by Cruz-Orive and Howard (1995). Virtual cycloids, suitable for optical sectioning on thick slices, were described by Gokhale, Evans, Mackes, and Mouton (2004), see also West (2012).

2.30 Ratios Based on Test Systems

2.30.1 Introduction

The foregoing results enable the derivation of a wide variety of stereological ratio identities which, in the classical abridged notation, and by virtue of Eq. (2.26.7), coincide formally with the corresponding identities derived in this chapter for single unbounded, or bounded, test probes.

For three-dimensional sets, a stereological ratio is often of the form $\gamma_V = \gamma(Y_2)/V(Y_1)$, where $Y_1 \subset \mathbb{R}^3$ is a reference domain of volume $V(Y_1)$, and $Y_2 \subset \mathbb{R}^3$ is a compact target set of finite measure $\gamma(Y_2)$. The sets Y_1, Y_2 may, or may not be disjoint. Usually $Y_2 \subset Y_1$ and, although it is not necessary, we stick to this condition below.

In design-based stereology the ultimate target is usually the global quantity $\gamma(Y_2)$, and γ_V is only a vehicle to access such quantity via the two-stage identity

$$\gamma(Y_2) = V(Y_1) \cdot \gamma_V. \tag{2.30.1}$$

The reference volume $V(Y_1)$ is usually accessed first by the Cavalieri design, see Section 2.25.3, independently from the ratio γ_V. Often, a multistage design is needed involving a chain of ratios (Section 4.2.3).

Next we consider a few examples in which the measurement functions observable in sections are counting measures. The justification of the different choices will be better appreciated when dealing with stereological estimation (Chapter 4).

2.30.2 Volume ratio

The target is $V_V \equiv V(Y_2)/V(Y_1)$. A practical design involves FUR Cavalieri planes, each plane bearing a FUR test system of points, the relevant identity being Eq. (2.25.14). For convenience, the latter test system may consist of a fundamental tile of area a, containing p_1 test points for Y_1, and p_2 test points, distinguishable from the p_1 ones, for Y_2. The result is a special case of a so-called "multipurpose test system", see Fig. 2.30.1(a) for an example. Thus,

$$V_V \equiv \frac{V(Y_2)}{V(Y_1)} = \frac{T \cdot (a/p_2) \cdot \mathbb{E}\{P(Y_2 \cap \Lambda_z \cap \Lambda_x)\}}{T \cdot (a/p_1) \cdot \mathbb{E}\{P(Y_1 \cap \Lambda_z \cap \Lambda_x)\}}, \tag{2.30.2}$$

and, setting $V_i \equiv V(Y_i)$ and $P_i \equiv P(Y_i \cap \Lambda_z \cap \Lambda_x)$, $i = 1, 2$, for short, we have

$$V_V \equiv \frac{V_2}{V_1} = \frac{p_1}{p_2} \cdot \frac{\mathbb{E}(P_2)}{\mathbb{E}(P_1)}. \tag{2.30.3}$$

A justification is that, if we have a rough idea of the value of γ_V, then, for reasons of statistical efficiency (Chapter 4), we should choose $p_1/p_2 \approx \gamma_V$, so that $\mathbb{E}(P_1)$ and

$\mathbb{E}(P_2)$ are not too different from each other. This recommendation usually applies to any other ratios.

An advantage of using ratios is that parameters such as T, a will cancel out. This advantage, however, should not distract us from the fact that Eq. (2.30.3), for instance, holds under a design involving FUR, (or IUR), Cavalieri planes combined with FUR (or IUR) point test systems.

2.30.3 Surface to volume, and surface ratios

Surface to volume ratio

Here $Y_2 \subset Y_1 \subset \mathbb{R}^3$ represents a surface of area S_2 in a reference set Y_1 of volume V_1. The target ratio is $S_V = S_2/V_1$. A suitable design is the ICav design. On each IUR planar section, an IUR multipurpose test system is superimposed with p test points and with a fundamental curve probe of length l, all included in a fundamental tile of area a, see Fig. 2.30.1(b). From Eq. (2.26.18) and Eq. (2.25.14) we obtain

$$S_V \equiv \frac{S_2}{V_1} = 2 \cdot \frac{p}{l} \cdot \frac{\mathbb{E}(I_2)}{\mathbb{E}(P_1)}, \tag{2.30.4}$$

where, for a realization of the ICav plane series, I_2 denotes the total number of intersections determined by the test curves in all the planar surface traces, whereas P_1 is the total number of test points hitting all the planar sections of the reference set, see Fig. 2.26.2(b), and compare with the last identity in Eq. (2.12.19).

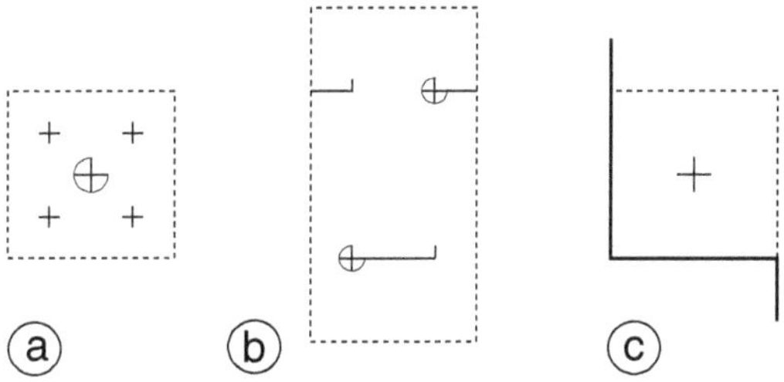

Fig. 2.30.1 (a) Fundamental tile of a multipurpose test system to access V_V, with $p_1 = 1$ and $p_2 = 4$, see Eq. (2.30.3). (b) Idem for S_V, see Eq. (2.30.4). (c) Idem for L_V, see Eq. (2.30.7), or N_V, see Eq. (2.30.8).

The preceding identity remains formally unaltered if each ICav section is first subsampled with a FUR test system of quadrats with fundamental quadrat area $a_0 \leq a$, and then an IUR test system is superimposed on each quadrat with fundamental area a_1, containing p test points and a fundamental test curve of length l in the fundamental tile. It is readily seen that the constants a, a_0, a_1 will cancel out (therefore they do not need to be known). The new expected counts of I_2 and P_1 will be a_0/a_1 times the earlier ones, hence the ratio of their means will remain constant.

The same identity will also hold for the combination of VCav planes with a FUR multipurpose test system containing test points and cycloids, with or without quadrat subsampling.

Surface ratio

If $Y_2 \subset Y_1 \subset \mathbb{R}^3$ are two surfaces of areas S_2 and S_1, respectively, then, with ICav planes sampled with an IUR test system of curves (no test points are needed), we have

$$S_S \equiv \frac{S_2}{S_1} = \frac{\mathbb{E}(I_2)}{\mathbb{E}(I_1)}, \tag{2.30.5}$$

with or without quadrat subsampling (compare with the second identity in Eq. (2.12.20)). The same identity holds for the vertical design with cycloids.

2.30.4 Length per unit volume

If $Y_2 \subset Y_1 \subset \mathbb{R}^3$ represents a curve of length L_2, then the target ratio is $L_V = L_2/V_1$, and a suitable probe is the ICav. On each ICav plane superimpose a FUR test system of points with a fundamental tile of area a containing p test points – in Fig. 2.30.1(c) we have $p = 1$. Then the combination of Eq. (2.26.17) and Eq. (2.25.14) yields

$$L_V \equiv \frac{L_2}{V_1} = 2 \cdot \frac{p}{a} \cdot \frac{\mathbb{E}(Q_2)}{\mathbb{E}(P_1)}, \tag{2.30.6}$$

where, for a realization of the ICav plane series, Q_2 denotes the total number of curve transects scored in the ICav planes with the forbidden line rule, see Fig. 2.30.1(c), and P_1 is as before.

If, besides the p test points, the fundamental tile of the preceding FUR test system contains a quadrat of area $a_0 \leq a$, and each ICav plane is subsampled with this test system, then,

$$L_V \equiv \frac{L_2}{V_1} = 2 \cdot \frac{p}{a_0} \cdot \frac{\mathbb{E}(Q_2)}{\mathbb{E}(P_1)}, \tag{2.30.7}$$

because the new $\mathbb{E}(Q_2)$ in the preceding identity is now equal to a_0/a times the old one without subsampling. Compare with the second identity in Eq. (2.12.21).

Note that the test points do not need to lie within the quadrats.

For two curves, their length ratio identity us analogous to Eq. (2.30.5).

2.30.5 Number per unit volume, and number ratio

Number per unit volume

If $Y_2 \subset Y_1 \subset \mathbb{R}^3$ represents a set of N_2 particles enclosed in a reference set of volume V_1, then the target ratio is $N_V = N_2/V_1$. For a design of FUR Cavalieri slab disectors of thickness $t > 0$ hitting Y_1, the combination of Eq. (2.25.21) and Eq. (2.25.14) yields

$$N_V \equiv \frac{N_2}{V_1} = \frac{p}{at} \cdot \frac{\mathbb{E}(Q_2^-)}{\mathbb{E}(P_1)}, \tag{2.30.8}$$

where Q_2^- denotes the total number of particles captured by all the slab disectors, whereas P_1 is the total number of points from a test system with p test points in a fundamental tile of area $a > 0$, see Fig. 2.30.1(c), hitting the planar sections determined in the reference set Y_1 by the reference (or by the look up) faces of the slabs.

Each slab may be subsampled with a FUR test system of bounded disectors with base area $a_0 > 0$ and thickness $t > 0$, in which case the combination of Eq. (2.25.29) and Eq. (2.25.14) yields the same identity (2.30.8) with the symbol a replaced with a_0 (compare with Eq. (2.15.14)).

Number ratio

If Y_2 and Y_1 are two different sets of N_2 and N_1 particles, respectively, and the disector thickness t is constant, then, whether the disectors are unbounded, or bounded, we have

$$N_N \equiv \frac{N_2}{N_1} = \frac{\mathbb{E}(Q_2^-)}{\mathbb{E}(Q_1^-)}, \tag{2.30.9}$$

with no constants (notably the disector thickness) involved.

Analogous identities hold for the Euler–Poincaré characteristic (Section 2.17).

2.30.6 Note

In design stereology, ratios are usually intermediate quantities used to estimate a global quantity: Eq. (2.30.1) corresponds to a simple design involving only one ratio – more general designs will be described in Chapter 4. Examples of such designs can be found in Weibel (1963, 1969). For further developments see for instance Cruz-Orive and Weibel (1981), Gundersen and Østerby (1981), Cruz-Orive (2009b), and Cruz-Orive, Insausti, Insausti, and Crespo (2004). For the adaptation of such designs to sets of particles, see the next section and Gundersen (1986).

2.31 Classical Ratio Designs for Mean Particle Size

Consider a population of N particles $Y = \{Y_1, Y_2, \ldots, Y_N\} \subset \mathbb{R}^3$. Common target parameters are $\mathbb{E}(V)$, $\mathbb{E}(S)$, namely the mean individual particle volume and surface area, respectively, see Eq. (2.19.10).

A classical setup is $Y \subset D \subset \mathbb{R}^3$, where D is a reference domain. The volume ratio $V_V = V(Y)/V(D)$ is available via Eq. (2.30.3) from FUR Cavalieri planes, whereas the number to volume ratio $N_V = N/V(D)$ is available separately via Eq. (2.30.8) from FUR Cavalieri slabs subsampled with bounded disectors. Then,

$$\mathbb{E}(V) = V_N = \frac{V_V}{N_V}. \tag{2.31.1}$$

The main drawback of the preceding design is that it involves test points not only in the particles, but also in the reference space. This may be circumvented by using FUR Cavalieri slabs directly, in which case the corresponding identity is basically the reciprocal of Eq. (2.30.8), namely,

$$\mathbb{E}(V) = V_N = \frac{at}{p} \cdot \frac{\mathbb{E}(P)}{\mathbb{E}(Q^-)}. \tag{2.31.2}$$

This design, however, requires that the point count P is scored in the particle transects determined by the flat reference (or look up) faces of the slab, or by an optical section.

If the target is $\mathbb{E}(S)$, then the ratio $S_V = S(Y)/V(D)$ is available via Eq. (2.30.4) from ICav planes, whereas N_V is available separately from FUR Cavalieri slabs as before, whereby

$$\mathbb{E}(S) = S_N = \frac{S_V}{N_V}. \tag{2.31.3}$$

The counterpart of Eq. (2.31.2) for mean particle surface area requires ICav slabs. Provided that the total intersections count I can be scored on the particle transect boundaries determined by the flat reference (or look up) faces of the slabs, or on an optical section, the combination of Eq. (2.30.4) and Eq. (2.30.8) yields,

$$\mathbb{E}(S) = S_N = \frac{2at}{l} \cdot \frac{\mathbb{E}(I)}{\mathbb{E}(Q^-)}. \tag{2.31.4}$$

2.32 Local Stereology for Particle Size

2.32.1 Number- and size-weighted particle sampling

Again, let $Y = \{Y_1, Y_2, \ldots, Y_N\} \subset \mathbb{R}^3$ represent a bounded aggregate of N particles with piecewise smooth boundaries. Although no special assumptions are necessary about particle shape, it is necessary to assume that any planar transect (whether connected, or not) of a particle must be unambiguously identifiable as belonging to the particle (resorting to adjacent serial sections if necessary). Let $\gamma_i \equiv \gamma(Y_i)$ denote a size parameter defined on the ith particle, where γ may stand for volume V, surface area S, integral of mean curvature M, mean caliper length H, or Euler–Poincaré characteristic χ. On the other hand, let $W_i \equiv W(Y_i) > 0$ denote a weighting factor for the ith particle. The individual mean W-weighted particle size is defined as

$$\mathbb{E}_W(\gamma) = \sum_{i=1}^{N} W_i \cdot \gamma_i \Big/ \sum_{i=1}^{N} W_i. \tag{2.32.1}$$

Hereafter the sampling probe is assumed to be an IUR test system $\Lambda \subset \mathbb{R}^3$ hitting the particle population Y. Isotropy is usually not necessary to sample particles, but it is so to apply local stereology. Then, $W_i = c \cdot \mathbb{E}\{\alpha(Y_i \cap \Lambda)\}$ where $c > 0$ is a known constant, and α represents the pertinent intersection measure. In the special cases considered below, W_i represents the mean number of times the particle Y_i is sampled with a realization of the test system – in other words, the multiplicity of γ_i in the sample.

If Λ consists of ICav disectors of thickness $t > 0$ and period $T \geq t$, see Fig. 2.32.1(a), then the number of times a particle is sampled is either 0, or 1, the mean number being $W_i = t/T$, namely the (constant) probability that the particle Y_i is sampled. Thus, in this case $\mathbb{E}_W(\gamma)$ becomes the ordinary number weighted mean

$$\mathbb{E}(\gamma) \equiv \mathbb{E}_N(\gamma) = \frac{1}{N} \sum_{i=1}^{N} \gamma_i. \tag{2.32.2}$$

Note that neither t nor T need to be known. With estimation purposes in mind, the preceding identity has to be expressed in terms of mean sample values. For each location and orientation of Λ relative to the particle population Y, consider the set of indices of the disector-sampled particles, namely,

$$\mathcal{S} = \{i \in \{1, 2, \ldots, N\} : \; Y_i \text{ is sampled}\}. \tag{2.32.3}$$

Now Eq. (2.25.20) becomes

$$\gamma(Y) \equiv \sum_{i=1}^{N} \gamma_i = \frac{T}{t} \cdot \mathbb{E} \sum_{i \in \mathcal{S}} \gamma_i, \tag{2.32.4}$$

the expectation being with respect to the joint probability element of the location and orientation of Λ. Further, by Eq. (2.25.21),

$$N = \frac{T}{t} \cdot \mathbb{E}(Q^-), \tag{2.32.5}$$

where Q^- is the random number of disector-sampled particles, namely the size of the set $\mathcal{S}$. Thus,

$$\mathbb{E}_N(\gamma) = \frac{\gamma(Y)}{N} = \frac{\mathbb{E} \sum_{i \in \mathcal{S}} \gamma_i}{\mathbb{E}(Q^-)}. \tag{2.32.6}$$

The next step is to express each γ_i by a convenient representation based on local probes, see Sections 2.32.2–2.32.7 below.

If Λ is a spatial grid of test points with fundamental tile volume $v > 0$ and $p = 1$ test points per tile, then by virtue of Eq. (2.29.1),

$$W_i = \mathbb{E}\{P(Y_i \cap A)\} = V_i/v \tag{2.32.7}$$

is the mean number of test points hitting the particle Y_i. Now $\mathbb{E}(\gamma)$ becomes the volume-weighted mean value of γ, namely,

$$\mathbb{E}_V(\gamma) = \sum_{i=1}^{N} V_i \cdot \gamma_i \Big/ \sum_{i=1}^{N} V_i = \sum_{i=1}^{N} \mathbb{E}(P_i) \cdot \gamma_i \Big/ \sum_{i=1}^{N} \mathbb{E}(P_i), \tag{2.32.8}$$

where $P_i \equiv P(Y_i \cap \Lambda)$, for short. Again, each γ_i has to be expressed as the mean value of a convenient local estimator evaluated at each of the P_i hitting test points, see Section 2.32.3 below.

Finally, suppose that Λ is a spatial grid of test lines with $l/v > 0$ test line length per volume. Then, by Eq. (2.29.4)

$$W_i = \mathbb{E}\{I(\partial Y_i \cap \Lambda)\} = (1/2)(l/v)S_i \tag{2.32.9}$$

is the mean number of times the particle Y_i is sampled, namely the mean number of intersections between the particle boundary and the test lines. Now $\mathbb{E}_W(\gamma)$ is the surface area weighted mean value of γ, namely,

$$\mathbb{E}_S(\gamma) = \sum_{i-1}^{N} S_i \cdot \gamma_i \Big/ \sum_{i=1}^{N} S_i = \sum_{i=1}^{N} \mathbb{E}(I_i) \cdot \gamma_i \Big/ \sum_{i=1}^{N} \mathbb{E}(I_i), \tag{2.32.10}$$

where $I_i \equiv I(Y_i \cap \Lambda)$. In this case, however, there is hitherto no general local estimator available for γ_i which is based on boundary intersections. Nonetheless, noting that $\mathbb{E}_N(S)/\mathbb{E}_N(V) \equiv S_V$, the mean surface area-weighted particle volume is expressible as follows,

$$\mathbb{E}_S(V) = \mathbb{E}_V(S)/S_V, \tag{2.32.11}$$

and $\mathbb{E}_V(S)$ is accessible with the invariator, see Section 2.32.5 below.

If both $\mathbb{E}_N(V)$ and $\mathbb{E}_V(V)$ are available then from the definitions

$$\begin{aligned} \mathbb{E}_V(V) &= \mathbb{E}_N(V^2)/\mathbb{E}_N(V), \\ \mathrm{Var}_N(V) &= \mathbb{E}_N(V^2) - \mathbb{E}_N^2(V), \end{aligned} \tag{2.32.12}$$

it follows that

$$\mathrm{CV}_N^2(V) = \frac{\mathbb{E}_V(V)}{\mathbb{E}_N(V)} - 1, \tag{2.32.13}$$

which shows that the knowledge of the ordinary coefficient of variation of particle volume does not require the knowledge of the volume distribution, but only disector- and point-sampled particles without reconstructions.

2.32.2 Mean volume of nucleolated particles with the nucleator

The target is the mean particle volume $\mathbb{E}_N(V)$ which, by Eq. (2.32.6), may be expressed as follows,

$$\mathbb{E}_N(V) = \frac{\mathbb{E}\sum_{i\in\mathcal{S}} V_i}{\mathbb{E}(Q^-)}. \tag{2.32.14}$$

The method is applicable whenever each particle Y_i contains a single, clearly distinguishable point $y_i \in Y_i$, called the nucleolus, which is adopted as the sampling unit – common examples are some types of neurons. Such particles are called nucleolated particles. Each ICav slab is swept by a plane parallel to the slab faces – in practice this is a sweeping optical section, see Fig. 2.32.1(b) and Section 2.15.2. As it meets a nucleolus y_i, the sweeping plane is effectively an IR pivotal plane through y_i, hence the pivotal nucleator, Eq. (2.18.8), or the integrated nucleator, Eq. (2.18.10), may be used to express the particle volume in the rhs of Eq. (2.32.14). In the former case (Fig. 2.32.1(c)),

$$\mathbb{E}_N(V) = \frac{4\pi \cdot \mathbb{E}\sum_{i\in\mathcal{S}} \mathbb{E}_\omega(f_i(\omega, u))}{\mathbb{E}(Q^-)}. \tag{2.32.15}$$

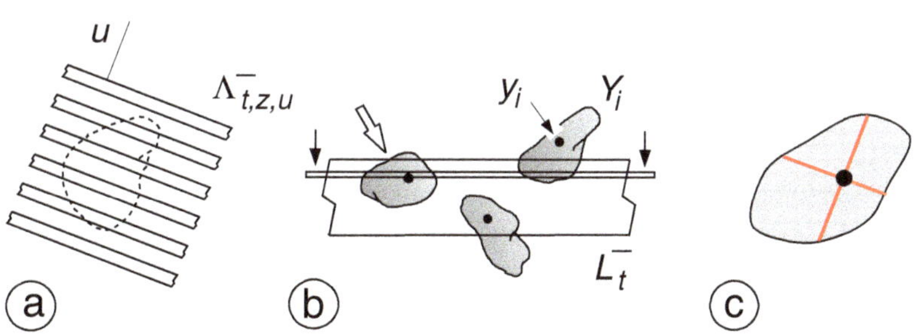

Fig. 2.32.1 Illustration of the nucleator method for individual mean particle volume. (a) A series of ICav slab disectors hitting a reference domain, to sample the particles contained in the domain with identical probabilities. (b) If the particles are nucleolated, then each slab may be swept by a thin optical section to sample a particle whenever its nucleolus is met for the first time. (c) Because the ICav series was isotropic, the corresponding transect may be analysed with the nucleator formula, see Eq. (2.32.15) (e.g. with four systematic IR rays of period $\pi/2$ in the figure).

2.32.3 Volume weighted mean particle volume from point-sampled intercepts

The target is the volume-weighted mean particle volume $\mathbb{E}_V(V)$ which, by Eq. (2.32.8) may be expressed as follows,

$$\mathbb{E}_V = \sum_{i=1}^{N} \mathbb{E}(P_i) \cdot V_i \Big/ \sum_{i=1}^{N} \mathbb{E}(P_i). \tag{2.32.16}$$

Consider an IUR spatial grid $\Lambda \equiv \Lambda_{x,u} \subset \mathbb{R}^3$ with fundamental tile $J_0 = [0, l)^3$, namely a cube of volume $v = l^3$, containing a fundamental probe consisting of the union of a test point at the origin plus the straight line segment $[0, l)$. Thus the grid is the union of an IUR fakir probe of parallel straight lines and a cubic grid of sampling points contained in the lines. The length l does not need to be known.

For each $x \in J_0$ and $u \in \mathbb{S}^2_+$, the grid Λ will generate $P_i \geq 1$ sampling points $\{x_{ij},\ j = 1, 2, \ldots, P_i\}$ hitting the sampled particle Y_i, $i \in \mathcal{S}$. By Eq. (2.19.8),

$$V_i = \pi \cdot \mathbb{E}_{P_i} \left\{ \frac{1}{P_i} \sum_{j=1}^{P_i} f_{0i}(x_{ij}, u) \right\} = \frac{\pi \cdot \mathbb{E} \sum_{j=1}^{P_i} f_{0i}(x_{ij}, u)}{\mathbb{E}(P_i)}, \tag{2.32.17}$$

where $\mathbb{E}_{P_i}$ denotes P_i-weighted expectation. Now Eq. (2.32.16) yields

$$\mathbb{E}_V(V) = \frac{\pi \cdot \sum_{i=1}^{N} \mathbb{E} \sum_{j=1}^{P_i} f_{0i}(x_{ij}, u)}{\sum_{i=1}^{N} \mathbb{E}(P_i)}. \tag{2.32.18}$$

Renumber the whole set of $P = P_1 + P_2 + \cdots + P_N$ functionals as $\{f_{0i},\ i = 1, 2, \ldots, P\}$. Then the preceding identity may be written in a shorthand form as

$$\mathbb{E}_V(V) = \frac{\pi \cdot \mathbb{E} \sum_{i=1}^{P} f_{0i}}{\mathbb{E}(P)}, \tag{2.32.19}$$

which reveals the fact that the sampling units are the sampling points, and not the individual particles. In practice, instead of an IUR spatial grid hitting the particle population, as considered above for convenience, it suffices to consider a test system of ICav planes (Fig. 2.32.2(a)) and to lay an IUR test system of parallel lines bearing equispaced test points on each ICav section, (Fig. 2.32.2(c)). It is also valid to consider a vertical plane, in which case the angle between the parallel lines and the vertical direction must be sine-weighted on each vertical transect.

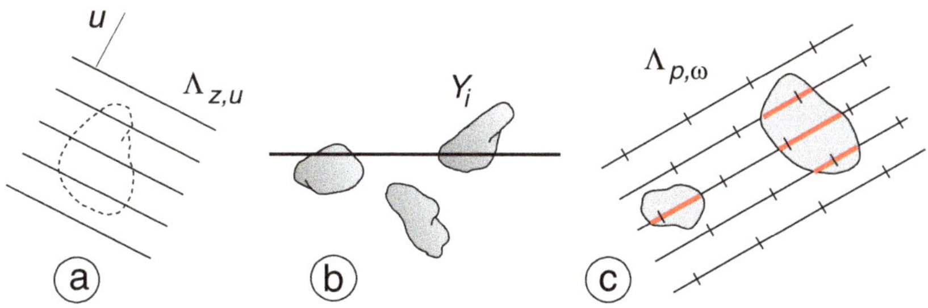

Fig. 2.32.2 Illustration of the point-sampled intercepts method for individual volume-weighted mean particle volume. (a) ICav planes hitting a reference domain containing particles. (b) A ICav section hits two particles. (c) Generation of point-sampled intercepts on the ICav section plane, see Eq. (2.32.19).

2.32.4 Mean particle volume with the selector

The target parameter is $\mathbb{E}_N(V)$, and the method is an alternative to the nucleator for particles that do not exhibit a nucleolus. The particle population Y is intersected by a serial stack of IUR parallel planes at a distance which does not need to be known. Systematic ICav disectors consisting of pairs of planes are used to sample particles, see Fig. 2.32.3(a). The largest disector thickness should not exceed the smallest particle caliper diameter in any direction, so that no particle can escape detection. Irrespective of disector thickness, each sampled particle will be captured by exactly one disector, and this implies that W_i will be constant for all $i = 1, 2, \ldots, N$, so that Eq. (2.32.14) holds unchanged. Each disector-sampled particle Y_i, see Fig. 2.32.3(b), is then hit by a series of Cavalieri planes parallel to the slab, extracted from the original stack, and independent from the slab faces, see Fig. 2.32.3(c)). On each serial plane hitting the same particle, superimpose an IUR test system of parallel lines bearing test points – in Fig. 2.32.3(d) the three Cavalieri sections hitting the particle have been brought to a common plane. Let P_i denote the total number of test points hitting the disector-sampled particle Y_i in its Cavalieri sections. For each hitting point x_{ij}, $j = 1, 2, \ldots, P_i$, compute the functional given by Eq. (2.19.8). Then the volume of Y_i is given by Eq. (2.32.17), which substituted into the rhs of Eq. (2.32.14) yields the selector identity,

$$\mathbb{E}_N(V) = \frac{\pi \cdot \mathbb{E} \sum_{i \in \mathcal{S}} \left\{ \mathbb{E} \sum_{j=1}^{P_i} f_{0i}(x_{ij}, u) \Big/ \mathbb{E}(P_i) \right\}}{\mathbb{E}(Q^-)}. \tag{2.32.20}$$

Because each sampled particle must be traceable on successive serial planes, the total stack height should be at least twice the largest particle diameter. In practice it may be convenient to record first the set of all serial sections of all the particles hit by the stack and then, a posteriori, identify the disector-sampled particles, and their corresponding serial transects, from the entire set. Not all serial sections, but a systematic subsample of them, may be used.

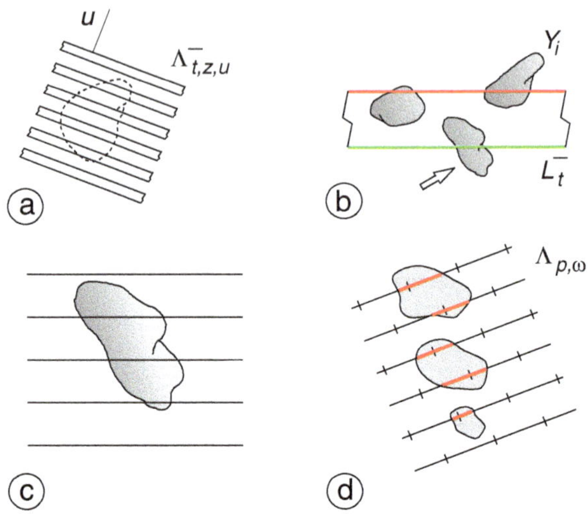

Fig. 2.32.3 Illustration of the selector method for the individual mean volume of not necessarily nucleolated particles. (a) ICav disectors (obtained from an original stack of serial sections) hitting a reference domain containing particles. (b) The arrowed particle is selected by a disector. (c) The disector-sampled particle is hit by independent Cavalieri sections from the original stack. (d) Point-sampled intercepts from the sections in (c). The parameters of the different test systems used need not be known, see Eq. (2.32.20).

By ignoring disectors, $\mathbb{E}_V(V)$ is also accessible from the same material via Eq. (2.32.19), whereby $\mathrm{CV}_N(V)$ is available from Eq. (2.32.13).

Alternatively, the stack of planes may be vertical, in which case the orientation of the test lines bearing test points should be sine-weighted with respect to the vertical direction, as usual.

2.32.5 Mean volume and surface area of nucleolated particles with the invariator

The invariator method is based on measurements made on an IR pivotal plane. Unlike the nucleator and the point-sampled intercepts designs, the invariator cannot be implemented on vertical sections. If each particle bears a single nucleolus, then the latter is adopted as the sampling unit for IUR disector sampling, as in the nucleator case, see Fig. 2.32.4(a,b). Moreover the nucleolus can be used as the pivotal point, whereby not only $\mathbb{E}_N(V)$, but also $\mathbb{E}_N(S)$, become available. The primary identity is Eq. (2.32.6) with $\gamma = V, S$. Next we incorporate the invariator identities for each individual V_i and S_i. The results are based on Sections 1.8 and 2.20.

A posteriori weighted test lines

On a pivotal plane $L^3_{2[0]} \equiv L^3_2(0, u)$, $u \sim \mathrm{UR}(\mathbb{S}^2_+)$, through a pivotal point $O \in Y$, generate an independent IUR grid $\Lambda_{z,\phi}$ of parallel test lines a distance $T > 0$ apart, see Fig. 2.32.4(c) and Eq. (2.26.8). Set

$$r_k = z + kT, \quad k \in \mathbb{Z}, \tag{2.32.21}$$

the signed distance from the pivotal point to the kth test line $L^2_1(r_k, \phi)$ of the grid. The joint probability element relevant to the present design is

$$\mathbb{P}(\mathrm{d}u, \mathrm{d}z, \mathrm{d}\phi) = \frac{\mathrm{d}u}{2\pi} \cdot \frac{\mathrm{d}z}{T} \cdot \frac{\mathrm{d}\phi}{\pi}. \tag{2.32.22}$$

For the ith particle, set

$$I_{ij} = I\left\{\left(\partial Y_i \cap L^3_{2[0]}\right) \cap L^2_1(r_{ij}, \phi)\right\}, \tag{2.32.23}$$

$$I_{ij} = L\left\{\left(Y \cap L^3_{2[0]}\right) \cap L^2_1(r_{ij}, \phi)\right\}. \tag{2.32.24}$$

The application of Eq. (2.26.6) to Eq. (1.8.1) and Eq. (1.8.2), yields the invariator identities for a posteriori weighted test lines applied to a single particle Y_i, namely,

$$S_i = 2\pi T \cdot \mathbb{E} \sum_{j \in \mathbb{Z}} |r_{ij}| \cdot I_{ij}, \tag{2.32.25}$$

$$V_i = \pi T \cdot \mathbb{E} \sum_{j \in \mathbb{Z}} |r_{ij}| \cdot L_{ij}, \tag{2.32.26}$$

respectively. Absolute distances are involved because, in Eq. (1.8.1) and Eq. (1.8.2), the orientation range of the test system was $[0, 2\pi)$, and $r > 0$, see also Section 2.20.1. Substitution of γ_i in turn with the preceding expressions of S_i, and then of V_i, in the rhs of Eq. (2.32.6), yield the required expressions for $\mathbb{E}_N(S)$ and $\mathbb{E}_N(V)$, respectively.

A priori weighted test lines

On the pivotal plane $L^3_{2[0]}$, generate a FUR grid of test points with a fundamental tile J_0 of area $a > 0$. The relevant joint probability element for the pivotal plane and the grid is

$$\mathbb{P}(\mathrm{d}u, \mathrm{d}x) = \frac{\mathrm{d}u}{2\pi} \cdot \frac{\mathrm{d}x}{a}, \quad u \in \mathbb{S}^2_+,\ x \in J_0. \tag{2.32.27}$$

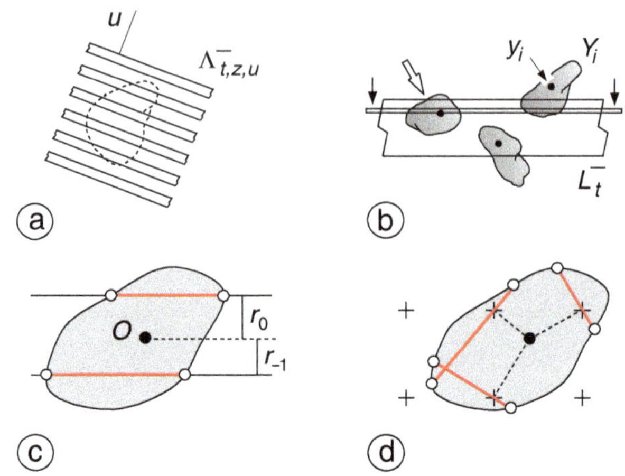

Fig. 2.32.4 Illustration of the invariator method for individual mean particle volume and surface area. (a,b) Same as in Fig. 2.32.1. (c) Pivotal section of a sampled particle analysed by a posteriori weighted test lines, see Eq. (2.32.25) and Eq. (2.32.26). (d) Idem by a priori weighted test lines, see Eq. (2.32.30) and Eq. (2.32.31).

For the ith particle, through each test point x_{ij} of the grid draw a test line $L^2_{1(x_{ij})}$ normal to the axis joining the pivotal point O with the test point, see Fig. 2.32.4(d), and set

$$I_i = \sum_{j \in \mathbb{Z}} I\left\{\left(\partial Y \cap L^3_{2[0]}\right) \cap L^2_{1(x_{ij})}\right\}, \tag{2.32.28}$$

$$L_i = \sum_{j \in \mathbb{Z}} L\left\{\left(Y \cap L^3_{2[0]}\right) \cap L^2_{1(x_{ij})}\right\}. \tag{2.32.29}$$

By the usual argument, the invariator identities for a priori weighted test lines applied to a single particle read,

$$S_i = 2a \cdot \mathbb{E}(I_i), \tag{2.32.30}$$

$$V_i = a \cdot \mathbb{E}(L_i), \tag{2.32.31}$$

and the corresponding identities for $\mathbb{E}(S)$ and $\mathbb{E}_N(V)$ are obtained from Eq. (2.32.6) similarly as above. Eq. (2.32.30) and Eq. (2.32.31) coincide formally with Eq. (2.26.20) and Eq. (2.25.16), respectively. This is not surprising because, under the invariator design, a priori weighted test lines are effectively IUR hitting an object directly in space. Note, however, that while Eq. (2.25.16) does not require isotropy, Eq. (2.32.31) does, because here the measurements are restricted to an IR pivotal plane.

2.32.6 Volume-weighted mean particle volume and surface area with the invariator

The development is analogous to that in Section 2.32.3. The particle population is hit by a series of ICav planes containing a FUR primary grid of points to sample particles with probabilities proportional to particle volume. Each hitting test point is adopted as a pivotal point, and the sectioning plane containing the point is effectively a pivotal plane through that point. Let P_i denote the number of sampling points hitting the particle Y_i. For the jth hitting point, a secondary FUR test grid of points (with tile area $a > 0$ and $p = 1$ test point per tile) is superimposed independently on the pivotal section. With a priori weighted test lines, let $\{I_{ij},\ j = 1, 2, \ldots, P_i\}$ denote the total number of intersections, as defined by Eq. (2.32.28), and $\{L_{ij},\ j = 1, 2, \ldots, P_i\}$ the total intercept lengths, as defined by Eq. (2.32.29), corresponding to the P_i hitting points. Then, by a method analogous to that leading to Eq. (2.32.17), we obtain

$$S_i = \frac{2a \cdot \mathbb{E} \sum_{j=1}^{P_i} I_{ij}}{\mathbb{E}(P_i)}, \qquad V_i = \frac{a \cdot \mathbb{E} \sum_{j=1}^{P_i} L_{ij}}{\mathbb{E}(P_i)}. \tag{2.32.32}$$

Renumber the whole set of $P = P_1 + P_2 + \cdots + P_N$ intersection numbers as $\{I_i,\ i = 1, 2, \ldots, P\}$, and the total intercept lengths as $\{L_i,\ i = 1, 2, \ldots, P\}$. Then, Eq. (2.32.16), and its analogue for surface area, yield

$$\mathbb{E}_V(S) = \frac{2a \cdot \mathbb{E} \sum_{i=1}^{P} I_i}{\mathbb{E}(P)}, \qquad \mathbb{E}_V(V) = \frac{a \cdot \mathbb{E} \sum_{i=1}^{P} L_i}{\mathbb{E}(P)}. \tag{2.32.33}$$

2.32.7 Mean particle volume and surface area with a combination of the selector and the invariator

The mean number-weighted volume and surface area of arbitrary particles which do not contain a distinguishable nucleolus is accessible by combining the selector and the invariator methods. It suffices to replace point-sampled intercepts with intersection numbers (for surface area), or with intercept lengths (for volume) as obtained for instance with a priori weighted invariator lines. With the notation introduced in the preceding section, instead of Eq. (2.32.20) we obtain

$$\mathbb{E}_N(S) = \frac{2a \cdot \mathbb{E} \sum_{i \in \mathcal{S}} \left\{ \mathbb{E} \sum_{j=1}^{P_i} I_{ij} \Big/ \mathbb{E}(P_i) \right\}}{\mathbb{E}(Q^-)},$$

$$\mathbb{E}_N(V) = \frac{a \cdot \mathbb{E} \sum_{i \in \mathcal{S}} \left\{ \mathbb{E} \sum_{j=1}^{P_i} L_{ij} \Big/ \mathbb{E}(P_i) \right\}}{\mathbb{E}(Q^-)}, \tag{2.32.34}$$

respectively.

2.32.8 Particle size distributions

Constant probability particle sampling – usually with disector probes – will yield particle size distributions in the number measure. For instance, for the population Y of N of particles considered so far, let $N(v)$ denote the total number of particles with volume less than or equal to v. Then, the number-weighted distribution function of particle volume is

$$F_N(v) = N(v)/N. \tag{2.32.35}$$

In general, W-weighted sampling will return particle size distributions in the same measure as W. For instance, let $V(v)$ denote the total volume of particles with volume less than or equal to v. Then, the volume-weighted distribution function of particle volume is

$$F_V(v) = V(v) \Big/ \sum_{i=1}^{N} v_i. \tag{2.32.36}$$

The volume of each sampled particle, however, has to be measured as accurately as possible – typically with a sufficient number of Cavalieri sections identifiable on serial stacks. Particle size distributions are expensive to obtain – if only a measure of the variation of particle volume is of interest, then Eq. (2.32.13) looks preferable.

2.32.9 Notes

1. Particle size weighting

Size-weighted particle sampling was studied for instance by Serra (1982), Gundersen and Jensen (1983), Jensen and Gundersen (1985), Gundersen (1986), Jensen (1991), Sørensen (1991) and Karlsson and Cruz-Orive (1997). Section 2.32.1 is partly based on the latter paper. Basic concepts, including the definition in Eq. (2.32.35), can be found in Miles (1978b).

2. Local particle size methods

As so often happens, the less simple concepts, see Section 2.32.3, appeared first, e.g. Miles (1979), and Gundersen and Jensen (1983, 1985). For early applications see Howard (1986) and Brüngger and Cruz-Orive (1987). The paper of Gundersen (1986) contains a wealth of new and useful information up to the date of its publication, including traditional methods (Section 2.31 above), and weighted particle size distributions, in more detail than in Section 2.32.8.

The selector method was proposed by Cruz-Orive (1987a) – for an application see McMillan and Sørensen (1992). Eq. (B.3) of the former paper inspired Gundersen (1988) to propose the nucleator (Section 2.32.2). For an application, see Tandrup

(1993). González-Villa, Cruz, and Cruz-Orive (2017) revisited the nucleator on isolated objects with a focus on its precision.

The invariator was reviewed by Cruz-Orive and Gual-Arnau (2015), who included the a posteriori variant for the first time (Section 2.32.5).

For further historical notes, see Cruz-Orive (2017).

2.33 Pappus–Guldin Identities for Volume

The background results for single probes are given in Section 2.18.4. The target is the volume V of a domain $Y \subset \mathbb{R}^3$. The relevant probes are also local, as above, but here we consider half-planes, or planes, containing a given axis, instead of rays, or lines, through a given point. If the target sets are particles, then the volume identities derived below may be used to obtain the mean number-weighted, or the volume-weighted mean particle volume.

2.33.1 The ray identity for coaxial planes

With the setup defined in Section 1.7.2 , here we consider a test system of $N \geq 1$ half-plane probes a fixed angle $2\pi/N$ apart, sharing a common rotation axis Ox_3, namely,

$$\begin{aligned} &\{L_{2+}^3(\phi_i),\ i = 0, 1, \ldots, N-1\}, \\ &\phi_i = \phi + 2\pi i/N, \quad \phi \sim \mathrm{UR}[0, 2\pi/N). \end{aligned} \tag{2.33.1}$$

The rotation axis does not need to hit the domain Y. Within the ith half-plane probe, $i = 0, 1, \ldots, N-1$, we lay a test system of parallel rays a distance $T > 0$ apart, normal to the rotation axis, namely,

$$\{L_{1+}^2(z_j, \phi_i),\ j \in \mathbb{Z}\}, \quad z_j = z + jT,\ z \sim \mathrm{UR}[0, T). \tag{2.33.2}$$

The relevant joint probability element reads,

$$\mathbb{P}(\mathrm{d}\phi, \mathrm{d}z) = \frac{\mathrm{d}\phi}{2\pi/N} \cdot \frac{\mathrm{d}z}{T}. \tag{2.33.3}$$

Provided that it is nonempty, the ijth ray intercept $L_{1+}^2(z_j, \phi) \cap Y$ will consist of $m_{ij} \geq 1$ separate segments. Let (l_{ijk-}, l_{ijk+}) denote respectively the closer and the farther distances from the Ox_3 axis of the end points of the kth intercept segment within the ijth ray intercept. From Eq. (1.7.17) we obtain,

$$V = \frac{\pi}{N} \cdot T \cdot \mathbb{E}\left\{ \sum_{i=0}^{N-1} \sum_{j \in \mathbb{Z}} \sum_{k=1}^{m_{ij}} (l_{ijk+}^2 - l_{ijk-}^2) \right\}, \tag{2.33.4}$$

the expectation being with respect to the joint probability element $\mathbb{P}(\mathrm{d}\phi, \mathrm{d}z)$.

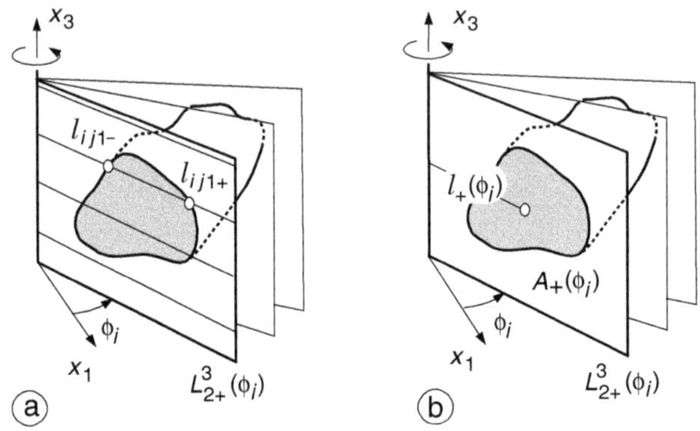

Fig. 2.33.1 Coaxial planes for particle volume. (a) The ray method, see Section 2.33.1. (b) The Pappus-Guldin method, see Section 2.33.3.

As a cursory check, suppose that Y is a ball centred at O, with Ox_3 as the rotation axis, and $N = 1$ half-plane probe only. If the jth ray intercept is nonempty, then $m_{0j} = 1$ and $l_{0j1} = 0$. Moreover, $l_{0j1+} \equiv r_j$, the radius of the disk $L_2^3(z_j, 0) \cap Y$ determined in the ball Y by a plane $L_2^3(z_j, 0)$ normal to the rotation axis through the point $(0, 0, z_j)$. Thus, Eq. (2.33.4) reduces to

$$V = T \cdot \mathbb{E} \sum_{j \in \mathbb{Z}} \pi r_j^2, \tag{2.33.5}$$

which is the Cavalieri identity (with Ox_3 as the sampling axis) for the ball.

In general, set $a(z_j) \equiv A(L_2^3(z_j, 0) \cap Y)$, the section area determined in the domain Y by the plane $L_2^3(z_j, 0)$, which may be regarded as a virtual Cavalieri plane. Within this plane, each coaxial half-plane determines a horizontal ray emanating from the point $(0, 0, z_j)$. Therefore the planar nucleator identity, see Eq. (1.7.4), may be applied to the corresponding virtual Cavalieri section, and we obtain

$$a(z_j) = \mathbb{E}\left\{ \frac{1}{N} \sum_{i=0}^{N-1} \sum_{k=1}^{m_{ij}} \pi \cdot (l_{ijk+}^2 - l_{ijk-}^2) \right\}, \tag{2.33.6}$$

the expectation being over $\phi \sim \text{UR}[0, 2\pi/N)$. Substitution into the rhs of Eq. (2.33.4) yields the alternative identity

$$V = T \cdot \mathbb{E} \sum_{j \in \mathbb{Z}} a(z_j), \tag{2.33.7}$$

which is the Cavalieri identity – the expectation is now over $z \sim \text{UR}[0, T)$.

2.33.2 The vertical rotator

Suppose that the set $Y \in \mathbb{R}^3$ is a particle containing a nucleolus $O \in Y$, and adopt a convenient vertical rotation axis Ox_3 through the nucleolus. This adaptation of the coaxial planes design is called the vertical rotator, and it is useful to access the mean volume $\mathbb{E}(V)$ of a given set of cells bearing a nucleolus. In this case, Eq. (2.32.14) holds for vertical slab disectors. For each particle whose nucleolus O has been met for the first time by a vertical scanning plane within a vertical slab disector, the vertical rotation axis is laid on this plane through the nucleolus, and Eq. (2.33.4) holds. For convenience, $N = 2$ is usually adopted, in which case the two relevant half-planes lie in the vertical plane, and the (horizontal) ray measurements can be made in the same plane on both sides of the vertical rotation axis. With reference to Eq. (2.33.7), it is easy to see that the rotator is a Cavalieri method, with the sections normal to the rotation axis, in which each transect area is obtained by means of two opposed planar nucleator rays.

2.33.3 The Pappus–Guldin identity

Based on the system $\{L_{2+}^3(\phi)\}$ of coaxial planes given by Eq. (2.33.1), application of Eq. (1.7.19) yields

$$V = \frac{2\pi}{N} \cdot \mathbb{E} \sum_{i=0}^{N-1} l_+(\phi_i) \cdot A_+(\phi_i), \tag{2.33.8}$$

where $A_+(\phi_i) \equiv A(Y \cap L_{2+}^3(\phi_i))$ and $l_+(\phi_i)$ denotes the distance from the rotation axis to the centroid of the coaxial section $Y \cap L_{2+}^3(\phi_i)$. The expectation is over $\phi \sim \text{UR}[0, 2\pi/N)$. The preceding identity also holds under the vertical rotator setup.

2.33.4 Note

The coaxial sections method was outlined in Cruz-Orive (1987d), and later developed by Cruz-Orive and Roberts (1993), who applied it to estimate the volume of a human bladder using ultrasound probes. Independently, Jensen and Gundersen (1993) coined the term rotator, developed the method, and applied it to nucleolated neurons on vertical sections.

Chapter 3
Model and Second-Order Stereology

3.1 Random Processes of Geometric Objects: Basic Concepts

3.1.1 Introduction

In the preceding chapters the approach was design-based, involving a bounded, non-random target set hit by a test probe equipped with a motion-invariant density. In the present chapter, however, the approach is model-based: the target is a random process $Y \subset \mathbb{R}^d$ of geometric objects of dimension $q \in \{0, 1, \dots, d\}$. For instance, $Y \subset \mathbb{R}^3$ may represent a point, curve, surface, or volume process. Henceforth, "$Y \subset \mathbb{R}^d$" means that the process Y has realizations in $\mathbb{R}^d$. In simple terms, the main aspects of model stereology considered here are the following.

1. The probes do not need to be equipped with a motion-invariant probability element because the randomness is attributed to the target object, which is modelled by a random process. This implies some degree of belief, which is often justified by 'inspection', especially in materials sciences. Roughly, a section of the material looks 'nearly the same' everywhere – this corresponds to the concept of stationarity or homogeneity, as described below.
2. Stationarity implies that the model is unbounded, with no appreciable inhomogeneities, nor natural boundaries typical in biological specimens. The design is simpler because it usually consists of sampling with arbitrarily located bounded probes.
3. Unboundedness implies that the target is usually a ratio, e.g., V_V, or S_V, which are interpreted as the volume, or the surface area, 'intensities' of the corresponding processes.
4. If stationarity holds, then the theorem of Joseph Mecke (1938–2014), outlined in Section 3.3.2 below, warrants that the classical ratio identities of design stereology hold formally unchanged.
5. Model stereology is founded on stochastic geometry. While design stereology is mainly concerned with object properties such as L, S, V, etc., called 'first-order properties', stochastic geometry supplies in addition tools for second-order

L. M. Cruz-Orive, *Stereology*, Interdisciplinary Applied Mathematics 59,
https://doi.org/10.1007/978-3-031-52451-6_3

properties, often addressed to characterize partial aspects of structural patterns, see e.g. Sections 3.6, 3.7. Means and variances studied in classical statistics may be regarded as first- and second-order properties, respectively.

6. Beyond the stationarity property, a common approach of stochastic geometry is to formulate models for point, line, or particle processes, well-known examples of which are described in the latter sections of this chapter. The idea is to adopt a model depending on a set of parameters which can be related to sample measures using statistical inference. In practice, realistic models are often intractable analytically, whereby inference usually requires intensive computing. These aspects transcend the scope of this book.

3.1.2 Definitions

The definition of a random q-dimensional set or 'manifold' process $Y \subset \mathbb{R}^d$, also called a random closed set, may be outlined as follows. Let Ω denote a family of closed subsets $y \in \mathbb{R}^d$, with the property that for any ball $D \subset \mathbb{R}^d$, the intersection $y \cap D$ is a closed q-dimensional set, piecewise smooth of class C^1, and of finite measure $\gamma(y \cap D)$. This property warrants the necessary regularity conditions on the sets of Ω. Now it only remains to define the corresponding measurable space, and a probability measure on it. Let $\mathcal{A}$ denote the σ-algebra of sets of Ω generated by all functions $f : y \to \gamma(y \cap T)$, where $T \subset \mathbb{R}^d$ is any Borel set – in the present context, T may be regarded as a fixed and bounded probe. Then, a q-dimensional manifold process is a random variable Y whose probability element $\mathbb{P}(\mathrm{d}y)$ is defined on the measurable space $(\Omega, \mathcal{A})$. In particular, the preceding construction ensures that $\gamma(y \cap T)$is a random variable.

The mean value of an $\mathcal{A}$-measurable function $f : \mathcal{A} \to [0, \infty)$ is

$$\mathbb{E}(f) = \int f(y)\mathbb{P}(\mathrm{d}y). \tag{3.1.1}$$

For instance, if $T \subset \mathbb{R}^d$ is a fixed and bounded r-dimensional probe, then

$$\mathbb{E}\{\alpha(Y \cap T)\} = \int \alpha(y \cap T)\mathbb{P}(\mathrm{d}y). \tag{3.1.2}$$

Remark on notation

As in the preceding chapters, the q-measure of a bounded portion of Y, the r-measure of a bounded probe T, and the $(q + r - d)$-measure of $Y \cap T$, will be denoted by γ, ν, and α, respectively, for short.

Example

If $Y \subset \mathbb{R}^3$ represents a surface process and $T \subset \mathbb{R}^3$ is a fixed and bounded curve probe, see Fig. 3.1.1, then

$$\mathbb{E}\{I(Y \cap T)\} = \int I(y \cap T)\mathbb{P}(\mathrm{d}y). \tag{3.1.3}$$

In the analogous design-based case, Y represented a bounded surface of area S, and T a test curve of length l equipped with the invariant probability element given by Eq. (2.11.5). The identity analogous to Eq. (3.1.3) was based on Eq. (1.16.7), and it yielded the explicit mean value given by Eq. (2.12.16). In contrast, in the model-based case the randomness is attributed to the surface process, and the probe may be arbitrarily fixed. However, Eq. (3.1.3) is open ended and, as such, it does not supply an explicit result directly. In order to implement stereology in the model case, a common resource is to assume that the random process is stationary, as described next.

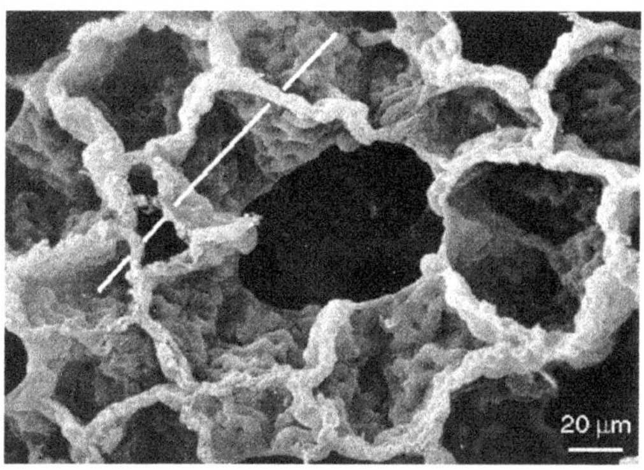

Fig. 3.1.1 A block of lung parenchyma of gnu, obtained by SEM (Scanning Electron Microscopy), with a needle probe hitting it, to illustrate Eq. (3.1.3). If the material can be modelled by a stationary and isotropic surface process, then the intensity of the latter is $S_V = 2I_L$, namely twice the mean number of intersections divided by the needle length, see Eq. (3.3.15). Courtesy of Ewald R. Weibel, modified from Gehr et al. (1981), with permission of Elsevier.

3.2 Stationarity and Isotropy

Let f denote a measurable function, as above. If the identity

$$\int f(y)\mathbb{P}(\mathrm{d}y) = \int f(y+x)\mathbb{P}(\mathrm{d}y) \tag{3.2.1}$$

holds for any translation vector $x \in \mathbb{R}^d$, then the process $Y \subset \mathbb{R}^d$ is said to be (first-order) stationary, see Fig. 3.2.1. In this case, if $C_d = [0,1]^d$ denotes the unit cube, then

$$\gamma_V \equiv \mathbb{E}\{\gamma(Y \cap C_d)\} = \int \gamma(y \cap C_d)\mathbb{P}(\mathrm{d}y) \tag{3.2.2}$$

is called the intensity of the stationary process Y. For an arbitrarily fixed domain probe $T \subset \mathbb{R}^d$ with $\dim(T) = d$ and volume $V(T) > 0$, we have

$$\gamma_V = \frac{\mathbb{E}\{\gamma(Y \cap T)\}}{V(T)}. \tag{3.2.3}$$

If $d = 3$ and $\dim(Y) = q$, then γ_V stands for $N_V, \chi_V, L_V, S_V, M_V, V_V$, according to whether $q = 0, 0, 1, 2, 2, 3$, respectively.

If the identity

$$\int f(y)\mathbb{P}(\mathrm{d}y) = \int f(u_d y)\mathbb{P}(\mathrm{d}y) \tag{3.2.4}$$

holds for any rotation $u_d \in G_{d[0]}$ (see Section 1.16.4), then the process $Y \subset \mathbb{R}^d$ is said to be isotropic, see Fig. 3.2.1(a,c).

If, for any special motion $g = (x, u_d)$, $x \in \mathbb{R}^d$, $u_d \in G_{d[0]}$, the identity

$$\int f(y)\mathbb{P}(\mathrm{d}y) = \int f(gy)\mathbb{P}(\mathrm{d}y) \tag{3.2.5}$$

always holds, then the process $Y \subset \mathbb{R}^d$ is said to be stationary and isotropic or, equivalently, motion-invariant, see Fig. 3.2.1(a).

Remarks

In the following sections we show that, if a process is motion-invariant, then the classical design-based stereological equations hold without formal change for an arbitrary fixed probe. This is because the relevant identities for a motion-invariant process are based on the motion-invariant densities of integral geometry, e.g. the kinematic density, see next. For instance, given that a point of a stationary point process hits a domain, the point is UR in the domain. Or, given that a line from a stationary and isotropic line process hits a compact set, the line is IUR hitting the set, etc. As far as sampling is concerned, if a process is stationary but not isotropic,

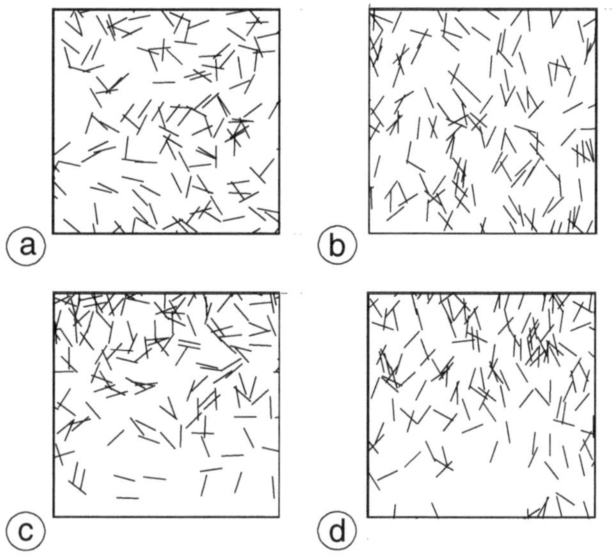

Fig. 3.2.1 (a) A realization of a stationary (or 'homogeneous') and isotropic straight line segment process in a quadrat, see Section 3.12.4. (b) Idem stationary and anisotropic. (c) Idem non-stationary and isotropic. (d) Idem non-stationary and anisotropic.

then for the pertinent equations to hold it suffices to fix the location of the AP of the probe, and to equip the latter with the isotropic probability element.

3.3 Motion-Invariant Process Hit By a Fixed Bounded Probe

3.3.1 Preliminary design-based identities

Consider a fixed compact set $Y \subset \mathbb{R}^d$ of dimension $q \in \{0, 1, \dots, d\}$ and measure $\gamma(Y)$, hit by a bounded probe $T \subset \mathbb{R}^d$ of dimension $r \in \{0, 1, \dots, d\}$, $q + r \geq d$, and measure $\nu(T)$, equipped with the kinematic density $\mu(\mathrm{d}g) = [\mathrm{d}x\, \mathrm{d}u_d]$. In the sequel we will use Eq. (1.14.1), namely,

$$\int \gamma(Y \cap gT)\mu(\mathrm{d}g) = c_2 \cdot \nu(T) \cdot \gamma(Y), \tag{3.3.1}$$

where $c_2 \equiv c_2(q, r, d)$ is a constant given by Eq. (1.16.11), $g(T)$ is the result of applying a rigid motion g to the probe, and $\mu(\mathrm{d}g)$ means the same as $\mathrm{d}T$ in our usual notation. If $\dim(T) = d$, then,

$$\int \gamma(Y \cap gT)\mu(\mathrm{d}g) = c_{20} \cdot V(T) \cdot \gamma(Y), \tag{3.3.2}$$

where $c_{20} \equiv c_{20}(d)$ is given by Eq. (1.16.12).

3.3.2 General equation. Mecke's theorem

Here $Y \subset \mathbb{R}^d$ represents a motion-invariant manifold process of dimension q, whereas $T \subset \mathbb{R}^d$ is a fixed and bounded probe of dimension r. Mecke's theorem states that the intensity of the process may be expressed as follows,

$$\gamma_V = \frac{c_{20}}{c_2} \cdot \frac{\mathbb{E}\{\alpha(Y \cap T)\}}{\nu(T)}, \tag{3.3.3}$$

where c_{20}/c_2 is given by Eq. (2.26.7). To prove the preceding result, first multiply both sides of Eq. (3.2.2) by $\nu(T)$ to obtain

$$\nu(T) \cdot \gamma_V = \nu(T) \int \gamma(y \cap C_d)\mathbb{P}(\mathrm{d}y), \tag{3.3.4}$$

and we want to show that this is equal to $(c_{20}/c_2)\mathbb{E}\{\alpha(Y \cap T)\}$. In Eq. (3.3.1) replace the fixed set Y with the set $y \cap C_d$, namely,

$$\nu(T) \cdot \gamma(y \cap C_d) = \frac{1}{c_2} \int \alpha(y \cap C_d \cap gT)\mu(\mathrm{d}g). \tag{3.3.5}$$

Multiplying both sides of the preceding identity by $\mathbb{P}(\mathrm{d}y)$, integrating, and using Eq. (3.3.4), we obtain

$$\begin{aligned} \nu(T) \cdot \gamma_V &= \frac{1}{c_2} \int \mathbb{P}(\mathrm{d}y) \int \alpha(y \cap C_d \cap gT)\mu(\mathrm{d}g) \\ &= \frac{1}{c_2} \int \mu(\mathrm{d}g) \int \alpha(gy \cap gT \cap C_d)\mathbb{P}(\mathrm{d}y) \\ &= \frac{1}{c_2} \int \mathbb{P}(\mathrm{d}y) \int \alpha(y \cap T \cap g^{-1}C_d)\mu(\mathrm{d}g), \end{aligned} \tag{3.3.6}$$

where the second identity follows by virtue of Eq. (3.2.5), and the third by the motion invariance of the kinematic density. In the lhs of Eq. (3.3.2), replace Y with $y \cap C_d$. By motion invariance we can write

$$\int \alpha(y \cap C_d \cap gT)\mu(\mathrm{d}g) = \int \alpha(y \cap T \cap g^{-1}C_d)\mu(\mathrm{d}g), \tag{3.3.7}$$

whereby the new set is $y \cap T$ (of dimension $q + r - d$) and the new probe is C_d. Because $\dim(C_d) = d$ and $\nu(C_d) = 1$, for each subset y Eq. (3.3.2) gives us

$$\int \alpha(y \cap T \cap g^{-1} C_d)\mu(\mathrm{d}g) = c_{20} \cdot \alpha(y \cap T). \tag{3.3.8}$$

Substitution of the preceding identity into the rhs of Eq. (3.3.6) yields

$$\nu(T) \cdot \gamma_V = \frac{c_{20}}{c_2} \int \alpha(y \cap T)\mathbb{P}(\mathrm{d}y) \tag{3.3.9}$$

$$= \frac{c_{20}}{c_2} \mathbb{E}\{\alpha(Y \cap T)\}, \tag{3.3.10}$$

which is Eq. (3.3.3), namely the general stereological equation applying to the model-based case for a bounded probe.

For two processes $Y_2, Y_1 \subset \mathbb{R}^d$ with intensities γ_V, β_V, respectively, it is possible to define the ratio $\gamma_\beta = \gamma_V/\beta_V$ and apply Mecke's theorem for numerator and denominator.

Remark

The bounded probe T may be connected, or may be a bounded portion of a test system, see next.

3.3.3 Special cases for a bounded probe

Planar case

Let $Y \subset \mathbb{R}^2$ denote a planar, motion-invariant curve process of intensity B_A hit by an arbitrary bounded curve probe $T \subset \mathbb{R}^2$ of fixed length $L(T)$, see Fig. 3.3.1(a). Thus, $d = 2$ and $q = r = 1$, whereby Eq. (3.3.3) yields

$$B_A = \frac{2\pi}{4} \cdot \frac{\mathbb{E}(I)}{L(T)} = \frac{\pi}{2} \cdot I_L. \tag{3.3.11}$$

If the process Y is stationary but anisotropic, then the probe must be isotropically oriented, see Fig. 3.3.1(b).

For a stationary planar particle process of intensity N_A sampled by a fixed bounded quadrat T, or by a bounded portion of a test system of quadrats, we have

$$N_A = \frac{\mathbb{E}(Q^-)}{A(T)} = Q_A^-, \tag{3.3.12}$$

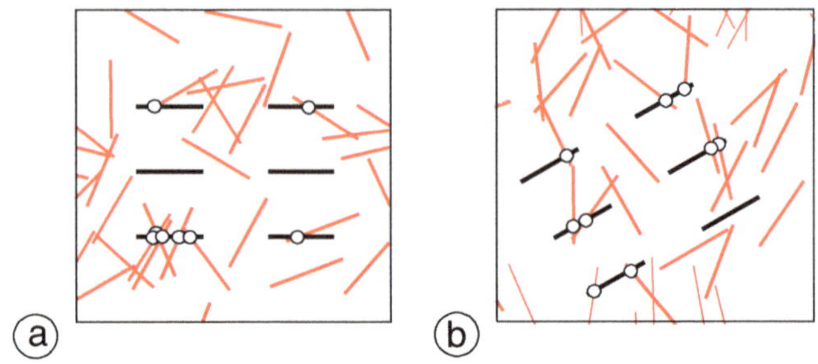

Fig. 3.3.1 (a) A motion-invariant segment process (in red) hit by a fixed probe consisting of a portion of a test system of line segments to illustrate Eq. (3.3.11). (b) If the process is anisotropic, then Eq. (3.3.11) holds if the probe is isotropically oriented.

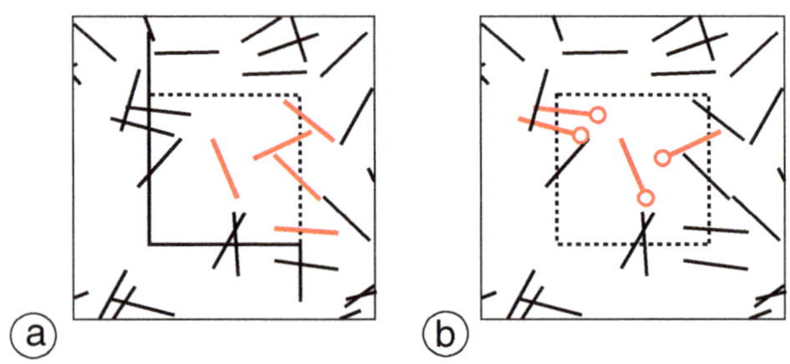

Fig. 3.3.2 (a) A stationary segment process hit by a fixed quadrat to apply Eq. (3.3.12) with the forbidden line counting rule. For the given realization, $Q^- = 5$ segments (in red) are counted. (b) Idem with the AP rule. The AP of a segment is defined here as its lowest end point, this yielding $Q^- = 4$. Alternatively both end points may be counted, in which case the total count must be halved – here $Q^- = 11/2 = 5.5$.

where Q^- is the number of particles captured by the probe with a proper counting rule that removes edge effects, see Section 2.15.2 and Fig. 3.3.2.

Eq. (3.3.12) extends to higher dimensions.

Spatial case

For a stationary volume process $Y \subset \mathbb{R}^3$, $(q = d = 3)$, we have $c_{20} = 8\pi^2$. For a fixed probe of dimension $r = 3, 2, 1, 0$, Eq. (3.3.3) yields the familiar identities,

$$V_V = A_A = L_L = P_P, \tag{3.3.13}$$

respectively.

For a motion-invariant surface process hit by a fixed and bounded surface probe, $(d = 3, q = r = 2)$, we have

$$S_V = \frac{8\pi^2}{2\pi^3} \cdot \frac{\mathbb{E}(B)}{A(T)} = \frac{4}{\pi} \cdot B_A. \tag{3.3.14}$$

If the probe is a fixed and bounded curve probe $(q = 2, r = 1)$, then

$$S_V = \frac{8\pi^2}{4\pi^2} \cdot \frac{\mathbb{E}(I)}{L(T)} = 2I_L. \tag{3.3.15}$$

Finally, for a motion-invariant curve process hit by a fixed and bounded surface probe, $(q = 1, r = 2)$,

$$L_V = \frac{8\pi^2}{4\pi^2} \cdot \frac{\mathbb{E}(Q)}{A(T)} = 2Q_A. \tag{3.3.16}$$

3.4 Motion-Invariant Process Hit By a Fixed r-Plane Probe

3.4.1 General equation for an r-plane probe

If a motion-invariant process $Y \subset \mathbb{R}^d$ of intensity γ_V is hit by a fixed r-plane $L^d_{r[0]}$, then the process $Y \cap L^d_{r[0]}$ is also motion-invariant because in Eq. (3.2.5) the special motion can be $g = (x, u_r)$, $x \in \mathbb{R}^r$, $u_r \in G_{r[0]}$. Let α_A denote the intensity of the process $Y \cap L^d_{r[0]}$, and consider the r-cube $C_r = [0, 1]^r \subset L^2_{r[0]}$. Then,

$$\alpha_A = \int \alpha(y \cap C_r)\mathbb{P}(\mathrm{d}y) = \frac{c_2}{c_{20}} \cdot \gamma_V, \tag{3.4.1}$$

where the first identity follows by the definition of intensity, and the second by Eq. (3.3.10). Thus,

$$\gamma_V = \frac{c_{20}}{c_2} \cdot \alpha_A, \tag{3.4.2}$$

which is the general model-based stereological equation for an unbounded r-plane probe.

3.4.2 Special cases for an unbounded probe

If a motion-invariant process $Y \subset \mathbb{R}^3$ is hit by a fixed plane, or a straight line, then, depending on the dimension of Y, Eq. (3.4.2) yields formally the same equations as for the bounded probe case, namely,

$$V_V = A_A = L_L, \; S_V = \frac{4}{\pi} B_A = 2I_L, \; L_V = 2Q_A, \; M_V = C_A. \tag{3.4.3}$$

The Delesse principle, expressed by the identity $V_V = A_A$, see Eq. (2.6.5), was originally conceived for a polished section of a rock, or a mineral. The closest model for this scenario is a stationary volume process hit by a fixed plane, as above.

Mean free path

Consider an invariant particle process $Y \subset \mathbb{R}^3$ with volume and surface area intensities V_V and S_V, respectively. If in Eq. (2.7.7) we let the reference domain D increase indefinitely, then the mean free path for the process Y becomes

$$\mathbb{E}(l) = 4 \cdot \frac{1 - V_V}{S_V}. \tag{3.4.4}$$

3.5 Intersection of Two Motion-Invariant Processes

3.5.1 General stereological equation

Consider two independent motion-invariant processes $Y, T \subset \mathbb{R}^d$ of dimensions q, r, with $q + r \geq d$, and intensities γ_V, ν_V, respectively. With probability one, their intersection $Y \cap T$ is a motion-invariant process of dimension $q + r - d$ and intensity α_V, say. By definition,

$$\begin{aligned} \alpha_V &= \mathbb{E}\{\alpha(Y \cap T \cap C_d)\} \\ &= \int \mathbb{P}(\mathrm{d}t) \int \alpha(y \cap t \cap C_d)\mathbb{P}(\mathrm{d}y), \end{aligned} \tag{3.5.1}$$

where the subset t is, for the process T, the analogue of the subset y for the process Y. By virtue of Eq. (3.3.9) with T replaced with $t \cap C_d$, we have

$$\int \alpha(y \cap t \cap C_d)\mathbb{P}(\mathrm{d}y) = \frac{c_2}{c_{20}} \cdot \nu(t \cap C_d) \cdot \gamma_V, \tag{3.5.2}$$

which, substituted into the rhs of Eq. (3.5.1), yields the intensity of the intersection process, namely,

$$\begin{aligned} \alpha_V &= \frac{c_2}{c_{20}} \cdot \gamma_V \cdot \int \nu(t \cap C_d)\mathbb{P}(\mathrm{d}t) \\ &= \frac{c_2}{c_{20}} \cdot \gamma_V \cdot \nu_V. \end{aligned} \tag{3.5.3}$$

3.5.2 Special cases

For $d = 2$ and $q = r = 1$, namely for two motion-invariant curve processes of intensities B_{A1} and B_{A2} in the plane, their intersection is a motion-invariant point process of intensity

$$I_A = \frac{2}{\pi} \cdot B_{A1} B_{A2}, \qquad (3.5.4)$$

see Fig. 3.5.1.

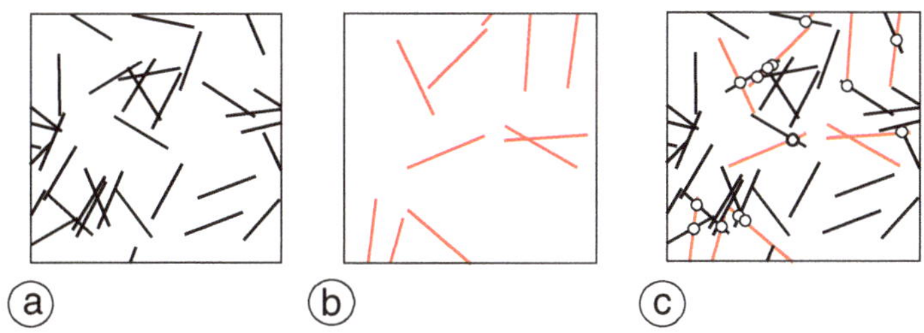

Fig. 3.5.1 (a) A realization in a quadrat of a motion-invariant segment process of length intensity B_{A1}. (b) Idem of length intensity B_{A2}. (c) Their intersection is a motion-invariant point process of intensity I_A given by Eq. (3.5.4).

The intersection of two motion-invariant surface processes in $\mathbb{R}^3$ is a motion-invariant curve processes of intensity

$$L_V = \frac{\pi}{4} \cdot S_{V1} S_{V2}. \qquad (3.5.5)$$

Finally, the intersection of two motion-invariant curve and surface processes of intensities L_V and S_V in $\mathbb{R}^3$, respectively, is a motion-invariant point processes of intensity

$$Q_V = \frac{1}{2} \cdot L_V S_V. \qquad (3.5.6)$$

3.5.3 Notes

1. Mecke's theorem

The material in Sections 3.1–3.5 is adapted from Mecke (1981). The Delesse principle $V_V = A_A$ was proposed by the geologist Delesse (1847) in the model-based context. The same can be said of Rosiwal's relation $V_V = L_L$ (Rosiwal, 1898), see Note 1 in Section 2.8.5, and Glagolev's $V_V = P_P$ (Glagolev, 1933), see Note 2 of Section 2.25.8. Eq. (3.4.4) was derived by Fullman (1953), Eq. (22).

2. Stochastic geometry

The general theory of random sets belongs to stochastic geometry – for early developments see Matérn (1960), Harding and Kendall (1974) and Matheron (1975). More recent references are Stoyan, Kendall, and Mecke (1995), Baddeley (1999), Beneš and Rataj (2004), and Schneider and Weil (2008).

3.6 Second-Order Functions

3.6.1 Preliminary comments

The intensity γ_V of a stationary process is a first-order property, because it is proportional to the first moment of the random variable $\gamma(Y \cap T)$, see Eq. (3.2.3). In the applications, the intensity may be a satisfactory descriptor of a material, but often it is advisable to seek higher-order properties which better discriminate among different materials. The common choices are second-order properties, which are the subject of this section. It should be borne in mind, however, that in general, higher-order properties do not characterize a random set completely.

For didactic purposes the ensuing results are restricted to $\mathbb{R}^3$ – the extensions to $\mathbb{R}^d$ are relatively straightforward, and usually available in the literature.

3.6.2 Non-centred covariance functions

Consider a volume process $Y \subset \mathbb{R}^3$, and its indicator function $1_Y(x)$, $x \in \mathbb{R}^3$. For any two distinct points or vectors $x_1, x_2 \in \mathbb{R}^3$, the vectorial centred covariance of the corresponding indicators is

$$\begin{aligned}\mathrm{Cov}\{1_Y(x_1), 1_Y(x_2)\} &= \mathbb{E}\{1_Y(x_1) \cdot 1_Y(x_2)\} - \mathbb{E}\{1_Y(x_1)\} \cdot \mathbb{E}\{1_Y(x_2)\} \\ &= \mathbb{P}(x_1 \in Y, x_2 \in Y) - \mathbb{P}(x_1 \in Y) \cdot \mathbb{P}(x_2 \in Y). \qquad (3.6.1)\end{aligned}$$

The corresponding coefficient of covariation is

$$\mathrm{CCV}\{1_Y(x_1), 1_Y(x_2)\} = \frac{\mathbb{P}(x_1 \in Y, x_2 \in Y)}{\mathbb{P}(x_1 \in Y) \cdot \mathbb{P}(x_2 \in Y)} - 1. \qquad (3.6.2)$$

In stochastic geometry and second-order stereology, however, non-centred versions are preferred for practical reasons. The vectorial non-centred covariance function of Y is

$$C(x_1, x_2) = \mathbb{P}(x_1 \in Y, x_2 \in Y), \quad x_1, x_2 \in \mathbb{R}^3. \qquad (3.6.3)$$

If Y is stationary, then the vectorial covariance function depends only on the vector $x_2 - x_1 = x$, that is,

$$C(x) = \mathbb{P}(0 \in Y, x \in Y), \quad x \in \mathbb{R}^3, \tag{3.6.4}$$

where $0 \in \mathbb{R}^3$ is a fixed point, and the point or vector x is arbitrary.

If the volume process $Y \subset \mathbb{R}^d$ is stationary of intensity V_V, then

$$C(0) = \mathbb{P}(Y \ni 0) = V_V. \tag{3.6.5}$$

In fact, if C_d denotes the unit cube, then

$$\begin{aligned} V_V &= \mathbb{E}\{V(Y \cap C_d)\} \\ &= \int_{C_d} \mathbb{E}\{1_Y(x)\}\, \mathrm{d}x \\ &= \int_{C_d} \mathbb{P}(Y \ni x)\, \mathrm{d}x \\ &= \mathbb{P}(Y \ni 0) \int_{C_d} \mathrm{d}x, \end{aligned} \tag{3.6.6}$$

the last identity following from the stationarity of Y.

Let (r, u_2), $r > 0$, $u_2 \in \mathbb{S}^2$ denote the spherical polar coordinates of the point $x \in \mathbb{R}^3$. If the volume process Y is motion-invariant, then the covariance function depends only on the modulus r of the vector x, and it becomes a scalar function independent of the orientation u_2 of the vector, namely,

$$C(r) = \mathbb{P}(0 \in Y,\ x = (r, u_2) \in Y), \quad r > 0,\ u_2 \in \mathbb{S}^2. \tag{3.6.7}$$

3.6.3 The K-function, and the pair correlation function

First we give a useful interpretation of the scalar covariance function $C(r)$. For a motion-invariant volume process $Y \subset \mathbb{R}^3$ we have

$$\begin{aligned} C(r) &= \mathbb{P}(x \in Y | 0 \in Y)\, \mathbb{P}(0 \in Y) \\ &= V_V \cdot \mathbb{P}(x \in Y | 0 \in Y). \end{aligned} \tag{3.6.8}$$

Consider a sphere $\partial D_0(r)$ of a given radius $r > 0$ centred at a point $0 \in Y$ of the volume process (hence the subscript '0'), called a 'typical point' of Y, see Fig. 3.6.1(a). The analogue in design stereology is a UR point in a domain. The area element at a point $x \in \partial D_0(r)$ on the corresponding spherical surface is $\mathrm{d}x = r^2\, \mathrm{d}u_2$, $u_2 \in \mathbb{S}^2$, whereby the mean surface area of the intersection between the sphere and the volume process, conditional on the event $0 \in Y$, is proportional to $C(r)$, namely,

$$
\begin{aligned}
\mathbb{E}\{S(Y \cap \partial D_0(r))\} &= \int_{\partial D_0(r)} \mathbb{P}(x \in Y | 0 \in Y)\,\mathrm{d}x \\
&= V_V^{-1} \int_{\partial D_0(r)} C(r)\,\mathrm{d}x \\
&= V_V^{-1} r^2 C(r) \int_{\mathbb{S}^2} \mathrm{d}u_2 \\
&= V_V^{-1} \cdot 4\pi r^2 C(r). \qquad (3.6.9)
\end{aligned}
$$

Therefore, the mean volume occupied by Y within a spherical shell $D_0(\mathrm{d}\rho)$ of radii ρ and $\rho + \mathrm{d}\rho$ centred at $0 \in Y$ is

$$
\begin{aligned}
\mathbb{E}\{V(Y \cap D_0(\mathrm{d}\rho))\} &= \mathbb{E}\{S(Y \cap \partial D_0(\rho))\}\,\mathrm{d}\rho \\
&= V_V^{-1} \cdot 4\pi \rho^2 C(\rho)\,\mathrm{d}\rho \\
&\equiv V_V \cdot \mathrm{d}K(\rho), \qquad (3.6.10)
\end{aligned}
$$

where $K(r)$, $r > 0$, is called the K-function of the volume process Y. Integration over ρ, shows that, for a volume process, $K(r)$ is V_V^{-1} times the mean total volume occupied by Y within the ball $D_0(r)$, see Fig. 3.6.1(b), namely,

$$
\begin{aligned}
\mathbb{E}\{V(Y \cap D_0(r))\} &= V_V \cdot K(r) \\
&= \frac{4\pi}{V_V} \int_0^r \rho^2 C(\rho)\,\mathrm{d}\rho, \quad r > 0. \qquad (3.6.11)
\end{aligned}
$$

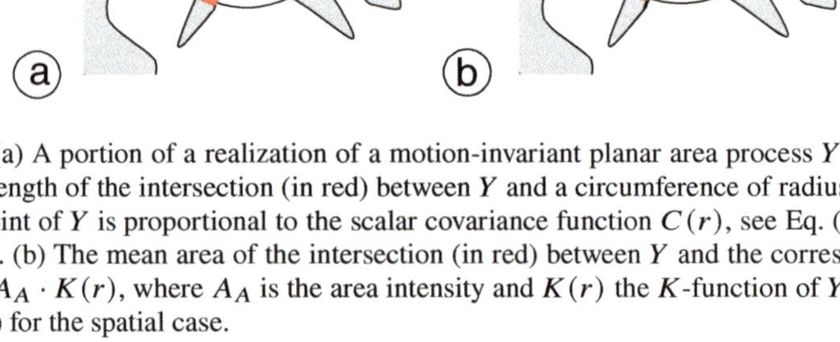

Fig. 3.6.1 (a) A portion of a realization of a motion-invariant planar area process Y in the plane. The mean length of the intersection (in red) between Y and a circumference of radius r centred at a typical point of Y is proportional to the scalar covariance function $C(r)$, see Eq. (3.6.9) for the spatial case. (b) The mean area of the intersection (in red) between Y and the corresponding disk is equal to $A_A \cdot K(r)$, where A_A is the area intensity and $K(r)$ the K-function of Y, see the first Eq. (3.6.11) for the spatial case.

Let $D(\mathrm{d}r)$ denote a spherical shell of radii r and $r + \mathrm{d}r$ centred at an arbitrary point $0 \in \mathbb{R}^3$, that is, not necessarily at a typical point of Y. Then, by Eq. (3.2.3) the mean volume of Y within this shell is $V_V \cdot 4\pi r^2\,\mathrm{d}r$ whereas, if the shell $D_0(\mathrm{d}r)$ is centred at $0 \in Y$, then the corresponding volume is $V_V \cdot \mathrm{d}K(r)$. By Eq. (3.6.10), the ratio of the latter to the former volume is

$$\frac{V_V \cdot \mathrm{d}K(r)}{V_V \cdot 4\pi r^2\,\mathrm{d}r} = \frac{C(r)}{V_V^2} \equiv c(r), \quad r > 0, \tag{3.6.12}$$

called the pair correlation function. By reference to Eq. (3.6.2) and Eq. (3.6.5), it is seen that $c(r)$ is the non-centred, scalar coefficient of covariation of a motion-invariant volume process covering two fixed points a distance r apart.

3.6.4 Extensions of the K-function

The scalar covariance function $C(r)$ given by Eq. (3.6.4) is defined for a motion-invariant volume process Y. The definition of the K-function, however, may be extended to the case in which $Y \subset \mathbb{R}^3$ represents a q-dimensional motion-invariant process, $q \in \{0, 1, 2, 3\}$. Thus,

$$K(r) = \gamma_V^{-1} \cdot \mathbb{E}\{\gamma(Y \cap D_0(r))\}, \tag{3.6.13}$$

namely γ_V^{-1} times the mean measure of Y in a ball $D_0(r) \subset \mathbb{R}^d$ centred at a typical point of Y. If $q = 0$, then the point event at the centre of $D_0(r)$ is not counted. Note that $K(r)$ is always a volume, irrespective of the dimension q of the process Y.

The spatial relationship between two processes Y_1, Y_2, of possibly different dimensions, may be described, at least in part, with the aid of the bivariate K-function,

$$K_{12}(r) = \gamma_V^{-1} \cdot \mathbb{E}\{\gamma(Y_2 \cap D_0(r))\}, \quad 0 \in Y_1. \tag{3.6.14}$$

Let $Y_1 = \{y_{1i}\}$, $Y_2 = \{y_{2i}\}$ denote two distinguishable point processes. If a ball of a given radius $r > 0$, centred at a point y_{1i}, captures a point y_{2j} then, for each pair (i, j), the same ball centred at the point y_{2j} will capture the point y_{1i}, whereby $K_{12}(r) = K_{21}(r)$. A similar argument may be used for any two processes of arbitrary dimensions by replacing points with curve, surface, or volume elements. The preceding result is analogous to the statistical property $\mathrm{Cov}(X_1, X_2) = \mathrm{Cov}(X_2, X_1)$ for a pair of random variables (X_1, X_2).

3.6.5 Sampling typical points

A typical point of a motion-invariant process Y of dimension q is a point of the process sampled with a probability proportional to the corresponding q-measure.

Planar process $Y \subset \mathbb{R}^2$

1. If Y is a stationary area process ($q = 2$) and $0 \in Y$, where 0 is a fixed point, then 0 is a typical point of Y. In fact, because Y is stationary, an area element $\partial y \subset Y$ is equipped with the translation-invariant density $\mathrm{d}y$, and therefore the translation-invariant measure of δy hitting 0 is the area element $\mathrm{d}y$. This is a consequence of Eq. (1.4.1).
2. If Y is a motion-invariant curve process ($q = 1$) and L_1^2 is an arbitrarily fixed straight line, then $Y \cap L_1^2$ is a stationary point process on the line consisting of typical points of Y. In fact, an arc element $\delta y \subset Y$ of length $\mathrm{d}b$ is equipped with the kinematic density $\mathrm{d}y\,\mathrm{d}\omega$, and therefore the motion-invariant measure of δy hitting a fixed linear element of L_1^2 is proportional to $\mathrm{d}b$, see Eq. (1.16.4) and Fig. 1.16.2.
3. If Y is a stationary process of disjoint particles, then any particles captured by a fixed quadrat according to an unbiased counting – or equivalently selecting – rule, are typical particles of the process. In fact, here sampling is number-weighted. For instance, the red coloured segments in Fig. 3.3.2 are typical segments of the process.

Spatial process $Y \subset \mathbb{R}^3$

By similar arguments, for a volume, surface, curve, or particle process in space, typical points (or particles in the latter case) may be selected with arbitrarily fixed points, lines, planes, or unbiased counting bricks, respectively. The underlying ideas, extensible to higher dimensions, are given in Section 2.32.1.

3.6.6 The spherical contact distribution function

Consider a motion-invariant volume process $Y \subset \mathbb{R}^d$ of intensity V_V, and let $\rho \geq 0$ denote the smallest distance to Y from a point 0 outside Y. Let $D_0(r) \subset \mathbb{R}^d$ denote a ball of radius $r \geq 0$ centred at $0 \notin Y$. Then, the distribution of ρ, called the spherical contact distribution function $H(r)$, is

$$\begin{aligned} H(r) &= \mathbb{P}(\rho \leq r | 0 \notin Y) \\ &= \mathbb{P}(Y \cap D_0(r) \neq \emptyset | 0 \notin Y) \\ &= 1 - \frac{\mathbb{P}(Y \cap D_0(r) = \emptyset,\ 0 \notin Y)}{1 - V_V}. \end{aligned} \tag{3.6.15}$$

Thus, for a given $r > 0$, $H(r)$ is the probability that the ball $D_0(r)$ ‘bites’ the process Y. If Y is not a volume process, then the preceding definition still applies with $V_V = 0$. Examples for special models are given in Sections 3.10–3.13 below.

3.6.7 Notes

1. The covariance function

The definition and implementation of the covariance function were pioneered by Matérn (1960) and Matheron (1967), see also Serra (1982).

2. The K-function

The K-function was initially developed mainly for point processes, e.g. Ripley (1977), see also Diggle (2003). For early applications in stereology, see Braendgaard and Gundersen (1986), Baddeley, Howard, Boyde, and Reid (1987) and Baddeley, Moyeed, Howard, and Boyde (1993). It was later adapted for curve, surface, and volume processes (Ripley (1981), Cruz-Orive (1989c), Stoyan et al. (1995)). Diggle (1986) proposed Eq. (3.6.14) for the union of two distinguishable point processes. Mattfeldt, Frei, and Rose (1993) applied Eq. (3.6.11) to compute the K-function for a volume process model of the human mammary gland.

3.7 Local Stereology for Second-Order Measures

3.7.1 The K-function from the nucleator

Let $Y \subset \mathbb{R}^3$ denote a motion-invariant volume process. An arbitrarily oriented ray emanating from the centre $0 \in Y$ of a ball $D_0(r)$ will determine, say, $m \geq 1$ separate intercepts in the bounded random set $Y \cap D_0(r)$. With the same notation as in Eq. (2.18.6), we have

$$\begin{aligned} V_V K(r) &= \mathbb{E}\{V(Y \cap D_0(r))\} \\ &= \frac{4\pi}{3} \cdot \mathbb{E} \sum_{i=1}^{m} (l_{i+}^3 - l_{i-}^3), \end{aligned} \tag{3.7.1}$$

which is the direct nucleator formula for the model-based case. With a proper interpretation of the intercept lengths, the same formal expression holds if an arbitrary radius is drawn in a great disk of the ball $D_0(r)$, namely in $d_0(r) = D_0(r) \cap L^3_{2[0]}$, where $L^3_{2[0]}$ is an arbitrary plane through $0 \in Y$. In this case we are implementing the pivotal nucleator in the model-based case, see Eq. (2.18.8).

3.7.2 The K-function from the invariator

Preliminary comments

The invariator yields not only the volume of a realization of a motion-invariant volume process Y within the ball $D_0(r)$ for each $r > 0$, but also the corresponding area of a motion-invariant surface process ∂Y. The line probes are contained in a pivotal plane through a typical point 0 of the process. Unfortunately, the invariator will not work if $0 \in \partial Y$, recall the paragraph following Eq. (2.32.10), because in this case the plane through 0 will not necessarily be isotropic relative to ∂Y anymore. For this reason the K-functions considered below are defined with respect to $0 \in Y$.

A posteriori line weighting

With a posteriori weighted test lines hitting the equatorial disk $d_0(r)$, see Fig. 3.7.1(a), application of Eq. (2.32.25) and Eq. (2.32.26) yields the K-functions of a motion-invariant volume process $Y \subset \mathbb{R}^3$ and of a motion-invariant surface process $\partial Y \subset \mathbb{R}^3$, with respect to a typical point $0 \in Y$, namely

$$S_V K(r) = 2\pi T \cdot \mathbb{E} \sum_{k \in \mathbb{Z}} |r_k| \cdot I_k(r), \tag{3.7.2}$$

$$V_V K(r) = \pi T \cdot \mathbb{E} \sum_{k \in \mathbb{Z}} |r_k| \cdot L_k(r), \tag{3.7.3}$$

respectively. The preceding equations are applied on the planar, bounded random sets $\partial Y \cap d_0(r)$ and $Y \cap d_0(r)$, respectively.

A priori line weighting

With a priori weighted test lines applied to the same random sets, see Fig. 3.7.1(b), direct application of Eq. (2.32.30) and Eq. (2.32.31) yields the corresponding K-functions with respect to a typical point $0 \in Y$, namely

$$S_V K(r) = 2a \cdot \mathbb{E}\{I(r)\}, \tag{3.7.4}$$

$$V_V K(r) = a \cdot \mathbb{E}\{L(r)\}, \tag{3.7.5}$$

respectively. For estimation details, see Section 4.26.1.

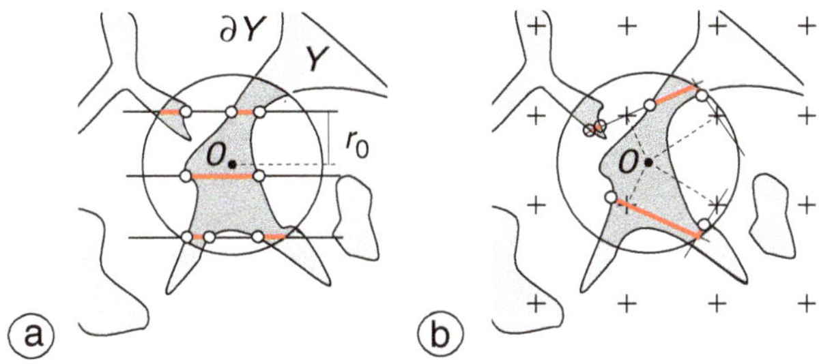

Fig. 3.7.1 A portion of a pivotal section through a typical point 0 of a motion-invariant volume process $Y \subset \mathbb{R}^3$ within a disk centred at 0. (a) Application of the invariator with a posteriori line weighting to obtain the K-functions of ∂Y and of Y, see Eq. (3.7.2) and Eq. (3.7.3), respectively. (b) Idem with a priory weighted test lines, see Eq. (3.7.4) and Eq. (3.7.5), respectively.

3.7.3 The K-function from a pivotal slab

The K-function of a motion-invariant process $Y \subset \mathbb{R}^3$ of point particles may be represented with the aid of a slab probe $L_t(0)$ of thickness $t > 0$, whose midplane contains a typical point particle $0 \in Y$. Let d_i denote the distance from 0 of a point particle $y_i \in Y \cap L_t(0) \cap D_0(r)$, namely of a point particle captured by the slab and contained in the ball $D_0(r)$, and let $\mathcal{S}_r \subset \{1, 2, \ldots, N_r\}$ denote the index set of the sampled point particles. Then, by Eq. (2.18.25),

$$N_V K(r) = \frac{2}{t} \mathbb{E} \sum_{i \in \mathcal{S}_r} d_i \text{ if } t < 2d_{(1)}, \tag{3.7.6}$$

where $d_{(1)}$ is the smallest observed distance. The method extends to arbitrary particles if the slab is used as a slab disector. For estimation details, see Section 4.26.2.

3.7.4 The pair correlation function from a pivotal section

Preliminary result: an intersection formula on the sphere

Consider a fixed domain Y of area $S(Y)$ on a fixed sphere $\partial D_0(r)$, and let $\partial d_0(r)$ denote a great circle of the sphere with rotation-invariant density du, $u \in \mathbb{S}_+^2$, which is that of the polar axis $L_1^3(0, u)$ normal to the disk $d_0(r)$. Then,

$$\int_{\mathbb{S}_+^2} B(Y \cap \partial d_0(r)) \, \mathrm{d}u = \frac{\pi}{r} \cdot S(Y), \tag{3.7.7}$$

see Note 2 in Section 3.7.5. Consider now a UR point $x \in \partial d_0(r)$, whereby,

$$\mathbb{P}(\mathrm{d}u, \mathrm{d}x) = \frac{\mathrm{d}u}{2\pi} \cdot \frac{\mathrm{d}x}{2\pi r}. \tag{3.7.8}$$

We have,

$$\begin{aligned}\mathbb{E}\{1_Y(x)\} &= \frac{1}{4\pi^2 r} \int_{\mathbb{S}^2_+} \mathrm{d}u \int_{\partial d_0(r)} 1_Y(x)\,\mathrm{d}x \\ &= \frac{1}{4\pi^2 r} \int_{\mathbb{S}^2_+} B(Y \cap \partial d_0(r))\,\mathrm{d}u \\ &= \frac{S(Y)}{4\pi r^2}.\end{aligned} \tag{3.7.9}$$

A formula for the pair correlation function

Consider a motion-invariant volume process $Y \subset \mathbb{R}^3$ and an arbitrary great circle $\partial d_0(r)$ of the sphere $\partial D_0(r)$ centred at a typical point $0 \in Y$. For an arbitrary point $x \in \partial d_0(r)$, define the intersection indicator

$$I_r(x) = \begin{cases} 1 \text{ if } Y \cap \partial d_0(r) \ni x \\ 0 \text{ otherwise.} \end{cases} \tag{3.7.10}$$

The combination of Eq. (3.7.9) (with the fixed domain Y replaced with the random domain $Y \cap \partial D_0(r)$) and of the last identity in Eq. (3.6.9) yields

$$\begin{aligned}\mathbb{E}\{I_r(x)\} &= \frac{\mathbb{E}\{S(Y \cap \partial D_0(r))\}}{4\pi r^2} \\ &= V_V^{-1} C(r),\end{aligned} \tag{3.7.11}$$

where the expectation is of the form given by Eq. (3.1.1). Finally, by Eq. (3.6.12),

$$c(r) = \frac{\mathbb{E}\{I_r(x)\}}{V_V}. \tag{3.7.12}$$

If $n \geq 1$ arbitrary points are laid on $\partial d_0(r)$ then, setting

$$I_I(r) \equiv \frac{1}{n} \cdot \mathbb{E} \sum_{i=1}^{n} I_r(x_i), \tag{3.7.13}$$

we obtain a simple identity for the pair correlation function, namely,

$$c(r) = \frac{I_I(r)}{V_V}. \tag{3.7.14}$$

In practice the points may be equispaced, e.g., in polar coordinates:

$$\{x_i = (r, 2\pi i/n),\ i = 0, 1, \ldots, n-1\}, \tag{3.7.15}$$

see Fig. 3.7.2.

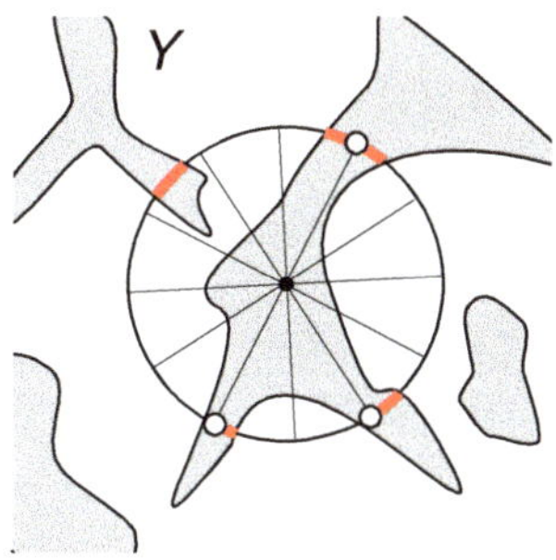

Fig. 3.7.2 Illustration of Eq. (3.7.13). Here $n = 12$ and $\sum_{i=1}^{12} I_r(x_i) = 3$.

3.7.5 Notes

1. Stereology of the K-function

Local stereology for the K-function, see Eq. (3.7.1), and for the pair correlation function, see Eq. (3.7.14), of a motion-invariant volume process, was proposed by Cruz-Orive (1989c). The local disector Eq. (3.7.6) for the function of a particle process was proposed by Gundersen et al. (1988) and by Evans and Gundersen (1989). A general scheme of local stereology for second-order properties was developed by Kiêu and Jensen (1993), see Jensen (1998) for a compendium and further references. The invariator versions Eq. (3.7.2)–(3.7.5) have apparently not been proposed before.

2. Stereology on the sphere

Eq. (3.7.7) is a particular Crofton formula on the sphere, see Eq. (20) of Santaló (1942), or the first Eq. (18.10) of Santaló (1976), which pertain to the unit sphere ($r = 1$).

3.8 Second-Order Measures for a Planar Domain

3.8.1 Geometric covariogram

Let $T \in \mathbb{R}^2$ denote a planar domain of area $A(T)$, and $x = (r, \omega) \in \mathbb{R}^2$ a point or vector of modulus $r \geq 0$ and orientation $\omega \in [0, 2\pi)$. The (vectorial) geometric covariogram of T is

$$g(x) = \int_{\mathbb{R}^2} 1_T(z) \cdot 1_T(z+x)\,\mathrm{d}z, \quad x \in \mathbb{R}^2, \tag{3.8.1}$$

and it has the following properties (the first of which is illustrated in Fig. 4.8.1(a)),

$$\begin{aligned} &g(x) = A(T \cap (T - x)), \\ &g(x) = g(-x), \\ &g(0) = A(T) \geq g(x) \geq g(\infty) = 0, \\ &\int_{\mathbb{R}^2} g(x)\,\mathrm{d}x = A^2(T). \end{aligned} \tag{3.8.2}$$

The mean of $g(x)$ with respect to the isotropic orientation is the isotropic covariogram $g(r)$. We may use $\mathbb{P}(\mathrm{d}\omega) = \mathrm{d}\omega/\pi$, $\omega \in [0, \pi)$, because $g(x) = g(-x)$, and therefore $g(r, \omega) = g(r, \omega + \pi)$. Thus,

$$g(r) = \int_0^{\pi} g(r, \omega)\mathbb{P}(\mathrm{d}\omega), \quad r \geq 0, \tag{3.8.3}$$

which is a scalar function.

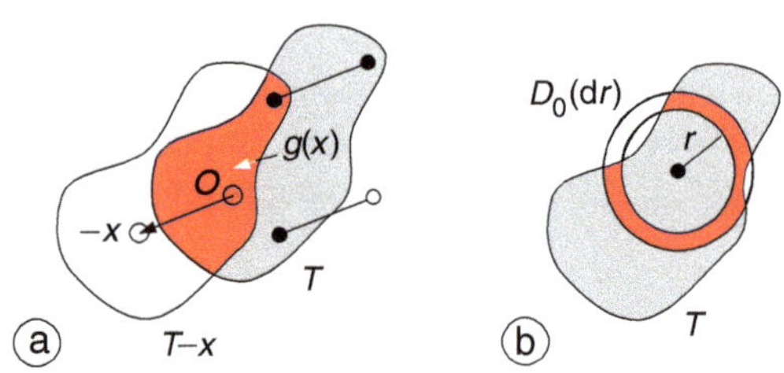

Fig. 3.8.1 (a) The vectorial geometric covariogram $g(x)$ of the planar domain T is the area of the red domain, see the first Eq. (3.8.2). (b) The point pair distance distribution concept, see Eq. (3.8.6).

3.8.2 Set covariance

Let $z \sim \mathrm{UR}(T)$ be a uniform random point in T, namely a typical point of T, with probability element $\mathbb{P}(\mathrm{d}z) = \mathrm{d}z/A(T)$, $z \in T \subset \mathbb{R}^2$. The vectorial set covariance of T is

$$
\begin{aligned}
C(x) &= \int_T \mathbb{P}(z + x \in T | z \in T)\mathbb{P}(\mathrm{d}z) \\
&= \frac{1}{A(T)} \int_{\mathbb{R}^2} 1_T(z) \cdot 1_T(z + x)\, \mathrm{d}z \\
&= \frac{g(x)}{A(T)}, \quad x \in \mathbb{R}^2,
\end{aligned}
\tag{3.8.4}
$$

namely a normalized version of the vectorial set covariogram.

The corresponding scalar, isotropic set covariance is

$$
C(r) = \frac{g(r)}{A(T)}, \quad r \geq 0. \tag{3.8.5}
$$

3.8.3 Point pair distance distribution

Let $r \geq 0$ denote the scalar distance between two independent typical points of T, called the point pair distance, and let $\mathbb{P}(\mathrm{d}r) = p(r)\,\mathrm{d}r$ be the corresponding probability element. Further, let $D_0(\mathrm{d}r) \subset \mathbb{R}^2$ denote a circular annulus of radii r and $r + \mathrm{d}r$ centred at a typical point of T, see Fig. 3.8.1(b). Then, by the two-dimensional analogue of the second identity from Eq. (3.6.10), we have

$$
\begin{aligned}
p(r)\,\mathrm{d}r &= \frac{\mathbb{E}\{A(T \cap D_0(\mathrm{d}r))\}}{A(T)} \\
&= \frac{2\pi r C(r)\,\mathrm{d}r}{A(T)} \\
&= \frac{2\pi r g(r)\,\mathrm{d}r}{A^2(T)}, \quad r \geq 0.
\end{aligned}
\tag{3.8.6}
$$

3.8.4 Variance of the area of intersection between a plate probe and a motion-invariant volume process

Let $Y \subset \mathbb{R}^3$ denote a motion-invariant volume process of intensity V_V, and $T \subset L_2$ a fixed one-sided plate probe of area $A(T)$ in an arbitrarily fixed plane L_2. The intersection $Y \cap L_2$ is a motion-invariant area process of intensity A_A, and by

Eq. (3.3.3),

$$V_V = \mathbb{E}\{A(Y \cap T)\}/A(T) = A_A. \tag{3.8.7}$$

As an application of the foregoing formulae, next we find an expression for $\mathrm{Var}(A_A)$. We have,

$$\begin{aligned}
\mathbb{E}\{A^2(Y \cap T)\} &= \mathbb{E}\left\{\int_{L_2}\int_{L_2} 1_T(z)1_T(z+x)1_Y(z)1_Y(z+x)\,\mathrm{d}z\,\mathrm{d}x\right\} \\
&= \int_{L_2}\left\{\int_{L_2} 1_T(z)1_T(z+x)\,\mathrm{d}z\right\}\cdot\mathbb{E}\{1_Y(z)1_Y(z+x)\}\,\mathrm{d}x \\
&= \int_{L_2} g_T(x)C_Y(r)\,\mathrm{d}x \\
&= \int_0^{2\pi}\mathrm{d}\omega\int_0^\infty g_T(r,\omega)C_Y(r)r\,\mathrm{d}r \\
&= \int_0^\infty 2\pi r g_T(r)C_Y(r)\,\mathrm{d}r \\
&= A^2(T)\int_0^\infty p_T(r)C_Y(r)\,\mathrm{d}r,
\end{aligned} \tag{3.8.8}$$

whereby,

$$\begin{aligned}
\mathrm{Var}(A_A) &= A^{-2}(T)\cdot\mathbb{E}\{A^2(Y \cap T)\} - V_V^2 \\
&= \int_0^\infty \{C_Y(r) - V_V^2\}\cdot p_T(r)\,\mathrm{d}r.
\end{aligned} \tag{3.8.9}$$

The so-called mixing property of a motion-invariant volume process Y establishes that, as $r \longrightarrow \infty$, the events $0 \in Y$ and $x = (r, u_2) \in Y$ tend to be independent. Because the mixing property usually applies, we may assume that

$$\lim_{r\longrightarrow\infty} C_Y(r) = V_V^2. \tag{3.8.10}$$

Thus, if the distance function $p_T(r)$ of the probe is concentrated at large values of r, then Eq. (3.8.9) suggests that $\mathrm{Var}(A_A)$ will be reduced. In this sense, using for instance a bounded portion of a square grid consisting of a sparse 10×10 array of unit quadrats will in principle be better than using a single quadrat of area 100.

3.8.5 Notes

1. The geometric covariogram

The geometric covariogram, see Section 3.8.1, was proposed by Matheron (1965, 1967) as a tool to study compact sets and to predict the estimation variance under Cavalieri sampling (Chapter 5).

2. The point pair distance distribution

Integral geometric results involving the distance between two UR points in a domain were classical in integral geometry; for details and early references, see Santaló (1976). The point pair distance distribution for a domain was used by Kellerer (1986), and with a different name by Enns and Ehlers (1978).

3. The variance of A

Eq. (3.8.9) was given without proof by Miles (1978a), who discussed its impact on sampling efficiency, as reflected in the paragraph following Eq. (3.8.10) above.

3.9 Stationary Point Processes. The Poisson Point Process

3.9.1 Notions on point processes relevant to stereology

Basic descriptors

In the applied sciences it is often useful to use a stationary point process model

$$Y = \bigcup_{i \in \mathbb{Z}} y_i \tag{3.9.1}$$

to study aggregates of disjoint particles in $\mathbb{R}^d$. The ith particle is replaced with an associated point y_i, which is called an 'event' of the process. The intensity $\lambda \equiv N_V > 0$ of the process, which represents the mean number of events per unit d-volume, is constant. To compare point processes, useful descriptors (illustrated in Fig. 3.9.1 for $d = 2$) are:

- The contact distribution $H(r)$, see Eq. (3.6.15), which in this case is the distribution function of the distance r from an arbitrary point of the containing space to the nearest event. Equivalently,

  $$H(r) = 1 - \mathbb{P}(Y \cap D_0(r) = \emptyset,\ 0 \notin Y), \tag{3.9.2}$$

  namely the probability that a ball $D_0(r)$ centred at a point which is not an event contains at least one event of the process.
- The nearest neighbour distribution $G(r)$, namely the distribution function of the distance r from an arbitrary event of the process to its nearest event. Thus,

  $$G(r) = 1 - \mathbb{P}(Y \cap D_0(r) = \emptyset,\ 0 \in Y), \tag{3.9.3}$$

namely the probability that a ball $D_O(r)$ centred at an arbitrary event contains at least one further event of the process.

- The K-function $K(r)$, namely λ^{-1} times the number of further events in a ball $D_O(r)$ centred at an arbitrary event.

An idea of the discriminating performance of $G(r)$ for different point processes is illustrated in Fig. 3.9.2.

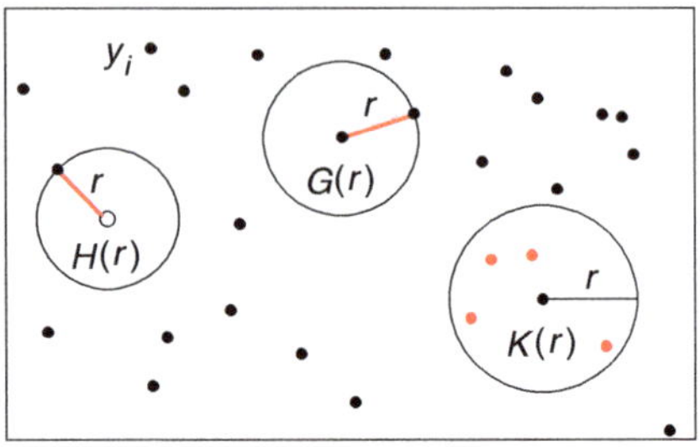

Fig. 3.9.1 Conventional illustrations of the contact distribution $H(r)$, the nearest neighbour distribution $G(r)$, and the K-function, for a planar point process, see Section 3.9.1.

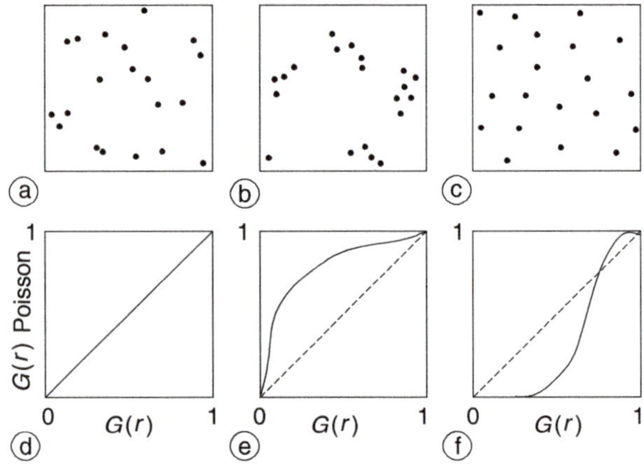

Fig. 3.9.2 Idea of the expected behaviour of the function $G(r)$ for a Poisson process, a cluster process, and a hard core process, respectively. The latter may be interpreted as a process of the centres of non-interpenetrating disks.

Point processes on the line

Consider a stationary process Y of disjoint points on a given axis, namely

$$Y = \{\cdots < y_{-2} < y_{-1} < y_0 < y_1 < y_2 < \cdots\} \subset \mathbb{R}. \tag{3.9.4}$$

The ith interval, $i \in \mathbb{Z}$, is $I_i = [y_i, y_{i+1})$. If the support is the continuous time axis, then the interval length $l_i = |y_{i+1} - y_i|$ is called the waiting time (from the event y_i to the next).

Let $F(l)$, $l > 0$, denote the distribution of the random length l of the typical interval. To sample typical intervals (that is, to sample them with identical probabilities), the unbiased stereological sampling rules (Section 2.15.2) apply. For all $i \in \mathbb{Z}$, adopt the left end point y_i as the associated point (AP) of the interval I_i. On the other hand, consider a unidimensional disector $L_t^-(x) = (x, x + t]$ of a fixed length $t > 0$. If for an arbitrary point $x \in \mathbb{R}$ there is an index $i \in \mathbb{Z}$ such that $L_t^-(x) \ni y_i$, then y_i, I_i are a typical point and a typical interval of the process, respectively, and the length l_i is a realization of the random variable l. In practice it may be convenient, and efficient (see the paragraph following Eq. (3.8.10)) to use a finite portion of a Cavalieri test system of period $T \geq t > 0$, consisting of n segment disectors, say, at an arbitrary location, namely,

$$\Lambda_t^- = \{L_t^-(z + kT),\ z \in [0, T),\ k = 1, 2, \ldots, n\}, \tag{3.9.5}$$

see Fig. 3.9.3.

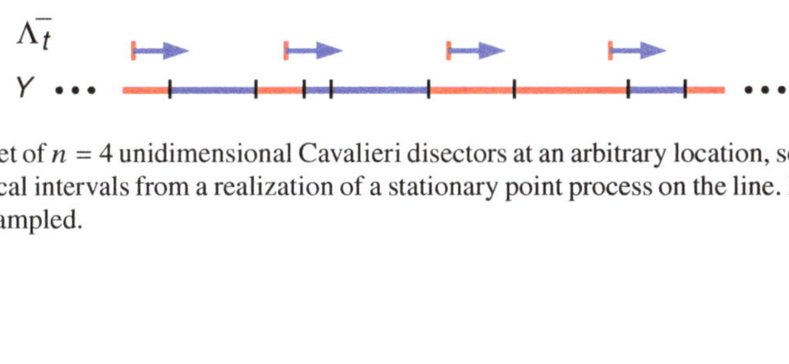

Fig. 3.9.3 A set of $n = 4$ unidimensional Cavalieri disectors at an arbitrary location, see Eq. (3.9.5), to sample typical intervals from a realization of a stationary point process on the line. Four intervals (in blue) are sampled.

If the interval lengths $\{l_i\}$ are mutually independent, then Y is called a renewal process. Consider an arbitrary test point $x \in \mathbb{R}$. Then, if there is an index $i \in \mathbb{Z}$ such that $I_i \ni x$, then l_i is length-weighted, and its distribution is given by Eq. (2.3.6). However, due to independence the next interval I_{i+1} will be a typical interval of the process, and l_{i+1} will be a realization of the random variable l. Note that this rule is not warranted for a stationary process whose intervals are not independent, whereas disector sampling requires stationarity only.

3.9.2 The Poisson point process on an axis

In the applied sciences, the stationary Poisson point process (henceforth called the Poisson point process, for short) is adopted as a model for 'complete spatial randomness' of points or 'events'. On a support axis, a Poisson point process Y of intensity $\lambda \equiv N_L > 0$ may be constructed as follows. On each elementary interval $(x, x + \mathrm{d}x]$, an event takes place with probability $\lambda\,\mathrm{d}x$, and no events with probability $1 - \lambda\,\mathrm{d}x$, with independence among disjoint intervals. As shown below, the number n of events in an interval of length L is a Poisson random variable with probability function

$$p(n) = \frac{(\lambda L)^n \exp(-\lambda L)}{n!}, \qquad n = 0, 1, \ldots, \tag{3.9.6}$$

whereby $\mathbb{E}(n) = \mathrm{Var}(n) = \lambda L$.

To verify the preceding result, split the unit interval into N subintervals of equal length $1/N$, with the condition $\lambda/N < 1$, and assign an event to each subinterval, independently, with probability λ/N. Then, the total number n of events generated in an interval of length L is a binomial random variable with NL trials, probability of success λ/N, and probability function

$$\begin{aligned}&\binom{NL}{n}\left(\frac{\lambda}{N}\right)^n\left(1-\frac{\lambda}{N}\right)^{NL-n}\\ &= \frac{NL\cdot(NL-1)\cdot\cdots\cdot(NL-n+1)}{N^n}\cdot\frac{\lambda^n}{n!}\left(1-\frac{\lambda}{N}\right)^{NL-n}.\end{aligned} \tag{3.9.7}$$

As $N \to \infty$, that is, as the length of each subinterval becomes infinitesimal, the limit of the preceding expression is the rhs of Eq. (3.9.6).

To generate Poisson events in a fixed interval $D \subset \mathbb{R}$ of length L or, more precisely, to generate a realization of $Y \cap D$, we may use the following procedure.

1. Generate a realization n from the Poisson random variable of mean λL.
2. Generate n independent UR events in D.

Consider the ordered abscissas $0 < y_1 < y_2 < \cdots$ of successive Poisson events from an arbitrary origin 0. Here y_k is called the kth (forward) neighbour from 0. The probability element of the abscissa $y \equiv y_k$, $k \in \{1, 2, \ldots\}$, of the kth neighbour is

$$\begin{aligned}\mathbb{P}(\mathrm{d}y) &= \mathbb{P}\{k-1 \text{ events in } (0, y)\}\cdot\mathbb{P}\{1 \text{ event in } (y, y+\mathrm{d}y)\}\\ &= \frac{(\lambda y)^{k-1}\exp(-\lambda y)}{(k-1)!}\cdot\lambda\,\mathrm{d}y\\ &= \frac{\lambda^k}{\Gamma(k)}\cdot y^{k-1}\mathrm{e}^{-\lambda y}\,\mathrm{d}y,\end{aligned} \tag{3.9.8}$$

which shows that

$$y_k \sim \mathrm{Gamma}(k, \lambda), \qquad k = 1, 2, \ldots, \tag{3.9.9}$$

with $\mathbb{E}(y) = n/\lambda$ and $\mathrm{Var}(y) = n/\lambda^2$. If we consider instead the ordered sequence $y_0 < y_1 < y_2 < \cdots$, where y_0 is now an arbitrary event of the process, then, due to interval independence, $y_k - y_0 \sim \mathrm{Gamma}(k, \lambda)$ as well. This generally does not hold if the point process is not Poisson. Further, setting $k = 1$ in Eq. (3.9.8) shows that either random variable $l \equiv y_1$, or $l \equiv y_1 - y_0$, are $\mathrm{Gamma}(1, \lambda)$, namely the exponential distribution, so that $H(l) = G(l)$ in this case. The corresponding density function of the typical interval is

$$h(l) = g(l) = \lambda \mathrm{e}^{-\lambda l}, \quad l > 0, \tag{3.9.10}$$

with $\mathbb{E}(l) = 1/\lambda$, logically the reciprocal of the mean number of events per unit length.

However, if $l > 0$ represents the random length of a Poisson interval $(y_{i-1}, y_i]$ hit by an arbitrary test point x independent from the process, then, by virtue of Eq. (2.3.6),

$$\mathbb{P}(\mathrm{d}l | \uparrow) = \lambda^2 l \mathrm{e}^{-\lambda l}, \quad l > 0, \tag{3.9.11}$$

which shows that $(l| \uparrow) \sim \mathrm{Gamma}(2, \lambda)$. This is actually the distribution of the second neighbour from a typical Poisson event. It means that the hitting test point x plays the role of an event interposed between the events y_{i-1} and y_i, whereby y_i is now the second neighbour of y_{i-1}. The test point x splits the hit interval into two typical Poisson intervals $(y_{i-1}, x]$ and $(x, y_i]$ with independent and exponentially distributed lengths of mean $1/\lambda$. A well-known property of the $\mathrm{Gamma}(k, \lambda)$ random variable is that it can be expressed as the sum of k independent exponential random variables of mean $1/\lambda$.

3.9.3 The Poisson point process in the plane, and in space

The extension of the preceding results to a Poisson point process $Y \subset \mathbb{R}^d$ is straightforward. For instance, the number n of Poisson events of intensity $\lambda \equiv N_A$ in the plane, within a bounded domain of area A, is a Poisson random variable with mean (and variance) λA, irrespective of the domain shape. Thus, Eq. (3.9.6) formally holds with L replaced with A for the planar case, or in general with the d-volume V_d of a bounded domain $D \subset \mathbb{R}^d$. Further, if $Y \cap D$ consists of n events then, conditional on n, the latter events are effectively independent UR in D.

From the preceding properties, the K-function of a planar Poisson point process $Y \subset \mathbb{R}^2$, for instance, is

$$K(r) = \pi r^2. \tag{3.9.12}$$

In fact, by definition $\lambda K(r)$ is the mean number of events in a disk of radius r, excluding the event at the centre of the ball, namely $\pi r^2 \lambda$.

The corresponding contact distribution function $H(r)$ is the probability that a disk $D_0(r)$ of radius r centred at a point $0 \notin Y$ contains at least one event, namely,

$$\begin{aligned} H(r) &= 1 - \mathbb{P}(\text{No events in } D_0(r)) \\ &= 1 - \exp(-\lambda \pi r^2), \quad r \geq 0. \end{aligned} \tag{3.9.13}$$

Consider the ordered distances $y_0 < y_1 < y_2 < \cdots$ of Poisson events in the plane, indistinctly from a fixed origin, or from an arbitrary event y_0, where y_k denotes the kth neighbour distance from y_0. Let A_k denote the area of the disk of radius y_k, with $\mathrm{d}A_k = 2\pi y_k \, \mathrm{d}y_k$, namely the area of the elementary annulus of radii $\{y_k, y_k + \mathrm{d}y_k\}$. Then, an argument analogous to that leading to Eq. (3.9.6) shows that

$$A_k \sim \text{Gamma}(k, \lambda), \quad k = 1, 2, \ldots. \tag{3.9.14}$$

Note that the preceding result holds for the area of the disk that contains the event y_k in its boundary, and not for the distance y_k itself.

Let $a_i = \pi(y_i^2 - y_{i-1}^2)$, $i = 1, 2, \ldots$, denote the area of the annulus of radii $\{y_{i-1}, y_i\}$. For $k = 1$ we have $a_1 = A_1$, hence the area a_1 is exponential of mean $1/\lambda$. It can be shown that the successive annuli areas $a_2, a_3, \ldots$ are also exponential of mean $1/\lambda$, and independent. Note that $a_1 + a_2 + \cdots + a_k = A_k$, whereby the additive property of the Gamma distribution again verifies Eq. (3.9.14).

The preceding results suggest that the properties considered here of the Poisson point process are formally the same for any space dimension, provided that the distinct elements involved are properly interpreted. For instance, a kth neighbour distance in $\mathbb{R}$ from a given point, or event 0, corresponds in $\mathbb{R}^d$ to the d-volume of a ball centred at 0, whose boundary contains the kth neighbour event. Or, the length of a Poisson interval in $\mathbb{R}$ corresponds in $\mathbb{R}^d$ to the d-volume of a spherical shell which contains two consecutive neighbour events in its inner and outer boundaries, respectively.

3.9.4 Notes

Point processes with realizations on $\mathbb{R}$ were studied toward the mid 20th century as a branch of stochastic processes. The textbook of Cox and Miller (1965) contains most of the concepts given here, and lists early references. For a modern treatment in the stochastic geometry context, with pertinent references, see Stoyan et al. (1995). For the statistics of point processes, see Diggle (2003) and Baddeley, Rubak, and Turner (2016).

3.10 Motion-Invariant Line Processes in the Plane

3.10.1 Construction

A special stationary and isotropic line process $Y \subset \mathbb{R}^2$, useful for didactic purposes, is an infinite union of straight lines in the plane,

$$Y = \bigcup_{i \in \mathbb{Z}} L_1^2(p_i, \phi_i), \tag{3.10.1}$$

with the following properties:

1. The signed distances $\{p_i \in (-\infty, \infty),\ i \in \mathbb{Z}\}$ from the origin 0 to the lines constitute a stationary process of separate points of intensity N_L on an axis.
2. For each realization p_i, $i \in \mathbb{Z}$, the corresponding orientation angle ϕ_i is an independent UR realization in $[0, \pi)$. Thus,
   (a) The process Y contains no parallel lines with probability 1.
   (b) At each vertex, the number of intersecting lines is equal to 2 with probability 1.
3. The process Y is ergodic, which means that the intensity N_L remains fixed for all realizations of Y.

The next properties follow from the Remarks in Section 3.2.

i. Given that a point p_i from a stationary point process on the line hits a fixed interval, the point is UR in that interval. Also, by Eq. (3.2.3), the mean number of points $\{p_i\}$ hitting a fixed interval of length $H > 0$ is equal to $N_L H$.
ii. Given that a line from a motion-invariant planar line process hits a fixed domain $D \subset \mathbb{R}^2$, the line is IUR hitting D. For instance, the corresponding mean length of a connected linear intercept is given by Eq. (2.4.13).
iii. A line process may be regarded as a countable union of line segments (and similarly a plane process as a countable union of platelets, etc.), whereby Mecke's results for motion-invariant curve and surface processes extend to motion-invariant line and plane processes, respectively.

3.10.2 First-order properties

Consider a fixed probe domain $T \subset \mathbb{R}^2$ of area $A(T) > 0$. Then, by Eq. (3.2.3),

$$\mathbb{E}(L(Y \cap T)) = L_A \cdot A(T), \tag{3.10.2}$$

where L_A denotes the length intensity of the invariant line process Y.

The mean of the total intercept length determined in a fixed disk D of diameter $H > 0$, perimeter B and area A, by the lines of the process, is equal to the mean number of lines hitting D, times the mean intercept length determined in the disk by an IUR line $L_1^2(p, \phi)$ hitting it. By the preceding properties (i), (ii),

$$\begin{aligned} L_A &= \frac{1}{A} \cdot \mathbb{E}\{N(Y \cap D)\} \cdot \mathbb{E}\{L(L_1^2(p, \phi) \cap D)\} \\ &= \frac{1}{A} \cdot N_L H \cdot \frac{\pi A}{B} \\ &= N_L. \end{aligned} \tag{3.10.3}$$

Thus, the identity

$$L_A = N_L \tag{3.10.4}$$

holds for the entire line process because it does not depend on the disk size. The identities in Eq. (3.10.3) also hold if D is an arbitrary convex set of parameters H, B, A.

The intersection $Y \cap L_1^2(0, 0)$ of the line process Y with a fixed line probe $L_1^2(0, 0)$ is a stationary intersection process of intensity I_L on the line. By Eq. (3.4.2) with $d = 2$,

$$I_L = \frac{2}{\pi} \cdot L_A, \tag{3.10.5}$$

which is a model-based version of Eq. (2.4.10).

The vertices of the line process Y, namely the self-intersections between pairs of lines, constitute a stationary point process of intensity I_A in the plane. By Remark (iii) above, Eq. (3.5.4) applies with the factor $2/\pi$ replaced with $1/\pi$, because each vertex is generated by two lines of the same process Y. Thus,

$$I_A = \frac{1}{\pi} \cdot L_A^2. \tag{3.10.6}$$

The random polygons formed by the lines of the process Y are convex, because they are intersections of half-spaces. A vertical sweeping line scanning the plane from left to right will almost surely (a.s.) meet each polygon for the first time at a unique vertex, which may be adopted as the associated point (AP) of the polygon. If follows that the polygon intensity N_A is identical to the vertex intensity, namely,

$$N_A(\text{polygons}) = I_A(\text{vertices}) = L_A^2/\pi. \tag{3.10.7}$$

The mean area of a polygon is therefore

$$\mathbb{E}\{A(\text{polygon})\} = 1/N_A(\text{polygons}) = \pi/L_A^2. \tag{3.10.8}$$

For each AP first met by the sweeping line, a.s. the corresponding two edges are also first met. Thus, the number intensity of edges is

$$N_A(\text{edges}) = 2I_A(\text{vertices}) = (2/\pi)L_A^2. \tag{3.10.9}$$

The mean perimeter length of a polygon is

$$\mathbb{E}\{B(\text{polygon})\} = \frac{L_A}{(1/2)N_A(\text{polygons})} = \frac{2\pi}{L_A}, \tag{3.10.10}$$

the factor $1/2$ stemming from the first identity because each edge is shared by two polygons.

The mean number of vertices, or of edges, of a polygon, may be expressed as the mean number of vertices per line length, times the mean polygon perimeter length, namely,

$$\begin{aligned}\mathbb{E}\{N(\text{polygon})\} &= I_L \cdot \mathbb{E}\{B(\text{polygon})\}\\ &= \frac{2}{\pi}L_A \cdot \frac{2\pi}{L_A}\\ &= 4.\end{aligned} \tag{3.10.11}$$

A cross-check. The mean polygon intercept length with an arbitrary line must be equal to the mean interval length between intersections along the line, namely $I_L^{-1} = (\pi/2)L_A^{-1}$, see Eq. (3.10.5). By Eq. (2.4.13), the mean intercept length determined in a convex set of area A and perimeter length B by an IUR line hitting it is equal to $\mathbb{E}(L|\uparrow) = \pi A/B$. Because the probability of hitting the set is proportional to B, the unconditional mean polygon intercept length is the two-dimensional version of Eq. (2.16.24), namely,

$$\mathbb{E}(L) = \frac{\pi\mathbb{E}\{A(\text{polygon})\}}{\mathbb{E}\{B(\text{polygon})\}} = \frac{\pi^2/L_A^2}{2\pi/L_A} = \frac{\pi}{2L_A}, \tag{3.10.12}$$

as we wanted to verify.

3.10.3 Notes

As stated in Remark (iii) of Section 3.10.1, motion-invariant line, or plane, processes may be treated for the preceding purposes as motion-invariant manifold processes. As indicated in Sections 3.3–3.5, general first-order properties relevant to the pertinent stereological equations were derived by Mecke (1981). For further details, see Stoyan et al. (1995).

3.11 The Motion-Invariant Poisson Line Process in the Plane

3.11.1 Construction

The Poisson line process $Y \subset \mathbb{R}^2$ is a motion-invariant line process in the plane with the distinct property that the point process of the $\{p_i,\ i \in \mathbb{Z}\}$, is Poisson of intensity N_L. With this additional property, higher-order properties can be obtained for Y.

3.11.2 Some properties of the Poisson line process

Besides the first-order properties given in Section 3.10.2, specific properties of the Poisson line process are given next.

The contact distribution function

The contact distribution function $H(r)$ of a Poisson line process $Y \subset \mathbb{R}^2$ is the probability that no line hits a disk $D_0(r)$ centred at a point $0 \in \mathbb{R}^2$, namely,

$$H(r) = 1 - \exp(-2N_L r), \quad r \geq 0. \tag{3.11.1}$$

The K-function and the pair correlation function

By definition, $L_A K(r)$ is the mean length of Poisson lines in a disk $D_0(r)$ centred at a point O from one of the lines. The line through O contributes the length $2r$ of a disk diameter, whereas the remaining lines contribute with a mean length $\pi r^2 L_A$ (which would not necessarily be the case if the line process was not Poisson, because the centre of the disk is not an arbitrary point of the plane). Thus, the K-function is

$$K(r) = \frac{2r + \pi r^2 L_A}{L_A} = \frac{2r}{L_A} + \pi r^2, \tag{3.11.2}$$

whereby the pair correlation function is

$$c(r) = \frac{K'(r)}{2\pi r} = \frac{1}{\pi r L_A} + 1. \tag{3.11.3}$$

Some properties of Poisson polygons

Let $D(H)$ denote the in-circle (of diameter $H > 0$) of a typical Poisson polygon P. The in-circle of a convex polygon is the largest circle contained in the polygon, and it is in general tangent to three of the sides of the polygon, see Fig. 3.11.1(a). The distribution function of H is

$$F(h) = \mathbb{P}(H \le h) = \mathbb{P}(\text{No line hits } D(h)) = 1 - \exp(-N_L h), \tag{3.11.4}$$

that is, $H \sim \text{Gamma}(1, N_L)$, the negative exponential distribution of mean $\mathbb{E}(H) = 1/N_L$.

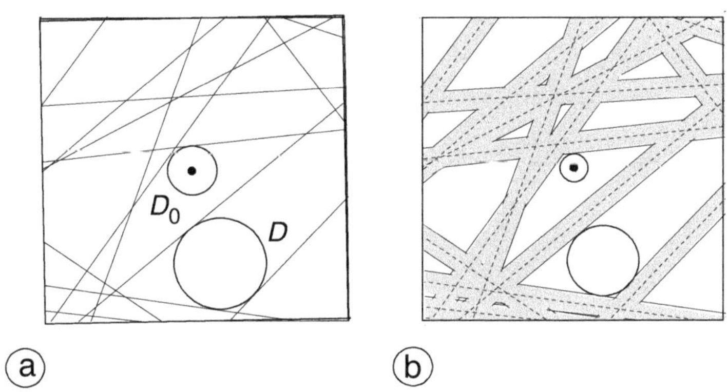

Fig. 3.11.1 (a) A realization of Poisson lines in a quadrat. D is the incircle of a Poisson polygon, and D_0 is the largest circle, with arbitrary centre, contained in a Poisson polygon. (b) Idem of the corresponding Poisson stripes. See text, Section 3.11.2.

Let $D_0(H_0)$ denote the largest circle of centre $0 \in \mathbb{R}^2$ (and diameter H_0) contained in the polygon $P_0 \ni 0$, see Fig. 3.11.1(a). We also have that $H_0 \sim \text{Gamma}(1, N_L)$, the same as H. This is an apparent contradiction, because $D_0(H_0)$ is in general a tangent circle, whereas $D(H)$ is an incircle, and therefore it seems that $H_0 < H$. The point is, however, that P_0 is not a typical, but an area-weighted polygon, whose mean area $\mathbb{E}(A_0)$ will therefore be larger than the mean area $\mathbb{E}(A) = \pi/N_L^2$ of P, recall Eq. (3.10.8) and Eq. (3.10.4).

To compute $\mathbb{E}(A_0)$ we apply Robbins' theorem. Let $X \subset \mathbb{R}$ denote a random domain. The random volume of X is

$$V(X) = \int_{\mathbb{R}^d} 1_X(x)\, dx. \tag{3.11.5}$$

Therefore,

$$\begin{aligned}\mathbb{E}\{V(X)\} &= \int_{\mathbb{R}^d} \mathbb{E}\{1_X(x)\}\,\mathrm{d}x \\ &= \int_{\mathbb{R}^d} \mathbb{P}(x \in X)\,\mathrm{d}x,\end{aligned} \tag{3.11.6}$$

which is Robbins' theorem. Thus,

$$\begin{aligned}\mathbb{E}(A_0) &= \int_{\mathbb{R}^2} \mathbb{P}(x \in P_0)\,\mathrm{d}x \\ &= \int_{\mathbb{R}^2} \mathbb{P}(\text{No line crosses the segment } 0x)\,\mathrm{d}x \\ &= \int_0^{2\pi} \mathrm{d}\omega \int_0^{\infty} \exp(-(2/\pi)N_L r)\,\mathrm{d}r \\ &= \frac{\pi^3}{2N_L^2} > \frac{\pi^2}{N_L^2} = \mathbb{E}(A).\end{aligned} \tag{3.11.7}$$

As a plus, we obtain,

$$\mathrm{Var}(A) = \mathbb{E}(A_0)\cdot\mathbb{E}(A) - \mathbb{E}^2(A) = \frac{\pi^2(\pi^2-2)}{2N_L^4}. \tag{3.11.8}$$

Poisson stripes

If each Poisson line is the midline of a stripe of width t, then the result is a Poisson stripe process Y_t, see Fig. 3.11.1(b). The distributions of the circle diameters H and of H_0, coincide with those of the Poisson line process Y. The same is the case for the properties of the polygons constituting the uncovered space. This is also an apparent paradox, dispelled by the fact that in the stripe process the smaller Poisson polygons tend to disappear, whereby there is a loss only of the number of polygons per unit area, not of their size. The area fraction covered by Y_t, namely the area intensity, is

$$\begin{aligned}A_A &= \mathbb{P}(Y_t \ni 0) \\ &= \mathbb{P}(Y \cap D_0(t) \neq \emptyset) \\ &= 1 - \exp(-tN_L),\end{aligned} \tag{3.11.9}$$

where $D_0(t)$ is a disk of diameter t centred at an arbitrary point $0 \in \mathbb{R}^2$.

3.11.3 Notes

The first detailed study of Poisson lines and stripes is due to Miles (1964), who derives higher-order moments, and some distributions, for Poisson polygons. Subsequent studies of Poisson line and plane processes abound, for references see for instance Stoyan et al. (1995) and Schneider and Weil (2008). The result in Eq. (3.11.6) is due to Robbins (1944).

3.12 Germ–Grain and Boolean Models

3.12.1 Basic notions on germ-grain processes

A general germ-grain model $Y \subset \mathbb{R}^d$ may be expressed as

$$Y = \bigcup_{i \in \mathbb{N}} (Y_i + y_i), \tag{3.12.1}$$

where the $\{Y_i\}$ are compact subsets called grains, whereas the $\{y_i\}$ are the corresponding associated points (APs), called germs, which constitute a point process in $\mathbb{R}^d$. The latter process need not be stationary, and the subsets need not be either isotropic or independent.

Without further assumptions, the properties of germ-grain models are difficult to study. The germ process is usually assumed to be stationary, in which case each grain Y_i may be assumed to be a realization of a random compact subset Y_0, called the typical grain, whose size and shape are random variables with a suitable joint distribution. In this case we may consider the following possibilities.

(a) The grains may overlap.
(b) The grains or particles cannot overlap, the result being a process of disjoint particles.
(c) The germs constitute a Poisson point process, hence the grains may overlap. The result is a Boolean model.

In each of the preceding three cases, if the grains are isotropically rotated, or their distribution is rotation-invariant, then the process is motion-invariant.

Explicit expressions for the mathematical properties of germ-grain models are scant, an exception being the Boolean model with grains of a simple shape. In most cases these properties have to be explored numerically. The usual goal is to fit a real structure with a particular model. This is more common in materials sciences than in biology, where the structures tend to be highly ordered. Whenever a model is found to fit a real structure reasonably, however, it should be borne in mind that the germ-grain mechanism used to construct the model may be very different from the mechanism (physical, or chemical) generating the real structure.

3.12.2 First-order properties

As in the line process case, the stereological equations involving a motion-invariant germ-grain model are given by Mecke's theorem, see Eq. (3.3.3), which applies to any motion-invariant process.

As a cursory illustration, consider a motion-invariant planar curve process

$$Y = \bigcup_{i \in \mathbb{N}} Y_i(y_i, \omega_i) \tag{3.12.2}$$

whose typical grain is a piecewise smooth planar curve $Y_0(0,0)$ of finite length $b > 0$ and arbitrary shape, with a well-defined AP placed at the origin. For simplicity, b is assumed to be fixed. The AP realizations $\{y_i\}$ constitute a stationary point process of intensity N_A in the plane, and the orientation angles $\{\omega_i\}$ of the curve realizations are independent UR in $[0, 2\pi)$. Curve shapes are also independent from the location of the APs.

The preceding curve process Y may be regarded as a stationary particle process because, although the segments are allowed to intersect, they are almost surely distinguishable. Therefore, the length intensity is

$$B_A = N_A b. \tag{3.12.3}$$

Consider a fixed test curve $T_1 \equiv T_1^2(0,0)$, piecewise smooth, of length $l > 0$ and arbitrary shape. Then, by Eq. (3.3.11),

$$\mathbb{E}\{I(Y \cap T_1)\} = \frac{2}{\pi} \cdot l B_A, \tag{3.12.4}$$

which is a model-based version of Eq. (2.12.10).

Similarly as in Eq. (3.5.4), the intensity of the stationary point process formed by the self-intersections of the curves of Y is

$$I_A = \frac{1}{\pi} \cdot B_A^2. \tag{3.12.5}$$

3.12.3 The Boolean model

Let Y denote a Boolean model with realizations in $\mathbb{R}^d$. The associated point process is stationary Poisson of intensity λ, and the corresponding grains are independent realizations of a fundamental random grain $Y_0 \subset \mathbb{R}^d$ with its AP at the origin, and d-volume $V(Y_0)$. Fig. 3.12.1 illustrates a realization of a planar Boolean model whose fundamental grain is a disk of a fixed radius.

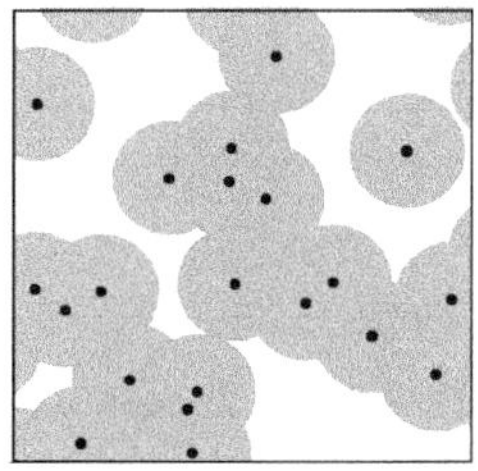

Fig. 3.12.1 A realization of a Boolean model of disks in a quadrat.

Hitting probability

The probability that an arbitrarily fixed compact test probe $T \subset \mathbb{R}^d$ hits Y is

$$\begin{aligned}\mathbb{P}(Y \cap T \neq \emptyset) &= 1 - \mathbb{P}(\text{no grain hits } T)\\ &= 1 - \mathbb{P}(\text{no AP is in } T \oplus \check{Y}_0)\\ &= 1 - \exp\{-\lambda \mathbb{E} V(T \oplus \check{Y}_0)\}, \end{aligned} \tag{3.12.6}$$

(recall Eq. (1.20.1)).

Volume intensity

In particular, if the probe T is a fixed test point $0 \in \mathbb{R}^d$ then, combining the preceding result with Eq. (3.6.5), the volume fraction occupied by the Boolean model is

$$V_V = \mathbb{P}(Y \ni 0) = 1 - \exp\{-\lambda \mathbb{E} V(Y_0)\}. \tag{3.12.7}$$

The spherical contact distribution function

The application of Eq. (3.12.6) to Eq. (3.6.15) yields

$$H(r) = 1 - \frac{\exp\{-\lambda \mathbb{E} V(D_0(r) \oplus \check{Y}_0)\}}{1 - V_V}, \tag{3.12.8}$$

with $V_V = 0$ if the Boolean process Y is not a volume process.

The covariance function

The non-centred scalar covariance function $C(r)$, $r > 0$, of a motion-invariant volume process Y allows the computation of useful items, for instance,

- The K-function, see Eq. (3.6.11).
- The variance of the ratio $V(Y \cap T)/V(T)$ generated by a fixed test probe T, see Eq. (3.8.8) for the special case of a platelet probe in $\mathbb{R}^3$.

Below we derive an expression for $C(r)$ for a motion-invariant Boolean volume process.

By Eq. (3.6.4), for $x \in \mathbb{R}^d$ the vectorial non-centred covariance function of a stationary volume process is

$$\begin{aligned} C(x) &= \mathbb{P}(0 \in Y, x \in Y) \\ &= 1 - \mathbb{P}(0 \notin Y, \text{ or } x \notin Y) \\ &= 1 - \mathbb{P}(0 \notin Y) - \mathbb{P}(x \notin Y) + \mathbb{P}(0 \notin Y, x \notin Y) \\ &= 1 - 2(1 - V_V) + \mathbb{P}(0 \notin Y, x \notin Y). \end{aligned} \tag{3.12.9}$$

Now, for a Boolean model,

$$\begin{aligned} \mathbb{P}(0 \notin Y, x \notin Y) &= \mathbb{P}\{\text{No AP in } \check{Y}_0 \cup (\check{Y}_0 + x)\} \\ &= \exp\{-\lambda \mathbb{E} V(\check{Y}_0 \cup (\check{Y}_0 + x))\}, \end{aligned} \tag{3.12.10}$$

and, in turn,

$$\begin{aligned} V(\check{Y}_0 \cup (\check{Y}_0 + x)) &= 2V(\check{Y}_0) - V(\check{Y}_0 \cap (\check{Y}_0 + x)) \\ &= 2V(Y_0) - V(Y_0 \cap (Y_0 - x)) \\ &= 2V(Y_0) - g_0(x), \end{aligned} \tag{3.12.11}$$

where $g_0(x)$ is the vectorial geometric covariogram of the typical grain, see the first Eq. (3.8.2). Combining the preceding results, and recalling Eq. (3.12.7), we finally obtain

$$C(x) = 2V_V - 1 + (1 - V_V)^2 \exp\{\lambda \mathbb{E} g_0(x)\}, \quad x \in \mathbb{R}^d. \tag{3.12.12}$$

If the grains are isotropic then, averaging their covariogram with respect to orientations, see Eq. (3.8.3) for the planar case, we obtain the scalar non-centred covariance of the Boolean model, namely,

$$C(r) = 2V_V - 1 + (1 - V_V)^2 \exp\{\lambda \mathbb{E} g_0(r)\}, \quad r > 0. \tag{3.12.13}$$

If the grains differ only in size, then the expectation of the scalar covariogram $g_0(r)$ in the preceding expression is over the grain size distribution.

3.12.4 Boolean model of straight line segments

As an illustration, a Boolean model Y of straight line segments in the plane may be represented by Eq. (3.12.2). The fundamental grain Y_0 is a straight line segment of length $b > 0$ which, for simplicity, is assumed to be constant. The midpoint y_0 of Y_0 is adopted as its AP. The segment APs, namely the $\{y_i\}$, constitute a Poisson point process of intensity N_A, and the corresponding segment orientations $\{\omega_i\}$ are independent UR in $[0, \pi)$. The result is a Poisson line segment process of length intensity $B_A = N_A b$ in the plane, see Fig. 3.2.1(a). Apart from the first-order properties given by Eq. (3.12.4) and Eq. (3.12.5), warranted solely by the motion-invariance of the segment process, specific higher-order properties stemming from the Poisson assumption are given next.

The K-function

Here the quantity $B_A K(r)$ is the mean length of Poisson segments in a disk $D_0(r)$ of radius $r > 0$ centred at a UR point within a segment, i.e., $0 \in Y$. If $r \geq b$, then the length contribution L of the hit segment is equal to b, whereas if $r < b$, then the contribution L is random, and its mean value may be computed as follows. Take a reference axis along the hit segment, and fix the origin so that the abscissa of the centre of the disk is equal to b. If $b \in (r, 2r)$, then

$$\begin{aligned} b\mathbb{E}(L) &= \int_0^{b-r} (r+x)\,\mathrm{d}x + \int_{b-r}^{r} b\,\mathrm{d}x + \int_r^b (r+b-x)\,\mathrm{d}x \\ &= 2rb - r^2. \end{aligned} \tag{3.12.14}$$

If $b > 2r$, then the corresponding integrals are different, but the result is the same. Thus,

$$K(r) = \begin{cases} \frac{1}{B_A}\left(2r - \frac{r^2}{b}\right) + \pi r^2, & r < b, \\ \frac{b}{B_A} + \pi r^2, & r \geq b. \end{cases} \tag{3.12.15}$$

Note that, if the segment process was not Poisson, then the second term would not necessarily be equal to πr^2.

If the segment length b is random with a pdf $f(b)$, say, then, when computing the mean value of the first term of $K(r)$, the pdf $bf(b)/\mathbb{E}(b)$ should be used, because the pdf of the length of the segment containing the typical point 0 is length-weighted.

The spherical contact distribution function

With the preceding notation, the area of the Minkowski sum $D_0(r) \oplus \check{Y}_0$ is in this case equal to $\pi r^2 + 2rb$, whereby the adapted Eq. (3.12.8) yields

$$H(r) = 1 - \exp\{-rN_A(\pi r + 2b)\}, \quad r > 0. \tag{3.12.16}$$

3.12.5 Notes

1. Germ–grain models

With no specific name, similar models were already studied by Matérn (1960), Section 2.7, also by Matheron (1967), Section 8. A specific compendium was provided by Hall (1988), see also Stoyan et al. (1995), and Schneider and Weil (2008).

2. Boolean models.

Germ-grain models with Poisson germs were also treated in the preceding references. Early formal treatments can be found in Giger and Hadwiger (1968), and Matheron (1975). See also Kellerer (1983), who elaborates on the pioneering work of Mack (1954). The stereology of a Boolean model in space intersected by a slab was treated by Miles (1976). Various authors studied special models with a segment, a disk, etc., as a model for the primary grain, see Stoyan et al. (1995).

3.13 Particle Size Weighting and Size Distributions

Let Y represent a stationary process of piecewise smooth, disjoint particles $\{Y_i\}$.

3.13.1 Number weighting

A particle sampled with a sweeping plane by the scanning principle (Section 2.15.2) is called a typical particle, and its analogue in design stereology is a disector-sampled particle, or a particle met for the first time by the sweeping plane. For particles sampled that way, the distribution of a size parameter γ (which may stand for V, S, M, H, or χ) is number-weighted, and denoted by $F_N(\gamma)$. For instance, let $\gamma \equiv V$, and let $N_V(v)$ denote the number intensity of particles with volume no greater than v. Then, analogously as in Eq. (2.32.35),

$$F_N(v) = N_V(v)/N_V, \tag{3.13.1}$$

where N_V is the number intensity of the process Y. Suppose that F_N admits a density function f_N, and let $N_V(\mathrm{d}v)$ denote the number intensity of particles with volume in the interval $(v, v + \mathrm{d}v]$. Then the corresponding probability element is

$$f_N(v)\,\mathrm{d}v = N_V(\mathrm{d}v)/N_V. \tag{3.13.2}$$

3.13.2 Volume weighting

Consider a fixed origin 0. If for a realization of the process Y there is a particle containing 0, then this particle is sampled with a probability proportional to its volume. To see this, let $Y(\mathrm{d}v)$ represent the process of the particles from Y whose volume lies in the interval $(v, v + \mathrm{d}v]$, and let $V_V(\mathrm{d}v)$ denote the corresponding volume intensity. Then, by virtue of Eq. (3.6.5),

$$\mathbb{P}(Y(\mathrm{d}v) \ni 0) = V_V(\mathrm{d}v) = vN_V(\mathrm{d}v) \propto v. \tag{3.13.3}$$

Let V_V denote the volume intensity of Y. By definition, the mean particle volume in number is $\mathbb{E}_N(V) = V_V/N_V$. If the probability density function f_V in the volume distribution exists, then it follows that the corresponding probability element is

$$f_V(v)\,\mathrm{d}v = \frac{V_V(\mathrm{d}v)}{V_V} = \frac{v \cdot N_V(\mathrm{d}v)}{N_V \cdot \mathbb{E}_N(V)} = \frac{v \cdot f_N(v)\,\mathrm{d}v}{\mathbb{E}_N(V)}, \tag{3.13.4}$$

as expected (compare with Eq. (2.3.2)). Thus, if a histogram of f_N is available (e.g. via the volume measurement of disector-sampled particles using serial sections, see Section 2.32.8), then a histogram of f_V may be constructed a posteriori numerically using the preceding identity. Conversely, if a histogram of f_V is available (e.g. via the volume measurement of point-sampled particles using serial sections), then a histogram of f_V may be recovered numerically by discretization of the following identity,

$$\begin{aligned} f_N(v) &= \mathbb{E}_N(V) \cdot v^{-1} f_V(v) \\ &= \frac{v^{-1} f_V(v)}{\int_0^\infty v^{-1} f_V(v)\,\mathrm{d}v}. \end{aligned} \tag{3.13.5}$$

3.13.3 Mean size-weighted particle size

In principle, the identities developed in Section 2.32 extend to the model-based case with few practical adaptations. For instance, in order to implement Eq. (2.32.19) for $\mathbb{E}_V(V)$, it suffices to superimpose arbitrarily on a section a portion of a test system consisting of a few lines bearing test points, see Fig. 2.32.2. In materials science, however, a limitation in the use of such methods is the presence of a large proportion of complicated non-convex particles, which may preclude the identification of transects consisting of several separate profiles as belonging to a given particle. This may in principle be controlled using serial section images, see next.

3.13.4 Notes

As stated in the preceding section, the available design-based methods extend to the model-based case. An informative, pioneering paper on size-weighted particle size distributions is that of Gundersen and Jensen (1983).

If the particles are not too complicated in shape, then $\mathbb{E}_V(V)$, for instance, is easy to obtain from independent sections. As shown by Gundersen (1986), this parameter usually enjoys a higher discrimination power than $\mathbb{E}_N(V)$, which is also more expensive to obtain. Moreover, because the method to access $\mathbb{E}_V(V)$ is distribution-free, it may (and indeed, it should) be used to check the performance of particle distribution models. On the other hand, parameters requiring sampling in 3D, notably disector sampling to access $\mathbb{E}_N(V)$, or even $f_N(v)$, for instance, may also be implemented in materials science by a suitable combination of serial grinding and polishing of the material with computer-aided section imaging, as already suggested in Fig. 6.1 from Gundersen (1986).

3.14 Band and Membrane Models

3.14.1 Model, mean thickness

Curved stripes, or bands, may be useful models for planar maps of roads, rivers, blood vessels on a flat support, etc. Loosely, a bounded band $Y \subset \mathbb{R}^2$ is a narrow portion of space confined between two non-intersecting, piecewise smooth curve boundaries ∂Y_1 (the reference face) and ∂Y_2.

Consider a simple band model example consisting of the union of a horizontal rectangle of base (or length) $l_1 = 10\,\text{m}$ and height (or width) $\tau_1 = 3\,m$, followed by an adjacent horizontal rectangle of length $l_2 = 40\,\text{m}$ and width $\tau_2 = 1\,\text{m}$. The ordinary average width of 2 m is not a fair measure of the mean width of the band. More correct is

$$\mathbb{E}_L(\tau) = \frac{3 \times 10 + 1 \times 40}{10 + 40} = 1.4\,\text{m}, \tag{3.14.1}$$

namely the length-weighted mean

$$\begin{aligned} &\mathbb{E}_L(\tau) = w_1\tau_1 + w_2\tau_2, \quad (w_1 + w_2 = 1), \\ &w_i = l_i/(l_1 + l_2), \quad i = 1, 2. \end{aligned} \tag{3.14.2}$$

Equivalently,

$$\mathbb{E}_L(\tau) = \frac{\text{Area of the band}}{\text{Length of the band}}. \tag{3.14.3}$$

Returning to the more general band model $Y \subset \mathbb{R}^2$, we define the thickness $\tau(y)$ of Y at a point $y \in \partial Y_1$ as the smallest distance from y to ∂Y_2. Let $\mathrm{d}b(y)$ be the length element of ∂Y_1 at y. As a natural extension of the preceding example, conventional definitions of the area and the mean thickness of Y are

$$A(Y) = \int_{y \in \partial Y_1} \tau(y)\,\mathrm{d}b(y), \tag{3.14.4}$$

$$\mathbb{E}_B(\tau) = \frac{A(Y)}{\frac{1}{2}\{B(\partial Y_1) + B(\partial Y_2)\}}, \tag{3.14.5}$$

respectively. Note that the integrand in the rhs of Eq. (3.14.4) is the length-weighted local thickness at y.

The preceding ideas are readily extensible to $\mathbb{R}^d$. Of interest is the motion-invariant membrane model $Y \subset \mathbb{R}^3$ of volume intensity V_V, with piecewise smooth boundary ∂Y of area intensity S_V, in which case,

$$\mathbb{E}_S(\tau) = 2V_V/S_V. \tag{3.14.6}$$

For a design-based applications, see Sections 4.13, 4.17, and 4.19.

3.14.2 Local sampling on sheet-like structures

Without loss of generality, consider a motion-invariant band process $Y \subset \mathbb{R}^2$, of area intensity A_A, with total piecewise smooth boundary ∂Y of length intensity B_A, in which case, $\mathbb{E}_B(\tau) = 2A_A/B_A$, as we have seen. An alternative version of $\mathbb{E}_B(\tau)$, generally different from the preceding one, may be accessed by boundary length-weighted observations of τ, which is assumed to be finite. This may be done in the following two steps.

1. Sample typical points $\{y_i\}$ determined by the intersections between ∂Y and a bounded portion, say, of an arbitrary system of test line segments.
2. Measure the local thickness τ_i at each typical boundary point y_i. Because the local band thickness sampled this way is boundary length-weighted (Section 3.6.5), as the number of observations increases the corresponding histogram converges to the probability density function of τ in the boundary length frequency measure, and therefore the mean of this distribution is the mean band thickness $\mathbb{E}_B(\tau)$.

In biological sheet-like objects like the brain cortex, it is often desired not only to measure thicknesses, but also to study internal structures, such as concrete neuron populations, by means of small blocks which are locally perpendicular to the reference boundary (e.g. to the pial surface of the brain in space, not to planar section traces). The problem is to sample the block locations properly, namely surface area-weighted. In theory, these locations should be the intersections determined in the reference surface by say a portion of a fakir probe – in Fig. 2.25.2(b) such locations are encircled. At this initial sampling step, non-invasive methods, based on say laser technology, might be tried to mark the prospective block sites on the reference surfaces. Later, blocks (e.g. tissue cylinders) would be cut out at these sites, locally perpendicular to the reference surface boundaries in 3D.

3.14.3 Membrane thickness distribution

A motion-invariant membrane process in 3D hit by a plane L_2^3 is a motion-invariant planar band process, as described in Section 3.14.2. Also, a membrane process hit by a test line L_1^3 (either directly in 3D, or in a planar section) yields a motion-invariant process of linear membrane intercepts along the line. Classical problems of stereology were (i) to identify the distribution in surface area of membrane thickness,

$$S_V(\tau) = S_V(y \in \partial Y : \tau(y) \leq \tau) \tag{3.14.7}$$

from the observable distribution in (trace) length of the apparent thickness (namely of band width) in a planar section,

$$B_A(r) = B_A(y \in \partial Y \cap L_2^3 : \tau(y) \leq r), \tag{3.14.8}$$

and (ii) to identify $S_V(\tau)$ from the observable distribution in boundary trace length of the linear intercept length,

$$B_A(l) = B_A(y \in \partial Y \cap L_1^3 : l(y) \leq l). \tag{3.14.9}$$

3.14.4 Notes

1. Membrane thickness

The latter problem was addressed in part by Weibel and Knight (1964) and solved by Gundersen, Jensen, and Østerby (1978) and Cruz-Orive (1979). The problem (i) was also solved in the latter paper and in Jensen, Gundersen, and Østerby (1979). The local membrane model consisted of an essentially planar surface element at a typical point $y \in \partial Y$ and a parallel plane a distance $\tau(y)$ apart. For a brief review, see Cruz-Orive (2017), p. 166.

2. Other unfolding problems

For the reasons given in Section 2.16.4, Note 2, unfolding methods for particle size distributions are not considered here – for a brief, critical review, see Cruz-Orive (2017), p. 160.

3. Anisotropy, orientation distributions

These subjects are important, but their treatment requires space that would distract from the basic purposes of this book. Moreover, at least in biomedicine the practical implementation of the available results is not straightforward. The main progress is taking place in materials science, see for instance Ohser and Schladitz (2009), Wirjadi, Schladitz, Easwaran, and Ohser (2016), etc. – for a brief survey, see Cruz-Orive (2017), pp. 166–67.

Chapter 4
Sampling and Estimation for Stereology

4.1 Estimation in Design Stereology: Basic Ideas

Chapter 2 covers geometric sampling and mean values of random variables observable on the intersection between a non-random target object and a single probe equipped with a motion-invariant probability element. As a simple example,

$$\mathbb{E}\{P(Y \cap x)\} = \frac{A}{a} \tag{4.1.1}$$

is a special case of Eq. (2.2.3). Note that Eq. (4.1.1) involves only non-random quantities, namely the mean value in the lhs, A, and a. Here A denotes the area of a planar domain Y contained in a reference square $D = [0, H)^2$ of area $a = H^2$, and x is a UR test point in D. The target parameter is the unknown quantity A. In this chapter we address estimation problems. With the purpose of estimating A, the sampling experiment consists in generating a UR test point x in D. If $Y \cap x \neq \emptyset$, then we score $P(Y \cap x) = 1$, whereas if $Y \cap x = \emptyset$, then $P(Y \cap x) = 0$. Now, if we replace the expected value $\mathbb{E}\{P(Y \cap x)\}$ with the random outcome $P(Y \cap x)$, then the mean value identity (4.1.1) can be exploited to obtain an unbiased estimator (UE) $\widehat{A}$ of A, namely,

$$\widehat{A} = a \cdot P(Y \cap x), \tag{4.1.2}$$

which is a random variable. In fact, by Eq. (4.1.1) the mean of $\widehat{A}$ with respect to the UR distribution of x in D is

$$\mathbb{E}(\widehat{A}) = a\mathbb{E}\{P(Y \cap x)\} = A, \tag{4.1.3}$$

which amounts to saying that $\widehat{A}$ is a UE of A. The random variable $P(Y \cap x)$, namely the random number of hits, is an example of a 'statistic', and an estimator is just a formula involving one or more statistics. In general a statistic is a function of the data from a sample. For instance, the sum of the data from a sample is a statistic, whereas that sum divided by the sample size is an estimator of the population mean under some conditions. If a statistic is replaced with an observed numerical value,

L. M. Cruz-Orive, *Stereology*, Interdisciplinary Applied Mathematics 59,
https://doi.org/10.1007/978-3-031-52451-6_4

then the estimator formula yields an 'estimate', which is also a numerical value. For instance, if we observe $P(Y \cap x) = 1$, then $\widehat{A} = a$, which is a known numerical value if a is known. For simplicity, the same notation ($\widehat{A}$ in this case) will be used for an estimator and for a particular estimate.

The foregoing sampling experiment is an example of the implementation of a sampling design. In design stereology, a sampling design defines the pertinent sampling probabilities, see Section 4.3.2. It is aimed at estimating a target parameter, namely a fixed and unknown geometric quantity defined on an object of interest.

Unbiasedness and precision are two different concepts, see Fig. 4.1.1. Here each dot impact may be interpreted as an estimate of the position of the bullseye. If the design is correct, e.g. if $x \sim \mathrm{UR}(D)$ in the above example, then the centroid of a cluster of infinitely many impacts will coincide with the bullseye (cases (c,d)), which means that the corresponding estimator will be unbiased. Otherwise the estimator may be biased, see cases (e,f). The bias is the distance of the cluster centroid from the bullseye. Bias may also arise if the estimator formula is incorrect, e.g. if the multiplying constant in the rhs of Eq. (4.1.3) is other than a.

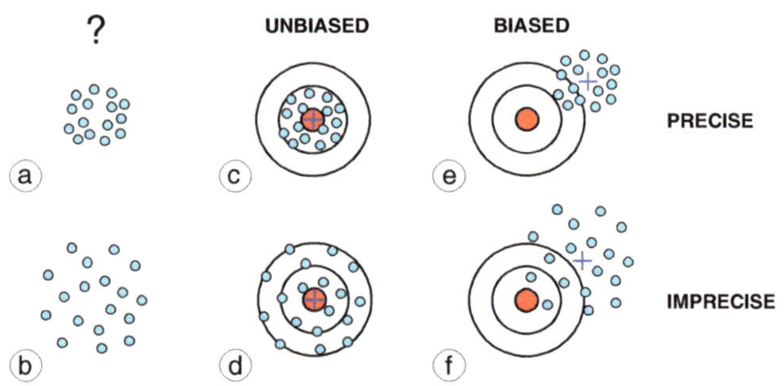

Fig. 4.1.1 The quality of the data in (a), (b) cannot be assessed because the target is unknown. Often the data in (a) is preferred, but if the estimator is biased, see (e), then (b) may be preferable, see (f). The estimator leading to the replications in (c) is fairly accurate, the one leading to (d) is unbiased but not as accurate. Only the mathematics underlying the sampling design can predict whether an estimator will be design unbiased, or not.

The estimator in Eq. (4.1.2) is unbiased, but it may take only two possible values (that is, it can yield only two possible estimates), namely a and 0, with probabilities $p = A/a$ and $1 - p$ respectively. Thus, a UE may not be useful if it comes from one, or very few observations. Here comes the question of precision, which can be increased arbitrarily by increasing the number of test points for each estimate, so that the scenario should change from case (d) to case (c). Precision can also be increased by improving the design, e.g., by using a test system. The effect of so doing is illustrated in Section 4.6.4, Note 2, for a special case. The estimates in the upper row of Fig. 4.1.1 are all precise, but only the estimator of case (c) is accurate (namely not only precise, but also unbiased). Case (a) is uncertain.

In the example, $\{P(Y \cap x)\}$ denotes a random sample of size 1. Suppose that we generate $n \geq 1$ independent UR test points $\{x_1, x_2, \ldots, x_n\}$ in the reference square. Then, $\{P(Y \cap x_i),\ i = 1, 2, \ldots, n\}$ represents a random sample of size n consisting of 0's and 1's, and

$$\widehat{A}_n = a \cdot \frac{1}{n} \sum_{i=1}^{n} P(Y \cap x_i), \tag{4.1.4}$$

represents the corresponding UE of A. If $n = 5$, say, then $\{0, 1, 1, 0, 1\}$ would represent a particular sample of size 5, (not "five samples"!), and the corresponding estimate, analogous to a dot impact, would be $\widehat{A}_5 = 3a/5$. Note that n is a property of each impact, and not the size of a cluster in Fig. 4.1.1. In statistics, each impact corresponds to a random replication (or realization) of an estimator (of A in this case).

The error of an estimator is the signed difference between the estimator and the target, and the bias is the mean error, see Eq. (A.1.52). Thus, unbiasedness means zero bias. In statistics, unbiasedness is not always a priority – it is often preferred to minimize the mean square error (MSE), see Eq. (A.1.53). In stereology, however, design-unbiasedness is a valuable property, rather easy to achieve (usually easier than "doing it wrong"!). On the other hand, increasing the sample size (e.g. using a denser test system) is no problem, hence the MSE can be decreased at will. The smaller the MSE, the more accurate is the estimator.

If the target is unknown, then resampling experiments (e.g., increasing arbitrarily the number of impacts per cluster in either Fig. 4.1.1(a,b)) can tell us something about precision (often called "reproducibility"), but cannot disclose whether bias is present, or not – only mathematics can tell whether a design will necessarily lead to unbiasedness, or not. Nonetheless, even if the sampling design is correct, unavoidable sources of bias are common in practice, arising from specimen preparation prior to sectioning, observation artifacts, etc. For this reason the estimators studied here may cautiously be called 'design-unbiased' – i.e., strictly unbiased only as far as the geometric sampling design is concerned.

It is important to distinguish between unbiasedness and consistency. A sufficient condition that an estimator is consistent is that its MSE (or its variance if the estimator is unbiased) tends to zero as the size of the sample used in the estimator tends to infinity. Thus, consistency bears the idea that the estimator gets closer and closer (but not necessarily "exactly equal") to the target parameter as the sample size (e.g., the value of n in Eq. (4.1.4)) increases. In contrast, the mean value of a UE over all possible replications (= 'dot impacts') is always exactly equal to the target parameter no matter the sample size (even if $n = 1$, see Eq. (4.1.3)).

The estimator $\widehat{A}_n + 1/n$ of A has a bias equal to $1/n$, hence it is not unbiased for any $n \geq 1$, but it is consistent because

$$\mathrm{MSE}(\widehat{A}_n + 1/n) = \mathrm{Var}(\widehat{A}_n) + 1/n^2 = O(1/n). \tag{4.1.5}$$

This is easy to see because the summation in the rhs of Eq. (4.1.4) follows a Binomial distribution with n independent trials and a constant hitting probability $p = A/a$

for each trial, hence $\mathrm{Var}(\widehat{A}_n) = a^2 p(1-p)/n = O(1/n)$. On the other hand, if an IUR square grid of test points is used, then an approximation of $\mathrm{Var}(\widehat{A})$ is $\mathrm{var}(\widehat{A}) = O(1/P^{3/2})$, where P is the total number of test points hitting the object Y, see Section 4.6.3. Thus, convergence to the target parameter as P increases is usually much faster with test systems than with independent probes.

In the example, the procedure used to obtain a UE of A works because Eq. (4.1.1) is linear in A. The mean value identities Eq. (2.1.5) and Eq. (2.1.6) are linear in the target parameter $\gamma \equiv \gamma(Y)$, and therefore the estimation technique extends easily: mean values are replaced with the corresponding statistics, and γ is replaced with its estimator $\widehat{\gamma}$, whereby,

$$\widehat{\gamma} = \frac{h_1(D)}{c_1} \cdot \alpha(Y \cap L_r) \tag{4.1.6}$$

for an unbounded probe, or

$$\widehat{\gamma} = \frac{h_2(D)}{c_2 \cdot \nu(T_r)} \cdot \alpha(Y \cap T_r) \tag{4.1.7}$$

for a bounded probe. In statistics this technique is known as ‘point estimation’. The main condition that $\widehat{\gamma}$ is unbiased is that the sampling experiment is carried out according to the motion-invariant probability element associated with the probe. This is equivalent to fixing the probe and assigning the corresponding probability element to the target object.

The model-based counterpart for a motion-invariant process Y with realizations in $\mathbb{R}^d$, hit by a fixed compact probe $T \subset \mathbb{R}^d$ is the UE

$$\widehat{\gamma}_V = \frac{c_{20}}{c_2} \cdot \frac{\alpha(Y \cap T)}{\nu(T)}, \tag{4.1.8}$$

which follows from Eq. (3.3.3).

4.2 Estimation of Global Quantities

4.2.1 Introduction

A global parameter $\gamma \equiv \gamma(Y)$ is the total of a fixed quantity defined in the target object $Y \subset \mathbb{R}^d$, which is regarded as a fixed compact set. In $\mathbb{R}^3$, γ may stand for

$$N\,\mathrm{cm}^0,\ \chi\,\mathrm{cm}^0,\ M\,\mathrm{cm},\ L\,\mathrm{cm},\ S\,\mathrm{cm}^2,\ \text{or } V\,\mathrm{cm}^3. \tag{4.2.1}$$

The exponent of the measurement unit expresses the physical dimension of the corresponding target parameter. This situation pertains to design-based stereology (or design stereology, for short), where the object is fixed and the test probe is endowed with a motion-invariant probability element (Section 2.1.2). The main fields of application of design stereology are biology and biomedicine, where the target object is typically an organ (e.g. a lung, or a brain), or a well-defined compartment of it (e.g. the entire alveolar surface of a lung, or the population of neurons of a given type in the gray matter of a brain). In these two cases, the target quantities may be total surface area, or total number, respectively.

The estimation of γ may be direct (Section 4.2.2), or indirect via ratios (Section 4.2.3), depending on the object size and on the magnification required to observe the target quantity on sections. The term 'magnification' will always mean 'final linear magnification'. Ratios are also natural target parameters in model stereology. In this chapter the adopted probes are usually test systems because, as hinted in the preceding section and in Section 2.24, they enjoy important advantages.

General estimators are given next, and particular cases are illustrated with real examples in subsequent sections.

4.2.2 Direct estimation for single objects

If the target parameter γ is, for instance, the total volume, or the total surface area of an organ, and the relevant sections can be observed and measured entirely at a given magnification, then γ may be estimated at a single sampling stage. We may consider the following cases.

FUR designs

FUR designs may be applied if at least one among the target parameter, or the test system, are of full dimension d, namely the dimension of the containing space.

For a FUR test system Λ_x with a bounded fundamental test probe $T_0 \subset J_0 \subset \mathbb{R}^d$, (e.g., a test point, a test curve, or a quadrat for $d = 2$), Eq. (2.25.3) yields the following UE,

$$\widehat{\gamma} = \frac{V(J_0)}{\nu(T_0)} \cdot \alpha(Y \cap \Lambda_x), \tag{4.2.2}$$

which is illustrated, for instance, in Sections 4.6 and 4.10. The preceding estimator also applies to FUR slabs (Section 4.9), to FUR slab disectors, and to FUR spatial grids.

For a FUR test system Λ_z of unbounded test probes, (e.g., parallel test lines, or planes, typical of Cavalieri designs), Eq. (2.25.5) yields

$$\widehat{\gamma} = V(J_0) \cdot \alpha(Y \cap \Lambda_z), \tag{4.2.3}$$

which is illustrated in Sections 4.8 and 4.9.

A section determined by an unbounded probe in a target object is usually called an extensive section of the object.

IUR designs

These apply if the dimensions of the target parameter, and of the test system, are both less than d (their sum however, being no less than d, see Section 1.3.2).

For an IUR test system Λ_{x,u_d} with a bounded fundamental test probe, (e.g., an IUR square grid to estimate curve length in the plane, see Section 4.6), Eq. (2.26.3) yields the following UE,

$$\widehat{\gamma} = \frac{c_{20}}{c_2} \cdot \frac{V(J_0)}{\nu(T_0)} \cdot \alpha(Y \cap \Lambda_{x,u_d}), \tag{4.2.4}$$

which applies also to IUR spatial grids.

For an IUR test system $\Lambda_{z,u}$ of unbounded test probes, (e.g., isotropic Cavalieri planes), Eq. (2.26.6) yields the following UE,

$$\widehat{\gamma} = \frac{c_{10}}{c_1} \cdot V(J_0) \cdot \alpha(Y \cap \Lambda_{z,u}), \tag{4.2.5}$$

which is illustrated in Section 4.12.

Vertical designs for a single object

The preceding formulae easily extend to the vertical designs described in Section 2.28 for $d = 3$. For details and illustrations, see Sections 4.16, 4.17, and 4.19.

4.2.3 Indirect estimation via ratios

Often the measurement of a relevant structure in an extensive section of a bounded object requires a relatively large magnification, and therefore the section cannot be measured directly. Fig. 4.2.1(b) shows an extensive section of the Dentate Gyrus (DG, regarded as a domain $Y_0 \subset \mathbb{R}^3$) within the hippocampus of a mouse – the design is revisited in Section 4.14. The target parameter was the number $N(Y)$ of a well-defined population of neurons $Y \subset Y_0$, whose observation requires a magnification $M \approx 20 \cdot M_0$, where M_0 is the magnification required to measure a Cavalieri section of Y_0. The section of the DG in Fig. 4.2.1(b) is about 1 mm wide. The magnification M required to count the neurons will be about 600×, whereby the entire magnified section would be around 60 cm wide! Here a strategy is to regard Y_0 as a reference compartment (usually called the "reference space"), and to estimate separately the reference volume $V(Y_0)$ at the magnification M_0, and then the ratio $N_V \equiv N(Y)/V(Y_0)$ at the magnification M. In the end, the target number $N(Y)$ is estimated indirectly using the design suggested by Eq. (2.30.1), namely,

$$\widehat{N}(Y) = \widehat{V}(Y_0) \cdot \widehat{N}_V. \tag{4.2.6}$$

At the magnification M, the ratio N_V may be estimated by subsampling each section with systematic optical disectors, see Fig. 4.2.1(d).

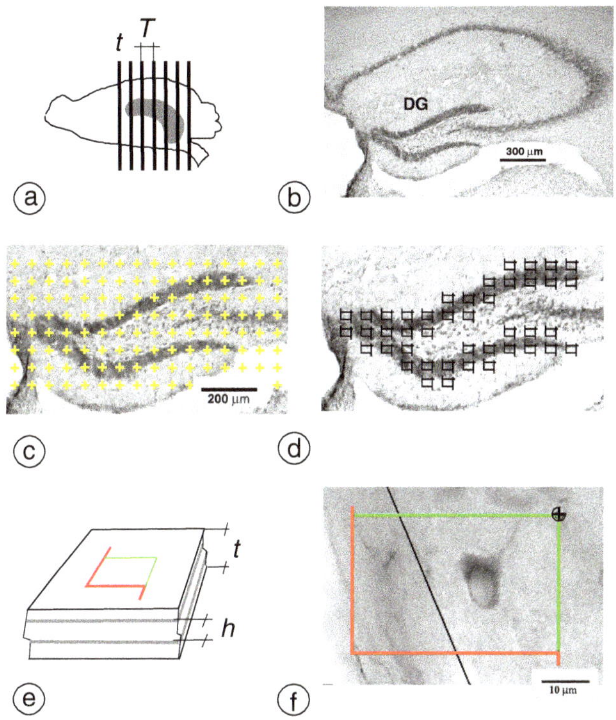

Fig. 4.2.1 (a) Sketch of a lateral projection of the hippocampus of a mouse hit by Cavalieri planes. (b) A Cavalieri section of the dentate gyrus (DG), which is adopted as the reference space Y_0. (c) Test points used to estimate section area. (d) Sketch (not scaled) of the systematic disectors used to estimate the ratio N_V, namely the number of target neurons per unit volume of Y_0. (e) Sketch of the optical disector tool. (f) A target neuron captured by an optical disector. Modified from Cruz-Orive et al. (2004), with permission of Oxford University Press.

Ratios are generally useful in model stereology (e.g. in materials science), but they can yield misleading conclusions in biology. For instance an increase (or decrease) in the ratio N_V may actually correspond to no change, or even to a decrease (or increase) of the total cell number N in the entire organ if the reference space has shrunken (swollen). In fact, the physical dimensions of N_V and N are quite different. A ratio $\gamma_V \equiv \gamma(Y)/V(Y_0)$, $Y \subset Y_0 \subset \mathbb{R}^3$, may stand for

$$N_V\,\mathrm{cm}^{-3},\ \chi_V\,\mathrm{cm}^{-3},\ M_V\,\mathrm{cm}^{-2},\ L_V\,\mathrm{cm}^{-2},\ S_V\,\mathrm{cm}^{-1},\ \text{or}\ V_V\,\mathrm{cm}^{0} \qquad (4.2.7)$$

(compare with Eq. (4.2.1)). The ratio γ_V, with the volume of the reference space Y_0 in the denominator, is often called the "density" of Y in Y_0 both in design and in model stereology. In the latter case, if two subsets $Y_1, Y_2 \subset \mathbb{R}^3$ satisfy the conditions $Y_1 \cap Y_2 = \emptyset$ and $Y_1 \cup Y_2 = \mathbb{R}^3$, then Y_1 and Y_2 are usually called the "phases" of a material. For instance, Y_2 may represent aluminium pores, or tungsten grains in a carbide, and Y_1 the corresponding embedding medium.

The design corresponding to Eq. (4.2.6) is a one-stage design. In more complex organs such as the lung, or liver, a multistage, nested, or "cascade" design has to be used involving different scales. For instance, if the target object is the contents of the capillary bed in a lung, then Fig. 4.2.2 suggests the following sequence of nested phases,

$$\text{lung} \supset \text{parenchyma} \supset \text{septa} \supset \text{capillaries}, \tag{4.2.8}$$

which have to be observed at increasing magnifications: the lung may be observed at less than 10× magnification, the parenchyma at less than 100×, the septa at less than 1000× by light microscopy (LM), or by low power electron microscopy (EM), and the fine structure of the capillaries at 5000 – 9000× by EM. In general, the magnification used at each stage should be the minimal one that allows unambiguous identification of the target structures. In this way, each image will contain as much relevant information as possible. Let Y denote the target object, or phase, (e.g. the capillaries), and Y_0 the entire organ (the lung in this case). A general s-stage design is based on the following sequence of nested phases,

$$Y_0 \supset Y_1 \supset \cdots \supset Y_{s-1} \supset Y, \quad s \geq 1 \tag{4.2.9}$$

The target parameter $\gamma(Y)$ may be expressed as follows,

$$\gamma(Y) = V_0 \cdot \frac{V_1}{V_0} \cdot \frac{V_2}{V_1} \cdots \frac{V_{s-1}}{V_{s-2}} \cdot \frac{\gamma(Y)}{V_{s-1}}, \tag{4.2.10}$$

where $V_i \equiv V(Y_i)$, $i = 0, 1, \ldots, s-1$.

For large organs, the reference volume V_0 is often measured by fluid displacement (Section 4.7). For smaller, or internal organs, V_0 is usually estimated from Cavalieri sections at a magnification M_0. The estimator of Eq. (4.2.6), for instance, requires a second magnification M_1 and, strictly, it is a two-stage design, but it is usual to regard it as a one-stage ratio estimator. Thus, the general convention is that an s-stage ratio design involves exactly s ratios, see Eq. (4.2.10) or, equivalently, exactly s nested phases or compartments in addition to the reference space Y_0, see Eq. (4.2.9).

The $s-1$ intermediate ratios in the rhs of Eq. (4.2.10) do not need to be volume ratios, but they usually are. Each ratio is usually estimated by superimposing test systems on systematic quadrats. The corresponding estimators are obtained as the pertinent ratios of the global estimators given in Section 4.2.2. For details, and practical recommendations, see Section 4.13.

The distinction between design- and model-based stereology is not always neat. For instance, the lung parenchyma observed in a bounded window (Fig. 4.2.2(b)) resembles a realization of a stationary and isotropic (volume, or surface) process, see also Fig. 3.1.1.

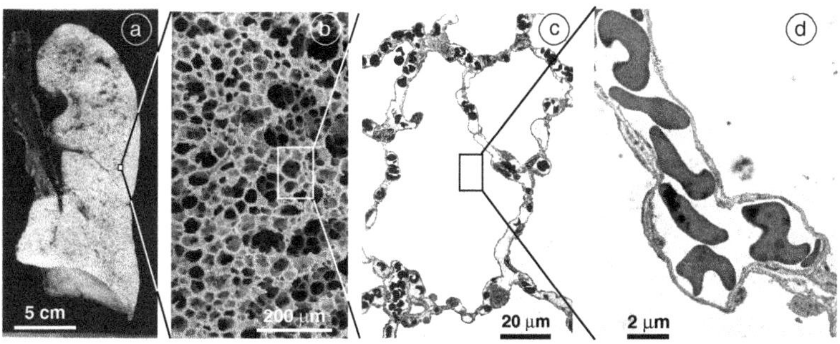

Fig. 4.2.2 Illustration of the nested design in Eq. (4.2.8). The target is the capillary compartment inside the lung alveolar septa (c), or even the erythrocytes within them (dark transects in (d)). (a) Extensive section of a human left lung. (b) Scanning electron microscopy (SEM) image of a slice of the parenchyma of a rabbit lung. (c) Transmission electron microscopy (TEM) image of an ultrathin section of a Grant gazelle lung parenchyma. Modified from Cruz-Orive and Weibel (1981), with permission of Wiley-Blackwell.

4.2.4 Notes

1. Ratios versus global quantities

In the early years, the classical stereological equations were motivated by problems arising in materials science, thereby involving ratios almost exclusively. The books of Saltykov (1958), DeHoff and Rhines (1968), Underwood (1970), and references therein, illustrate this. The general treatment of Miles (1972) was also concerned mainly with ratios. As a consequence, ratios also pervaded the early applications of stereology in biology. A better understanding of the real biological problems was often motivated by the need to correlate structure and function, and some biologists began to regard ratios as mere vehicles to estimate the relevant parameters, namely global quantities pertaining to complete organs. A good illustration of this is Chapter V from Weibel (1963). Shortly afterwards, the classical multistage design was made more explicit by Weibel (1969), p. 271, later formalized by Cruz-Orive and Weibel (1981), and applied by Weibel et al. (1981).

2. Caveat on nested designs

The primary reference volume V_0 is usually the first factor in a nested, multistage design, see Eq. (4.2.10). The next factor is a ratio bearing V_0 in the denominator, and it is necessary that the definition of V_0 is the same for both factors, so that V_0 cancels out. This extends to the subsequent ratios of the chain. Such conditions are not always easy to fulfill, because the definition of a given structure on sections usually varies at different magnifications – this could cause the so-called “laboratory bias”, discussed in Cruz-Orive and Weibel (1981), p. 239, and in Weibel et al. (1981), Figs. 2 and 3. At least as important is tissue shrinkage, or swelling, which may vary among stages due to the different tissue fixation and embedding techniques used. Tissue deformation is addressed for instance in Dorph-Petersen, Nyengaard, and Gundersen (2001), and in Howard and Reed (2005). The fractionator design, however, usually circumvents deformation artifacts, see Section 4.11.

3. The 'reference trap'

The term 'reference trap' was coined by Braendgaard and Gundersen (1986) in the neuroscience context to illustrate the fact that ratios may behave in the opposite way as the global parameter of interest, see also Gundersen (1992). Haug (1985) showed that old human brains shrink less than young ones after laboratory fixation, whereby the routine neuron density estimates in the cortex were the larger in younger brains. Shrinkage corrections dispelled the theory of neuron loss with age (briefly discussed in Cruz-Orive (2017) – see also Baddeley and Jensen (2005), p. 5).

4.3 Discrete Sampling

4.3.1 Purpose and definitions

The sampling designs discussed so far are based on motion-invariant probability elements, see Section 2.1.2 for single probes, or their adaptation to systematic sampling in Sections 2.25–2.26. Thus, we have considered sampling in a continuous domain. Here we consider sampling and estimation on a discrete domain. Consider for instance the estimation of the volume V of an object Y. Instead of using the Cavalieri method, say, we may split Y exhaustively into N fragments or 'blocks' (of arbitrary size and shape), labelled $\{1, 2, \ldots, N\}$, of corresponding unknown volumes $\{V_1, V_2, \ldots, V_N\}$, adding up to the total volume V. In this way we have constructed a discrete population of N elements (the fragments) corresponding to N quantities (the fragment volumes). In the example the target parameter is V. The idea is to replace a continuous sampling domain (such as the interval $[0, T)$ of the Cavalieri design) with the discrete one $\{1, 2, \ldots, N\}$, and the purpose is to estimate V in this scenario. If the elements are particles, then the target may be the total number, or the mean particle volume, etc. – see also Section A.1.6, where some of the concepts given below are sketched.

Prior to sampling it is necessary to define a sampling frame, namely the population of sampling units. A sampling unit is an element of the population, or a subset of the element which satisfies the fundamental condition that it is unambiguously observable. For instance, a group of people may be represented as projections on a planar image, and the target may be their total number. Entire bodies may not be fully observable due to overlapping artifacts, but head projections may, in which case the adopted sampling units may be the human heads (for references see Note 2 in Section 4.10.4). Unobservable heads are ignored. For a population of neurons exhibiting a unique, prominent nucleolus, a convenient sampling unit may be the nucleolus, etc. Thus, the population available to sampling is the population of sampling units, which is denoted by the set of labels $\{1, 2, \ldots, N\}$, and it corresponds 1:1 with the observable elements of the target population. The corresponding population of element values, e.g. $\{V_1, V_2, \ldots, V_N\}$ in the above example, is denoted by $\{y_1, y_2, \ldots, y_N\}$ in general. Depending on the context, the target parameter may be the population total, the population mean, etc.

A sample of size $n \geq 1$ is a subset of n element values from the population, drawn according to a sampling design with the purpose of estimating the target parameter. For instance, if the sampling units labelled $\{6, 19, 12\}$ are drawn, then $\{y_6, y_{19}, y_{12}\}$ is the corresponding sample (of size 3). For simplicity, a set $\mathcal{S}$ indexing a sample, and the sample $\{y_i,\ i \in \mathcal{S}\}$ itself, will indistinctly be called 'a sample', the distinction being easy from the context.

In sampling without replacement each unit can be sampled at most once – this is, for instance, the case if particles are sampled with disectors. In sampling with replacement each unit can be sampled more than once, e.g. if particles are sampled with test points.

4.3.2 Discrete sampling design

A sampling design is a law that assigns a probability to each of the possible samples that can arise under a well-defined procedure. The sum of the probabilities of all possible samples must be equal to 1. Explicit expressions for such probabilities may or may not be accessible, but this is of no consequence in practice. A convenient way to characterize a sampling design is to define a priori the corresponding inclusion probability for each sampling unit, which must be positive, namely,

$$\pi_i = \mathbb{P}(\text{The } i\text{th unit is selected}) > 0, \quad i = 1, 2, \ldots, N. \tag{4.3.1}$$

Unlike the sample probabilities, the individual inclusion probabilities $\{\pi_i\}$ generally do not add to 1, but to the mean value of the sample size n. To see this, define the following indicator random variable,

$$1_i = \begin{cases} 1 \text{ if the } i\text{th unit is selected,} \\ 0 \text{ otherwise.} \end{cases} \tag{4.3.2}$$

In sampling without replacement, $1_1 + 1_2 + \cdots + 1_N = n$, namely the sample size, and we have

$$\sum_{i=1}^{N} \pi_i = \sum_{i=1}^{N} \mathbb{E}(1_i) = \mathbb{E}\left\{\sum_{i=1}^{N} 1_i\right\} = \mathbb{E}(n). \tag{4.3.3}$$

More generally, suppose that sampling is with replacement, and let $m_i \geq 0$ denote the random number of times that the ith unit is included in the sample, namely its inclusion number. If particles are sampled with test points, or with test lines, then m_i will represent the number of test points hitting the particle, or of intersections determined in its boundary, respectively. The corresponding mean,

$$\mu_i = \mathbb{E}(m_i) > 0, \quad i = 1, 2, \ldots, N, \tag{4.3.4}$$

is fixed and positive. For instance, if a particle of volume V is hit by a UR spatial grid of volume v per test point, then $\mathbb{E}(P) = V/v$, etc., see Section 2.29. In this case, $m_1 + m_2 + \cdots + m_N = n$ is the total number of test points, intersections, etc., scored in the particles, and

$$\sum_{i=1}^{N} \mu_i = \sum_{i=1}^{N} \mathbb{E}(m_i) = \mathbb{E}\left(\sum_{i=1}^{N} m_i\right) = \mathbb{E}(n). \tag{4.3.5}$$

4.3.3 The Horvitz–Thompson estimator of a population total

Let Y represent a finite population of N sampling units with fixed associated values $\{y_1, y_2, \ldots, y_N\}$, and let the target parameter γ be the population total, namely,

$$\gamma = y_1 + y_2 + \cdots + y_N. \tag{4.3.6}$$

Let $\mathcal{S} \subset \{1, 2, \ldots, N\}$ denote a sample of units without replacement, and impose the condition that the inclusion probabilities $\{\pi_i,\ i \in \mathcal{S}\}$ of the sampled units are always known. Then, the Horvitz–Thompson estimator

$$\widehat{\gamma} = \sum_{i \in \mathcal{S}} \frac{y_i}{\pi_i}, \text{ or, equivalently, } \widehat{\gamma} = \sum_{i=1}^{N} \frac{1_i y_i}{\pi_i}, \tag{4.3.7}$$

is unbiased for γ. In fact,

$$\mathbb{E}(\widehat{\gamma}) = \sum_{i=1}^{N} \frac{\mathbb{E}(1_i) y_i}{\pi_i} = \sum_{i=1}^{N} \frac{\pi_i y_i}{\pi_i} = \sum_{i=1}^{N} y_i = \gamma. \tag{4.3.8}$$

Likewise, if sampling is with replacement, then

$$\widehat{\gamma} = \sum_{i=1}^{N} \frac{m_i y_i}{\mu_i}, \tag{4.3.9}$$

is also unbiased for the population total γ.

4.3.4 Notes

1. Sampling in the discrete and in the continuum

In statistics, "sampling" traditionally refers to sampling on a discrete domain, and it constitutes an old, well-developed discipline motivated mainly by human censuses and surveys – for general reference see Cochran (1977). A basic purpose of stereology was to adapt the classical sampling ideas to continuous domains by the incorporation of integral geometry and geometric probability (Davy & Miles, 1977).

2. The Horvitz–Thompson estimator

One of the most useful tools incorporated by stereology from sampling theory is the estimator of Horvitz and Thompson (1952). Uniform sampling, see below, is also a classical topic of sampling theory, especially simple random sampling.

4.4 Uniform Sampling

4.4.1 Definition

A sampling design in which all the inclusion probabilities are identical, that is,

$$\pi_1 = \pi_2 = \cdots = \pi_N = \mathbb{E}(n)/N, \tag{4.4.1}$$

is called a uniform sampling design. However, UR sampling is not unique, as shown in the sequel.

It is worth noting that, if the population of units is a UR permutation of the set $\{1, 2, \ldots, N\}$, then any arbitrary subset $\mathcal{S}_n$ of size n from the population is effectively a simple random sample (srs, see next). Thus, in this case any design is effectively an srs design. This scenario is analogous to that arising in model stereology, in which the object is modelled by a stationary random set, whereby the probe may be arbitrary (Chapter 3).

4.4.2 Simple random sampling (srs)

This is uniform sampling without replacement, usually known as the 'lottery method'. The size $n \in \{1, 2, \ldots, N\}$ of a simple random sample $\mathcal{S}_n$ is fixed a priori. In our context, the order of the sampled units within $\mathcal{S}_n$ is irrelevant. Each draw is performed uniformly and independently at random among all the units remaining in the population by means of a reliable random number generator.

The inclusion probability π_i is equal to the total number of those samples which include the ith unit, divided by the total number of all possible samples, namely,

$$\pi_i = \binom{N-1}{n-1} \Big/ \binom{N}{n} = \frac{n}{N}, \quad i = 1, 2, \ldots, N, \tag{4.4.2}$$

which agrees with Eq. (4.4.1) for fixed n, and therefore verifies that srs is uniform sampling.

Application of the Horvitz–Thompson estimator yields a UE of the population total γ, namely,

$$\widehat{\gamma} = \sum_{i \in \mathcal{S}_n} \frac{y_i}{n/N} = N\bar{y}, \tag{4.4.3}$$

which is not useful unless N is known. In contrast, the sample mean

$$\bar{y} = \frac{1}{n} \sum_{i \in \mathcal{S}_n} y_i \tag{4.4.4}$$

is a directly accessible UE of the population mean $\mu = \gamma/N$.

4.4.3 Systematic sampling on a discrete population

Fix a finite sampling period $T \in \{1, 2, \ldots\}$ and select every Tth unit with a UR start $z \in \{1, 2, \ldots, T\}$. The corresponding sample of units is

$$\mathcal{S}_z = \{z + iT \leq N, \ i = 1, 2, \ldots\}. \tag{4.4.5}$$

Although it is not strictly necessary, in practice one chooses $T \leq N$, otherwise the sample may be empty. Note that this sampling design is the discrete version of Cavalieri sampling.

There are T possible samples $\{\mathcal{S}_1, \mathcal{S}_2, \ldots, \mathcal{S}_T\}$, mutually disjoint, and their union is the population. Because the starting unit z is UR in the set $\{1, 2, \ldots, T\}$, all the T possible samples are equally likely, namely,

$$\mathbb{P}(\mathcal{S}_z) = \mathbb{P}(z) = \frac{1}{T}, \quad z = 1, 2, \ldots, T. \tag{4.4.6}$$

Moreover, because the samples constitute a partition of the population, each unit will belong to exactly one sample, and therefore the inclusion probability is constant, namely,

$$\pi_i = \frac{1}{T}, \quad i = 1, 2, \ldots, N. \tag{4.4.7}$$

This confirms that systematic sampling is uniform sampling.

The sample size n is random unless N is a multiple of T. By virtue of Eq. (4.3.3),

$$\mathbb{E}(n) = \sum_{i=1}^{N} \pi_i = \sum_{i=1}^{N} \frac{1}{T} = \frac{N}{T}. \tag{4.4.8}$$

By the Horvitz–Thompson estimator, the sampling period times the sample total, namely

$$\widehat{\gamma} = \sum_{i \in \mathcal{S}_z} \frac{y_i}{1/T} = T \sum_{i \in \mathcal{S}_z} y_i, \tag{4.4.9}$$

is a UE of the population total γ. Thus, unlike srs, systematic sampling is convenient to estimate γ. It is also easy to implement because any sample depends on a single UR integer z. However, the sample mean, see Eq. (4.4.4), is usually not unbiased for the population mean because the sample size n is generally random in systematic sampling. It is however ratio-unbiased because $\mathbb{E}(\sum y_i)/\mathbb{E}(n) = (\gamma/T)/(N/T) = \gamma/N$.

4.4.4 The Murthy–Gundersen (MG) smooth arrangement

In contrast to srs, the precision of $\widehat{\gamma}$ under systematic sampling depends on the arrangement of the sampling units prior to sampling. This dependence is analogous to that of Cavalieri sampling with respect to the smoothness properties of the measurement function (Section 5.2.3). In Section 5.2.5 it is shown that an approximation of the error variance $\mathrm{Var}(\widehat{\gamma})$ is given by its extension term, namely,

$$\mathrm{Var}_E(\widehat{\gamma}) = \frac{T^2}{12} \sum_{i=0}^{N} (y_{i+1} - y_i)^2, \tag{4.4.10}$$

with $y_0 = y_{N+1} = 0$. The preceding approximation suggests that, in order to reduce the error variance, the jumps $\{y_{i+1} - y_i\}$ should be as small as possible. A convenient choice is the Murthy–Gundersen arrangement Y_{MG}, which may be implemented as follows. Sort the population values in non-decreasing order of magnitude, namely

$$y_{(1)} \le y_{(2)} \le \cdots \le y_{(N-1)} \le y_{(N)}, \tag{4.4.11}$$

where $y_{(i)}$, $i = 1, 2, \ldots, N$ is called the ith order statistic. Then, if N is even,

$$Y_{\mathrm{MG}} := \{y_{(1)}, y_{(3)}, \ldots, y_{(N-1)}, y_{(N)}, y_{(N-2)}, \ldots, y_{(2)}\}, \tag{4.4.12}$$

and analogously if N is odd. For a numerical illustration, see the paragraph including Eq. (4.4.15) below.

4.4.5 Cluster sampling

Fig. 4.4.1 represents an object Y exhaustively split into $N = 7$ blocks in an arbitrary, non-random manner. The ith block contains y_i point particles – e.g., associated points (APs) of arbitrary particles – and the target parameter is the total number γ of such particles, defined as in Eq. (4.3.6). As in classical sampling theory, the subset of APs contained in a block is called a cluster. Thus, in the present context the ith cluster would be the union of the y_i particles associated with the ith block. In cluster sampling, the sampling units are not the particles, but the clusters themselves.

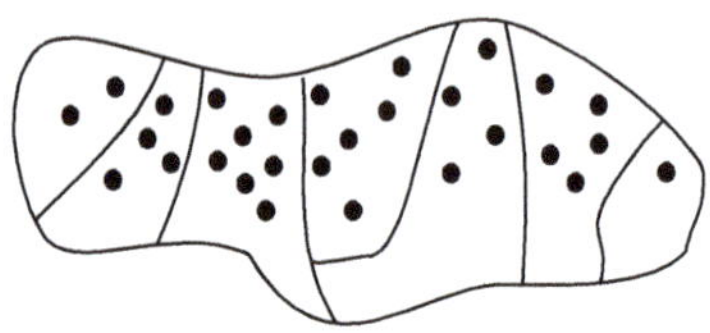

Fig. 4.4.1 A domain exhaustively split into 7 blocks, each containing a cluster of particles. The target is the total number of particles in the domain, and the primary sampling units are the clusters.

Systematic cluster sampling

The population consists of $N = 7$ clusters with particle numbers $Y = \{2, 4, 7, 6, 4, 5, 1\}$, listed in their original order. The target parameter is the total particle number $\gamma = 29$. With a systematic sampling period $T = 3$, there are three possible cluster samples, namely $\{2, 6, 1\}$, $\{4, 4\}$, and $\{7, 5\}$, respectively. By Eq. (4.4.9), the corresponding estimates of γ are $\widehat{\gamma}_1 = 27$, $\widehat{\gamma}_2 = 24$, and $\widehat{\gamma}_3 = 36$ and each estimate is obtained with a constant probability of $1/3$, see Eq. (4.4.7). In addition, each particle is contained in exactly one cluster, and therefore the particle inclusion probability is effectively $1/3$ as well. We readily check that $\mathbb{E}(\widehat{\gamma}) = 87/3 = 29 = \gamma$, which verifies the general fact that $\widehat{\gamma}$ is unbiased for γ. Further, the variance among the three possible estimates is $\mathrm{Var}_{syst}(\widehat{\gamma}) = 26$, whereby $\mathrm{CE}_{syst}(\widehat{\gamma})\% = 100\sqrt{26}/29 \approx 17.6\%$.

Systematic versus simple random sampling (srs) of clusters

Consider srs with a sample size $n = 3$ of clusters from the population of $N = 7$ clusters. As shown in Section 4.4.2, the estimator $\widehat{\gamma} = N\bar{y}$, where $\bar{y}$ denotes the sample mean of the number of particles per cluster, is unbiased for γ. This means that the average of $\widehat{\gamma}$ over the 35 possible estimates of γ will be equal to $\widehat{\gamma}$. Moreover, under srs the exact variance of $\widehat{\gamma}$ over the 35 possible estimates is

$$\mathrm{Var}_{srs}(\widehat{\gamma}) = N^2 \mathrm{Var}_{srs}(\bar{y}) = N^2 \cdot \frac{N-n}{N-1} \cdot \frac{\mathrm{Var}(y)}{n}, \tag{4.4.13}$$

where $N = 7$, $n = 3$ and $\mathrm{Var}(y) \approx 3.8367$ is the variance among the 7 cluster totals in the population Y. With these data, we obtain $\mathrm{CE}_{srs}(\widehat{\gamma})\% \approx 22.3\%$. Thus, in this example the relative efficiency of systematic versus srs is

$$\mathrm{eff}(\mathrm{syst}, \mathrm{srs}) = \frac{\mathrm{CE}^2_{srs}(\widehat{\gamma})}{\mathrm{CE}^2_{syst}(\widehat{\gamma})} \approx 1.61. \tag{4.4.14}$$

To achieve a relative efficiency of less than 1, the sample size with srs should be $n \geq 4$.

For a given sampling period T, the $\mathrm{CE}_{syst}(\widehat{\gamma})$ can often be decreased by adopting the smooth arrangement for the population of clusters prior to sampling, see Eq. (4.4.12). In the example, $Y_{MG} := \{1, 4, 5, 7, 6, 4, 2\}$ (or its reverse), and with $T = 3$ the three possible estimates of γ are $\widehat{\gamma}_1 = 30$, $\widehat{\gamma}_2 = 30$, and $\widehat{\gamma}_3 = 27$, whereby $\mathrm{CE}_{syst}(\widehat{\gamma})\% \approx 4.9\%$, and now,

$$\mathrm{eff}(\mathrm{syst}, \mathrm{srs}) \approx 20.9, \tag{4.4.15}$$

a dramatic increase. Even if the sample size under srs was of $n = 6$, that is, if each sample consisted of nearly the entire population, systematic cluster sampling with $T = 3$ would still be 2.6 times more efficient than srs in this example.

4.4.6 Notes

1. Exact systematic sampling

As observed by Mattfeldt (1989), if an isosceles triangle of area A is sampled with an even number of Cavalieri lines normal to its base, then $\widehat{A} = A$, i.e., $\mathrm{Var}(\widehat{A}) = 0$ – thus, the results at the end of the preceding example are not that surprising. The isosceles triangle is a special case of a class of measurement functions yielding zero variance under systematic sampling with an even number of observations (Tinajero-Bravo, Eslava-Gómez, & Cruz-Orive, 2014).

2. The Murthy–Gundersen (MG) smooth arrangement

Systematic sampling based on the arrangement of Eq. (4.4.12) was proposed by Murthy (1967) under the name 'balanced systematic sampling'. The idea was to arrange the population as closely as possible to an isosceles triangle. Gundersen (1986, 2002b) rediscovered the idea and adapted it to stereology, see Section 4.11 for an example. Cluster sampling in a geometric vein, with applications in agriculture and ecology, was also proposed by Jolly (1979).

3. Variance reduction. The proportionator

Particles will be unobservable when embedded in opaque materials, hence the MG arrangement can seldom be implemented exactly; it can be useful to increase estimation precision, however, if the number of particles in each cluster is roughly proportional to the apparent size of the cluster itself. An alternative is the proportionator method (Gardi, Nyengaard, and Gundersen (2008), Andersen, Hahn, and Jensen (2015), Boyce and Gundersen (2018)) whereby the number of particles in a cluster is known to be approximately proportional to some auxiliary variable, (e.g., the intensity of tissue stain when counting cells).

4. Further recommendations on variance reduction

In order to reduce the estimation variance, Tinajero-Bravo et al. (2014) suggest: (a) In discrete systematic sampling, besides using the MG arrangement, both the number N of clusters and the number n of sampled clusters should be even, and $N/n = T$ should be an integer. (b) In Cavalieri sampling of sections with period T, the error variance may be reduced if T is chosen so that the mean number of sections $H(u)/T$, see Eq. (2.25.12), is an integer. This requires that the caliper length $H(u)$ of the object along the sampling axis can be readily measured.

The foregoing recommendations are just intended to reduce estimation variance – remember that uniform sampling is design-unbiased anyway.

4.5 Planning and Optimizing a Stereological Design

4.5.1 Introduction

At the outset, a design involving stereology in biology is like an ordinary experimental design, inasmuch as the animals, or the biological items, constituting a primary sample or "group" are selected assuming that the sample will be effectively a srs. Within each item, however, the target will be a geometric parameter estimated by systematic sampling.

For simplicity we consider the common problem of comparing two population means, the null hypothesis being that the two means are equal. In our usual notation, we also assume that the relevant variable defined on each sampling unit is a global geometric quantity γ. The control population consists of untreated animals kept under standard conditions, and the mean of γ in this population is a fixed unknown quantity μ_1, say. The treated population consists of animals affected by a disease, or by some treatment, the mean of γ in this population being also a fixed unknown quantity μ_2.

In materials sciences the relevant variable is usually a density γ_V, and the control and the treatment populations are modelled by stationary and ergodic random sets of fixed intensities μ_1 and μ_2, respectively.

4.5.2 Optimal sample size

A common goal of a biomedical experiment is to test the null hypothesis $\delta \equiv \mu_2 - \mu_1 = 0$ against the alternative $\delta \neq 0$. To achieve this, simple random samples consisting of n_1 and n_2 animals are taken from the control and the treatment population, respectively. If the biological variance $\mathrm{Var}_b(\gamma)$ of γ between animals is the same in both populations, then one may take $n_1 = n_2 = n$, say – we assume this in order to circumvent well-known technical details.

Let $\{\gamma_1, \gamma_2, \ldots, \gamma_n\}$ represent a sample of true global quantities which constitute either the control, or the treatment group. The following analysis requires that, for each $i = 1, 2, \ldots, n$, and for each group, each target quantity γ_i is estimated without bias as $\widehat{\gamma}_i$ by means of a suitable stereological design.

The basic task is to optimize the primary sample size n. This is facilitated if the scientist is content with detecting a relative difference between the two population means of at least δ_0/μ_1 (e.g., of 15%), a smaller difference being deemed unimportant a priori for the given experiment. In addition, one may fix the significant level α, (i.e., the probability of rejecting the null hypothesis when true), and the power $1-\beta$, (namely the probability of detecting the desired relative difference between means). Then, n may be chosen to be the ceiling of

$$n_0 = 2c^2 \cdot \mathrm{Var}_b(\widehat{\gamma})/\delta_0^2, \tag{4.5.1}$$

- $\mathrm{Var}_b(\widehat{\gamma})$ is the true, observable variance of the unbiased stereological estimator $\widehat{\gamma}$ between animals.
- The factor 2 applies if the pertinent statistical test is a two-sample t-test. For a paired t-test, or for a one-sample test, the factor 1 applies.
- $c = z(1-\alpha/2) + z(1-\beta)$, where $z(p)$ is the percentile of the normal $N(0, 1)$ distribution corresponding to a probability p. With the usual choices $\alpha = 0.05$, $\beta = 0.90$, we get $c^2 \approx 10.5$. This value is useful if the estimator $\widehat{\gamma}$ is normally distributed, at least approximately, in the population of animals.

4.5.3 Variance decomposition to optimize a design

Eq. (4.5.1) suggests that, as soon as $\mathrm{Var}_b(\widehat{\gamma}) \geq \delta_0^2$, the required sample size n may become unduly large. If increasing δ_0 (i.e. decreasing the "resolution" of the experiment) cannot be contemplated, then the only strategy left is to decrease $\mathrm{Var}_b(\widehat{\gamma})$. From the variance decomposition formula, see Eq. (A.1.44), and recalling that $\widehat{\gamma}$ is unbiased for γ, namely that $\mathbb{E}_w(\widehat{\gamma}|\gamma) = \gamma$, we have

$$\begin{aligned}\mathrm{Var}_b(\widehat{\gamma}) &= \mathrm{Var}\{\mathbb{E}_w(\widehat{\gamma}|\gamma)\} + \mathbb{E}_b\{\mathrm{Var}_w(\widehat{\gamma}|\gamma)\} \\ &= \mathrm{Var}_b(\gamma) + \mathbb{E}_b\{\mathrm{Var}_w(\widehat{\gamma}|\gamma)\}. \end{aligned} \tag{4.5.2}$$

The subscripts 'b' and 'w' denote 'between' and 'within' animals, respectively. Thus, the first term in the rhs of Eq. (4.5.2) is the true biological variance, whereas the second term is the mean over animals of the true error variance within animals, namely the mean of the error variance due to the estimation of the individual $\{\gamma_i\}$ by stereology. In order to decide which of the two terms should be decreased, it is necessary to have some knowledge of them, either from previous experience from similar experiments, or from a pilot sample from the current experiment. Suppose that a sample of estimates $\{\widehat{\gamma}_1, \widehat{\gamma}_2, \ldots, \widehat{\gamma}_n\}$ is available. Then the estimates of each of the terms in Eq. (4.5.2) may be computed as follows.

- If the sample of animals is a srs, then the observable variance in the lhs of Eq. (4.5.2) may be estimated by the ordinary sample variance, namely,

$$\mathrm{var}_b(\widehat{\gamma}) = \frac{1}{n-1}\sum_{i=1}^{n}(\widehat{\gamma}_i - \bar{\gamma})^2, \tag{4.5.3}$$

  where $\bar{\gamma}$ denotes the sample mean of the $\{\widehat{\gamma}_i\}$.
- An estimator of the second term in the rhs of Eq. (4.5.2) is

$$\mathrm{mean}_b\{\mathrm{var}_w(\widehat{\gamma})\} = \frac{1}{n}\sum_{i=1}^{n}\mathrm{var}_w(\widehat{\gamma}_i), \tag{4.5.4}$$

  where $\mathrm{var}_w(\widehat{\gamma}_i)$ denotes an estimator of the stereological error variance of $\widehat{\gamma}_i$ (Chapter 5).
- An estimator of the biological variance may now be estimated by difference, namely,

$$\mathrm{var}_b(\gamma) = \mathrm{var}_b(\widehat{\gamma}) - \mathrm{mean}_b\{\mathrm{var}_w(\widehat{\gamma}|\gamma)\}. \tag{4.5.5}$$

The preceding analysis may be opportune if the observed difference between group means is no less than δ_0, but it is not statistically significant at the level α. In this case, one may consider the following strategies.

1. To redefine the target parameter into another one which may exhibit a lower biological variation.
2. To decrease the error variance within animals by increasing in each animal, or item, the number of sections, quadrats, or counts (in that order), bearing the corresponding costs in mind.

Usually the main variance component is the biological one. Many laboratory experiments involve pure animal strains, or even clones, to warrant a low biological variance, whereby groups of 6–10 animals may yield significant results of scientific value because the stereological error variance can be controlled. If the treatment under investigation is to be applied to a natural population, however, then the situation may be quite different. Here, no matter how much stereological work is invested per individual, hundreds, or even thousands of them may be required per group to obtain conclusive results, mainly because the biological variance will be high, and α has to be small for security reasons. This is a familiar scenario in clinical trials for pharmacology, for instance.

A snag of the preceding analysis is that the design used to estimate each γ_i involves systematic sampling, and thereby no estimators exist of $\text{Var}_w(\widehat{\gamma}_i)$ that are unbiased in general – only approximations are available whose accuracy is shape-dependent. The approximation formulae are easy to apply, but their theoretical derivation requires advanced mathematics. This topic is covered in Chapter 5 for the most common variance predictors.

4.5.4 The case examples

In the sequel, a variety of stereological methods are described to estimate γ for single objects and, whenever possible, to predict the corresponding estimator variance from a single sample. The methods are illustrated by means of worked examples involving real material. For short, the words "Estimation of" are omitted from the section headings.

For simplicity, the sample sizes used are often relatively small, hence the corresponding coefficients of error may occasionally lie around 15%. In practice, however, the target is a group mean μ, which is estimated from a sample of n animals, or primary units, as

$$\widehat{\mu} = (\widehat{\gamma}_1 + \widehat{\gamma}_2 + \cdots + \widehat{\gamma}_n)/n. \tag{4.5.6}$$

If the primary units are drawn independently, and the stereological estimators $\{\widehat{\gamma}_i\}$ are unbiased, then $\widehat{\gamma}$ is unbiased for μ, and

$$\text{Var}_b(\widehat{\mu}) = \frac{1}{n}\text{Var}_b(\gamma) + \frac{1}{n}\mathbb{E}_b\{\text{Var}_w(\widehat{\gamma}|\gamma)\}. \tag{4.5.7}$$

This means that the mean over animals of the stereological error variance within animals is reduced by a factor of $1/n$ when the target is the group mean.

The data collection is illustrated mainly with manual methods, this obeying to the following considerations.

1. Sampling has to be man-designed, and it is more important than measuring.
2. Once the 'know-how' of the measurements is well understood, the latter can be incorporated into automatic, or semiautomatic devices.
3. In spite of current advances, automatic pattern recognition is effective only in special cases – it is usually not in TEM for cell biology, say. Semiautomatic (i.e., man-supervised computer assisted measurements) are often a reasonable option.

4.5.5 Notes

1. Optimizing sample sizes

Eq. (4.5.1) is well known from classical statistics, see for instance Snedecor and Cochran (1980), Section 6.1. Shay (1975) considered the optimization of the sample sizes in a nested design involving organs, blocks, micrographs and measurements. Later, Gundersen and Østerby (1981) adapted these ideas to stereology, with the main conclusion that, in most biological studies, the biological variation among organs matters more than the error variance within organs. For an illustration in neuroscience, see Cruz-Orive et al. (2004).

2. Manual versus automatic stereology

Faced with a new experiment, machines alone generally cannot devise a stereological design. Measurements may occasionally be automatized, but not always without problems – see the Introduction in Cruz, Gómez, and Cruz-Orive (2015).

4.6 Planar Area and Boundary Length With a Square Grid

4.6.1 Purpose, material, and method

The purpose is to estimate the area A, cm^2, and the boundary length B, cm, of the flat projected image of a leaf (Fig. 4.6.1(a)). The leaf stem is ignored.

The targets A, B will be estimated manually by point and intersection counting, respectively, with an IUR square grid whose fundamental tile is shown in Fig. 4.6.1(b)). In practice, isotropic uniform randomness is attempted by superimposing the grid "at random", without looking at the image. From Eq. (2.25.6) and Eq. (2.26.12), we obtain the following UEs of A, B,

$$\widehat{A} = \frac{a}{p} \cdot \frac{1}{M^2} \cdot P = h^2 M^{-2} P, \qquad (4.6.1)$$

$$\widehat{B} = \frac{\pi}{2} \cdot \frac{a}{l} \cdot \frac{1}{M} \cdot I = \frac{\pi}{4} \cdot h \cdot M^{-1} I, \qquad (4.6.2)$$

respectively. The gap length h, cm, is measured directly on the grid. It is advisable to measure the total length of 10 gaps, say, in both the horizontal and the vertical direction, and divide by 20. The final linear magnification M is the actual length (mm) of the calibration ruler supplied with the image (Fig. 4.6.1(a)) divided by 50 mm. In general, M is the observed length of the calibration ruler accompanying the image, divided by the real length represented by the ruler (both lengths in the same units). Incidentally, unless the target is dimensionless:

$$\text{No calibration ruler} = \text{No stereology.} \tag{4.6.3}$$

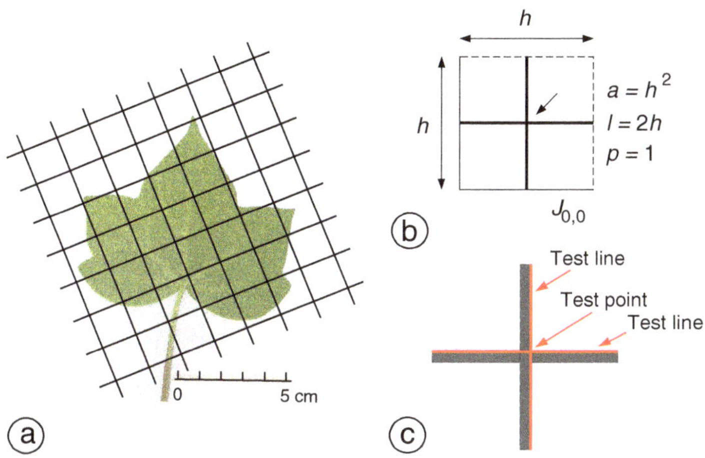

Fig. 4.6.1 Estimation of planar area and boundary length by point and intersection counting, see Section 4.6. Modified from Sagaseta (2005).

4.6.2 Results

In Fig. 4.6.1(a) the observed scores are $P = 20$ test points hitting the leaf image, and $I = 26$ intersections with its boundary. The definitions of a dimensionless test point and a one-dimensional test line are illustrated in Fig. 4.6.1(c)). For a grid with $h = 1.368\,\text{cm}$, and $M = 0.87$, the estimates obtained were $\widehat{A} = 49\,\text{cm}^2$, $\widehat{B} = 32\,\text{cm}$ (with two significant digits each, because the raw data P, I consist of two digits).

4.6.3 Error variance predictors

From Eq. (5.11.2), a predictor of the square coefficient of error of $\widehat{A}$, namely of $\text{CE}^2(\widehat{A}) = \text{Var}(\widehat{A})/A^2$, is

$$\text{ce}^2(\widehat{A}) = 0.07284 \cdot \frac{\widehat{B}}{\sqrt{\widehat{A}}} \cdot \frac{1}{P^{3/2}}. \tag{4.6.4}$$

Besides uniform randomness of the test system relative to the object, the preceding predictor also requires isotropic orientation. Here $\widehat{B}/\sqrt{\widehat{A}} = 4.57$, and $\text{ce}(\widehat{A})\% = 6.1\%$.

The available predictor formula for $\text{CE}^2(\widehat{B})$ is more complicated, see Eq. (5.7.5), but easy to program. The raw data are the ordered, individual intersection numbers with the horizontal, and the vertical test lines, namely $\{2, 4, 2, 2, 2, 2\}$ and $\{2, 2, 2, 2, 2, 2\}$, respectively. The result is $\text{ce}(\widehat{B})\% = 5.9\%$.

The relative contributions to the total $\text{ce}^2(\widehat{B})$ of the variation among the two orthogonal linear projected lengths of the leaf boundary in two perpendicular directions, and among the intersection numbers for each direction, are of 14.2% and 85.8%, respectively – see the last paragraph of Section 5.7.1.

4.6.4 Notes

1. Sources

The preceding case example is a modification from Sagaseta (2005). Eq. (4.6.4) is based on Matheron's transitive theory, see Gundersen and Jensen (1987). The numerical factor was improved by Kiêu and Mora (2009), see also Cruz-Orive (2013) and Section 5.11.2.

2. Efficiency of the square grid relative to independent test point sampling

Suppose that the leaf image is included in a reference rectangle of area A_0, and set $p = A/A_0 \in (0, 1)$. Generate n independent UR test points in the rectangle, and let $P \geq 1$ denote the total number of test points hitting the leaf. The distribution of P is Binomial $b(n, p)$, hence $\mathbb{E}(P) = np$ and $\widehat{A} = A_0 P/n$ is unbiased for A. Further $\text{Var}(P) = np(1 - p)$, hence an estimator of $\text{CE}^2_{ind}(\widehat{A})$ under independent point replications is

$$\text{ce}^2_{ind}(\widehat{A}) = (1 - p)/P. \tag{4.6.5}$$

The area of the smallest horizontal rectangle containing the leaf is about $A_0 = 96\,\text{cm}^2$, hence we may take $p = 49/96 \approx 0.51$ and, for a fair comparison, $P = 20$, as obtained above with the square grid. Then, $\text{ce}_{ind}(\widehat{A})\% = 100\sqrt{(1 - 0.51)/20} = 15.7\%$, whereby,

$$\text{eff}(\text{syst}, \text{indep}) = 0.157^2/0.061^2 \approx 6.6, \tag{4.6.6}$$

which implies that 20 hitting test points scored with the square grid achieve a similar precision as about $20 \cdot 6.6 = 132$ independent test points hitting the leaf image. To generate 132 hitting points we would need to generate about $n = 132/p = 259$ independent UR test points in the reference rectangle.

3. Precision of the boundary length estimator

Eq. (5.7.5) to predict $\mathrm{CE}^2(\widehat{B})$ was derived by Cruz-Orive and Gual-Arnau (2002), and its performance was studied by Gómez et al. (2016). For grids other than Buffon's (parallel lines), or Buffon–Steinhaus' (square grid), to our knowledge no suitable variance approximation exists.

4.7 Volume By Fluid Displacement

4.7.1 Purpose, material, and method

The volume V_0 of an isolated object Y_0 (regarded as a reference object, hence the subscript '0') may often be estimated by the non-invasive method of fluid displacement. Fig. 4.7.1 illustrates the following protocol for a banana.

1. On a weighing scale, read the weigh W_1 of a suitable jar containing enough water.
2. Introduce the object entirely in the water, taking care that it does not touch the jar whatsoever – for the banana this was easily achieved by holding it with a needle (Fig. 4.7.1(b)). Read the weigh W_2 of the union of jar, water, and object.

By the Archimedes principle, the difference $W_2 - W_1$ is the weight of a "banana of water". If the specific gravity of the water used is $\rho = 1$, then $W_2 - W_1$ is the volume of the banana.

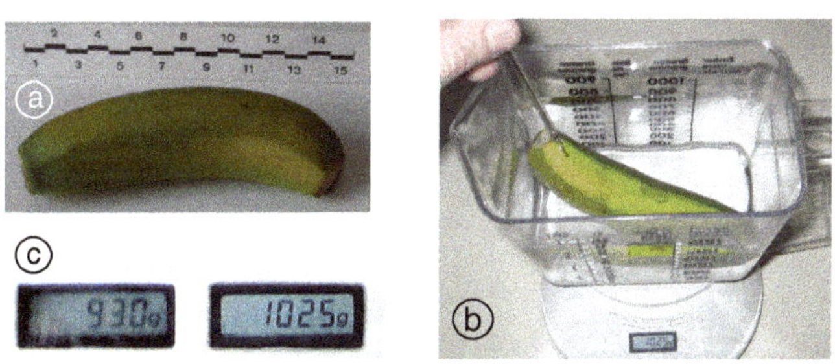

Fig. 4.7.1 Illustration of the fluid displacement method to estimate volume, see Section 4.7.

4.7.2 Result

From the readings (in grams) of Fig. 4.7.1(c), and assuming the value $\rho = 1\,\mathrm{g/cm}^3$, we get,

$$V_0 = (W_2 - W_1)/\rho = (1025 - 930)/1 = 95\,\mathrm{cm}^3. \tag{4.7.1}$$

4.7.3 Notes

1. The Archimedes principle

The first Eq. (4.7.1) is found in most elementary physics textbooks. The method was adapted to biomedicine by Scherle (1970), see also Weibel (1979). The banana example is adapted from Sagaseta (2005).

2. Practical aspects

If the object is a biological organ, then the fluid should be a proper buffer of a known specific gravity ρ. If the object is an organ such as a lung lobe, containing air spaces connected with the exterior through some opening, then the latter should be occluded (with the aid of a forceps, say) prior to submersion in the fluid. The extra volume of the submerged part of the forceps may be measured separately using the same method.

4.8 Volume From Cavalieri Sections

4.8.1 Purpose, material, and method

The problem is to estimate the volume V_0 of the banana Y_0 considered in the preceding section, and the volume V of its pulp $Y \subset Y_0$, using an invasive version of the Cavalieri design. Note that internal structures like the pulp cannot be directly estimated by fluid displacement.

The object was set stable on a bed of plastiline laid flat on a ruled mat. A simple aluminium device was used as a guide for a knife to produce parallel sections normal to the mat a constant distance T apart, see Fig. 4.8.1(a). The first cut was placed at a UR distance in the interval $[0, T)$ from the tip of the object. After each successive cut, the device was moved backwards T cm using the parallel rules of the mat as a reference. For a total of n cuts, $n+1$ blocks are produced (Fig. 4.8.1(b)). The rightmost block, say, is discarded, whereby the n Cavalieri sections are the rhs faces of the remaining n blocks (Fig. 4.8.1(c)). Let $\{A_1, A_2, \ldots, A_n\}$ denote the corresponding ordered sequence of n section areas, and suppose that the latter are measured directly with a negligible error. Then, by Eq. (2.25.10) the Cavalieri estimator

$$\widehat{V} = T \sum_{i=1}^{n} A_i \qquad (4.8.1)$$

is unbiased for V. Alternatively, for each $i = 1, 2, \ldots, n$ a UE $\widehat{A}_i$ of A_i may be obtained by subsampling with a test system of points using Eq. (4.6.1), see Fig. 4.8.1(d). If the constants T and a/p are given at the object scale (that is, corrected for magnification), then

$$\widetilde{V} = T \cdot \frac{a}{p} \cdot \sum_{i=1}^{n} P_i \qquad (4.8.2)$$

is an unbiased Cavalieri estimator of V with subsampling, based on Eq. (2.25.14).

Fig. 4.8.1 Illustration of the Cavalieri method to estimate volume (with section areas estimated by point counting), see Section 4.8.

4.8.2 Results

For a fairly smooth object such as the banana, 4–6 Cavalieri sections should yield a reasonable estimation precision. As the longitudinal diameter of the banana was about 15 cm long, a sectioning period $T = 3\,\mathrm{cm}$ was adopted. In turn, a square grid with $h = 0.87\,\mathrm{cm}$ (at the banana scale) was superimposed on each section – thus, $a/p = h^2$. The grid was IUR (i.e., not just UR) to enable the estimation of the variance contribution of point counting (recall Section 4.6.3). The observed point count sequences were $\{12, 11, 11, 8\}$ for the banana, and $\{5, 7, 7, 4\}$ for its pulp. These data yielded

$$\widetilde{V}_0 = 3 \cdot 0.87^2 \cdot 42 = 95\,\mathrm{cm}^3,$$
$$\widetilde{V} = 3 \cdot 0.87^2 \cdot 23 = 52\,\mathrm{cm}^3, \qquad (4.8.3)$$

respectively. For the banana, the Cavalieri volume estimate fortuitously coincided with the fluid displacement measurement, see Eq. (4.7.1), up to two digits.

4.8.3 Error variance predictors

To predict $\mathrm{CE}^2(\widetilde{V}_0)$ we apply Eq. (5.3.10), which corresponds to sections of zero thickness, and to section areas estimated by point counting with an IUR square grid. The smoothness constant q is taken to be equal to 1, because the section area function of the banana is expected to be continuous – therefore we adopt $\alpha(1) = 1/240$. For the point counting contribution to the variance, the shape constant φ, see Eq. (5.3.8), is taken to be equal to 3.628 for every section. For simplicity, this value corresponds to the banana section 1, see Fig. 4.8.1(d), in which $P_1 = 12$, $I_1 = 16$. With this, and with the known parameters T, h, we obtain $\mathrm{ce}(\widetilde{V}_0)\% = 4.9\%$ for the banana volume. The relative contributions to $\mathrm{ce}^2(\widetilde{V}_0)$ of sectioning and point counting are of 19.2% and 80.8%, respectively. In turn, for the pulp we obtain $\mathrm{ce}(\widetilde{V})\% = 7.1\%$, and the aforementioned contributions are of 3.9% and 96.1%, respectively. Thus, to increase the estimation precision in this example, it is more efficient to decrease h (that is, to choose a denser grid in order to score more test points) than to increase the number of sections.

4.8.4 Notes

1. The Cavalieri estimator of volume

Early references to Eq. (4.8.1) are given in Section 2.25.8, Note 3. The estimator was named in honor of B. Cavalieri (1598–1647), see Section 1.4.4, Note 3. The example is adapted from Sagaseta (2005).

2. Fluid displacement

Fluid displacement may be a convenient choice to measure the volume of an isolated object, but not of internal compartments. Here the Cavalieri estimator is a convenient choice, and its precision can be easily controlled. The sections may be invasive, as in the example, or non-invasive. In the latter case, Roberts et al. (1993) used magnetic resonance imaging (MRI) for human body compartments, whereas Pache et al. (1993) used computed tomography (CT) for the human lung. For manual invasive sectioning, a variety of devices have been developed, see for instance Michel and Cruz-Orive (1988), or Howard and Reed (2005).

3. On the precision of the Cavalieri estimator

If the section areas were measured exactly, then, using the approximate relative section contribution of 19.2% to the total $\text{ce}^2(\widehat{V}_0)$ found above,

$$\text{ce}(\widehat{V}_0) = \sqrt{0.192} \times \text{ce}(\widetilde{V}_0) = 0.021, \tag{4.8.4}$$

namely 2.1% only. This is not surprising: for an ellipsoid, Eq. (5.2.7) yields 2.0% with $m = 4$ Cavalieri sections.

4.9 Volume From Cavalieri Slabs

4.9.1 Purpose, material, and method

The purpose is to illustrate a non-invasive version of the Cavalieri method to estimate the volume of grey matter in a human brain. The basic data set consists of 183 consecutive coronal MRI slices of 1 mm thickness encompassing the entire brain, henceforth called the primary slices.

Fig. 4.9.1(a) represents a lateral projection of a computer reconstruction of the grey matter. The sampling axis is along the horizontal direction. The projection of a Cavalieri sample of $n = 4$ primary slices of period $T = 45$ mm, is also represented. The corresponding slice projections are displayed in Fig. 4.9.1(b) with false colours, see the inset of Fig. 4.9.1(c). The cerebrum is defined as the union of grey and white matter.

In Fig. 4.9.1(c), the ordinate of a curve at a point of abscissa p along the sampling axis is the measurement function, which in this case is the mean primary slice area

$$A_t(p) = V_t(p)/t \tag{4.9.1}$$

where t is the slab thickness, see Eq. (2.10.6). For $t = 1$ mm, the volume $V_t(p)$ of each primary slice was measured automatically by pixel counting (the pixel side length was of 1 mm) using the Analyze software (Mayo Foundation, MN, USA). The abscissa p of a slab is that of its reference (lhs) face. Each curve is polygonal, joining the available 183 data points with straight line segments.

Fig. 4.9.1(d) is analogous to Fig. 4.9.1(a) for Cavalieri slabs of $t = 27$ mm thickness. Here each slice is the union of 27 primary slices. If the range of the abscissa of a primary slice is (p_1, p_{183}), say, then the range of the abscissa of a slab of thickness t hitting the brain is $(p_1 - t, p_{183})$, see Fig. 2.10.1 and Fig. 2.10.2. Thus, to compute the volume $V_t(p)$ determined in the brain by a hitting slab of t mm thickness, t zeros have to be incorporated to the left of the vector of 183 primary slice volumes. The corresponding area functions are displayed in Fig. 4.9.1(e). The smoothing effect discussed in Section 2.10.3 is apparent when compared with the corresponding curves in Fig. 4.9.1(c).

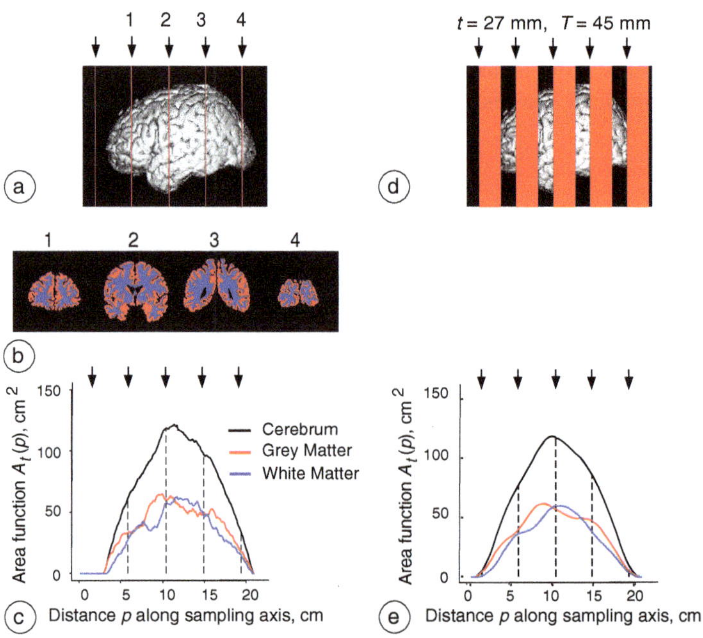

Fig. 4.9.1 (a–c) Estimation of brain compartment volumes with MRI, 1 mm thick Cavalieri slabs. (d,e) Idem with 27 mm thick slabs. See Section 4.9. Modified from McNulty et al. (2000), with permission of Lippincott Williams & Wilkins, Inc.

Set $V_{ti} \equiv V_t(p_i)$, $A_{ti} \equiv A_t(p_i)$, and relabel the nonempty Cavalieri slices with the indexes $\{1, 2, \ldots, n\}$ in their natural order. By Eq. (2.25.20), the estimator

$$\widehat{V} = \frac{T}{t} \sum_{i=1}^{n} V_{ti} = T \sum_{i=1}^{n} A_{ti}, \qquad (4.9.2)$$

is unbiased for the volume V of the brain compartment considered.

4.9.2 Results

For the Cavalieri sample with $t = 0.1$ cm, $T = 4.5$ cm, illustrated in Fig. 4.9.1(a), the mean primary slice areas (rounded to the third digit) corresponding to the grey matter (red coloured curve in Fig. 4.9.1(c)) are

$$\{A_{ti}\} = \{33.6, 60.9, 49.4, 18.8\}\ \text{cm}^2. \qquad (4.9.3)$$

Direct application of Eq. (4.9.2) yields $\widehat{V} = 732.2\,\text{cm}^3$. The ‘true’ grey matter volume given by the sum of the 183 primary slice volumes was $V = 724.0\,\text{cm}^3$.

For the corresponding Cavalieri sample with $t = 2.7$ cm, $T = 4.5$ cm, we have,

$$\{A_{ti}\} = \{4.1, 40.4, 57.4, 47.6, 6.0\}\ \text{cm}^2. \qquad (4.9.4)$$

whereby $\widehat{V} = 699.8\,\text{cm}^3$. Note that an additional Cavalieri slice has emerged.

4.9.3 Error variance predictors

Cavalieri sample of Eq. (4.9.3)

The error variance of the corresponding Cavalieri volume estimator is estimated by Eq. (5.4.5) with $\tau = t/T = 1/45$, $\widehat{\gamma}_{ti}$ replaced with A_{ti}, and no local errors ($vn = 0$). The value adopted for the smoothness constant of the measurement function is $q = 0.42$, whereby the relevant constant factor for the variance predictor becomes $\alpha(0.42, 1/45) = 0.0273$, whereby $\text{ce}(\widehat{V})\% = 4.3\%$. For further details, see Note 2 below.

Cavalieri sample of Eq. (4.9.4)

Similarly, with $\tau = 27/45$ and $q = 0.42$ we obtain $\alpha(0.42, 27/45) = 0.00691$, whereby $\text{ce}(\widehat{V})\% = 2.5\%$.

4.9.4 Notes

1. Choice of method

The data of the example comes from the study of McNulty et al. (2000). While the estimator of Eq. (4.9.2) is more accurate than the point count one of Eq. (4.8.2), the gain in precision will often not compensate the amount of work involved in digitizing the material. In most cases using point counting on non-invasive section images will be a reasonable choice.

2. Choice of the smoothness constant

The value $q = 0.42$ for the smoothness constant was proposed by García-Fiñana and Cruz-Orive (2004), (see Section 5.2.6, Note 3), who checked its performance on a basic data set of 183 slices of 1 mm thickness consisting of the union of the grey matter data used here, plus the cerebellum of the same brain (for this reason, their Fig. 4a differs slightly from Fig. 4.9.1(c)). Later, Cruz-Orive (2006) performed analogous resampling experiments (with the grey matter basic data set used here) for $t = 1, 3, 9$ and 27 mm, to check the performance of Eq. (5.4.5) without local errors, and Fig. 4.9.2 is an extract of Fig. 2 from the latter paper.

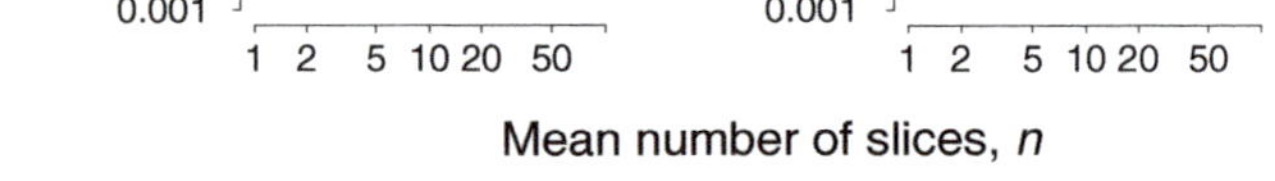

Fig. 4.9.2 Performance of the error predictor (red curves) of the estimator of grey matter volume from Cavalieri slabs (Eq. (5.4.5) without local errors). The black curve corresponds to the empirical coefficient of error obtained by resampling. Modified from Cruz-Orive (2006), Fig. 2, with permission of Wiley-Blackwell.

3. Performance of the error variance predictor given by Eq. (5.4.5)

The empirical coefficient of error computed from the 45 possible samples with $\tau = 1/45$ was $\mathrm{CE}(\widehat{V})\% = 5.4\%$. On the other hand, the error variance predictor computed with Eq. (5.4.5) with $\tau = t/T = 1/45$, $\upsilon_n = 0$, and $\alpha(0.42, 1/45) = 0.027326$, averaged over the 45 possible samples, yields $\mathrm{ce}(\widehat{V})\% = 5.1\%$. On the other hand, with $\tau = 27/45$, $\upsilon_n = 0$, and $\alpha(0.42, 27/45) = 0.006915$ the corresponding results were $\mathrm{CE}(\widehat{V})\% = 2.5\%$ and $\mathrm{ce}(\widehat{V})\% = 2.3\%$, respectively. Reasonable agreement was also obtained for the white matter and for the cerebrum.

4. Early error variance prediction with $q = 0, 1$

McNulty et al. (2000) also carried out resampling experiments for $t = 1, 3, 9$ and 27 mm, and for each brain compartment (grey and white matter, and cerebrum) to check the performance of the variance predictor given by Eq. (5.4.5) for $q = 0$ and $q = 1$, obtained earlier by Gual-Arnau and Cruz-Orive (1998). For a brief historical account of the evolution of the variance prediction problem for Cavalieri sampling, see Cruz-Orive (2017).

5. Confidence intervals

Exact confidence intervals for the target parameter from Cavalieri data are usually not available because the sampling distribution of estimators obtained under systematic sampling is generally unknown. Moreover, such distributions usually do not converge to the normal. García-Fiñana (2006) obtained approximate confidence intervals in this context.

6. Further applications with non-invasive Cavalieri sections

For early applications of the Cavalieri design with MRI sections, see for instance Roberts et al. (1993), Roberts, Garden, Cruz-Orive, Whitehouse, and Edwards (1994), Roberts, Cruz-Orive, et al. (1997) and García-Fiñana, Cruz-Orive, Mackay, Pakkenberg, and Roberts (2003).

4.10 Number and Mean Size of Planar Particles From Systematic Quadrats

4.10.1 Purpose, material, and method

Fig. 4.10.1 represents an approximately transverse section of a portion of rat Soleus muscle (distal end). For didactic purposes the fibre transects are regarded as two-dimensional particles, without reference to their three-dimensional meaning. The interfibrillar space is of no interest here.

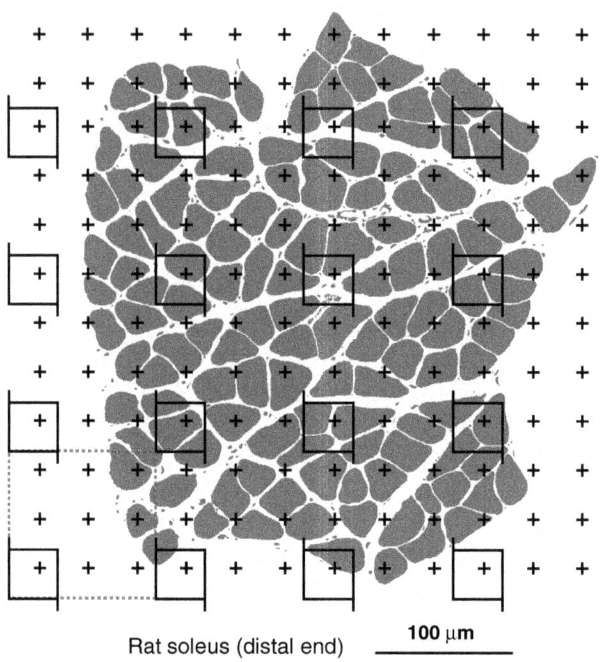

Fig. 4.10.1 Estimation of the total number of an aggregate of planar particles, and of the individual mean particle area and boundary length, see Section 4.10.

The purpose is to estimate the number N of particles, the mean particle area $\mathbb{E}_N(a) = A/N$, where A is total particle area, and the mean boundary length $\mathbb{E}_N(b) = B/N$, where B is total particle boundary length, by means of the multipurpose test system illustrated in the figure. The fundamental tile (lower left corner) is a square of area $a = T^2$, containing a quadrat of side length $t = T/3$, and $p = 9$ test points. It is assumed that the test system is IUR relative to the particle aggregate.

By Eq. (2.25.28), a UE of N is

$$\widehat{N} = \frac{a}{a_0} \cdot Q, \tag{4.10.1}$$

where $a_0 = a/9$ is the quadrat area, and Q denotes the number of particles captured by the quadrats according to the forbidden line rule, see Fig. 2.15.3(c). On the other hand,

$$\bar{a}_N = \widehat{A}/\widehat{N}, \quad \bar{b}_N = \widehat{B}/\widehat{N}, \tag{4.10.2}$$

estimate $\mathbb{E}_N(a)$ and $\mathbb{E}_N(b)$, respectively. In turn, $\widehat{A}$ is computed via Eq. (4.6.1). To compute $\widehat{B}$ we adapt the upper horizontal side of each quadrat as the fundamental test line, and apply the first Eq. (4.6.2) with $a/l = a/t$.

The estimators given by Eq. (4.10.2) are not strictly unbiased, but ratio-unbiased, because their denominator is a random variable. For instance, $\mathbb{E}(\bar{a}_N) = \mathbb{E}(\widehat{A}/\widehat{N}) \neq A/N$ hence $\bar{a}_N$ is not unbiased in general, but it is ratio-unbiased because

$$\mathbb{E}(\widehat{A})/\mathbb{E}(\widehat{N}) = A/N = \mathbb{E}_N(a), \tag{4.10.3}$$

and similarly for $\bar{b}_N$, see also Section A.2.3.

4.10.2 Results

In Fig. 4.10.1 we count $Q = 15$ particles with the forbidden line rule, (Fig. 2.15.3), and therefore $\widehat{N} = 9 \cdot 15 = 135$ particles. The true number is $N = 131$.

The remaining targets are not dimensionless, and therefore they require the working magnification M. In the working image the calibration bar measured 33 mm, and therefore $M = 33/0.1 = 330$. In the same image, $t = 12$ mm and $T = 36$ mm.

We score $P = 57$ test points in the particles, and $I = 21$ intersections in their boundaries, whereby

$$\begin{aligned} \widehat{A} &= \frac{36^2}{9} \cdot \frac{1}{330^2} \cdot 57 = 0.0754\,\text{mm}^2, \\ \widehat{B} &= \frac{\pi}{2} \cdot \frac{36^2}{12} \cdot \frac{1}{330} \cdot 21 = 10.80\,\text{mm}. \end{aligned} \tag{4.10.4}$$

Finally,

$$\bar{a}_N = 559\,\mu\text{m}^2, \quad \bar{b}_N = 80.0\,\mu\text{m}. \tag{4.10.5}$$

Because the particles are not far from convex, by Eq. (2.4.28), the mean particle caliper diameter may be approximated by $\bar{h}_N = 80.0/\pi = 25\,\mu$m directly.

4.10.3 Error variance predictors

The $\mathrm{CE}(\widehat{N})$ may be predicted from the matrix of quadrat counts, namely,

	[, 1]	[, 2]	[, 3]
[1,]	2	2	1
[2,]	2	2	3
[3,]	0	1	1
[4,]	0	1	0

The matrix columns correspond to Cavalieri stripes of thickness t. The stripe sampling fraction is $\tau = t/T = 1/3$, whereby Eq. (5.6.2) with

$$\alpha(0, 1/3) = \frac{1}{6} \cdot \frac{(1 - 1/3)^2}{2 - 1/3} = \frac{2}{45}, \tag{4.10.6}$$

(see Eq. (5.4.11)), yields $\mathrm{ce}(\widehat{N}) = 0.087$ for the above matrix, and $\mathrm{ce}(\widehat{N}) = 0.134$ with its transpose. We adopt the average,

$$\mathrm{ce}(\widehat{N}) = \{(0.087^2 + 0.134^2)/2\}^{1/2} = 0.113, \tag{4.10.7}$$

namely $\mathrm{ce}(\widehat{N})\% = 11.3\%$. The relative between and within stripes contributions to the total $\mathrm{CE}^2(\widehat{N})$, see Section 5.6.1, are estimated as 60% and 40% respectively.

Eq. (4.6.4) yields $\mathrm{ce}^2(\widehat{A}) = 0.0816^2$, and by Eq. (A.2.16),

$$\mathrm{ce}^2(\bar{a}_N) \approx \mathrm{ce}^2(\widehat{A}) + \mathrm{ce}^2(\widehat{N}) = 0.139^2, \tag{4.10.8}$$

that is, $\mathrm{ce}(\bar{a}_N)\% = 13.9\%$.

4.10.4 Notes

For an application of Section 4.10.1 to nerve cross sections, see J. O. Larsen (1998).

Cruz et al. (2015), Cruz and González-Villa (2018) and Gómez, Cruz, and Cruz-Orive (2019) used automatic Monte Carlo resampling to check the performance of Eq. (5.6.2), see also Section 5.6.3. The authors developed the free software Countem (`http://countem.unican.es`) for the semiautomatic estimation of a population total using the present design. The value of $\mathrm{ce}(\widehat{N})$ is also computed via Eq. (5.6.2).

4.11 Number By the Fractionator Method

4.11.1 Purpose, material, and method

The problem is to estimate the total number N of lymphatic valves in the left lower lobe of an infant lung, see Fig. 4.11.1(a). For background details, see Note 1 of Section 4.11.4 below. The chosen method is the fractionator, which is basically multistage systematic sampling. Next we illustrate a 3-stage fractionator design.

Stage 1. The lobe is exhaustively cut into slices, which are laid flat in their natural order (Fig. 4.11.1(b)). In turn, each slice is exhaustively cut into stripes, from which a systematic sample (Section 4.4.3) is drawn with period T_1, (Fig. 4.11.1(c)). If the sampled stripes contain a random number n_1 of valves, then by Eq. (4.4.8) we have

$$T_1 \mathbb{E}(n_1) = N. \tag{4.11.1}$$

Stage 2. Each sampled stripe is exhaustively cut into blocks, from which a systematic sample is drawn with period T_2, (Fig. 4.11.1(d)). If the sampled blocks contain n_2 valves, then

$$T_2 \mathbb{E}(n_2|n_1) = n_1, \tag{4.11.2}$$

and taking expectations with respect to the distribution of n_1,

$$T_1 T_2 \mathbb{E}\{\mathbb{E}(n_2|n_1)\} = T_1 \mathbb{E}(n_1) = N. \tag{4.11.3}$$

Stage 3. Each sampled block is exhaustively cut into serial sections, from which a systematic sample of disectors is drawn with period T_3. That is, from the ordered pool of all serial sections from all blocks, every T_3th section, and the next (Fig. 4.11.2), are taken to constitute a final sample of disectors. If a total of n_3 valves are counted in all the sampled disectors, then

$$T_3 \mathbb{E}(n_3|n_2) = n_2, \tag{4.11.4}$$

and proceeding similarly as above, it follows that the fractionator estimator,

$$\widehat{N} = T_1 T_2 T_3 n_3 \tag{4.11.5}$$

is unbiased for N.

Note that $\widehat{N}$ is a genuine 0-dimensional estimator – it does not depend on either slab thickness, stripe width, block size, or section thickness. However, the latter parameters, and the sampling periods, will affect the precision of $\widehat{N}$. In the example, the disectors were physical (Fig. 4.11.2). In neuroscience, for instance, optical disectors reduce the workload.

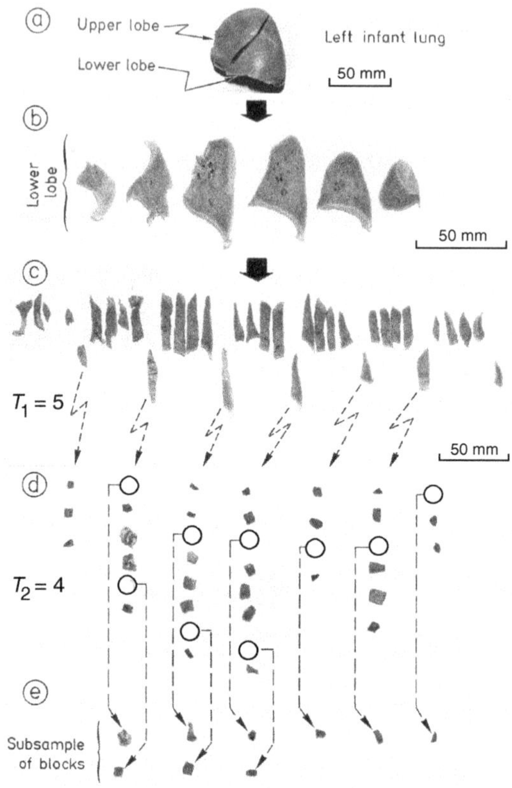

Fig. 4.11.1 Illustration of a three-stage fractionator method to estimate the number of lymphatic valves in an infant lung lobe, see Section 4.11. Modified from Ogbuihi and Cruz-Orive (1990), with permission of Wiley-Blackwell.

4.11.2 Results

The primary slices were about 8 mm thick, and the stripes about 8 mm wide. The periods were $T_1 = 5$, $T_2 = 4$ and $T_3 = 400$. The blocks sampled at the second stage were embedded in paraffin and exhaustively cut (on a Reicher Bio-Cut^{TM} rotary microtome) into serial sections of 5 μm nominal thickness. Possible tissue shrinkage and deformation artifacts had no impact here because the target was number and only valve counts were used. Every 400th section, the next, and the second next, were mounted on the same slide (Fig. 4.11.2(b,c)). Each section pair was used as a double disector, that is, valves were counted if present in the first but not in the second section, and also if present in the second but not in the first section. A total of 34 double disectors of 5 μm nominal thickness were sampled in total. A third

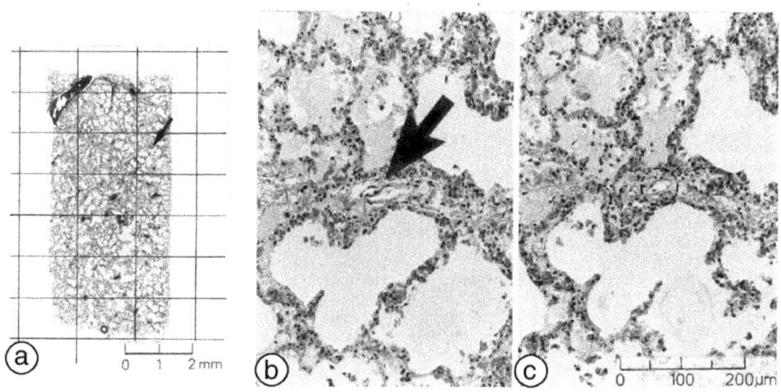

Fig. 4.11.2 In the third fractionator stage, systematic primary disectors were subsampled from each second stage tissue block (Fig. 4.11.1(e)). (a) Each primary disector was exhaustively examined using a grid of subdisectors. (b), (c). A subdisector capturing a lymphatic valve, present in (b), (arrowed), but not in (c). Modified from Ogbuihi and Cruz-Orive (1990), with permission of Wiley-Blackwell.

section was reserved in each disector for valve identification purposes whenever required. Each reference section was entirely examined under LM across a partition of quadrats of $1.8 \times 1.2\,\text{mm}^2$ real size (Fig. 4.11.2(a)). Valves were rarely captured by the disectors – only 4 valves were counted in total, whereby their total number N in the entire lobe was estimated with Eq. (4.11.5) as

$$\widehat{N} = 5 \times 4 \times 400 \times (4/2) = 16000 \text{ valves.} \tag{4.11.6}$$

The factor $1/2$ stems from the fact that the 4 valves were scored in double disectors.

4.11.3 Error variance prediction

Hitherto no general variance predictor exists for the fractionator. If smooth arrangements (Section 4.4.4) are implemented at each stage, however, then experience suggests that $\text{Var}(\widehat{N})$ can be rather small, even for quite small sample sizes. The original study involved 3 control and 4 SIDS (sudden infant death syndrome) complete infant lungs, respectively. The respective mean valve number estimates were 65500 ($\text{cv}(\widehat{N})\% = 0.9\%$) and 63100 ($\text{cv}(\widehat{N})\% = 8.9\%$), whereby no significant difference was detected among the two group means. The $\text{cv}(\widehat{N})$ among the pool of 7 lungs was of 6.5%, which includes the natural variation among lungs plus the average fractionator error variance within lungs. Thus, in this experiment the error variance was probably quite low, in spite of the fact that the total numbers of valves counted in the 3 control lungs and in the 4 SIDS lungs were 43 and 74 only, respectively.

The slice ordering was naturally smooth. However, the fractionator precision could still be improved if the pool of stripes was rearranged into a smooth series prior to the first sampling stage, and similarly for the pool of blocks prior to the second stage. These considerations would apply provided that the number of valves in a lung fragment was always proportional (at least approximately) to the observable size of the fragment (Section 4.4.6, Note 3).

4.11.4 Notes

1. Source

For details on the original study see Ogbuihi and Cruz-Orive (1990). The scheme illustrated in Fig. 4.11.1, however, was prepared a posteriori, and the corresponding data were not included in the study.

2. Error variance prediction

For a design similar to the one in the example, a modification of the variance predictor used in Section 4.10.3 could be tried by regarding the blocks sampled from a stripe as quadrats sampled within a flat stripe. In cases such as the present one, however, the idea could hardly be implemented, because most blocks yielded zero counts. The prevailing attitude with the fractionator is: "Try to ensure a very low error variance by using an MG smooth arrangement at each stage, and do not worry about estimating the variance itself".

3. The fractionator is not only for number

For a general parameter γ, a k-stage fractionator estimator would read

$$\widehat{\gamma} = T_1 \cdot T_2 \cdot \cdots \cdot T_k \cdot \widehat{\gamma}_k, \tag{4.11.7}$$

where $\widehat{\gamma}_k$ is estimated from a final sample of blocks by a suitable method. For instance, Artacho-Pérula, Roldán-Villalobos, and Cruz-Orive (1999) applied a 4-stage fractionator design to illustrate the unbiased estimation of the total length $\gamma \equiv L$ of the capillary bed in the soleus muscle of a Wistar rat, see Section 4.18.4, Note 4. For interesting applications in agronomy, see Wulfsohn, Sciortino, Aaslyng, and García-Fiñana (2010) and Wulfsohn, Aravena Zamora, Potin Téllez, Zamora Lagos, and García-Fiñana (2012).

4.12 Curve Length in Space From ICav Sections

4.12.1 Purpose, material, and method

The general problem is to estimate the total finite length L of a bounded set $Y \subset \mathbb{R}^3$ of smooth curves by means of ICav planes a distance T apart, see Section 2.26.3 and Fig. 4.12.1. Let $\{Q_1, Q_2, \ldots, Q_n\}$ denote the ordered transect numbers determined in the curves by the Cavalieri planes. Based on Eq. (2.26.17), a UE estimator of L is

$$\widehat{L} = 2T \sum_{i=1}^{n} Q_i. \tag{4.12.1}$$

To illustrate the method, three cooked spaghetti of a known length were arbitrarily curved and embedded into a ball of plastiline, taking care to avoid stretching them. The ball was first rotated, then spun around its vertical diameter (to comply, at least approximately, with Eq. (2.11.5)), and finally cut into Cavalieri sections by the practical procedure described in Section 4.8.1, see Fig. 4.12.2.

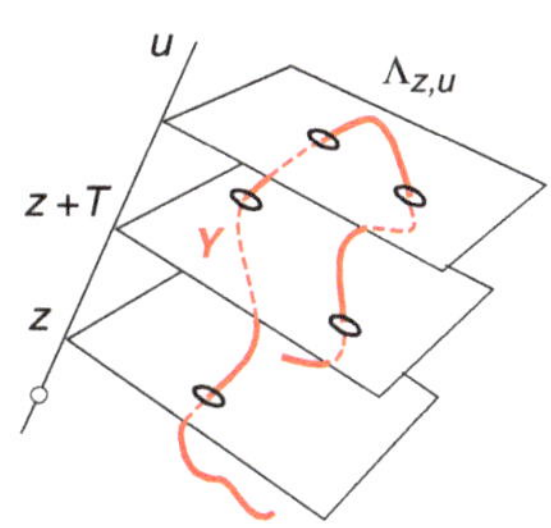

Fig. 4.12.1 Sketch of a curve in 3D hit by a test system of isotropic Cavalieri (ICav) planes, see also Fig. 2.26.2.

4.12.2 Results

In the plastiline ball, Cavalieri cuts with period $T = 1\,\text{cm}$ generated an exhaustive series of 8 extensive slices, whose left faces constituted $n = 7$ sections. Deformation of the latter would not affect the transect counts. The corresponding numbers of spaghetti transects, in their natural order, were $\{2, 7, 10, 9, 10, 11, 0\}$, whereby,

$$\widehat{L} = 2 \times 1 \times 49 = 98\,\text{cm}. \tag{4.12.2}$$

Prior to embedding, each spaghetti measured 33.3 cm, (see Fig. 4.12.2(a)), this making a total true length of $L = 100\,\text{cm}$.

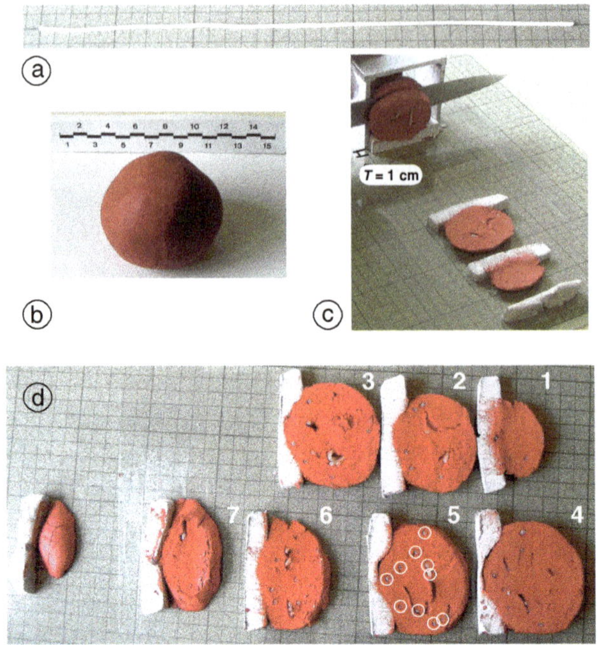

Fig. 4.12.2 (a) One of the three cooked spaghettis (on a 1 cm ruled mat) which were bent and embedded in a plastiline ball, see (b). (c) The ball was exhaustively cut into 1 cm thick ICav slices. (d) The first slice was discarded, and the lhs face of each of the remaining 7 slices was an ICav section used to estimate the total spaghetti length, see Eq. (4.12.2). In the 5th section, the relevant transects ($Q_5 = 10$) are encircled. Modified from Sagaseta (2005).

4.12.3 Error variance prediction

For the given sectioning orientation, the observed transect counts represent Cavalieri observations from an integer-valued function without local errors. Thus, we may apply Eq. (5.2.10) with $\alpha(0) = 1/12$ to obtain $\text{ce}(\widehat{L})\% = 7.2\%$, which does not include the orientation contribution.

4.12.4 Notes

1. Source

The example was adapted from Sagaseta (2005).

2. In practice, curvilinear features are cylinders, or tubes

The spaghetti of the example were solid twisted cylinders. Will this induce bias in the length estimator? Gundersen (1979) showed that the main source of bias is due to the presence of cylinder ends, being negligible if their number is relatively low. See also Gundersen (2002a).

4.13 Surface Area, Volume, and Mean Barrier Thickness By a Ratio Design

4.13.1 Purpose, material, and method

The target object is the septum of a mammalian lung, and the target parameters are the total septum volume $V(s)$, the total alveolar surface area $S(\text{alv})$ (which is equivalent to the external surface area of septum), and the mean septum thickness, which by Eq. (3.14.6) is conventionally defined as

$$\mathbb{E}_S(\tau(\text{s})) = \frac{V(\text{s})}{\frac{1}{2}S(\text{alv})} \equiv 2V_S(\text{s}, \text{alv}). \tag{4.13.1}$$

The septa are the tissue barriers separating the alveoli (which are air spaces) from each other, see Fig. 3.1.1. The union of all septa and alveoli constitute the lung parenchyma, see also Fig. 4.2.2(b). The union of all air spaces in the lung is simply connected. The symbol $\mathbb{E}_S(\cdot)$ means that the expectation is over the surface weighted distribution, see Section 3.14. Thus, the relative frequency of septum thickness in the interval $[\tau, \tau + \mathrm{d}\tau)$ is proportional to the surface area of septum with thickness in that interval.

The relevant phases are lung $\supset$ parenchyma $\supset$ septa, which in the sequel are written l $\supset$ par $\supset$ s for short. The corresponding sampling design (Fig. 4.13.1) leads to the following estimator,

$$\widehat{V}(\text{s}) = \widehat{V}(\text{l}) \cdot \frac{\widehat{V}(\text{par})}{\widehat{V}(\text{l})} \cdot \frac{\widehat{V}(\text{s})}{\widehat{V}(\text{par})}, \tag{4.13.2}$$

which is a particular instance of Eq. (4.2.10). In a more customary notation,

$$\widehat{V}(\text{s}) = \widehat{V}(\text{l}) \cdot \widehat{V}_V(\text{par}, \text{l}) \cdot \widehat{V}_V(\text{s}, \text{par}), \tag{4.13.3}$$

and similarly for $\widehat{S}(\text{alv})$.

An alternative, shortcut design may be adopted: Instead of estimating $V(\text{par})$ via the product $V(\text{l}) \cdot \widehat{V}_V(\text{par}, \text{l})$, one can do it directly by the Cavalieri method using Eq. (4.8.2), whereby

$$\widehat{V}(\text{s}) = \widehat{V}(\text{par}) \cdot \widehat{V}_V(\text{s}, \text{par}), \tag{4.13.4}$$

and similarly for $\widehat{S}(\text{alv})$. It is nonetheless advisable to estimate $V(\text{l})$ as well, e.g. by fluid displacement (Section 4.7).

If Eq. (4.13.3) is used, then $\widehat{V}_V(\text{par}, \text{l})$ may be obtained by point counting on Cavalieri sections at a low magnification (Fig. 4.13.1(b)) by adapting Eq. (4.13.6) below. At the second sampling stage, small systematic blocks (of about 2 mm diameter) are sampled in the parenchyma from Cavalieri slices (Fig. 4.13.1(c)). Because the parenchyma tissue is fairly homogeneous throughout the lung, it may be locally modelled by a realization of a stationary and isotropic random set, whereby model stereology applies. Thus, it suffices to take one section per block (Fig. 4.13.1(d)), and either one extensive quadrat per section if LM is used (Fig. 4.13.1(e)), or several systematic quadrats per section if EM is used. Edge effects may be ignored.

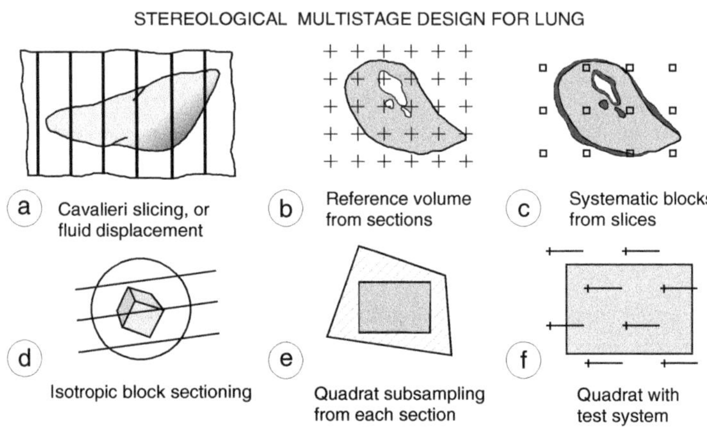

Fig. 4.13.1 Illustration of the multistage design corresponding to Eq. (4.13.3).

Fig. 4.13.2(a) represents a quadrat from a block section of lung parenchyma, with a test system whose fundamental tile is shown in Fig. 4.13.2(b), upper panel. Suppose that n block sections (which are effectively IUR hitting the parenchyma) are sampled from the entire lung. For the pool of all quadrats subsampled from the ith section, $i = 1, 2, \ldots, n$, let $I_i(\text{alv})$, $P_i(\text{par})$ denote respectively the total number of intersections between the test lines and the alveolar traces, and the total number of test points (encircled) in the reference space (parenchyma, which fills the entire quadrat). For instance, if the quadrat shown in the figure was the only quadrat observed in the first section, then $I_1(\text{alv}) = 16$, $P_1(\text{par}) = 18$.

Based on Eq. (2.30.4), a ratio-unbiased estimator of $S_V(\text{alv}, \text{par})$ is

$$\widehat{S}_V(\text{alv}, \text{par}) = 2 \cdot \frac{p}{l} \cdot M \cdot \frac{\sum_{i=1}^{n} I_i(\text{alv})}{\sum_{i=1}^{n} P_i(\text{par})}, \tag{4.13.5}$$

where $p = 2$ is the number of test points, and l the total test line length, in the fundamental tile; in this case, p/l is just the reciprocal of the length of a single test segment. Further, M is the final linear magnification of the quadrat image. Note that $l \cdot P_i(\text{par})$ is a UE of the total length of test line in the reference space in the ith section – recall Fig. 2.12.1(a,b) and Eq. (2.12.5).

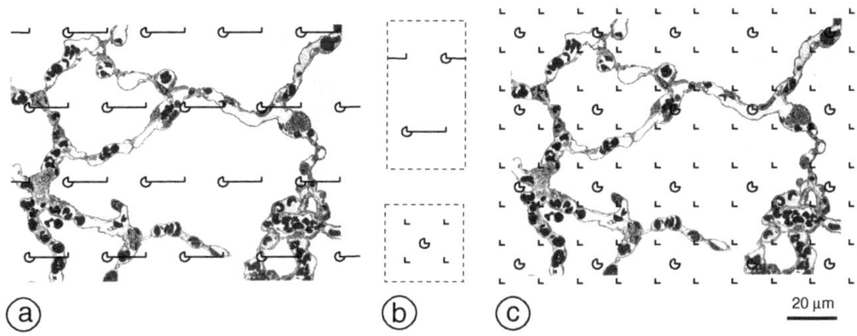

Fig. 4.13.2 (a) Estimation of the ratio of alveolar surface area to parenchymal volume by Eq. (4.13.5). (b) Fundamental tiles of the test systems used in (a,c). (c) Estimation of the ratio of septum to parenchymal volume by Eq. (4.13.6). Plain micrograph courtesy of E. R. Weibel, from Gehr et al. (1981) with permission of Elsevier.

Similarly, for the quadrat in Fig. 4.13.2(c),

$$\widehat{V}_V(\text{s}, \text{par}) = \frac{p_1}{p_2} \cdot \frac{\sum_{i=1}^{n} P_i(\text{s})}{\sum_{i=1}^{n} P_i(\text{par})}, \tag{4.13.6}$$

is, by virtue of Eq. (2.30.3), a ratio-unbiased estimator of $V_V(\text{s}, \text{par})$. Here $p_1 = 1$, $p_2 = 4$ are the numbers of test points per tile destined to hit the parenchyma (encircled) and to hit the septa, respectively, see Fig. 4.13.2(b), lower panel. In the figure, $P_1(\text{s}) = 11$, $P_1(\text{par}) = 20$. Finally, $\mathbb{E}_S(\tau(\text{s}))$ may be estimated by

$$\bar{\tau} = \frac{\widehat{V}_V(\text{s}, \text{par})}{\frac{1}{2}\widehat{S}_V(\text{alv}, \text{par})}. \tag{4.13.7}$$

4.13.2 Results

Fig. 4.13.2 constitutes a simplification for didactic purposes. In practice it is convenient and efficient to combine both test systems into a multipurpose one, see Fig. 4.13.3. The entire lung of a male albino rat (body weight = 300 g) from the study cited in Note 1 below, was fixed by airway instillation, embedded in agar and exhaustively cut into alternatively thin (1 mm) and thick (5 mm) Cavalieri slices.

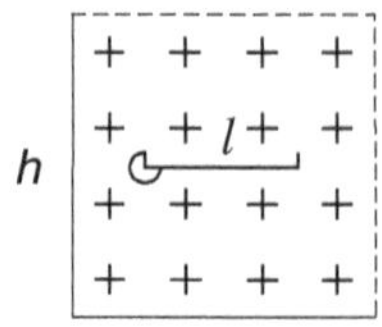

Fig. 4.13.3 Fundamental tile of a multipurpose test system designed to increase data collection efficiency with respect to the two test systems of Fig. 4.13.2.

The design corresponding to Eq. (4.13.4) was adopted, but nonetheless the total lung volume $\widehat{V}(\mathrm{l})$ was estimated first by fluid displacement (Section 4.7). Because the lung was small, $V(\mathrm{par})$ was estimated directly from extensive Cavalieri sections for LM (Fig. 4.13.1(a,b)). Relatively large airway and blood vessel transects exceeding about 0.5 mm in diameter were excluded. The result was $\widehat{V}(\mathrm{par}) = 7.54\,\mathrm{cm}^3$. From the thicker slices, systematic tissue blocks were cut, and a final sample of $n = 4$ blocks was retained. From each block an (effectively IUR) section of $1\,\mu$m nominal thickness was taken for LM, and from each section a rectangular quadrat of about $0.50 \times 0.33\,\mathrm{mm}^2$ was photographed. Each quadrat image was analysed at a final magnification $M = 415$, and the side length of the fundamental tile (Fig. 4.13.3) of the test system superimposed on each quadrat image was $h = 3.966\,\mathrm{cm}$, so that $l = 1.983\,\mathrm{cm}$. With the coarse test points (encircled in Fig. 4.13.3) we obtained $\{P_i(\mathrm{par})\} = \{18, 18, 17, 17\}$. With the finer test points (crosses), $\{P_i(\mathrm{s})\} = \{39, 65, 51, 59\}$, and with the test line segments, $\{I_i(\mathrm{s})\} = \{21, 23, 26, 27\}$, respectively. With these data, Eq. (4.13.6) and Eq. (4.13.5) yield the following estimates,

$$\widehat{V}_V(\mathrm{s}, \mathrm{par}) \frac{1}{16} \cdot \frac{214}{70} = 0.191,$$
$$\widehat{S}_V(\mathrm{alv}, \mathrm{par}) = 2 \cdot \frac{1}{1.983} \cdot 415 \cdot \frac{97}{70} = 580\,\mathrm{cm}^{-1}, \tag{4.13.8}$$

respectively. Thus, rounding to 3 significant digits, Eq. (4.13.4), its analogue for $\widehat{S}(\mathrm{alv})$, and Eq. (4.13.8) yield, respectively,

$$\widehat{V}(\mathrm{s}) = 7.54 \cdot 0.191 = 1.44\,\mathrm{cm}^3$$
$$\widehat{S}(\mathrm{alv}) = 7.54 \cdot 580 = 4370\,\mathrm{cm}^2,$$
$$\bar{\tau}(\mathrm{s}) = (2 \cdot 1.44/4370) \cdot 10^4 = 6.59\,\mu\mathrm{m}. \tag{4.13.9}$$

4.13.3 Error variance predictions

Because $V(\text{par})$ and $V_V(\text{s}, \text{par})$ are estimated independently, and $V(\text{s})$ is the product of both, Goodman's formula (Eq. (A.2.16)) may be applied to yield the following approximation,

$$\text{ce}^2(\widehat{V}(\text{s})) = \text{ce}^2(\widehat{V}(\text{par})) + \text{ce}^2(\widehat{V_V}(\text{s}, \text{par})), \tag{4.13.10}$$

and similarly for $\text{ce}^2(\widehat{S}(\text{alv}))$. Here, $\text{ce}(\widehat{V}(\text{par})) = 0.062$ was computed as in Section 4.8.3. In turn, Cochran's formula (Eq. (A.2.13)) may be used for the ratio estimators, because the tissue blocks were small and relatively distant from each other, hence the corresponding quadrats may be assumed to be independent. Note, however, that Cochran's formula takes into account the correlation between $P_i(\text{s})$ and $P_i(\text{par})$, or between $I_i(\text{alv})$ and $P_i(\text{par})$, within each block. Thus, setting $X_i \equiv P_i(\text{par})$, $Y_i \equiv P_i(\text{s})$ and $n = 4$, Eq. (A.2.13) yields

$$\text{ce}^2(\widehat{V_V}(\text{s}, \text{par})) = \frac{4}{3}\left(\frac{11828}{214^2} + \frac{1226}{70^2} - \frac{2 \cdot 3742}{214 \cdot 70}\right) = 0.01184, \tag{4.13.11}$$

and similarly, $\text{ce}^2(\widehat{S_V}(\text{alv}, \text{par})) = 0.005263$. Now Goodman's formula yields $\text{ce}(\widehat{V}(\text{s}))\% = 12.5\%$ and $\text{ce}(\widehat{S}(\text{alv}))\% = 9.5\%$. On the other hand, $\text{ce}(\bar{\tau}(\text{s}))$ should be better computed with Cochran's formula because Eq. (4.13.7) is the following ratio,

$$\bar{\tau}(\text{s}) = (l/p_2) \cdot M^{-1} \sum_{i=1}^{n} P_i(\text{s}) \Big/ \sum_{i=1}^{n} I_i(\text{alv}). \tag{4.13.12}$$

Thus,

$$\text{ce}^2(\bar{\tau}(\text{s})) = \frac{4}{3}\left(\frac{11828}{214^2} + \frac{2375}{97^2} - \frac{2 \cdot 5233}{214 \cdot 97}\right) = 0.008670, \tag{4.13.13}$$

whereby $\text{ce}(\bar{\tau}(\text{s}))\% = 9.3\%$.

4.13.4 Notes

1. Sources

The material used in Section 4.13.2 comes from the study of Bur, Bachofen, Gehr, and Weibel (1985). The plain micrograph used in Fig. 4.13.2 is a low power TEM image from a Grant's gazelle (modified from Gehr et al. (1981), Fig. 5a).

2. Biasing artifacts

For the design given by Eq. (4.13.3) and Eq. (4.13.4) to work, it is necessary that the definition of parenchyma is the same at both sampling stages. If $V(\text{par})$ is estimated on fresh tissue (Fig. 4.13.1(b)), whereas the second stage ratios are estimated on fixed and embedded tissue blocks, then shrinkage artifacts may arise; to minimize this, plastic embedding is often used.

3. The isector circumvents the isotropy assumption

If the parenchyma is modelled by an invariant random set, then a section from a tissue block may be regarded as an IUR section. Without the isotropy assumption, however, each block may be embedded in a plastic ball, which is randomly rotated before submitting it to the microtome blade – this is the isector method, see Nyengaard and Gundersen (1992).

4. Stratified design

Often it is convenient to partition a target object Y into r non-overlapping, non-random fragments or strata $\{Y_1, \ldots, Y_r\}$. In the lung, natural strata are the lung lobes. Let γ represent the target quantity defined on Y (e.g., either $V(\text{s})$, or $S(\text{alv})$, in the lung example), and let γ_i denote the target quantity for the ith stratum. The idea is to estimate γ_i (as described, for instance, in the preceding sections), independently for each stratum, whereby γ is estimated by the sum

$$\widehat{\gamma} = \widehat{\gamma}_1 + \widehat{\gamma}_2 + \cdots + \widehat{\gamma}_r. \tag{4.13.14}$$

Moreover, by independence, $\text{Var}(\widehat{\gamma})$ may be estimated by

$$\text{var}(\widehat{\gamma}) = \sum_{i=1}^{r} \text{var}(\widehat{\gamma}_i), \tag{4.13.15}$$

where $\text{var}(\widehat{\gamma}_i)$ denotes the stereological error variance estimator within the ith stratum. The stratified design will be more efficient than the global one if the quantities $\{\gamma_i\}$ are fairly different among each other and the within strata variances among the pertinent blocks, or sections, are small. This is well known in sampling theory, see e.g. Cochran (1977). For an application to lungs, see Michel and Cruz-Orive (1988).

If the estimation precision within strata is predicted by the square coefficients of error, then,

$$\text{var}(\widehat{\gamma}) = \sum_{i=1}^{r} \text{var}(\widehat{\gamma}_i) = \sum_{i=1}^{r} \text{ce}^2(\widehat{\gamma}_i) \cdot \widehat{\gamma}_i^2. \tag{4.13.16}$$

5. Statistical models for ratio estimation

Here we examine the possibility of obtaining unbiased ratio estimators of minimum variance in the model stereology context.

Model I

An example of this is the nesting s $\subset$ par considered above. More generally, the model for the target object (Section 4.13.1) is based on a motion-invariant process $Y \subset \mathbb{R}^3$ of intensity γ_V, see Section 3.3. A quadrat $T \subset \mathbb{R}^3$, see Fig. 4.13.1(e), of area $A(T)$, is regarded as a platelet probe. For convenience, we rewrite Eq. (3.3.3) as

$$\gamma_V = \frac{\mathbb{E}\{\alpha(Y \cap T)\}}{A(T)}, \tag{4.13.17}$$

where $\alpha \equiv A$ if $\gamma \equiv V$, and $\alpha \equiv (4/\pi)B$ if $\gamma \equiv S$. To estimate γ_V, we use n quadrats $\{T_i\}$ of controlled areas $\{x_i \equiv A(T_i)\}$. The corresponding observations $\{Y_i \equiv \alpha(Y \cap T_i)\}$ are random variables. Set $R \equiv \gamma_V$. By Eq. (4.13.17),

$$\mathbb{E}(Y_i|x_i) = R \cdot x_i, \quad i = 1, 2, \ldots, n. \tag{4.13.18}$$

In order to obtain a minimum variance-unbiased estimator $\widehat{R}$ of R we may use the following proportional regression model,

$$Y_i = Rx_i + e_i, \quad i = 1, 2, \ldots, n, \tag{4.13.19}$$

(Cruz-Orive, 1980a, 2009b), where the $\{e_i\}$ are random deviations which, if the quadrats are sufficiently far apart, may be assumed to be independent. Further,

$$\mathbb{E}(e_i|x_i) = 0, \quad \text{Var}(e_i|x_i) = a \cdot x_i^b, \tag{4.13.20}$$

where $a > 0$ and b are constants that can be estimated from data. Set $R_i \equiv Y_i/x_i$. The estimator

$$\widehat{R} = \sum_{i=1}^{n} w_i R_i, \tag{4.13.21}$$

is unbiased for R given any set of weights $\{w_i\}$ whose sum is unity. The idea is to find these weights by minimizing $\text{Var}(\widehat{R})$ under the above model. The result is

$$w_i = x_i^{2-b} \Big/ \sum_{j=1}^{n} x_j^{2-b}. \tag{4.13.22}$$

Moreover, a UE of the variance is

$$\text{var}(\widehat{R}) = \frac{1}{n-1} \sum_{i=1}^{n} w_i (R_i - \widehat{R})^2. \tag{4.13.23}$$

The preceding model was checked as follows (Cruz-Orive, 1980a). From the lung parenchyma of a female suni (body weight = 3.6 kg, $V(1) = 210\,\text{cm}^3$), $n = 12$ IUR blocks of about 2 mm diameter were sampled, and one $2\,\mu$m thick section was taken per block for LM. On the ith section, $i = 1, 2, \ldots, 12$, seven non-overlapping square quadrats $\{T_{i1}, T_{i2}, \ldots, T_{i7}\}$ were observed of areas $x_j \equiv A(T_{ij}) = x_1 \cdot 2^{j-1}$, $x_1 = 0.011\,\text{mm}^2$, see Fig. 4.13.4(a). Thus, for each of the 7 quadrats we had 12 independent replications. The corresponding quantities $Y_{ij} \equiv A(\text{s} \cap T_{ij})$, or $Y_{ij} \equiv B(\text{alv} \cap T_{ij})$, were measured automatically with a Quantimet 720 (Cambridge Instruments). For the septum section areas, the data points $\{(x_j, Y_{ij})\}$ were well-fitted by weighted regression lines through the origin, see Fig. 4.13.4(b). This only confirms that an invariant volume process is a reasonable model for the lung septa, whereby Eq. (4.13.17) applies. The constant b in the rhs of the second Eq. (4.13.20) was estimated by the slope of the regression line fitted to the points

$$\{(\log(x_j), \log(\text{var}(Y_{ij}|x_j)), \quad i = 1, 2, \ldots, 12\}, \tag{4.13.24}$$

yielding $b = 1.47$ for the septa, and $b = 1.16$ for the alveoli. For a given specimen, the value of b may vary depending on the dimension of the probe.

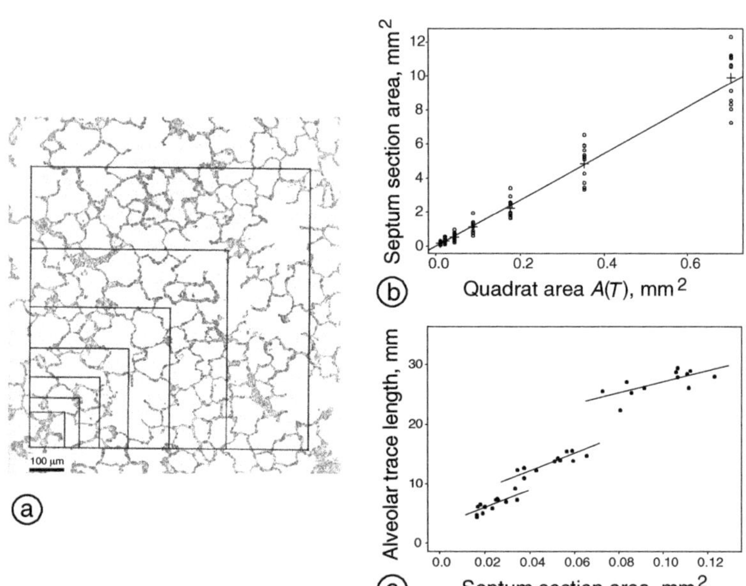

Fig. 4.13.4 Ratio estimation by a proportional regression model, Eq. (4.13.19). (a) Relative sizes of the quadrats used to check the model. (b) If the abscissa is controlled (quadrat area), then the data are compatible with the model. (c) If the abscissa is random (septum section area), then the model does not fit the data.

Model II

In this model the reference phase is no longer the parenchyma, but the septum (s) within the parenchyma, whereas the object phase is the alveolar surface (alv). Cruz-Orive (1980a) tried Eq. (4.13.19) also for this model, and noted that the pooled data points

$$\{(x_{ij} \equiv A(\mathrm{s} \cap T_{ij}), Y_{ij} \equiv B(\mathrm{alv} \cap T_{ij}))\} \tag{4.13.25}$$

were apparently well-fitted by a regression line through the origin. Here, however, the septum section areas $\{x_{ij}\}$ are random variables, whereby the proportional model need no longer hold. In fact, Jensen and Sundberg (1986b) studied the same data and showed that different quadrat areas lead to different regression lines which do not meet the origin, see Fig. 4.13.4(c).

Conclusions

Eq. (4.13.21)–(4.13.22) may be an option: (a) if the data are compatible with Model I, and (b) if the quadrats are of different sizes, (which is rarely the case). Besides, the sample sizes should be large enough to allow the estimation of the constant b from the data pairs in Eq. (4.13.24). The classical ratio-unbiased estimator $\widehat{R} = \sum_{i=1}^{n} Y_i / \sum_{i=1}^{n} x_i$ corresponds to Eq. (4.13.21) with $b = 1$, in which case Eq. (4.13.23) leads to a formula similar to Cochran's (Cruz-Orive, 1980a, Eq. (17)).

4.14 Neuron Number With the Optical Disector

4.14.1 Purpose, material, and method

The purpose is to estimate of the total number $N(\mathrm{neu})$ of calbindine stained neurons in the Dentate Gyrus (DG, which is adopted as the reference space, of volume $V(\mathrm{ref})$) from a single hemisphere of the hippocampus of a control mouse, see Fig. 4.2.1(a,b).

We consider two alternative estimation methods based on the ratio and the fractionator designs, respectively.

Ratio design

This design was discussed in Section 4.2.3, and it leads the following estimator,

$$\widehat{N}(\mathrm{neu}) = \widehat{V}(\mathrm{ref}) \cdot \widehat{N}_V(\mathrm{neu}, \mathrm{ref}). \tag{4.14.1}$$

The hippocampus hemisphere was exhaustively cut into serial coronal slices of $t = 50\,\mu$m thickness. Every fifth slice was stained with 0.25% thionin in order to exhibit the reference space properly, see Fig. 4.2.1(b), and was used to estimate $V(\text{ref})$ by the Cavalieri method with point counting using Eq. (4.8.2). An IUR square grid was superimposed on the upper face (that is, the uppermost optical section which is in focus) of each Cavalieri slice, see Fig. 4.2.1(c). Thus,

$$\widehat{V}(\text{ref}) = T \cdot \frac{a}{p} \cdot \sum P_i(\text{ref}), \tag{4.14.2}$$

where $T = 5 \cdot 50 = 250\,\mu\text{m}$ is the sampling period, a/p the area per test point at the specimen scale, and $P_i(\text{ref})$ denotes the number of test points hitting the reference space in the upper face of the ith slice. Here, and in the sequel, the summations are over Cavalieri slices.

Alternate slices of the Cavalieri series were stained with calbindine to exhibit the desired neuron subpopulation. UR systematic optical disectors were subsampled, see Fig. 4.2.1(d,e), with the aid of a CAST Grid system (Olympus Denmark). Based on Eq. (2.30.8), the ratio $N_V(\text{neu, ref})$ was estimated as follows,

$$\widehat{N}_V(\text{neu, ref}) = \frac{\sum Q_i^-(\text{neu})}{(a_0/p_0) \cdot \sum h_i P_i(\text{ref})}\ \text{mm}^{-3}, \tag{4.14.3}$$

where the denominator estimates the disector volume. Thus,

- a_0 denotes the area of each counting frame at the specimen scale, and p_0 the number of associated test points – here $p_0 = 1$, see Fig. 4.2.1(f).
- Q_i^-: Number of neurons counted in all the optical disectors hitting the ith calbindine slice. The theory is treated in Section 2.15.2. The nucleus of each neuron was adopted as the sampling unit. An optical section bearing an unbiased counting frame swept each slice by a fixed depth, see Fig. 4.2.1(e), and a neuron was counted if its nucleus was in focus and the corresponding optical transect was sampled by the frame according to the forbidden line rule, see Fig. 4.2.1(f).
- h_i: Optical disector thickness for the ith slice (Fig. 4.2.1(e)), corrected for tissue shrinkage. Recall that $t = 50\,\mu\text{m}$ was the original slice thickness, Let t_i' denote the thickness of the ith shrunken slice. The thickness of the optical disector within each shrunken slice was $h' = 10\,\mu\text{m}$. If we can assume that $h_i/t = h_i'/t_i'$, then $h_i = th'/t_i' = 500/t_i'$. For a discussion, see Section 4.14.4, Note 1.
- $P_i(\text{ref})$: Number of test points (sitting at the upper right corner of each unbiased frame, see Fig. 4.2.1(f)), hitting the reference domain (DG) in the ith section. Note that the relevant area of the intersection between all the frames and the reference domain in the ith section is estimated by $(a_0/p_0) \cdot P_i(\text{ref})$, and not by (a_0/p_0) times the number of frames hitting the domain, see Fig. 2.12.1(d) and Eq. (2.12.6).

Fractionator design

The target number $N(\text{neu})$ may also be estimated using the fractionator design. Let $N_i(\text{neu})$ represent the true number of neurons sampled by the ith Cavalieri slice disector, so that $N(\text{neu}) = \sum N_i(\text{neu})$. The fractionator estimator of $N_i(\text{neu})$ is $(T/h_i) \cdot (a_1/a_0) \cdot Q_i^-(\text{neu})$, where a_1 and a_0 denote respectively the fundamental tile area used for the disector subsampling and the area of the counting frame in that tile, see Fig. 4.2.1(d). Therefore $N(\text{neu})$ may be estimated by

$$\widehat{N}(\text{neu}) = T \cdot \frac{a_1}{a_0} \cdot \sum \frac{Q_i^-(\text{neu})}{h_i}. \tag{4.14.4}$$

The corrected disector thicknesses $\{h_i\}$ are random, whereby, apart from the aspects discussed in Note 2 from Section 4.14.4, the preceding estimator is not strictly unbiased in this case.

4.14.2 Results

Ratio design

To estimate the DG volume, a total of $n = 10$ Cavalieri slice faces of period $T = 0.250$ mm were subsampled with an IUR square grid with $a/p = 0.011525\,\text{mm}^2$ of area per test point at the specimen scale, yielding the following point counts in sequential order,

$$\{P_i(\text{ref})\} = \{4, 5, 10, 12, 9, 9, 8, 21, 28, 16\}. \tag{4.14.5}$$

Now Eq. (4.8.2) yields $\widehat{V}(\text{ref}) = 0.352\,\text{mm}^3$.

To estimate the number of calbindine neurons per unit reference volume by Eq. (4.14.3), nine out of the ten slices were suitable, yielding the following raw data,

$$\begin{aligned} &\{Q_i^-(\text{neu})\} = \{2, 0, 12, 24, 11, 21, 11, 41, 12\},\\ &\{P_i(\text{ref})\} = \{2, 1, 16, 23, 14, 24, 13, 36, 19\},\\ &\{h_i, \mu\text{m}\} = \{26.5, 26.8, 27.5, 24.7, 26.3, 23.8, 24.1, 24.7, 28.3\}. \end{aligned} \tag{4.14.6}$$

Each h_i was estimated as explained in Section 4.14.1 via $500/t_i'$, (μm). In turn, each t_i', (μm), was the average of three thickness measurements made at three different points of the upper plane with the microcator of the CAST Grid^{TM} system. The counting frame area per test point was $a_0/p_0 = a_0 = 0.001107\,\text{mm}^2$ at the specimen scale. Now Eq. (4.14.3) yields $\widehat{N}_V(\text{neu}, \text{ref}) = 32360\,\text{mm}^{-3}$, and by Eq. (4.14.1),

$$\widehat{N}(\text{neu}) = 0.352 \times 32360 = 11400. \tag{4.14.7}$$

Fractionator design

The area of the fundamental tile used to subsample the counting frames was $a_1 = 0.008348\,\mathrm{mm}^2$ at the specimen scale. Direct application of Eq. (4.14.4) yields

$$\begin{aligned}\widehat{N} &= 250 \times \frac{0.008348}{0.001107} \times \left(\frac{2}{26.5} + \frac{0}{26.8} + \cdots + \frac{12}{28.3}\right)\\ &= 10040. \end{aligned} \tag{4.14.8}$$

4.14.3 Error variance prediction

Ratio design

If $\widehat{N}(\mathrm{neu})$ is computed via Eq. (4.14.1), then its coefficient of error may be predicted by Goodman's formula, see Eq. (A.2.16).

First we compute $\mathrm{ce}^2(\widehat{V}(\mathrm{ref}))$ via Eq. (5.3.10), assuming zero section thickness because section areas, and no slice volumes, were measured. The shape factor for the DG sections was $\bar{B}/\sqrt{\bar{A}} = 5.0$. The smoothness constant was computed via Eq. (5.2.15) to yield $q = 0.6890$, whereby Eq. (5.2.13) returns $\alpha(0.6890) = 0.0125$. With this, and with the point counts given in Eq. (4.14.5), we get $\mathrm{ce}(\widehat{V}(\mathrm{ref}))\% = 3.5\%$.

To predict the coefficient of error of $\widehat{N}_V(\mathrm{neu}, \mathrm{ref})$ we may apply Cochran's formula to the data vectors $\{Q_i^-\}$ and $\{h_i P_i(\mathrm{ref})\}$, to obtain $\mathrm{ce}(\widehat{N}_V(\mathrm{neu}, \mathrm{ref}))\% = 10.1\%$.

Finally, Goodman's formula yields $\mathrm{ce}(\widehat{N}(\mathrm{neu}))\% = 10.6\%$.

Fractionator design

For practical purposes it suffices to adopt the Poisson model for the total number $Q^- = \sum Q_i^-(\mathrm{neu}) = 134$ of disector sampled neurons, whereby

$$\mathrm{ce}(\widehat{N}(\mathrm{neu}))\% = \frac{100}{\sqrt{Q^-}} = \frac{100}{\sqrt{134}} = 8.6\%. \tag{4.14.9}$$

For a derivation of the preceding formula, see Note 3 from Section 4.14.4 below.

4.14.4 Notes

1. The optical disector: advantages and limitations

a. Strictly, the optical disector is more a practical than a mathematical tool. The unambiguous identification and counting of the sampling units depends on the quality of focusing, and this in turn requires that the shape of the units is friendly for the purpose – cell nuclei and nucleoli are usually convex, in which case the optical disector can be an efficient tool.
b. The primary Cavalieri slices may undergo deformations during tissue processing – for a review, see Dorph-Petersen, Nyengaard, and Gundersen (2001). In order to avoid artifacts near the slice faces, the optical disector is usually placed near the midplane of the slice (Fig. 4.2.1(e)). If no cell number gradients are present within the slab, then one is implicitly assuming stationarity under the model-based approach (Chapter 3). If gradients are present, however, then unbiasedness could only be preserved – but hardly implemented – with Cavalieri optical disectors, as shown in Fig. 6b from Cruz-Orive and Geiser (2004).
c. Another important snag is the assessment of the effective optical disector thickness h under slice shrinkage.

2. Ratio versus fractionator designs

The fractionator has the advantage of not requiring the reference volume, but slice shrinkage unfortunately imposes the need to measure the optical disector thickness h. On the other hand, the fractionator requires that no neuron count is missed – it is necessary to keep a careful control of every disector hitting the reference space (including the grazing ones). Failure to achieve this may have a lesser impact on the estimation of $N_V(\text{neu}, \text{ref})$ because ratio estimators tend to be fairly stable.

3. Fractionator precision under the Poisson model

Consider a Poisson point process of intensity N_V in space. The number Q^- of Poisson points in a compartment of volume V has a Poisson distribution with

$$\mathbb{E}(Q^-) = \text{Var}(Q^-) = N_V \cdot V, \tag{4.14.10}$$

see Section 3.9.3. Thus,

$$\text{CV}^2(Q^-) = \frac{1}{\mathbb{E}(Q^-)}. \tag{4.14.11}$$

If the $\{h_i\}$ in Eq. (4.14.4) are constant, then $\widehat{N}(\text{neu}) = c \cdot Q^-$, where c is a constant. Thus, $\text{CE}\{\widehat{N}(\text{neu})\} = \text{CV}(Q^-)$, and its estimator, based on a total count of Q^- neurons in the compartment, is

$$\text{ce}(\widehat{N}(\text{neu}))\% = \frac{100}{\sqrt{Q^-}}, \qquad (4.14.12)$$

which is Eq. (4.14.9).

4.15 Connectivity of Trabecular Bone

4.15.1 Purpose, material, and method

The problem is to estimate $\chi_V(\text{trab}, \text{ref})$, namely the Euler–Poincaré characteristic (EPC) of trabeculae ('trab') per unit volume in a small block (in fact, a necropsy) from the iliac crest bone of a human subject. Rather than the entire bone, the reference space ('ref') is the compartment where the trabeculae sit, see Fig. 4.15.1(a). The location of the block was not UR in the iliac crest, but fixed in advance, and the target was the EPC density (mm^{-3}), and not a total number. Thus, we may adopt the model-based approach in which the target is the intensity χ_V of a stationary process $Y \subset \mathbb{R}^3$ of trabeculae. Isotropy is not necessary (Section 1.10.4, Note 3). Consider a box probe $T \subset \mathbb{R}^3$ of volume $V(T)$. Then, by Eq. (3.2.3),

$$\chi_V = \frac{\chi^-(Y \cap T)}{V(T)}, \qquad (4.15.1)$$

where $\chi^-(Y \cap T)$ could in principle be estimated by the three-dimensional shell method (Section 2.17); however, in this case the procedure may be simplified as follows. The tissue block was a cylinder core traversing a thin bone plate from the iliac crest. After methacrylate embedding, a series of $t = 10\,\mu$m thick slices were cut parallel to the cylinder axis. For some i, Fig. 4.15.2(a,b) shows the ith and the $(i+3)$th slices of such series, constituting a modification of a bounded disector probe T of $h = 3t = 30\,\mu$m thickness. In the picture, the core axis is horizontal, and the disector has natural bone boundaries on both sides, which are replaced with straight line segments to simplify the analysis. The reference face area $A(\text{ref})$ confined between the two line segments is estimated by counting, say, $P(\text{ref})$ test points with a test system of area a/p per test point. The disector volume is then estimated by

$$\widehat{V}(T) = h \cdot (a/p) \cdot P(\text{ref}), \qquad (4.15.2)$$

whereby Eq. (4.15.1), in combination with Eq. (1.12.7), or Eq. (2.17.1), yields the following ratio-unbiased estimator for a single disector,

$$\widehat{\chi}(\text{trab, ref}) = \frac{(1/2)(I - B + H)}{h \cdot (a/p) \cdot P(\text{ref})}. \tag{4.15.3}$$

The factor 1/2 stems from the fact the double disector rule is adopted: in Fig. 4.15.2(c) the two slice projections in Fig. 4.15.2(a,b) are perfectly registered and superimposed, whereby each of the relevant events (labelled I, B, H) appear either in green, or in red, this facilitating their identification. In the picture we observe $B = 18, I = H = 2$.

The EPC is closely related to a conventional definition of trabecular number, which is more intuitive. In an entire bone, the trabeculae constitute a connected, rigid network without enclosed cavities. The latter are rare and may be ignored for the present purpose, whereby Eq. (1.12.6) reduces to

$$\chi(\text{trab}) = 1 - N(\text{extra connections}). \tag{4.15.4}$$

As illustrated in Fig. 4.15.1(b), a conventional definition of the number of trabeculae is

$$N(\text{trab}) = 1 + N(\text{extra connections}) = 2 - \chi(\text{trab}). \tag{4.15.5}$$

For the number density the term '2' may be omitted, whereby

$$N_V(\text{trab, ref}) = -\chi_V(\text{trab, ref}). \tag{4.15.6}$$

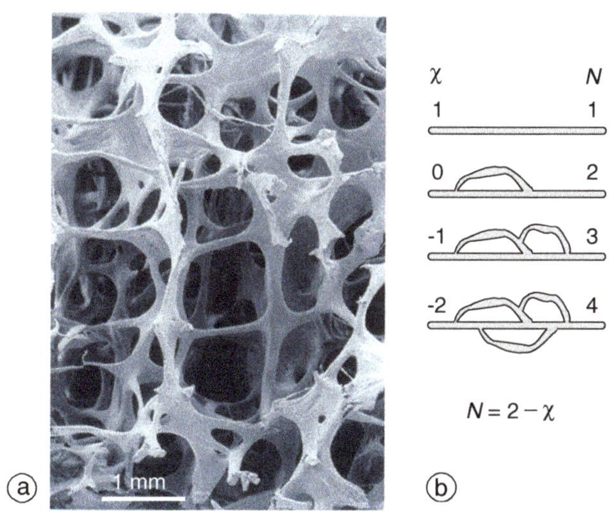

Fig. 4.15.1 (a) SEM image of a block of human trabecular bone. Plain micrograph courtesy of Lis Mosekilde, modified from Mosekilde (1990), with permission of Elsevier. (b) The number N of trabeculae is directly related to the Euler–Poincaré characteristic χ.

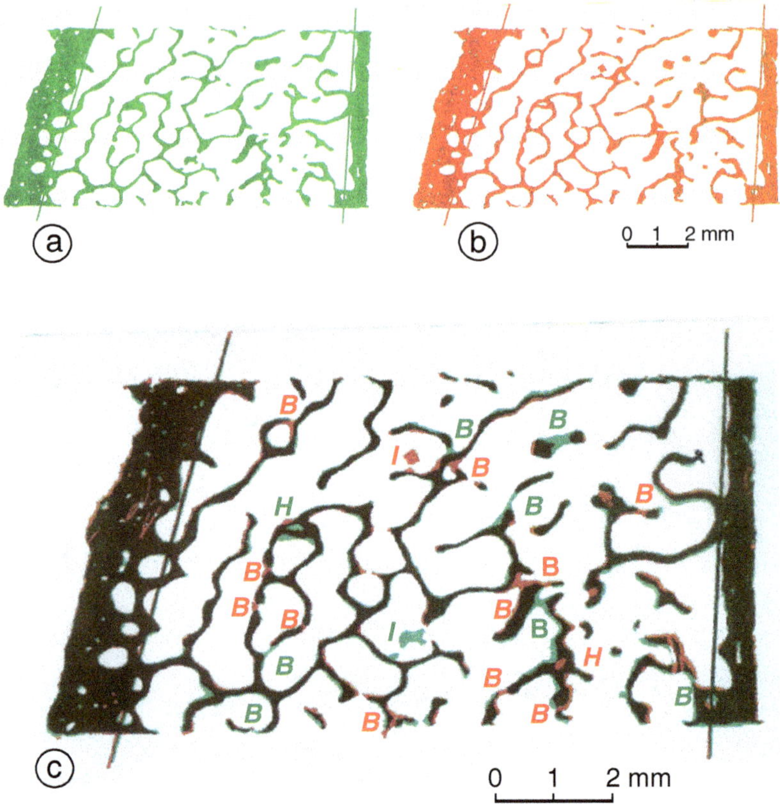

Fig. 4.15.2 (a,b) A pair of Cavalieri slices of a human iliac bone necropsy, to constitute a disector. (c) After superposition, the relevant events I, B, H involved in Eq. (4.15.3) can be identified.

4.15.2 Results

When analyzing a biopsy, or a necropsy, for diagnostic purposes, one section, or disector, may suffice. For didactic purposes, in this example $n = 4$ disectors were sampled every 400 μm along the slice series. The disector illustrated in Fig. 4.15.2 was the first one. The corresponding counts were:

$$\{(I_i, B_i, H_i)\} = \{(2, 18, 2), (1, 18, 2), (0, 16, 2), (1, 9, 1)\}. \tag{4.15.7}$$

To estimate $A(\text{ref})$, a test system of points (with a square fundamental tile of 28.3 mm side length containing a test point) was superimposed uniformly at random on the first face of each disector. The final magnification was $M = 15.3$, whereby $a/p = 3.42\,\text{mm}^2$ at the specimen scale. The total numbers of test points counted in the reference space were $\{P_i(\text{ref})\} = \{15, 20, 19, 18\}$, respectively. With these data the EPC density estimate was

$$\widehat{\chi}_V(\text{trab, ref}) = \frac{(1/2)\sum_{i=1}^{4}(I_i - B_i + H_i)}{h \cdot (a/p) \cdot \sum_{i=1}^{4} P_i(\text{ref})} = -3.38\,\text{mm}^{-3}, \tag{4.15.8}$$

and by Eq. (4.15.6), $N_V(\text{trab, ref}) = 3.38\,\text{trab/mm}^3$.

4.15.3 Error variance prediction

If we assume independence between disectors, then Cochran's Eq. (A.2.13) may be used with $Y_i \equiv |I_i - B_i + H_i|$, and $X_i \equiv P_i(\text{ref})$, $i = 1, 2, \ldots, n$, with $n = 4$. We obtain $\text{ce}(\widehat{N}_V(\text{trab, ref}))\% = 15.4\%$.

Alternatively, we may ignore the variation of $\widehat{V}(T)$ and use the fact that the disectors were Cavalieri with a period of 400 μm. Then we may apply Eq. (5.5.12) with $n = 2$ splitting subsets. The formula uses the net counts $\{Y_i\} = \{14, 15, 14, 7\}$ defined above. The two subset totals are in this case the odd and the even numbered net counts, namely $Q_1 = 14+14 = 28$ and $Q_2 = 15+7 = 22$, respectively. Further, the within error contribution may be estimated via the Poisson model, see Eq. (4.14.12), with the absolute counts $\{22, 21, 18, 11\}$, namely 72 in total. Finally, the required sampling fraction for the Cavalieri slabs is $\tau = 30/400$. We obtain,

$$\begin{aligned}\text{ce}^2(\widehat{N}_V(\text{trab, ref})) &= \frac{(11 - 3/40)^2}{3 - 6/40}\left[\left(\frac{28 - 22}{28 + 22}\right)^2 - \frac{1}{72}\right] + \frac{1}{72} \\ &= 0.119^2,\end{aligned} \tag{4.15.9}$$

namely $\text{ce}(\widehat{N}_V(\text{trab, ref}))\% = 11.9\%$, probably more realistic than Cochran's (15.4%) because the splitting formula takes the dependence among Cavalieri slices into account. Note that $\text{ce}^2(\widehat{\gamma})$ is obtained by dividing the rhs of Eq. (5.5.12) by the square of the rhs of Eq. (5.5.11).

4.15.4 Note

The example is based on an exercise prepared by Thomas Youngs and Hans J. G. Gundersen for the 1st Canadian-ISS Stereology Course given at The Banff Centre, Alberta, Canada in 1994. The material comes from the study of Youngs et al. (1994), kindly offered by the authors, and the method, called the conneulor, is due to Gundersen et al. (1993). Roberts, Reed, and Nesbitt (1997) apply the conneulor using a superposition technique analogous to that of Fig. 4.15.2 on digitized images of a porous medium. Wulfsohn, Knust, Ochs, Nyengaard, and Gundersen (2010) apply the method to count ventilatory units in mice lung.

4.16 Surface Area From Vertical Cavalieri Sections

4.16.1 Purpose, material, and method

This example complements the one illustrated in Section 4.8 (concerning the volumes V_0 and V of a banana Y_0, and of its pulp Y, respectively) with the estimation of the surface areas $S_0 \equiv S(\partial Y_0)$ and $S \equiv S(\partial Y)$ of the boundaries ∂Y_0 and ∂Y of banana and pulp, respectively.

As a plus, we also estimate the mean thickness $\mathbb{E}_S(\tau)$ of the peel, namely of the set difference $Y_0 \backslash Y$, which, similarly as in Eq. (4.13.1), is defined as follows,

$$\mathbb{E}_S(\tau) = \frac{V_0 - V}{\frac{1}{2}(S_0 + S)}. \tag{4.16.1}$$

The design used to estimate S_0 and S is that of vertical Cavalieri (VCav) sections for an isolated object, described in Section 2.28.2. The 5 blocks produced by the Cavalieri cuts (Fig. 4.8.1(b)) were rotated systematically with a period of $180°/5$, namely at angles

$$\{\phi_i = (U + i - 1) \cdot 36°,\ i = 1, 2, \ldots, 5\}, \quad U \sim \mathrm{UR}[0, 1), \tag{4.16.2}$$

(see Fig. 4.16.1(a) for the second block) and laid flat on the table (which was adopted as the horizontal plane), and stabilized on plastiline, see Fig. 4.16.1(b). Then, by a similar procedure as that illustrated in Fig. 4.8.1(a), Cavalieri sections a distance $T = 2$ cm apart, perpendicular to the table, were cut, which, in fact, constituted $n = 9$ vertical sections (Fig. 4.16.1(c)) – their isotropic orientation around the vertical, and their uniform location, being warranted by the angles $\{\phi_i\}$ and by the Cavalieri design, respectively.

On each vertical section, a cycloid test system was superimposed uniformly at random with the minor axes of the cycloids parallel to the vertical axis (Fig. 4.16.1(e)). The fundamental tile is shown in Fig. 4.16.1(d). For S, let I_i denote the number of intersections between the trace of $\partial Y$and the cycloids counted on the ith non-empty vertical section. Then, the corresponding estimator is based on Eq. (2.28.4), namely,

$$\widehat{S} = 2 \cdot T \cdot \frac{a}{l} \cdot \sum_{i=1}^{n} I_i, \tag{4.16.3}$$

and similarly for $\widehat{S_0}$.

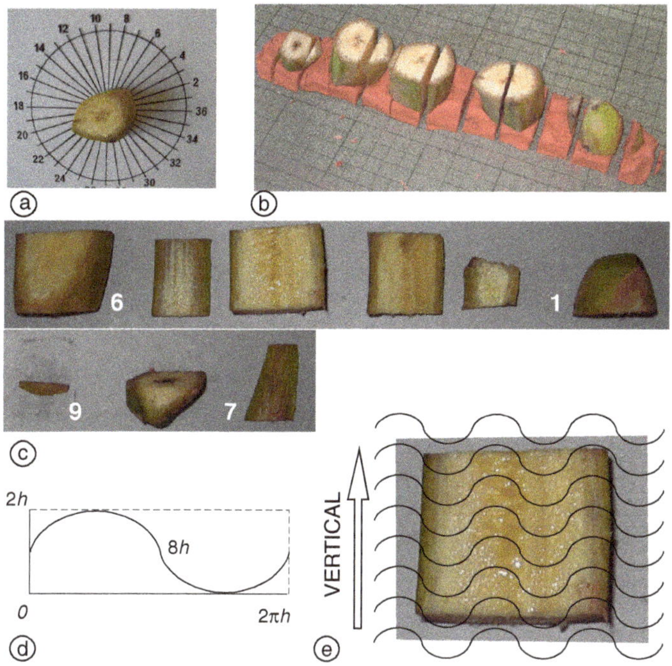

Fig. 4.16.1 Illustration of the vertical design to estimate the surface area of an isolated object using VCav sections and superimposing a UR vertical cycloid test system on each section, see Section 4.16. Modified from Sagaseta (2005).

4.16.2 Results

We apply Eq. (4.16.3) with Cavalieri period $T = 2\,\text{cm}$ (which was aimed at obtaining about 8–10 non-empty sections). Further, from Fig. 4.16.1(d) we have $a/l = 4\pi h^2/(8h) = \pi h/2$, where $2h = 0.49\,\text{cm}$ at the specimen scale. Thus, for this example,

$$\widehat{S} = \pi T h \sum_{i=1}^{n} I_i, \ (\text{cm}^2). \tag{4.16.4}$$

For the banana surface, the ordered intersection counts on the $n = 9$ vertical sections shown in Fig. 4.16.1(c) were $\{I_i\} = \{8, 7, 12, 11, 10, 10, 11, 7, 2\}$. For the pulp surface, $\{I_i\} = \{4, 7, 11, 12, 4, 10, 2, 4, 0\}$. Now Eq. (4.16.4) yields

$$\begin{aligned} \widehat{S}_0 &= 0.49\pi \cdot 78 = 120\,\text{cm}^2, \\ \widehat{S} &= 0.49\pi \cdot 54 = 83\,\text{cm}^2. \end{aligned} \tag{4.16.5}$$

Substitution of the preceding results, and those in Eq. (4.8.3), into Eq. (4.16.1), yields the mean peel thickness,

$$\bar{\tau} = \frac{95 - 52}{\frac{1}{2}(120 + 83)} = 0.42\,\text{cm}, \tag{4.16.6}$$

namely 4.2 mm, which sounds realistic.

4.16.3 Error variance prediction

To predict $\text{Var}(\widehat{S})$ we try Eq. (5.2.10), which includes neither the orientations, nor the intersections contributions. Another negative aspect is that, in the Cavalieri series shown in Fig. 4.16.1(b) there are two empty sections. This is of no consequence to estimate S_0 and S, but it has to be taken into account to predict their error variances by Matheron's formula. For the banana surface, the ordered data are $\{I_i\} = \{8, 7, 0, 12, 0, 11, 10, 10, 11, 7, 2\}$ which, with the smoothness constant $q = 0$ (because, unlike the banana itself, the union of the blocks is not smooth), namely with $\alpha(0) = 1/12$, Eq. (5.2.10) yields $\text{ce}(\widehat{S_0})\% = 11.2\%$. For the pulp surface, the corresponding data are $\{I_i\} = \{4, 7, 0, 11, 0, 12, 4, 10, 2, 4, 0\}$, and $\text{ce}(\widehat{S})\% = 18.5\%$. Thus, the error predictions are unduly large, and it is therefore important to avoid empty sections by reducing the space between the blocks as much as possible before cutting. The problem is basically open.

4.16.4 Notes

1. Vertical design with the ratio method

The design illustrated in Fig. 4.16.1(b) follows the example in Fig. 9 from Baddeley et al. (1986). The design adopted there, however, used the ratio method, which is also a reasonable alternative. In the preceding notation,

$$S_0 = V_0 \cdot \frac{S_0}{V_0} \equiv V_0 \cdot S_V(\partial Y_0, Y_0), \tag{4.16.7}$$

and the ratio was estimated with the analogue of Eq. (4.13.5) using the test system of Fig. 4.19.1(f), which bears test points in addition to cycloids. To estimate $\text{CE}(\widehat{S_V})$, Gundersen and Jensen (1987, Section 7) adapted Cochran's formula for a ratio of Cavalieri estimators, see also Pache et al. (1993, Section 6.3), and Section A.2.4.

2. Direct estimation of surface area

The direct estimator of surface area by the VCav design, see Eq. (4.16.3), was proposed by Gundersen (1986, Eq. 2.8), who hinted that the problem of predicting the corresponding error variance is basically open. See also Gundersen and Jensen (1987) and Michel and Cruz-Orive (1988).

4.17 Cortical Surface Area and Mean Thickness From Digitized Vertical Sections of a Human Brain

4.17.1 Purpose, material, and methods

High resolution, three-dimensional (3D) MR images of a human brain were obtained for a healthy female volunteer. The pixel side of a cubic voxel was 1 mm. For the union of both hemispheres, the target structures were the outer (pial) surface ∂Y_0, and the inner (subcortical) surface ∂Y, which is the boundary between the cortex and the underlying white matter. The purpose is to estimate the corresponding surface areas S_0 and S, plus the volume and mean thickness of the cortex. Similarly as in Eq. (4.16.1), the latter thickness is defined as follows,

$$\mathbb{E}_S(\tau(\text{cortex})) = \frac{V(\text{cortex})}{\frac{1}{2}(S_0 + S)}. \qquad (4.17.1)$$

The VCav design is adopted, see Fig. 4.17.1. With the aid of FreeSurfer software, the 3D MR data yielded triangulated representations of the surfaces ∂Y_0 and ∂Y. The former, for instance, consisted of 467426 triangles of a maximal side length of 5.899 mm. The corresponding total areas, which will be regarded as true areas for checking purposes, were $S_0 = 1863\,\text{cm}^2$ and $S = 1529\,\text{cm}^2$. Besides, the true cortical volume was $V(\text{cortex}) = 432.5\,\text{cm}^3$.

The intersection between a vertical plane and a 3D triangulation was a polygonal curve trace obtained with the aid of StereoTool, a software developed by Julius Gelšvartas. Each trace consisted of straight links with known end point coordinates. In addition, StereoTool enabled the automatic identification of the intersection points between a polygonal trace and a test curve, or a test system of curves such as the cycloids used to analyze the vertical sections.

The relevant estimator based on a UR cycloid test system superimposed on each VCav section is given by Eq. (4.16.3). If digitized polygonal traces are available, however, then an alternative estimator, called the W-method, is possible based on Eq. (2.28.2), namely,

$$\widehat{S} = T \sum_{i=1}^{n} \widehat{W}_i, \qquad (4.17.2)$$

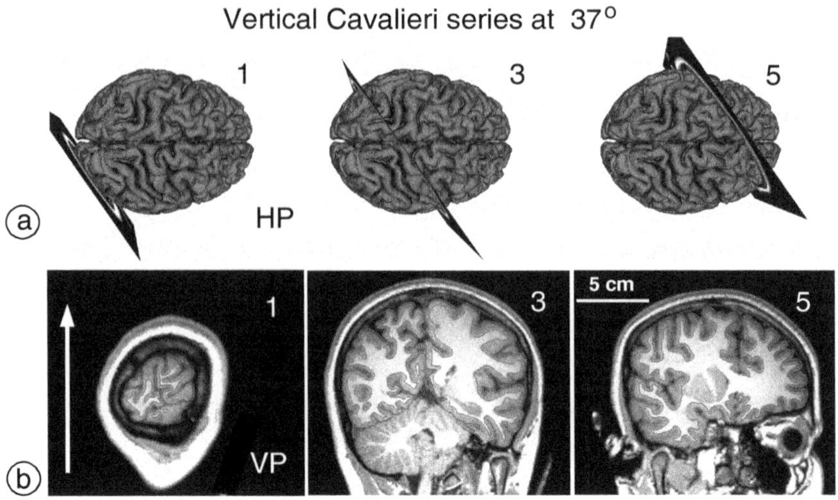

Fig. 4.17.1 (a) VCav sections (a distance $T = 5$ cm apart) of a human brain, following the design in Fig. 2.28.1(a). (b) Corresponding digitized sections, with the pial surface traces in red and the subcortical traces in blue. Modified from Cruz-Orive et al. (2014), with permission of Wiley-Blackwell.

where $\widehat{W}_i$ is a discretized approximation of the functional given by Eq. (1.6.4) for the ith VCav section, and n is the number of sections. The polygonal trace in the ith section will consist of, say, m_i straight line links $\{y_{ij},\ i = 1, 2, \ldots, n,\ j = 1, 2, \ldots, m_i\}$. Let h_{ij}, v_{ij} denote the unsigned horizontal and vertical orthogonal projected lengths, respectively, of the link y_{ij}. Then, $b_{ij} = \left(h_{ij}^2 + v_{ij}^2\right)^{1/2}$ is the length of the link, and $\psi_{ij} = \cos^{-1}(v_{ij}/b_{ij}) \in [0, \pi/2)$ is the angle between the vertical axis and the link, see Fig. 1.6.1(c), whereby,

$$\widehat{W}_i = \sum_{j=1}^{m_i} [\sin(\psi_{ij}) + (\pi/2 - \psi_{ij})\cos(\psi_{ij})] \cdot b_{ij}. \tag{4.17.3}$$

Note that the preceding expression is unchanged if the link projected lengths are signed and $\psi_{ij} \in [0, \pi)$.

4.17.2 Results

VCav sections analyzed with cycloids

In the original paper, see Section 4.17.4 below, two vertical, mutually perpendicular Cavalieri series were generated with a period of 2.5 cm. Here we only consider the odd numbered sections of the first series, see Fig. 4.17.1, whereby the period is $T = 5$ cm. The traces of the target surfaces in each vertical section were digitized as explained above, and a UR cycloid test system was superimposed automatically with the minor axes parallel to the vertical axis (arrow in Fig. 4.17.2(a)). With reference to Fig. 4.16.1(d), in the original paper the relevant constant of the test system was $a/l = \pi h/2$, where $2h = 0.70$ cm at the brain scale. For clarity, the test system in Fig. 4.17.2(a) is coarser ($2h = 1.50$ cm).

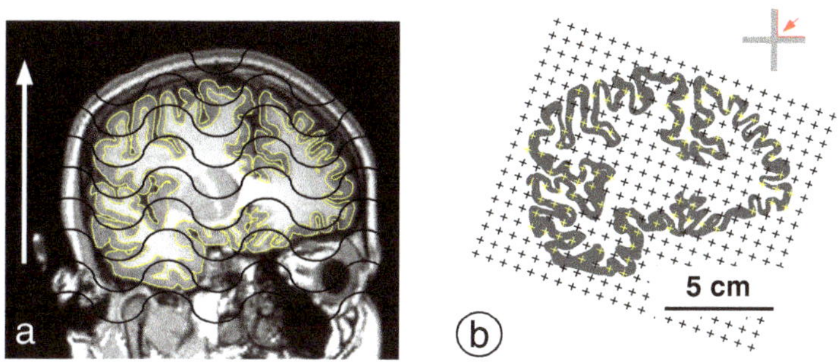

Fig. 4.17.2 (a) Digitized VCav section number 5 with the target surface traces approximated by polygonal curves (in yellow). A digital test system of cycloids is superimposed to estimate the corresponding surface areas by automatic intersection counting. (b) Digitized cortical trace with a grid to estimate cortical volume by point counting. Modified from Cruz-Orive et al. (2014), with permission of Wiley-Blackwell.

For the pial surface ∂Y_0, the automatically scored intersection numbers were $\{I_{0i}\} = \{54, 131, 156\}$. For the subcortical surface ∂Y, the corresponding counts were $\{I_i\} = \{24, 137, 114\}$. With $T = 5$ cm and $h = 0.35$ cm, Eq. (4.16.4) yields

$$\widehat{S_0} = 1.75\pi \cdot 341 = 1875\,\text{cm}^2,$$
$$\widehat{S} = 1.75\pi \cdot 275 = 1512\,\text{cm}^2. \tag{4.17.4}$$

Fortuitously, the preceding estimates are closer to the corresponding true values, see Section 4.17.1, than the estimates reported in the original study with 6 sections.

Cortex volume by point counting, and mean cortical thickness

The cortical volume was estimated by point counting via Eq. (4.8.2) with an IUR square lattice of 0.70 cm side length, whereby the test area per test point was $a = 0.49\,\text{cm}^2$ at the brain scale, see Fig. 4.17.2(b). The isotropy of the test lattice allowed the prediction of the error variance due to point counting. The number of test points hitting the cortex in the three VCav sections were $\{P_i\} = \{18, 77, 66\}$ respectively, whereby,

$$\widehat{V}(\text{cortex}) = T \cdot a \cdot \sum_{i=1}^{n} P_i = 5 \cdot 0.49 \cdot 161 = 394.5\,\text{cm}^3. \tag{4.17.5}$$

With the preceding results, Eq. (4.17.1) yields the following estimate of the mean cortical thickness,

$$\bar{\tau}(\text{cortex}) = \frac{\widehat{V}}{\frac{1}{2}(\widehat{S}_0 + \widehat{S})} = \frac{394.5}{\frac{1}{2}(1875 + 1512)} = 0.233\,\text{cm}, \tag{4.17.6}$$

As reported in the original paper, with pixel counting we obtained $V(\text{cortex}) = 432.5\,\text{cm}^3$ which, combined with the true surface areas, (Section 4.17.1), yielded $\bar{\tau}(\text{cortex}) = 0.258\,\text{cm}$.

VCav sections analyzed by the W-method

The relevant approximations calculated automatically for each VCav section via Eq. (4.17.3) were $\{\widehat{W}_i\} = \{55.8, 151.1, 149.3\}$ (cm) for the pial surface, and $\{\widehat{W}_i\} = \{33.8, 133.2, 131.4\}$ (cm) for the subcortical surface, respectively. Thus, Eq. (4.17.2) yields,

$$\begin{aligned} \widehat{S}_0 &= 5 \cdot 356.2 = 1781\,\text{cm}^2, \\ \widehat{S} &= 4 \cdot 298.4 = 1492\,\text{cm}^2. \end{aligned} \tag{4.17.7}$$

4.17.3 Error variance predictions

VCav sections analyzed with cycloids

The coefficient of error of the estimators $\widehat{S}_0$ and $\widehat{S}$, conditional on the given orientation of the vertical sections, and excluding the contribution due to the cycloid intersection counting, may be predicted via Eq. (5.2.10) with a common smoothness constant $q = 0.36$, as established in the original paper, Table 3. Thus $\alpha(0.36) = 0.0324$, and we obtain $\text{ce}(\widehat{S}_0)\% = 9.4\%$ and $\text{ce}(\widehat{S})\% = 10.2\%$.

Cortex volume by point counting, and mean cortical thickness

For the coefficient of error (conditional on the given orientation) of the estimator $\widehat{V} \equiv \widehat{V}(\text{cortex})$, we use Eq. (5.3.10). We adopt $q = 0.62$, (original paper, Table 6), whereby $\alpha(0.62) = 0.0155$. For the point counting contribution we need the individual cortex boundary length of each section (also given in the latter table), namely $\{70.6, 223.5, 218.9\}$ (cm), and the section area estimates $\{8.82, 37.73, 32.34\}$ (cm^2) obtained from the $\{P_i\}$ given to compute Eq. (4.17.5). Thus, $\varphi = \bar{B}/\sqrt{\bar{A}} = 33.3$, and we thereby obtain $\text{ce}(\widehat{V})\% = 7.9\%$. Sections contribute 67% to the total variance, and point counting the remaining 33%.

VCav sections analyzed by the W-method

For $\widehat{S_0}$ and $\widehat{S}$ we use Eq. (5.2.10) with $\alpha(0.45, 0) = 0.0253$ and $\alpha(0.41, 0) = 0.0283$, respectively (original paper, Table 1). We obtain $\text{ce}(\widehat{S_0})\% = 7.6\%$ and $\text{ce}(\widehat{S})\% = 8.9\%$.

4.17.4 Note

The material used in this exercise comes from Cruz-Orive, Gelšvartas, and Roberts (2014). The design was anticipated in Fernández-Viadero, González-Mandly, Verduga, Crespo, and Cruz–Orive (2008). In Section 4.3 of the former paper, Eq. 2.26 from Cruz-Orive (1993) was used to predict the contribution of intersections with cycloids to the total error variance of the surface area estimators.

4.18 Curve Length From Vertical Projections

4.18.1 Purpose, material, and method

The purpose is to estimate the length L of a twisted wire from total vertical projections (TVPs). The wire was measured before bending it, yielding $L = 90\,\text{cm}$.

The estimation method is based on Eq. (2.28.8). The twisted wire was fixed by introducing one of its ends in the upper face a cylindrical cork, whose axis was adopted as the vertical axis, see Fig. 4.18.1. The cork was then rotated at $n \in \mathbb{N}$ systematic angles around the vertical direction with period $T = 180°/n$, namely,

$$\{\phi_i = (U + i - 1) \cdot T,\ i + 1, 2, \ldots, n\},\ U \sim \text{UR}[0, 1). \tag{4.18.1}$$

A cycloid test system was superimposed uniformly at random onto each TVP with the major axes parallel to the vertical axis, and the corresponding total numbers $\{I_1, I_2, \ldots, I_n\}$ of intersections were counted. Because the wire projection and the cycloids have a thickness, an intersection is scored if the right-hand side border of a cycloid hits the axis or 'spine' of the wire projection. This was not difficult to decide in the present example. However, for some real features such as neuron dendrites, plant roots, etc., the definition of the spine may be ambiguous, in which case we recommend scoring intersections with both borders of the feature projection and divide the total by 2.

The estimator

$$\widehat{L} = 2 \cdot \frac{a}{l} \cdot \frac{1}{M} \cdot \frac{1}{n} \cdot \sum_{i=1}^{n} I_i \qquad (4.18.2)$$

is unbiased for L, where M is the final magnification of the TVPs, and a/l the test area per test length of the cycloid test system (here the adopted fundamental tile was that in Fig. 4.16.1(d)).

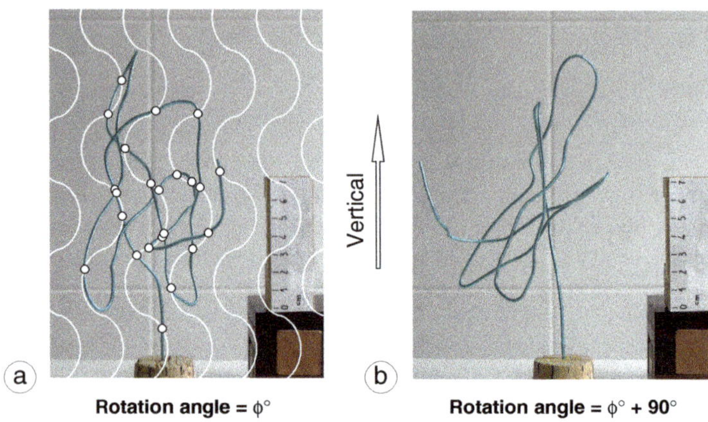

Fig. 4.18.1 (a) Total vertical projection (TVP) of a twisted wire by an angle $\phi \sim \text{UR}[0, 90°)$, with a UR cycloid test system superimposed on it to estimate the length of the 3D wire by intersection counting using Eq. (4.18.3). (b) The companion TVP.

4.18.2 Results

Here we took $n = 2$ TVPs, namely $T = 90°$. On the working images the length of the accompanying 70 mm ruler was 75 mm, whereby $M = 75/70 = 15/14$. For the cycloid test system used, the relevant constant is $a/l = \pi h/2$, namely half the length of the major axis of a cycloid arc. The total length of a chain of 6 major cycloid axes was 24.0 cm, hence $a/l = 24.0/12 = 2.0$ cm. The two intersection counts were $I_1 = 26$ and $I_2 = 21$. With these data, Eq. (4.18.2) yields

$$\widehat{L} = 2 \cdot \frac{a}{l} \cdot \frac{1}{M} \cdot \frac{1}{2} \cdot (I_1 + I_2) = 88.0\,\text{cm}. \qquad (4.18.3)$$

(A funny comment could be that the 'missing' 2 cm were inside the cork...).

4.18.3 Error variance prediction

The estimator $\widehat{L}$ is proportional to the mean $\bar{I} = (1/2)(I_1 + I_2)$, and I_1, I_2 are two systematic observations on the semicircle $[0, \pi)$. We may estimate $\text{CE}(\widehat{L})$ via Eq. (5.7.5) with $\widehat{v}_2 = 0$, that is, ignoring the contribution of the intersection counts within TVPs. Thus,

$$\text{ce}(\widehat{L})\% = \frac{100}{\sqrt{6}} \cdot \frac{|I_1 - I_2|}{I_1 + I_2} = 4.3\%. \tag{4.18.4}$$

For a discussion of the preceding formula, see Section 5.7.3, Note 1.

4.18.4 Notes

1. Sources

The example is adapted from Sagaseta (2005). It was first demonstrated at the 20th International Biometric Conference, University of California at Berkeley, WA, USA, in July 2000, see Cruz-Orive (2000). For the original reference on the method, see Section 1.19.7, Note 3, and for early applications, see Section 2.28.6, Note 2.

2. Capillary length from vertical slice projections

Batra et al. (1995) estimated total capillary length in the left ventricle of a rat heart. A ratio design was used based on the following identity,

$$L(\text{cap}) = V(\text{ref}) \cdot L_V(\text{cap}, \text{ref}), \tag{4.18.5}$$

or $L = V \cdot L_V$, for short. The ventricle volume V was estimated first as its weight to specific gravity ratio, adopting for the latter the value of $1.06\,\text{g/cm}^3$. On the other hand, the estimator of the ratio L_V of capillary length to ventricle volume is due to Gokhale (1990), see Section 2.28.4. Vertical Cavalieri slices of a convenient period T (which does not enter into the ratio estimator) and thickness $t < T$ were cut by a design analogous to that of Fig. 4.16.1 and, on the orthogonal projection of each slice onto the observation plane, a cycloid test system was superimposed uniformly at random with the major axes parallel to the vertical axis. The fundamental tile bore p/l test points per cycloid test length at working magnification M. If t is given at the specimen scale, the estimator is

$$\widehat{L}_V = \frac{2}{t} \cdot \frac{p}{l} \cdot M \cdot \sum_{i=1}^{n} I_i \Big/ \sum_{i=1}^{n} P_i, \tag{4.18.6}$$

(Gokhale, 1990), where I_i is half the number of intersections between the cycloids and both borders of the capillary projections in the ith section, P_i the corresponding number of test points hitting the heart ventricle, and n the number of vertical slices. Because V was measured with a negligible error, here $\mathrm{CE}^2(\widehat{L}) \approx \mathrm{CE}^2(\widehat{L}_V)$, which may be predicted by Cochran's formula, Eq. (A.2.13).

A similar design was implemented by Insausti, Cruz-Orive, Jáuregui, Manrique, and Insausti (1999) and Insausti, Insausti, Cruz-Orive, and Manrique (2000) to estimate astrocyte process length in the cochlear nuclei of a monkey.

3. Dendrite length per neuron

Howard, Cruz-Orive, and Yaegashi (1992) applied Eq. (4.18.2) to estimate the dendrite length of a reconstructed neuron from its TVPs. In addition, they considered the estimation of the mean dendrite length per neuron, $\mathbb{E}_N(L) = L_V/N_V$, in a layered neuron population contained in a suitable embedding medium. For simplicity the approach may be model-based, assuming stationarity but not isotropy. A convenient vertical axis (VA) is adopted. The idea is to cut a vertical slice and to scan it along a given depth t (which, as shown below, does not need to be known) with an optical plane parallel to the slice faces. The sweeping plane bears in focus (a) an unbiased counting frame to estimate N_V by the optical disector method, and (b), a test system of cycloids with the major axes parallel to the VA to estimate L_V. The frame should be the union of an integer number of complete tiles of the test system. Then, dividing the rhs of Eq. (4.18.6) by that of Eq. (4.14.3) (with $a_0/p_0 \equiv M^{-2}a/p$ and $h_i = t$ for all $i = 1, 2, \ldots, n$), the disector thickness t cancels out and

$$\widehat{L}_N = 2 \cdot \frac{a}{l} \cdot \frac{1}{M} \cdot \sum_{i=1}^{n} I_i \Big/ \sum_{i=1}^{n} Q_i^- \tag{4.18.7}$$

is a ratio-unbiased estimator of $\mathbb{E}_N(L)$, where n denotes the number of vertical slices. For the ith slice, I_i denotes half the number of intersections observed in focus inside the counting frame between projected dendrite borders and cycloids, and Q_i^- the number of neurons counted by the optical disector. The scanning depth t, which should be the same for the n vertical slices, does not enter into the estimator, but it is proportional to the sample size and therefore affects the estimation precision.

Under isotropy, an arbitrary system of test lines may be used with constant a/l, and the estimator $\widehat{L}_N$ is formally unchanged.

4. Capillary length from vertical slice projections and the fractionator

Artacho-Pérula et al. (1999) estimated the total capillary length in the soleus muscle of a rat with the fractionator, using vertical Cavalieri designs with known periods in two stages, then quadrat subsampling with a known period, and finally quadrat slice projections analysed with cycloids, as above, at the last stage. See also Section 4.11.4, Note 3, and Wulfsohn et al. (1999).

4.19 Stereology of Articular Cartilage From Local Vertical Sections: I. Global Size Properties

4.19.1 Purpose, material, and method

The articular cartilage tissue (hereafter denoted by "c") attached to the surface of the medial femoral condyle of a rabbit may be regarded as a thin cap of about 10 mm in diameter, and up to 0.5 mm thickness, see Fig. 4.19.1(a,b). The outer cap surface is a smooth patch of area $S(\mathrm{c})$. The tissue itself is the union of lacunae (denoted by "lac"), namely the rigid voids, or capsules, containing the chondrocyte cells, and the interlacunar matrix.

Instead of the global quantities $V(\mathrm{lac})$, $S(\mathrm{lac})$ defined in the cap, we use $S(\mathrm{c})$ as a reference, and consider the ratios

$$V_S \equiv V(\mathrm{lac})/S(\mathrm{c}), \text{ and } S_S \equiv S(\mathrm{lac})/S(\mathrm{c}), \tag{4.19.1}$$

namely the mean total volume, or surface area of lacunae, per unit area of the outer cartilage surface.

The ratios V_S, S_S may be estimated from local vertical sections, namely sections which are locally perpendicular to the outer cartilage surface, and isotropically rotated around the local normal (vertical) direction. In the sketch of Fig. 4.19.1(a), the arrows represent local vertical directions.

If the target object is the entire cap, then the design (which is extensible to bounded plate- or sheet-like structures in general) runs as follows.

1. The idea is to split the cap into blocks, as sketched in Fig. 4.19.1(a,b), and to draw a systematic sample of them. In the original study (see Section 4.19.4), the fixed condyle was first cut into about 2 mm thick slices (Fig. 4.19.1(b)) with a diamond saw. Each slice from a systematic sample was embedded in Epon and cut into wedge-shaped blocks (Fig. 4.19.1(c)). Finally, each block from a systematic subsample was re-embedded, cut into local vertical sections of 1 μm nominal thickness, and stained with a cationic dye (RHT) for light microscopy (LM), as sketched in Fig. 4.19.1(d).
2. Fig. 4.19.1(e) shows a micrograph of a local vertical section. Its upper boundary is the trace of the outer cartilage surface. The lower trace is more irregular, and it represents the transition between cartilage and subchondral bone. The white profiles are lacunae transects, whereas the matrix transect appears in black.

In the original study, a goal was to correlate stereological quantities in concrete cartilage regions with their mechanical function – for instance, the more weight bearing, the thicker the cartilage region. Entire cap properties would not be that useful. For didactic purposes, in this exercise only the vertical section shown in Fig. 4.19.1(e) is used. The reference space "c" is just a block constituting the region of interest, see Fig. 4.19.1(d).

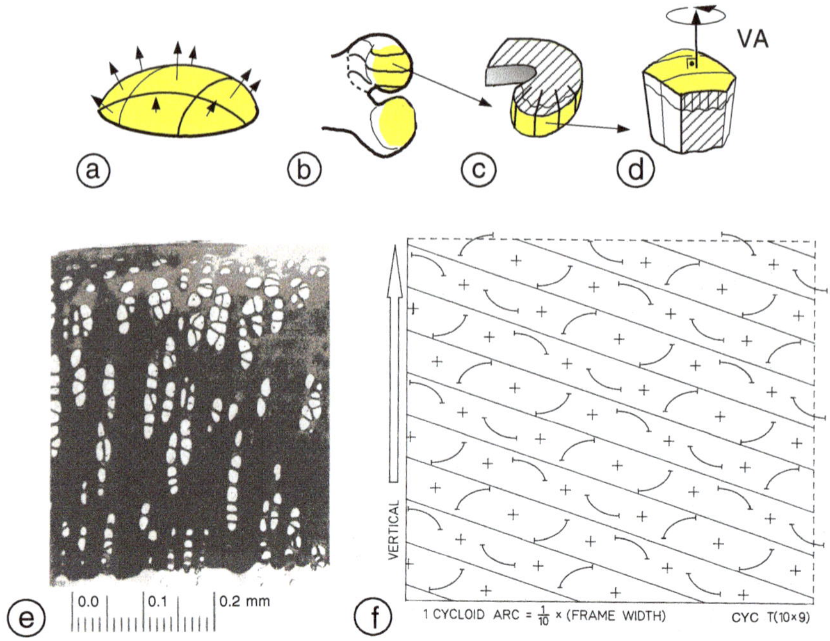

Fig. 4.19.1 Sampling design to generate a local vertical section for LM of the articular cartilage of a rabbit, see text, Section 4.19.1.

Indirect estimation of V_S, S_S

In this example, the conventional definition of mean thickness is

$$\mathbb{E}_S(\tau) = V(\mathrm{c})/S(\mathrm{c}), \tag{4.19.2}$$

which, in combination with Eq. (4.19.1), yields

$$\begin{aligned} V_S &= V_V(\mathrm{lac}, \mathrm{c}) \cdot \mathrm{E}_S(\tau), \\ S_S &= S_V(\mathrm{lac}, \mathrm{c}) \cdot \mathbb{E}_S(\tau). \end{aligned} \tag{4.19.3}$$

The volume ratio $V_V(\mathrm{lac}, \mathrm{c})$ is estimated on the vertical section by point counting using a test system whose fundamental tile contains $p_1 = 2$ coarse test points to score $P(\mathrm{c})$, and $p_2 = 16$ fine test points to score $P(\mathrm{lac})$, see Fig. 4.19.2. Thus,

$$\widehat{V}_V(\mathrm{lac}, \mathrm{c}) = \frac{p_1}{p_2} \cdot \frac{P(\mathrm{lac})}{P(\mathrm{c})}. \tag{4.19.4}$$

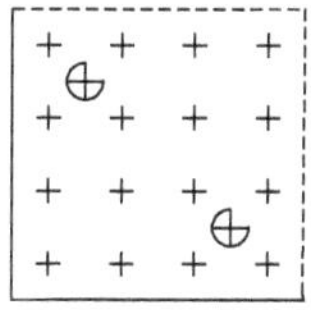

Fig. 4.19.2 Fundamental tile of the test system used to estimate $V_V(\text{lac, c})$.

On the other hand, the ratio $S_V(\text{lac, c})$ is estimated by point and intersection counting using the cycloid test system shown in Fig. 4.19.1(f), for which $l/p = 2.0\,\text{cm}$ (namely the length of a cycloid arc in this case), at a working magnification $M = 340$. Thus,

$$\widehat{S}_V(\text{lac, c}) = 2 \cdot \frac{p}{l} \cdot M \cdot \frac{I(\text{lac})}{P(\text{c})}. \tag{4.19.5}$$

Eq. (4.19.5) coincides with Eq. (4.13.5), because properly oriented UR cycloids on vertical sections 'do the same' as IUR test lines on IUR planar sections.

Lastly, the mean cartilage thickness $\mathbb{E}_S(\tau)$ may be estimated by the sample mean $\bar{\tau}$ of the vertical cartilage intercept lengths measured from the intersections between the tilted straight lines of the cycloid test system and the upper border of the vertical section, down to the subchondral trace, see Section 3.14.2. Such lines are also intended to facilitate the follow up of test points and cycloid arcs in the counting process. The test systems should be UR, always covering the entire cartilage section.

Direct estimation of S_S

Based on its definition in Eq. (4.19.1), S_S may be estimated directly with the cycloid test system as the ratio of the number $I(\text{lac})$ of intersections between the cycloids and the lacunae profile boundaries, to the corresponding number $I(c)$ of intersections between the cycloids and the upper border of the section, namely,

$$\widehat{S}_S(\text{lac, c}) = I(\text{lac})/I(\text{c}). \tag{4.19.6}$$

In this example $I(\text{c})$ is too low and highly variable, in which case it should be replaced with the mean $\bar{I}(\text{c})$ over no less than 3 independent superimpositions of the cycloid test system.

4.19.2 Results

With the test point system whose fundamental tile is shown in Fig. 4.19.2, we scored $P(\text{lac}) = 17$ and $P(\text{c}) = 16$. With the cycloid test system shown in Fig. 4.19.1(f), $I(\text{lac}) = 49$ and $P(\text{c}) = 26$. The mean cartilage thickness measured at the (three) intersection points between the tilted lines and the upper border of the section was $\bar{\tau} = 0.450$ mm at the specimen scale. Now, using the estimators given by Eq. (4.19.4) and Eq. (4.19.5), plus the ones based on Eq. (4.19.3) we get,

$$\widehat{V_S}(\text{lac}, \text{c}) = \frac{1}{8} \cdot \frac{17}{16} \cdot 0.450 = 0.0598\,\text{mm}^3/\text{mm}^2,$$
$$\widehat{S_S}(\text{lac}, \text{c}) = 2 \cdot \frac{1}{20} \cdot 340 \cdot \frac{49}{26} \cdot 0.450 = 28.8\,\text{mm}^2/\text{mm}^2. \quad (4.19.7)$$

On the other hand, in three independent UR superimpositions of the cycloid test system, a total of $I(\text{c}) = 5$ intersections were counted with the upper border of the section, namely with the outer cartilage trace. Thus, Eq. (4.19.6) yields the alternative estimate,

$$\widehat{S_S}(\text{lac}, \text{c}) = \frac{I(\text{lac})}{\bar{I}(\text{c})} = \frac{49}{5/3} = 29.4\,\text{mm}^2/\text{mm}^2. \quad (4.19.8)$$

4.19.3 Error variance prediction

No error variances can be estimated from a single section. If $n \geq 2$ vertical sections were analysed from a given region of interest, then $\widehat{V_V}$, $\widehat{S_V}$ and the direct estimator $\widehat{S_S}$ would involve ratios of sums, as in Section 4.13.1. If the sections were approximately independent, then the precision of the latter estimates could be predicted with Cochran's formula. On the other hand, $\bar{\tau} = (\tau_1 + \tau_2 + \cdots + \tau_m)/m$ may be treated as the mean of an srs, whereby $\text{ce}(\bar{\tau}) = \text{cv}(\tau_1)/m$. If m intercepts are measured on, say, n vertical sections, then the preceding formulae should be used with m and not with n, because intercept sampling is surface area-weighted, see Section 3.14.2. The coefficient of error of the indirect estimates $\widehat{V_S}$ and $\widehat{S_S}$ could be predicted with Goodman's formula, similarly as in Section 4.13.3.

4.19.4 Note

The pictures were kindly donated by the authors of Eggli, Hunziker, and Schenk (1998).

4.20 Stereology of Cartilage From Local Vertical Sections: II. Lacunae Number and Mean Size

4.20.1 Purpose, material, and method

From the material described in the preceding section, the first purpose is to estimate the mean number of lacunae per unit area of the outer cartilage surface in a given region, namely

$$N_S(\text{lac}, \text{c}) = N_V(\text{lac}, \text{c}) \cdot \mathbb{E}_S(\tau). \tag{4.20.1}$$

The new ratio needed is therefore the mean number of lacunae per unit cartilage volume, $N_V(\text{lac}, \text{c})$. To estimate this we use a physical disector bounded by the 1st and the 3rd slabs of a series. The thickness of each slab is $t = 1\,\mu\text{m}$, whereby the disector thickness is $h = 3t = 3\,\mu\text{m}$, see Fig. 4.20.1(a). In order that every lacuna has the same probability to be sampled by the disector, however, it is necessary to ensure that h is less than the smallest individual caliper length of the lacunae in a direction normal to the slabs. The two section images corresponding to the upper and lower slabs of the disector were printed on transparent films, superimposed onto each other, and carefully registered, see Fig. 4.20.1(b). The result is a double disector in which a lacuna is counted only if it is observable as a grey profile. Under this rule, a lacuna is effectively counted if it is hit by the upper face of the lower slab, and not by the upper face of the upper slab, as shown diagrammatically in Fig. 4.20.1(a)– and similarly if the role of the slabs is interchanged. This shows that a disector of thickness h bounded by two slabs does the same as one bounded by two planar faces a distance h apart. White profiles are not scored because they violate the preceding rule. No unbiased frame is needed because the disector has natural boundaries at the top and at the bottom. To facilitate counting, 4 vertical stripes are used whose union is the reference area of the disector. In each stripe, a grey profile is scored if it lies entirely within the stripe, or if it hits the right- but not the left-hand side edge of the stripe. Let Q_i^- denote the total number of grey profiles counted in the ith stripe, and set $Q^- = Q_1^- + \cdots + Q_4^-$, the total count. Based on Eq. (2.30.8), the following estimator

$$\widehat{N}_V(\text{lac}, \text{c}) = \frac{(1/2) \cdot Q^-}{h \cdot (a/p) \cdot M^{-2} \cdot P(\text{c})}, \tag{4.20.2}$$

is ratio-unbiased for $N_V(\text{lac}, \text{c})$. The numerator is half the total number of lacunae sampled by the double disector, whereas the denominator is a UE of the disector volume. The point count $P(\text{c})$ is scored within the reference area with the test system shown in Fig. 4.20.1(b). For simplicity, a test point is the inner corner of a wedge.

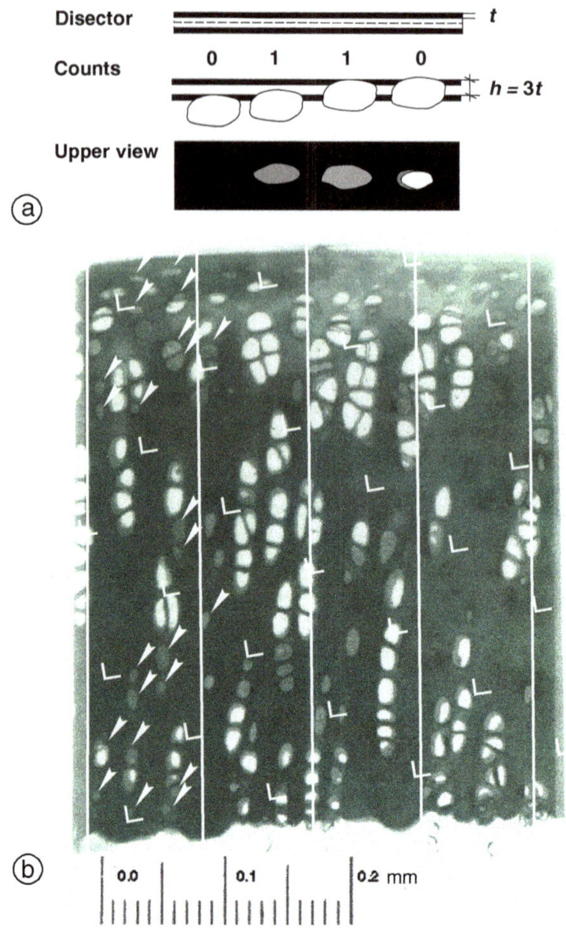

Fig. 4.20.1 (a) Physical disector to count chondrocyte lacunae in articular cartilage. The lacunae are transparent, the matrix opaque. The sketch shows why a lacuna is counted if it appears as a grey profile on an upper view of the disector, with both disector faces superimposed and perfectly registered. In (b), the lacunae counted that way in the first stripe are arrowed. Then, Eq. (4.20.2) estimates lacunae number per unit cartilage volume.

The second purpose is to estimate the (number-weighted) mean individual volume $\mathbb{E}_N(V)$, and surface area, $\mathbb{E}_N(S)$ of the lacunae. The design is based on Eq. (2.31.1) and Eq. (2.31.3), respectively. The corresponding estimators are,

$$\begin{aligned} \bar{v}_N(\text{lac}) &= \widehat{V}_V(\text{lac}, \text{c})/\widehat{N}_V(\text{lac}, \text{c}), \\ \bar{s}_N(\text{lac}) &= \widehat{S}_V(\text{lac}, \text{c})/\widehat{N}_V(\text{lac}, \text{c}). \end{aligned} \tag{4.20.3}$$

Of interest in this context is also the mean matrix volume per lacuna, whose estimator is,

$$\bar{v}_N(\text{mat}) = \frac{1 - \widehat{V}_V(\text{lac, c})}{\widehat{N}_V(\text{lac, c})}. \tag{4.20.4}$$

It is also possible to estimate the mean individual caliper length of the lacunae, $\mathbb{E}_N(H)$, in a direction normal to the disector. For the reference slab of thickness t of the disector, the classical equation for N_V in this context reads

$$N_V = \frac{Q_A}{\mathbb{E}_N(H) - t}, \tag{4.20.5}$$

namely the second Eq. (2.16.9) with $-t$ in the place of $+t$. This is because, in the present example, the slab probe $L_t(p)$ is opaque whereas the lacunae are transparent. As a consequence, Eq. (2.10.1) for a single lacuna of caliper H in a given direction should be replaced with

$$\mathbb{P}(\text{d}p) = \frac{\text{d}p}{H - t}, \quad p \in [0, H - t). \tag{4.20.6}$$

Thus, directly from Eq. (4.20.5) a ratio-unbiased estimator of $\mathbb{E}_N(H)$ is

$$\bar{h}_N = \widehat{Q}_A / \widehat{N}_V(\text{lac, c}) + t. \tag{4.20.7}$$

A convenient alternative is obtained bearing in mind that $\widehat{Q}_A = Q/\widehat{A}(\text{c})$ and $\widehat{N}_V = (Q^-/2)/(h \cdot \widehat{A}(\text{c}))$, where $\widehat{A}(\text{c})$ denotes an estimator of the cartilage transect area and Q^- is the double disector count. Thus,

$$\bar{h}_N = h \cdot \frac{Q}{Q^-/2} + t. \tag{4.20.8}$$

Here $Q = Q_1 + \cdots + Q_4$, and Q_i denotes the number of hit lacunae, observed as white profiles, in the ith stripe from the reference face of the disector (shown in Fig. 4.19.1(e)).

4.20.2 Results

Lacunae number

The numbers of lacunae (grey profiles) sampled by the double disector of Fig. 4.20.1(b) in each of the four stripes is $\{Q_i^-\} = \{23, 18, 17, 9\}$, hence $Q^- = 67$. At the working magnification the 0.2 mm long calibration ruler measured 76.5 mm, hence $M = 382.5$. At this magnification, the size of the square grid used was $h = 31.5$ mm, hence $a/p = 31.5^2$ mm^2. The number of test points of this grid hitting the reference space is $P(\text{c}) = 24$. Direct application of Eq. (4.20.2) now yields

$$\widehat{N}_V(\text{lac, c}) = \frac{67/2}{0.003 \cdot 31.5^2 \cdot 382.5^{-2} \cdot 24} = 68600\,\#/\text{mm}^3, \tag{4.20.9}$$

and now the estimation version of Eq. (4.20.1) yields,

$$\widehat{N}_S(\text{lac, c}) = \widehat{N}_V(\text{lac, c}) \cdot \bar{\tau} = 68600 \cdot 0.450 = 30900\,\#/\text{mm}^2. \tag{4.20.10}$$

Mean individual lacuna size

With the data given in Section 4.19.2, the estimators given by Eq. (4.19.4) and Eq. (4.19.5) yield $\widehat{V}_V(\text{lac, c}) = 0.133$ and $\widehat{S}_V(\text{lac, c}) = 64.1\,\text{mm}^{-1}$ respectively. Now, direct application of Eq. (4.20.3) and Eq. (4.20.4) yield,

$$\begin{aligned}
\bar{v}_N(\text{lac}) &= \frac{0.133 \cdot 10^9}{68600} = 1940\,\mu\text{m}^3,\\
\bar{s}_N(\text{lac}) &= \frac{64.1 \cdot 10^6}{68600} = 934\,\mu\text{m}^2,\\
\bar{v}_N(\text{mat}) &= \frac{(1 - 0.133) \cdot 10^9}{68600} = 12600\,\mu\text{m}^3.
\end{aligned} \tag{4.20.11}$$

Finally, in order to estimate the mean horizontal caliper length of a lacuna by Eq. (4.20.8) we need the total number of lacunae profiles in the reference section (Fig. 4.19.1(e)). The observed counts are: $\{Q_i\} = \{45, 46, 51, 43\}$, whereby $Q = 185$. Thus,

$$\bar{h}_N(90^\circ) = 3 \cdot \frac{185}{67/2} + 1 = 17.6\,\mu\text{m}. \tag{4.20.12}$$

4.20.3 Error variance prediction

If $n \geq 2$ vertical sections are available from a region of interest, then Eq. (4.20.2) and Eq. (4.20.8) are ratios of sums over sections, and Cochran's formula may be used. For Eq. (4.20.3) and Eq. (4.20.4), Goodman's formula may be used.

4.20.4 Note

The methods illustrated here, and in Section 4.19, constitute a subset of those applied in Cruz-Orive and Hunziker (1986) to growth cartilage lacunae.

4.21 Volume-Weighted Mean Nuclear Volume From Point-Sampled Intercepts

4.21.1 Purpose, material, and method

The purpose is to estimate the volume weighted mean volume $\mathbb{E}_V(V)$ of nuclei in a malignant melanoma and in a benign tumour of human skin. The volume-weighted mean volume of nuclei, or cells, is useful to assist in pathology evaluations because it is more effective than the number-weighted mean volume $\mathbb{E}_N(V)$ to detect the presence of unusually large nuclei. Besides, the estimation of the latter requires disectors, the former does not.

Natural sections to sample skin tissue are vertical Cavalieri sections normal to the outer skin surface. The skin surface is adopted as the horizontal plane (HP), whereas its normal is the vertical axis. The images in Fig. 4.21.1 correspond to quadrats from 2 μm thick vertical LM sections from resin-embedded skin tissue.

On each quadrat, a smaller rectangular frame is superimposed to avoid edge effects. In turn, a test system of parallel lines bearing test points (see also Fig. 2.32.2) is superimposed on the frame making a sine-weighted angle with the VA. If n quadrats are available from a given specimen, then a systematic design may be adopted to generate the sine-weighted angles $\{\theta_1, \theta_2, \ldots, \theta_n\}$ for the lines of the test system. The cosines of these angles constitute a systematic sample of period $T = 2/n$ in the range $[-1, 1)$, hence the range of the angles is $[0, 180°)$. Thus,

$$\{\theta_i = \cos^{-1}(1 - (i - U) \cdot T),\ i = 1, 2, \ldots, n\}, \quad U \sim \mathrm{UR}[0, 1). \tag{4.21.1}$$

Let $P \geq 1$ denote the (random) total number of test points hitting nuclear transects, and let $\{l_{0i},\ i = 1, 2, \ldots, P\}$ represent the corresponding line intercepts (in this example, all the observed intercepts were connected segments). Then, by Eq. (2.32.19),

$$\bar{v}_V = \frac{\pi}{3} \cdot \frac{1}{P} \sum_{i=1}^{P} l_{0i}^3 = \frac{\pi}{3} \cdot \overline{l_0^3}, \tag{4.21.2}$$

is a ratio-unbiased estimator of $\mathbb{E}_V(V)$. Note that the sampling units are the test points, and not the nuclei.

4.21.2 Results

In Fig. 4.21.1(a), corresponding to malignant melanoma, $P = 11$ point-sampled intercepts (marked in red) are observed. Using a 20 mm ruler, an intercept is manually classified into class c if it measures between $c - 1$ and c mm. The observed class order numbers are

$$\{c_i\} = \{5, 10, 7, 12, 10, 11, 11, 15, 11, 11, 10\}. \tag{4.21.3}$$

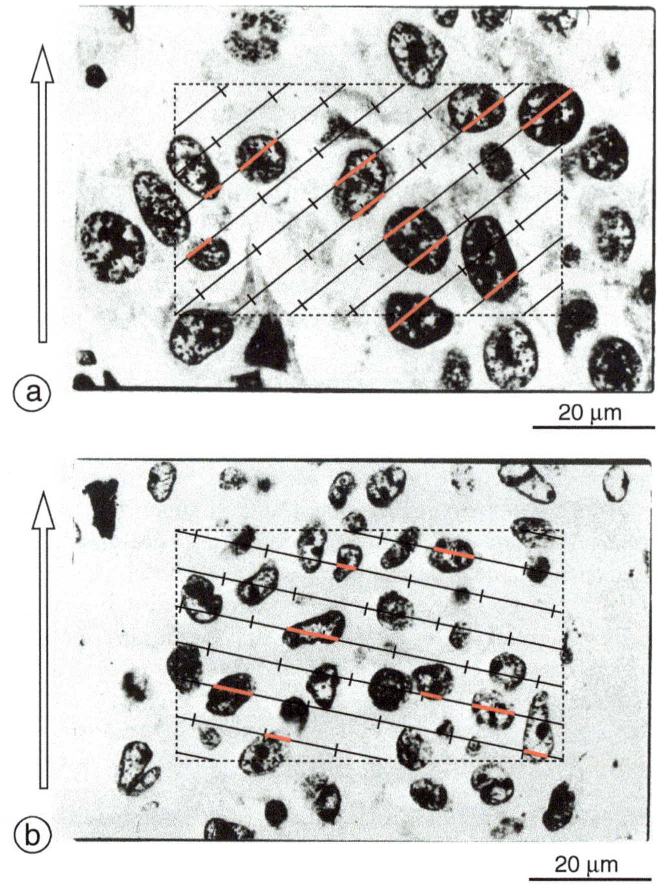

Fig. 4.21.1 Illustration of the point-sampled intercepts method on vertical sections to estimate the volume-weighted mean nuclear volume of (a) malignant and (b) benign human skin tumours. The arrows indicate the vertical axes. Plain micrographs courtesy of C. V. Howard, see Section 4.21.4.

That is, the first intercept length belonged to the 5th class, the second to the 10th, etc. The calibration ruler measured 27 mm at the working magnification, hence $M = 27000/20 = 1350$. For the observed ith intercept length we adopt the following value (at the specimen scale),

$$l_{0i} = (c_i - 1/2) \cdot (1000/M)\,\mu\text{m}, \quad i = 1, 2, \ldots, P, \tag{4.21.4}$$

whereby Eq. (4.21.2) yields

$$\bar{v}_V(\text{melan}) = \frac{\pi}{3} \cdot \left(\frac{1000}{1350}\right)^3 \cdot \frac{1}{11} \sum_{i=1}^{11} (c_i - 0.5)^3$$
$$= 470\,\mu\text{m}^3. \tag{4.21.5}$$

Likewise, for the benign tumour image of Fig. 4.21.1(b) we observe $P = 8$ point-sampled intercepts with class numbers

$$\{c_i\} = \{9, 5, 12, 5, 10, 8, 6, 6\}, \tag{4.21.6}$$

whereby,

$$\bar{v}_V(\text{benign}) = 219\,\mu\text{m}^3. \tag{4.21.7}$$

In the original exercise, $n = 3$ images were available for each of the two groups. For the melanoma group, we generated a UR number $U = 0.445$ in the interval $[0, 1)$, whereby Eq. (4.21.1) with $T = 2/3$ yielded the following sine-weighted angles of the test system with the vertical axis: $\{\theta_i\} = \{51°, 92°, 135°\}$. The image Fig. 4.21.1(a) was the first one in the series. A total of $P = 40$ point-sampled intercepts were measured, which yielded $\bar{v}_V(\text{melan}) = 385\,\mu\text{m}^3$.

For the benign group, $U = 0.188$, whereby $\{\theta_i\} = \{63°, 102°, 151°\}$. The image Fig. 4.21.1(b) was the second one in the series. A total of $P = 41$ point-sampled intercepts were measured, which yielded $\bar{v}_V(\text{melan}) = 211\,\mu\text{m}^3$.

4.21.3 Error variance prediction

Set $x_i \equiv (c_i - 0.5)^3$, $i = 1, 2, \ldots, P$. Then, assuming independence among point-sampled intercepts we may tentatively predict the square coefficient of error of $\bar{v}_V$ as follows,

$$\text{ce}^2(\bar{v}_V) = \frac{\text{cv}^2(x_1)}{P}. \tag{4.21.8}$$

With the data corresponding to the single pictures in Fig. 4.21.1, we get $\text{ce}(\bar{v}_V(\text{melan}))\% = 19.9\%$ and $\text{ce}(\bar{v}_V(\text{benign}))\% = 32.1\%$, respectively. To obtain moderate coefficients of error, about $P = 100$ point-sampled intercepts should be measured.

Suppose that n quadrats, or sections, are available, and let P_i, $\bar{v}_{Vi}$ denote the number of point-sampled intercepts and the volume-weighted mean volume estimator corresponding to the ith section, respectively. Set $P = P_1 + P_2 + \cdots + P_n$, and let l_{0ij} denote the jth point-sampled intercept from the ith section. The ith section is P_i-weighted – for instance, a section contributing 20 intercepts should count four times as much as one contributing 5 intercepts. Consequently, the final estimator becomes

$$\bar{v}_V = \sum_{i=1}^{n} \frac{P_i}{P} \cdot \bar{v}_{Vi} = \frac{\pi}{3} \cdot \frac{1}{P} \sum_{i=1}^{n} \sum_{j=1}^{P_i} l_{0ij}^3, \tag{4.21.9}$$

(see also Section 2.32.3), which may be formally rewritten as Eq. (4.21.2). This illustrates the fact that the sampling units are the point-sampled intercepts, not the sections.

With the aforementioned data corresponding to three quadrats per group with $P = 40$ and $P = 41$ we obtained $\mathrm{ce}(\bar{v}_V)\% = 13.0\%$ in either case.

A non-parametric t-test could now be used to compare both groups, namely,

$$t = \frac{|385 - 211|}{0.13 \cdot (385^2 + 211^2)^{1/2}} \approx 3.05, \tag{4.21.10}$$

large enough to reject the null hypothesis that $\mathbb{E}_V(V)$ is the same in both groups.

4.21.4 Note

The material of this exercise comes from the study of Howard (1986). The exercise was designed by H. J. G. Gundersen and C. V. Howard for the 10th ISS European Stereology Course at CS-Skalský Dvůr in 1986. For early applications in dermatology see Brüngger and Cruz-Orive (1987) and Sørensen (1989). Valuable practical details can also be found in Sørensen (1991). For manual analysis, the sine-weighted orientation of the test system was efficiently generated with a protractor, see Fig. 2 and Fig. 3 of the latter paper. The algorithm in Eq. (4.21.1) is conceived for computer-aided work.

4.22 Mean Neuron Volume With the Optical Nucleator

4.22.1 Purpose, material, and method

The purpose is to estimate the number-weighted mean volume $\mathbb{E}_N(V)$ of neuron bodies in a well-defined population, see Note 1 below. As described in Section 2.32.2, the basic design consists of the following two stages.

1. Systematic UR optical disectors are sampled from the compartment of interest. The disectors may be either IUR, see Fig. 2.32.1(a), or vertical. The first goal is to sample, say, Q^- neurons (usually between 50 and 100) with identical probabilities. If each neuron contains exactly one nucleolus, as in this case, then it is efficient to adopt it as the sampling unit, see Fig. 2.32.1(b).
2. If the optical disector is used then, whenever a nucleolus is observed in focus for the first time in the unbiased frame, the nucleator method is applied to the corresponding neuron transect in order to obtain a UE of the neuron volume. If a cell contains more than one nucleolus, then the pivotal point should be a nucleolus selected with the same probability as the others (not necessarily the first one encountered!). Let $\{v_i\ i = 1, 2, \ldots, Q^-\}$ denote the n unbiased volume

estimators so obtained. Then, based on Eq. (2.32.14),

$$\bar{v}_N = \frac{1}{Q^-} \cdot \sum_{i=1}^{Q^-} v_i \tag{4.22.1}$$

is a ratio-unbiased estimator of $\mathbb{E}_N(V)$ (note that the sample size Q^- will be random).

Next we illustrate the implementation of two widely used nucleator versions.

The pivotal nucleator

In Fig. 4.22.1, an IUR optical disector captures two neurons whose nucleolus is in focus. The pivotal nucleator, see Fig. 1.7.2(d), is applied on the pivotal transect of each neuron with 4 systematic rays of period $T = 90°$ emanating from the nucleolus at the following angles,

$$\{\alpha_i = (U + i - 1) \cdot 90°,\ i = 1, 2, \ldots, 4\}, \quad U \sim \mathrm{UR}[0, 1), \tag{4.22.2}$$

see also Fig. 2.32.1(c). The random number U may be generated independently for each neuron.

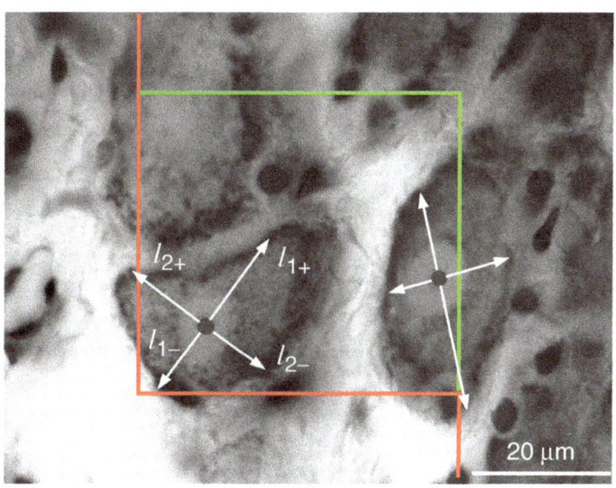

Fig. 4.22.1 Illustration of the pivotal nucleator to estimate the volumes of two neurons sampled with a sweeping isotropic optical section of about 0.5 μm thickness, using the nucleoli as sampling units. Plain micrograph courtesy of C. Avendaño, see Section 4.22.4, Note 1.

The vertical nucleator

In Fig. 4.22.2(a,c), a vertical optical disector captures one neuron whose nucleolus is in focus. The vertical nucleator, see Fig. 1.7.2(b), is applied with two sine-weighted directions which may be sampled in the following two alternative ways.

Design (1, 2). Here the vertical section corresponds to a vertical plane with UR orientation $\phi \sim \mathrm{UR}(0°, 360°)$ around the vertical axis. Then, two systematic sine-weighted rays emanating from the nucleolus are generated by means of Eq. (4.21.1) with $T = 2/2 = 1$, namely,

$$\theta_1 = \cos^{-1}(U), \quad \theta_2 = \cos^{-1}(U - 1), \quad U \sim \mathrm{UR}[0, 1), \tag{4.22.3}$$

see Fig. 4.22.2(b). The notation (1, 2) means that the design uses one longitude and two colatitude angles.

Design (2, 1). Here the vertical section is regarded as the union of two vertical half-planes with systematic orientations

$$\phi_1 = U \cdot 180°, \quad \phi_2 = (U + 1) \cdot 180°, \tag{4.22.4}$$

around a vertical axis through the nucleolus. The corresponding sine-weighted rays make an angle

$$\theta = \cos^{-1}(U), \quad U \sim \mathrm{UR}[0, 1), \tag{4.22.5}$$

with the vertical axis, see Fig. 4.22.2(d). The notation (2, 1) means that two longitude and one colatitude angles are generated.

The nucleator estimator of volume

For a given cell of volume V, and for either the pivotal, or the vertical nucleator, the primary ray intercept lengths are denoted by $\{l_{1+}, l_{2+}\}$, and their antipodal ray lengths by $\{l_{1-}, l_{2-}\}$, respectively. From Eq. (2.18.8), a UE of is

$$v = \frac{4\pi}{3} \cdot \frac{1}{4} \cdot \sum_{i=1}^{2} (l_{i+}^3 + l_{i-}^3) \equiv \frac{4\pi}{3} \cdot \overline{l^3}. \tag{4.22.6}$$

Based on Eq. (2.32.15), the preceding estimator is inserted into the rhs of Eq. (4.22.1) to estimate the mean neuron volume in the population.

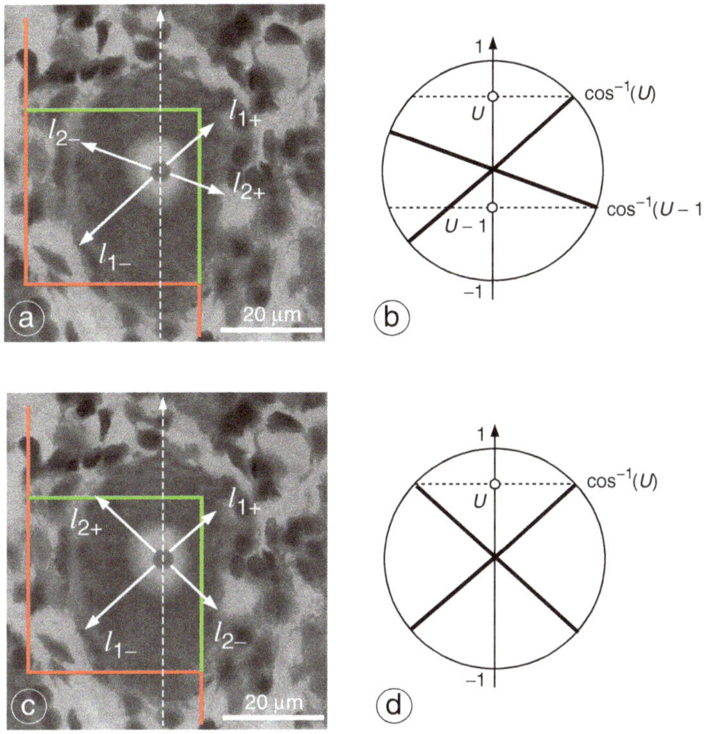

Fig. 4.22.2 Illustration of the vertical nucleator on a neuron sampled with a sweeping vertical optical section of about 0.5 μm thickness, using the nucleolus as the sampling unit. (a,b) Design (1,2), see text. (c, d) Design (2,1). Plain micrograph courtesy of C. Avendaño, see Section 4.22.4, Note 1.

4.22.2 Results

The pivotal nucleator

For the two neurons (left and right, respectively), in Fig. 4.22.1, Eq. (4.22.6) yields the following volume estimates,

$$
\begin{aligned}
v_1 &= \frac{4\pi}{3} \cdot \frac{1}{4}(15.5^3 + 10.8^3 + 13.2^3 + 10.6^3) = 8860\,\mu\text{m}, \\
v_2 &= \frac{4\pi}{3} \cdot \frac{1}{4}(10.5^3 + 7.7^3 + 12.6^3 + 19.5^3) = 11500\,\mu\text{m}^3. \qquad (4.22.7)
\end{aligned}
$$

The vertical nucleator

For the design (1,2), see Fig. 4.22.2(a,b),

$$v = \frac{4\pi}{3} \cdot \frac{1}{4}(13.9^3 + 22.1^3 + 16.4^3 + 12.9^3) = 21100\,\mu\text{m}^3. \tag{4.22.8}$$

The alternative estimate of the same cell volume obtained with the design (2,1), see Fig. 4.22.2(c,d) is

$$v = \frac{4\pi}{3} \cdot \frac{1}{4}(13.9^3 + 22.1^3 + 17.9^3 + 14.3^3) = 23200\,\mu\text{m}^3. \tag{4.22.9}$$

4.22.3 Error variance prediction

The pivotal nucleator

For a nucleator transect, set $F_i = l_{i+}^3 + l_{i-}^3$, $i = 1, 2$. If the transect is isotropic, then we may predict the coefficient of error of the corresponding nucleator volume estimator using the analogue of Eq. (4.18.4), namely,

$$\text{ce}(v)\% = \frac{100}{\sqrt{6}} \cdot \frac{|F_1 - F_2|}{F_1 + F_2}. \tag{4.22.10}$$

From the two isotropic neuron transects in Fig. 4.22.1 we obtain $\text{ce}(v_1)\% = 7.3\%$ and $\text{ce}(v_2)\% = 28.9\%$, respectively.

The vertical nucleator

For a vertical nucleator transect we may use

$$\text{ce}(v)\% = \frac{100}{\sqrt{3}} \cdot \frac{|F_1 - F_2|}{F_1 + F_2}, \tag{4.22.11}$$

see Note 3 below. For the single neuron of the vertical disector in Fig. 4.22.2 we obtain $\text{ce}(v)\% = 20.1\%$ and $\text{ce}(v)\% = 12.9\%$ for the designs (1,2) and (2,1), respectively.

Remarks

1. Eq. (4.22.11) is based on pseudosystematic sampling on the unit sphere, and it therefore takes the variation among coaxial vertical sections into account. However, Eq. (4.22.10) is based on systematic sampling on the circle (Section

5.7.3, Note 1), and it therefore does not include the variation among isotropic pivotal sections. See also Note 3 below.

2. The nucleator is rather conceived to estimate the population mean cell volume $\mathbb{E}_N(V)$ by Eq. (4.22.1). If the individual volume estimators can be regarded as independent, then the variance of $\bar{v}_N$ may be estimated by the usual formula pertaining to simple random sampling, namely,

$$\mathrm{var}(\bar{v}_N) = \frac{\mathrm{var}(v_1)}{Q^-} = \frac{1}{Q^-(Q^- - 1)} \cdot \sum_{i=1}^{Q^-} (v_i - \bar{v}_N)^2. \tag{4.22.12}$$

For this purpose, the error predictors in Section 4.22.3 are not needed.

4.22.4 Notes

1. Neuron micrographs

The images were kindly provided by Professor Carlos Avendaño (Universidad Autónoma, Madrid), and were included in the PhD thesis of Dr. Alfonso Lagares. They correspond to primary sensory neurons from the trigeminal ganglion of a monkey. The neurons were sampled with vertical optical disectors with the vertical axis along the longer axis of the ganglion. In the example, however, the pivotal transects in Fig. 4.22.1 are treated as isotropic for didactic purposes. The optical sections were at most $0.5\,\mu$m thick, taken within a $40\,\mu$m thick slice using oil immersion with a 100× planapochromatic objective (numerical aperture of 1.4). The tissue was embedded in nitrocellulose, and Nissl-stained.

2. Nucleator properties

The vertical designs (1,2) and (2,1) are illustrated in Fig. 3 of González-Villa et al. (2017). Tandrup (1993) proposed the design (1,2). Their relative performance depends on object shape, and on the vertical axis chosen, see Fig. 5 of the former paper. In their Fig. 4 the authors show that the distribution of the nucleator estimator can be very skewed. Thus, individual nucleator estimators based on a single section are usually unstable, but the population mean estimator $\bar{v}_N$ may be reasonably stable if computed from about $Q^- \approx 100$ disector sampled particles (Gundersen, 1988).

3. Prediction of the error variance of the nucleator

Gual-Arnau and Cruz-Orive (2002) obtained variance predictors for a vertical nucleator design based on coaxial sections. For a single section, Eq. (4.22.11) emerged as a special case. The performance of their variance predictors was checked by

González-Villa et al. (2017) using automatic Monte Carlo resampling on the volume rendering of the right brain hemisphere of a rat, and the conclusions were not encouraging. For three or more coaxial sections the variance of the nucleator estimator decreases rapidly, but the practical implementation of the corresponding design is precluded in ordinary stereological studies.

4.23 Volume and Surface Area of an Isolated Object With the Invariator

4.23.1 Purpose, material, and method

The purpose is to estimate the volume V and the external surface area S of the brain of an adult Wistar rat by the invariator method. The brain was split into its two hemispheres. Based on Section 2.20, the sampling protocol for an isolated object consists of the following two stages (complementary details are given in the original study, see Note 1 in Section 4.23.4 below).

1. The pivotal point is defined as the centre of a solid ball containing the object. The two brain hemispheres were embedded (approximately perpendicular to each other) in agar, in the interior of a hollow rubber ball of 34.8 mm interior diameter. The ball was rotated on the table, submitted to MRI, and exhaustively cut into a series of 30 virtual serial sections of 1.2 mm thickness, which were expected to be IUR. The 19th ball section was the largest, and its centre was therefore adopted as the pivotal point.
2. On the transect image corresponding to the pivotal section, two alternative test systems were superimposed consisting of a priori, and a posteriori weighted test lines (Sections 2.20.2 and 2.20.1, respectively), whose implementation is described next.

A priori weighted invariator test lines

On the pivotal image, superimpose a UR square grid of points of tile area a. Through each test point, a test line is drawn perpendicular to the ray joining the test point with the pivotal one, see Fig. 4.23.1(a). Such lines are effectively IUR in space. Let L (at the specimen scale) and I denote the total intercept length and the total number of intersections determined by the test lines in the brain transect and on its boundary. Based on Eq. (2.32.31) and Eq. (2.32.30), the estimators

$$v = aL, \tag{4.23.1}$$

$$s = 2aI, \tag{4.23.2}$$

are unbiased for V and S, respectively.

A posteriori weighted, parallel test lines

On the pivotal section image, superimpose an IUR test system of parallel lines a distance T apart, see Fig 4.23.1(b). Such lines are not IUR in space because, although they are IUR in the pivotal plane, the latter is IR but not IUR. To estimate V and S, however, it suffices to weight the density of each line with its distance from the pivotal point – see the factor r in Eq. (2.20.2) and Eq. (2.20.3). For the ith test line hitting the pivotal transect (counted from top to bottom, say), let r_i, L_i, I_i denote the distance of the test line from the pivotal point, the total intercept length determined in the transect, and the total number of intersections determined in the transect boundary, respectively. Then, based on Eq. (2.32.26) and Eq. (2.32.25), the estimators

$$v = \pi T \cdot \sum |r_i| \cdot L_i, \tag{4.23.3}$$

$$s = 2\pi T \cdot \sum |r_i| \cdot I_i, \tag{4.23.4}$$

are unbiased for V and S, respectively. The summations are over hitting test lines.

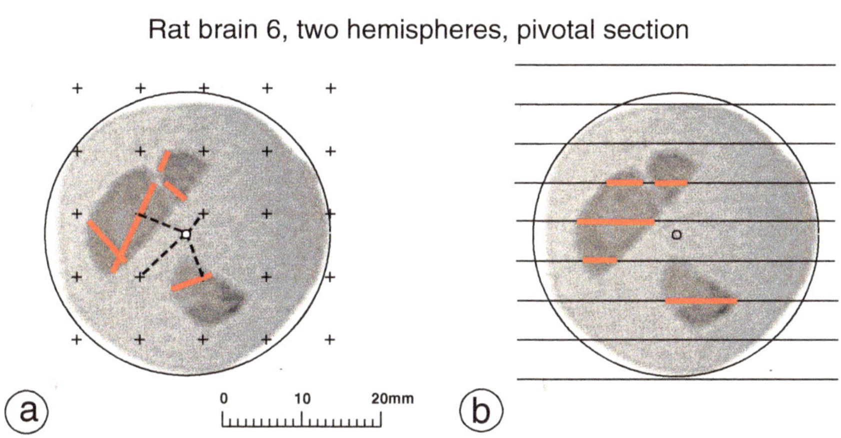

Fig. 4.23.1 Illustration of the invariator method to estimate the volume and the surface area of the union of the two hemispheres of a Wistar rat brain. (a) A priori weighted invariator test line method. (b) A posteriori weighted, parallel test lines method. See Section 4.23 and Note 1 from Section 4.23.4.

4.23.2 Results

A priori weighted invariator test lines

The tile area for the UR system of test points in Fig. 4.23.1(a) is $a = 64.0\,\text{mm}^2$ (at the specimen scale). The lengths (at the specimen scale) of the 5 individual intercepts observed in the same figure are

$$\{L_i\} = \{3.28, 12.39, 2.84, 7.01, 5.37\}\,\text{mm}, \tag{4.23.5}$$

the total length being $L = 30.89$ mm. For each intercept, two intersections with the transect boundaries are observed, hence $I = 10$. Thus,

$$\begin{aligned} v &= 64.0 \cdot 30.89 = 1977\,\text{mm}^3, \\ s &= 2 \cdot 64.0 \cdot 10 = 1280\,\text{mm}^2. \end{aligned} \tag{4.23.6}$$

A posteriori weighted, parallel test lines

In the IUR test system of Fig. 4.23.1(b), the parallel lines are $T = 5$ mm apart (at the specimen scale). The relevant measurements (at the specimen scale) are

$$\begin{aligned} \{(|r_i|, L_i, I_i)\} = \{&(6.79, 8.81, 4), (1.79, 9.70, 2), \\ &(3.21, 4.48, 2), (8.21, 5.97, 2)\}, \end{aligned} \tag{4.23.7}$$

and direct application of Eq. (4.23.3) and Eq. (4.23.4) yields

$$\begin{aligned} v &= \pi \cdot 5.0 \cdot (6.79 \cdot 8.81 + \cdots + 8.21 \cdot 5.97) = 2208\,\text{mm}^3, \\ s &= 2 \cdot \pi \cdot 5.0 \cdot (6.79 \cdot 4 + \cdots + 8.21 \cdot 2) = 1683\,\text{mm}^2, \end{aligned} \tag{4.23.8}$$

respectively.

Remark

In the original study the brain considered here was number 6. The corresponding 'true' values of V and S obtained by pixel counting on the entire set of digitized serial sections (regarded as isotropic Cavalieri sections with 1.2 mm period) were $1919\,\text{mm}^3$ and $1419\,\text{mm}^2$, respectively.

4.23.3 Error variance prediction

A priori weighted invariator test lines

A naive predictor of $\mathrm{CE}(v)$ assuming independence among intercept lengths is the sample coefficient of error of the sum L of the data vector in Eq. (4.23.5), namely,

$$\mathrm{ce}(v)\% = \mathrm{ce}(L)\% = 100 \cdot \frac{\mathrm{cv}(L_1)}{\sqrt{5}} = 27.9\%. \tag{4.23.9}$$

If the target object is a sphere with its centre at the pivotal point, then for a given grid position the invariator intercepts coincide with the fakir ones, see Fig. 4.23.2, in which case Eq. (5.12.2) could be applied. For an arbitrary object this will generally not be the case. Tentatively, Eq. (5.12.2) with S replaced with its estimator $s = 1280\,\mathrm{mm}^2$, and with the tile side length of the grid of test points on the pivotal plane, namely 8 mm, yields

$$\mathrm{ce}(v) = \frac{\mathrm{sd}(v)}{v} = \frac{(0.02430 \cdot 1280 \cdot 8^4)^{1/2}}{1977} = 0.1805, \tag{4.23.10}$$

namely $\mathrm{ce}(v)\% = 18.1\%$, which takes the dependence among intercepts into account.

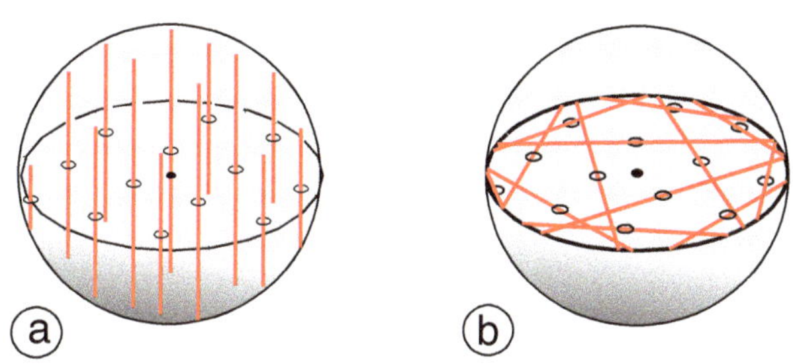

Fig. 4.23.2 Illustration of the equivalence between (a) isotropic fakir linear intercepts, and (b) invariator ones, for a ball. Modified from Cruz-Orive et al. (2010), with permission of Wiley-Blackwell.

To predict $\mathrm{CE}(s)$ we may adopt the rather conservative Poisson model for I, similarly as in Eq. (4.14.9), that is,

$$\mathrm{ce}(s)\% = \frac{100}{\sqrt{10}} = 31.6\%. \tag{4.23.11}$$

Indirect error variance prediction for a priori weighted invariator test lines

In the original study, 'true' values of V and S (obtained with the isotropic Cavalieri series) on the one hand, and the corresponding estimates v and s (obtained with invariator lines on the pivotal section) on the other, were obtained for each of seven Wistar rat brains. The results are displayed in Fig. 4.23.3. From these data, a UE of the mean error variance within animals can be obtained via Eq. (4.5.5) as follows,

$$\begin{aligned}\text{mean}_b\{\text{var}_w(v|V)\} &= \text{var}_b(v) - \text{var}_b(V)\\ &= 355206 - 32815 = 322391.\end{aligned} \tag{4.23.12}$$

On the other hand, $\text{mean}_b(V) = 1959$, whereby,

$$\text{mean}_b\{\text{ce}_w(v|V)\} = 322391^{1/2}/1959 = 0.290. \tag{4.23.13}$$

Similarly, $\text{mean}_b(S) = 1425$ and

$$\text{mean}_b\{\text{ce}_w(s|S)\} = 0.319. \tag{4.23.14}$$

The tile area of the square grid used in the original study, however, was of 37.21 mm^2 instead of the 64 mm^2 used in the example. Consequently, the totals L and I for brain number 6 were of 61 mm and 24 intersections instead of the 31 mm and 10 intersections of the example, respectively. Thus, in spite of their good agreement, the predictions of Eq. (4.23.9) and Eq. (4.23.11) are not strictly comparable with the preceding ones.

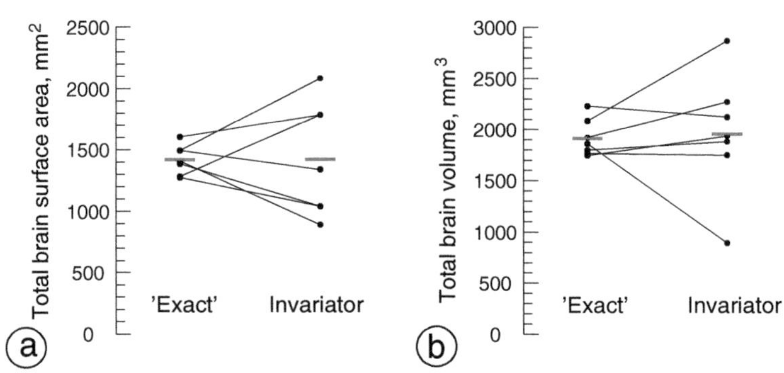

Fig. 4.23.3 For a group of seven rat brains, the 'exact' brain volumes and surface areas vary much less than the corresponding invariator estimates, but the respective means (red horizontal segments) are very similar – as expected, because the invariator estimators are unbiased. Modified from Cruz-Orive et al. (2010), with permission of Wiley-Blackwell.

Parallel invariator lines

Set $x_i = |r_i| L_i$ and $y_i = |r_i| I_i$. Then, using the data in Eq. (4.23.7), and treating them tentatively as independent,

$$\begin{aligned} \mathrm{ce}(v)\% &= 100 \cdot \mathrm{cv}(x_1)/\sqrt{4} = 32.4\%, \\ \mathrm{ce}(s)\% &= 100 \cdot \mathrm{cv}(y_1)/\sqrt{4} = 40.0\%. \end{aligned} \tag{4.23.15}$$

Indirect predictors analogous to those given above for the invariator lines are not available from the original study because the estimators corresponding to parallel invariator lines were unknown at the time. As suggested by the preceding results, however, it seems likely that the latter estimators are the less precise because the intercepts generated by a priori weighted invariator lines incorporate different IR orientations on each transect, whereas the parallel invariator lines use a single IR orientation.

4.23.4 Notes

1. Source

The material of this exercise comes from Cruz-Orive et al. (2010). In the Appendix of this paper, an easily programmable algorithm is given to plot a test system of invariator test lines within a disk. The 'pivotal tessellation' formed by a priori weighted lines, see Fig. 4.23.2(b), was studied in Cruz-Orive (2009a).

2. Efficiency of the invariator relative to the nucleator

Using a ball as a test object, Cruz-Orive (2008) showed that, for comparable intercept lengths, the invariator estimator v from a priori weighted test lines is more efficient to estimate V than the nucleator estimator, unless the pivotal point is close to the centre of the ball, see the lower panel of Fig. 7 from that paper. That figure also verifies that, for the invariator, $\mathrm{Var}(v)$ is constant for any position of the pivotal point. This is a general property of the invariator, because the distribution of the IUR linear intercepts through a set depends only on the size and shape of the set, and not on the position of the pivotal point used to generate the intercepts. In contrast, nucleator intercepts are not IUR, and the corresponding estimator is proportional to the intercept lengths raised to the power 3, which increases the estimator variance. In fact, for the union of the two hemispheres of rat brain number 5 from the original study, González-Villa et al. (2017) showed by automatic Monte Carlo resampling that the coefficient of error of the nucleator estimator was 99.2% for the nucleator design (1,2), and 120.0% for the design (2,1). By comparison with Eq. (4.23.13), the invariator was about 10 to 16 times more efficient than the nucleator for the mentioned objects.

On the other hand, at least for the ball the invariator estimator s of S is more efficient than the surfactor estimator for any position of the pivotal point, see the upper panel of Fig. 7 from Cruz-Orive (2008).

3. The invariator is designed for population mean particle size

Although, as hinted above, the invariator will usually be more efficient than the nucleator, its precision will still be too low for single objects – the current example was designed for didactic purposes. The fact that it is unbiased, and allows the simultaneous estimation of particle volume and surface area, however, makes the invariator potentially useful for particle populations. Remember, however, that, unlike the nucleator, the invariator cannot be implemented on vertical sections.

Variance splitting techniques analogous to that used in Eq. (4.23.12) were developed for stereological particle analyses by Pawlas, Nyengaard, and Jensen (2009).

4.24 Mean Neuron Volume and Surface Area With the Invariator

4.24.1 Purpose, material, and method

The purpose is to implement the invariator method on the two neuron transects used in Fig. 4.22.1 to illustrate the pivotal nucleator. Recall that each transect is regarded as an isotropically oriented pivotal transect with the corresponding nucleolus as pivotal point. The ultimate goal would be to estimate the mean neuron volume $\mathbb{E}_N(V)$ in a well-defined population by Eq. (4.22.1), where v_i would now represent the invariator estimator, and similarly for the mean neuron surface area $\mathbb{E}_N(S)$. At the first sampling stage Q^- neurons, say, are sampled with IUR optical disectors using the nucleolus as sampling unit, see Fig. 2.32.4(a,b). At the second stage we adopt the a priori weighted invariator test lines method on those transects whose nucleolus is captured by the unbiased frame, as described in Section 4.22.1 – see also Fig. 2.32.4(d). Thus, for each disector-sampled neuron we apply the estimators given Eq. (4.23.1) and Eq. (4.23.2), whereby

$$\bar{v}_N = \frac{1}{Q^-}\sum_{i=1}^{Q^-} v_i = \frac{a}{Q^-}\sum_{i=1}^{Q^-} L_i, \tag{4.24.1}$$

$$\bar{s}_N = \frac{1}{Q^-}\sum_{i=1}^{Q^-} s_i = 2\frac{a}{Q^-}\sum_{i=1}^{Q^-} I_i, \tag{4.24.2}$$

where L_i, I_i denote the total intercept length, and the total number of intersections, respectively, determined in the ith disector-sampled neuron by invariator lines generated upon a grid of points of tile area a.

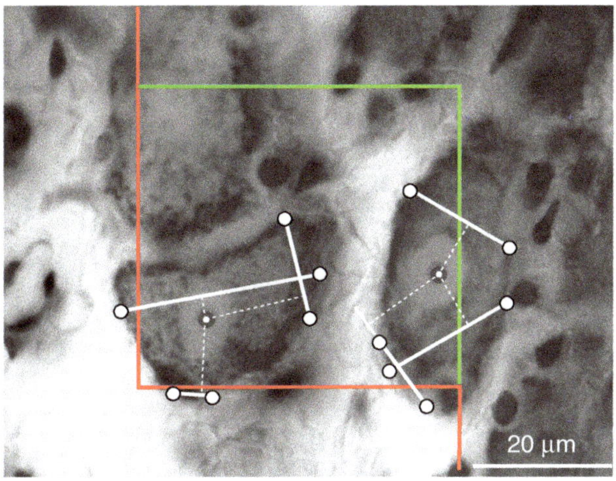

Fig. 4.24.1 Illustration of the invariator with a priori weighted test lines on the two neurons of Fig. 4.22.1, in order to estimate their volume and surface area. The UR square grid used to generate the test lines is omitted for clarity, see Fig. 2.32.4(d) and Fig. 4.23.1(a).

As a bonus, the number of neurons per unit reference volume, $N_V(\text{neu}, \text{ref})$, can in principle be estimated as

$$\widehat{N}_V = \widehat{V}_V / \bar{v}_N, \text{ or } \widehat{N}_V = \widehat{S}_V / \bar{s}_N, \tag{4.24.3}$$

which, unlike the approaches followed in Section 4.14, circumvent the need to measure the optical disector thickness. Recall, however, that the estimation of V_V requires at least FUR sections, and that of S_V requires IUR, or vertical sections.

Because the volume and the surface area of each disector-sampled cell can be estimated simultaneously with the invariator, mean ratios such as $\mathbb{E}_N(S/V)$, or the dimensionless one $\mathbb{E}_N(S^{3/2}/V)$, may be estimated respectively as

$$\overline{(s/v)}_N = \frac{2}{Q^-} \sum_{i=1}^{Q^-} I_i / L_i, \tag{4.24.4}$$

$$\overline{(s^{3/2}/v)}_N = \frac{2\sqrt{2a}}{Q^-} \sum_{i=1}^{Q^-} I_i^{3/2} / L_i, \tag{4.24.5}$$

In these cases, the tile area a should be chosen so that all the L_i are positive.

4.24.2 Results

In the first neuron transect, the observed intercept lengths (at the specimen scale) are $\{L_i\} = \{14.9, 28.3, 5.7\}\,\mu$m, the total intercept length being $L = 48.9\,\mu$m. The tile area of the UR square grid superimposed on the pivotal section is $a = 210.6\,\mu\text{m}^2$, whereby

$$v_1 = 210.6 \cdot 48.9 = 10300\,\mu\text{m}^3. \tag{4.24.6}$$

Similarly, for the second neuron transect $\{L_i\} = \{16.6, 11.7, 19.5\}\,\mu$m. Here $L = 47.8\,\mu$m, and

$$v_2 = 210.6 \cdot 47.8 = 10100\,\mu\text{m}^3. \tag{4.24.7}$$

For the surface areas of the neuron bodies, the total number of intersections is 6 for each transect, whereby

$$s_1 = s_2 = 2 \cdot 210.6 \cdot 6 = 2530\,\mu\text{m}^2. \tag{4.24.8}$$

4.24.3 Error variance prediction

The relevant error variance is usually that of the estimator of the population mean particle volume $\bar{v}_N$, and it may be predicted by Eq. (4.22.12), and similarly for $\bar{s}_N$.

4.24.4 Note

The methods used in this example were proposed in Cruz-Orive (2005), Section 4.1. For a biomedical application, see Noorhafshan, Omidi, Karbalay-Doust, Aliabadi, and Dehghani (2011).

4.25 Volume-Weighted Mean Grain Volume and Surface Area With the Invariator

4.25.1 Purpose, material, and method

The purpose is to estimate the volume-weighted mean volume $\mathbb{E}_V(V)$, and surface area $\mathbb{E}_V(S)$, of the hard tungsten carbide grains of a cemented carbide, see Fig. 4.25.1(a). The embedding material is a ductile cobalt matrix. The grains are approximately convex polyhedra.

The sampling design is model-based. The grains are modelled by separate nonvoid particles with piecewise smooth boundary, and their union constitutes a stationary, isotropic and ergodic particle process Y with realizations in $\mathbb{R}^3$, of fixed volume and surface intensities V_V and S_V, respectively. The latter two intensities may be estimated by point and intersection counting, see Section 3.3.3.

Here we adopt the invariator method with a priori weighted lines (Section 2.32.6). The sampling protocol consists of the following two stages.

1. On a quadrat from a ground, polished, and etched section of the material, superimpose a portion of a test system of points (e.g. an array of 2×3 test points in Fig. 4.25.1(a)) to sample grains. The sampling principle is that a UR point in 3D may be generated as a UR point in a UR, or IUR, plane, see Section 1.2.4. A suitable guard area (outside the inner rectangle in Fig. 4.25.1(a)) should be provided to avoid edge effects. For convenience and efficiency, the sampling points should be sufficiently spaced so that each grain is hit by at most one sampling point. The figure shows three point sampled grains.
2. A UR point hitting a UR grain transect samples the grain with a probability proportional to its volume. Moreover, if the transect is IUR the sampling point is effectively a realization of a pivotal point, and the section plane is a realization of a pivotal plane. For each test point hitting a transect, a test system of invariator test lines is superimposed independently, based on a UR square grid of points of a constant tile area a, see Fig. 4.25.1(b). Let $\{(L_i, I_i),\ i = 1, 2, \ldots, P\}$ denote the total intercept length, and the total number of intersections, corresponding to the ith hitting test point. Based on Eq. (2.32.33), the estimators

$$\begin{aligned}\bar{v}_V &= \frac{a}{P}\sum_{i=1}^{P} L_i = a\bar{L},\\ \bar{s}_V &= \frac{2a}{P}\sum_{i=1}^{P} I_i = 2a\bar{I},\end{aligned} \tag{4.25.1}$$

are ratio-unbiased for $\mathbb{E}_V(V)$ and $\mathbb{E}_V(S)$, respectively. Analogously as in Eq. (4.21.2), the sampling units are the pivotal points, and not the grains. In Fig. 4.25.1, each transect is hit by a single test point but, in general, the total number P of hitting points may exceed the number of point-sampled grains.

4.25.2 Results

In Fig. 4.25.1(b,c,d), the total intercept lengths determined by the invariator test lines (of tile area $a = 10.85\,\mu\text{m}^2$) for each of the $P = 3$ pivotal points are $L_1 = 8.24$, $L_2 = 63.53$, and $L_3 = 4.71\,\mu$m, respectively. The corresponding intersection numbers are $I_1 = 4$, $I_2 = 24$ and $I_3 = 4$. Now Eq. (4.25.1) yields

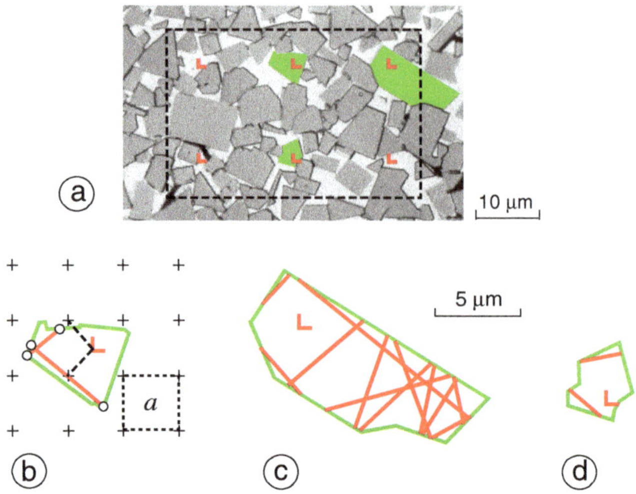

Fig. 4.25.1 Illustration of the invariator method to estimate the volume-weighted mean volume and surface area of the grains of a cemented carbide, see Section 4.25.

$$\bar{v}_V = \frac{a}{3}\sum_{i=1}^{3} L_i = 277\,\mu\text{m}^3,$$

$$\bar{s}_V = \frac{2a}{3}\sum_{i=1}^{3} l_i = 231\,\mu\text{m}^2. \tag{4.25.2}$$

4.25.3 Error variance prediction

We may tentatively treat the $\{L_i\}$ as independent, whereby, similarly as in Eq. (4.21.8), $\text{ce}^2(\bar{v}_V) = \text{cv}^2(L_1)/P$, and similarly for $\text{ce}^2(\bar{s}_V)$. We obtain $\text{ce}(\bar{v}_V)\% = 74.7\%$ and $\text{ce}^2(\bar{s}_V)\% = 62.5\%$, too large because the grain size is highly variable and the sample size ($P = 3$) is too small – as usual, P should be no less than 100.

4.25.4 Note

1. Source

The material used in this exercise comes from the study of Karlsson and Cruz-Orive (1997).

2. Advantages

The invariator is an alternative to the point-sampled intercept method (Section 4.21) to estimate $\mathbb{E}_V(V)$ from sections that are IUR with respect to the specimen. For convenience, the implementation of the invariator should be computer-aided (Section 4.23.4, Note 1). An advantage of the invariator is the simultaneous estimation of $\mathbb{E}_V(S)$.

3. Technical note

The fundamental tile of the UR grid of test points used to generate the a priori weighted test lines should be fixed throughout for a given material. If no test lines hit a point sampled particle transect, then one sets $L_i = I_i = 0$ for that transect in Eq. (4.25.1), but the corresponding sampling point contributes to the total P.

4.26 Estimation of the K-Function for Volume, Surface Area, and Number

4.26.1 K-function for volume and surface area with the invariator

Consider an invariant volume process Y of intensity V_V with realizations in $\mathbb{R}^3$. Its boundary $\partial Y$is a piecewise smooth, invariant surface process of intensity S_V.

Fig. 4.26.1(a) illustrates the estimation of the K-functions of Y, and of ∂Y, using the invariator method. Let $O \in Y$ be a typical point of Y (generated, for instance, similarly as in Fig. 4.25.1(a)), and let $D_0(r)$ represent a ball of finite radius $r \geq 0$ centred at O. Then, the function $K(r)$ for Y, or for ∂Y, is defined by Eq. (3.6.13). To estimate a histogram of $K(r)$ in each case it suffices to estimate $K(\Delta r) = K(r+\Delta r)-K(r)$, namely V_V^{-1}, or S_V^{-1}, times the contents of the intersection between Y, or ∂Y, respectively, and the shell $D_0(\Delta r) \equiv D_0(r + \Delta r)\backslash D_0(r)$ for any desired radius $r \geq 0$ and $\Delta r > 0$. Here Δr represents the width of the histogram class $(r, r + \Delta r]$. The intersection of $D_0(\Delta r)$ with a pivotal plane section through O is an annulus $d_0(\Delta r)$ of radii r and $r + \Delta r$. Based on O, and on a UR square grid of points of tile area a in the pivotal plane, construct a grid $\Lambda_0(a)$ of invariator test lines. Then, by virtue of Eq. (3.7.4) and Eq. (3.7.5), the estimators

$$V_V\widehat{K}(\Delta r) = a \cdot L(Y \cap d_0(\Delta r) \cap \Lambda_0(a)), \tag{4.26.1}$$

$$S_V\widehat{K}(\Delta r) = 2a \cdot I(\partial Y \cap d_0(\Delta r) \cap \Lambda_0(a)), \tag{4.26.2}$$

are unbiased for $V_V K(\Delta r)$ and $S_V K(\Delta r)$ respectively. When counting intersections, care should be taken to ignore those lying on the circular boundaries of the annulus $d_0(\Delta r)$.

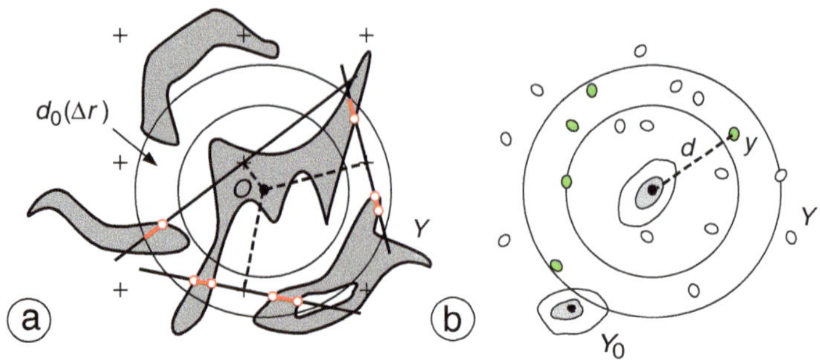

Fig. 4.26.1 (a) Estimation of the K-function of a stationary and isotropic (i.e. invariant) volume process Y, or of its surface ∂Y, by the invariator method. The linear intercepts (in red) are for Y, see Eq. (4.26.1), whereas the relevant intersections (open circles) are for ∂Y, see Eq. (4.26.2). (b) Estimation of the K-function (in number) of an invariant particle process Y with respect to another particle process Y_0 using a disector, see Section 4.26.2. In all cases, if the target process is stationary but not isotropic, then the pivotal section, or disector, should be isotropic.

4.26.2 K-function for particle number from a pivotal slab disector

Consider the union of two invariant point particle processes with realizations in $\mathbb{R}^3$, namely a reference process Y_0 and a target one Y of intensity N_V. For instance, Y_0 and Y may consist of the nucleoli of two different cell populations. The purpose is to estimate the K-function of Y in balls centred at the nucleoli of Y_0 with the idea of revealing the spatial relationship between Y and Y_0. We assume that the number of point particles of either process in any bounded set is finite.

The estimation approach is similar to that used in the preceding section, but instead of a pivotal plane we now use a pivotal slab disector $L_t(0)$ of known thickness $t > 0$, whose midplane contains a typical nucleolus $O \in Y_0$. Let $\mathcal{S}$ denote the finite set of indexes of the point particles in the bounded set $L_t(0) \cap D_0(\Delta r)$, where $D_0(\Delta r)$ is as defined in Section 4.26.1. Fig. 4.26.1(b) represents the projection onto the midplane from $L_t(0)$ of the five disector-sampled particles from Y, (in green), which are captured in the shell $D_0(\Delta r)$. Let $\{d_i,\ i \in \mathcal{S}\}$ denote the distances of such particles from O. Based on Eq. (3.7.6), and considering just the case in which all such distances exceed t, the estimator

$$N_V\widehat{K}(\Delta r) = \frac{2}{t}\sum_{i\in\mathcal{S}} d_i \tag{4.26.3}$$

is unbiased for $N_V K(\Delta r)$. If the particles are not points, as in the figure, then a disector-sampled particle is also sampled in the shell $D_0(\Delta r)$ if it lies inside the shell, or hits the internal, but not the external sphere of the shell.

4.26.3 Notes

1. Implementation of the K-function

Pertinent references on the theory are given in Section 3.7.5.

In a 3D study, Karlsson and Liljeborg (1994) estimated the spatial distributions G, H, K, see Section 3.9, for the centroids of aluminium pores using serial section stacks obtained by confocal scanning laser microscopy.

Stark, Gundersen, Gardi, Pakkenberg, and Hahn (2011) proposed the 'saucor' method to increase the efficiency in the estimation of the K-function for particle number, and illustrated it for glial cells, analogous to the smaller particles in Fig. 4.26.1(b), around neurons (larger particles).

2. Edge effects

For simplicity of exposition, no edge effect corrections have been included. For the (generally difficult) treatment of such corrections for estimating the K-function of invariant point processes in the plane, and in space, see respectively Jensen (1998), and Baddeley et al. (1993), and references therein.

3. Particle shape

Volume tensor analysis, see Jensen and Kiderlen (2017), has been implemented on sections to approximate particles by ellipsoids, this helping to compare particle aggregates by their aspect ratio. Ziegel, Nyengaard, and Jensen (2015) and Rafati, Ziegel, Nyengaard, and Jensen (2016) describe and apply the method to arbitrarily oriented neurons, whereas N. Y. Larsen, Ziegel, Nyengaard, and Jensen (2019) do it for neurons whose major axes tend to be parallel to a common direction.

Chapter 5
Variance Predictors for Systematic Sampling

5.1 Introduction

The purpose of this chapter is to derive error variance predictors for stereological estimators which include the applications described in Chapter 4. The estimators of the target parameters follow directly from the mean value identities derived in Chapters 2 and 3, and they are generally unbiased. Under systematic sampling with test systems, however, predicting the error variance of these estimators is generally nontrivial because the observations are correlated in a manner which depends on the shape and the relative size of both the object and the test system. The variance predictors given here are easy to use, but they are approximations rather than unbiased estimators, and their derivation (which is outlined here for completeness) requires relatively advanced mathematical methods.

The material may be classified as follows:

1. Variance prediction under Cavalieri sampling along a fixed axis, in the following cases.
   a. Cavalieri planes with exact measurements (Section 5.2). The basic technique is developed in Section 5.2.3.
   b. Cavalieri planes with estimated measurements (Section 5.3). The special case of Cavalieri areas estimated by point counting is considered in Section 5.3.2.
   c. Cavalieri slabs with slice contents affected by local errors (Section 5.4).
   d. Splitting design with Cavalieri slabs (Section 5.5). The special case of estimating particle number in the plane with systematic quadrats is considered in Section 5.6.
   e. Estimation of planar curve length with a square grid (Section 5.7). Sampling in the semicircle is involved (Section 5.7.2).
2. Variance prediction for systematic sampling with IUR test systems of geometric probes to estimate planar area with test lines (Section 5.9), points (Section 5.11), stripes (Section 5.14), quadrats (Section 5.16) and segments (Section 5.17), or volume with test planes (Section 5.10), lines (Section 5.12), points (Section 5.13),

L. M. Cruz-Orive, *Stereology*, Interdisciplinary Applied Mathematics 59,
https://doi.org/10.1007/978-3-031-52451-6_5

and slabs (Section 5.15). The practical predictors have a common, simple form. The basic techniques are described in Sections 5.9.2 and 5.14.2.
3. Apparent paradoxes in geometric sampling (Section 5.18).

5.2 Cavalieri Sampling With Section Areas Measured Exactly

5.2.1 Examples

The following two examples illustrate the relative complexity of the error variance under systematic sampling, even for simple geometric models (Fig. 5.2.1, full lines). The examples serve also as an introduction to the main purpose of this chapter, which is to derive approximations for a smooth, usually monotonic, 'trend' of the true variance (Fig. 5.2.1, broken lines), with only mild regularity assumptions for the measurement function.

Variance of the Cavalieri estimator of segment length

Consider a segment $Y = [0, L]$ of unknown length $L > 0$, which we want to estimate by Cavalieri sampling. By virtue of Eq. (2.25.8), the Cavalieri estimator

$$\widehat{L} = T \sum_{k \in \mathbb{Z}} 1_Y(z + kT) \equiv T \cdot P, \quad z \sim \mathrm{UR}[0, T), \tag{5.2.1}$$

is unbiased for L. The exact mean of the random number $P \equiv \widehat{L}/T$ of test points hitting Y is $m \equiv \mathbb{E}P = L/T$, and P can take only two possible values with the following probabilities,

$$\begin{aligned} \mathbb{P}(P = m + 1) &= \Delta, \\ \mathbb{P}(P = m) &= 1 - \Delta, \end{aligned} \tag{5.2.2}$$

where $\Delta = m - [m]$ denotes the fractional part of m. The exact variance and square coefficient of error of $\widehat{L}$ are, respectively,

$$\mathrm{Var}(\widehat{L}) = \Delta(1 - \Delta)T^2, \text{ and } \mathrm{CE}^2(\widehat{L}) = \Delta(1 - \Delta)m^{-2}, \tag{5.2.3}$$

which involve the unknown length L via Δ. Fig. 5.2.1(a), full line, reveals the oscillating behaviour (called the 'Zitterbewegung' effect) of $\mathrm{CE}(\widehat{L})$. This behaviour is typical of systematic sampling, and it is inherited here from the oscillating behaviour of the factor $\Delta(1 - \Delta)$ as a function of $m = L/T$.

The naive attempt at estimating $\mathrm{Var}(\widehat{L})$ directly from a single observation of P fails, because P is an integer-valued random variable and $\widehat{\Delta} = P - [P] = 0$. However, if Δ is assumed a priori to be $\mathrm{UR}[0, 1)$, then

$$\text{Mean}\,\text{Var}(\widehat{L}) = \frac{1}{6}T^2, \tag{5.2.4}$$

see Fig. 5.2.1(a), broken line.

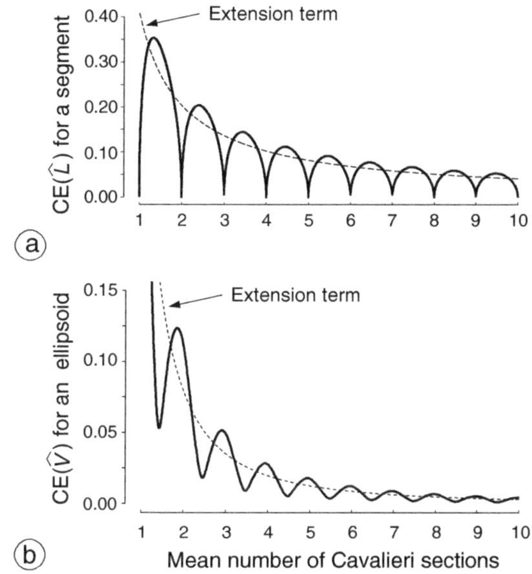

Fig. 5.2.1 (a) Exact $\text{CE}(\widehat{L})$ for a straight line segment computed via the second Eq. (5.2.3), and extension term (broken line) computed via Eq. (5.2.4). (b) Exact $\text{CE}(\widehat{V})$ for an ellipsoid, see Eq. (5.2.6), and extension term, see Eq. (5.2.7). In either case, the oscillations obey to the Zitterbewegung effect.

Variance of the Cavalieri estimator of ellipsoid volume

Consider an arbitrary ellipsoid $Y \subset \mathbb{R}^3$ of volume $V > 0$, which is the target parameter. By virtue of Eq. (2.25.10), the Cavalieri estimator

$$\widehat{V} = T\sum_{k\in\mathbb{Z}} A(z+kT), \quad z \sim \text{UR}[0,T), \tag{5.2.5}$$

is unbiased for V. By Eq. (2.25.12), the mean number of non-empty sections is $\mathbb{E}n = H/T$, where H is the length of the orthogonal projection of Y onto the sampling axis. For each $z \in [0,T)$ the ratio $\widehat{V}/V$, and thereby $\text{CE}^2(\widehat{V})$, do not change if the ellipsoid is replaced with a ball of radius $R = H/2$ (but $\text{Var}(\widehat{V})$ will change, because V will generally depend on ellipsoid shape for a given H), see Section 5.2.6, Note 4. Set $m \equiv \mathbb{E}n$, for short, and $\Delta = m - [m]$. It can be verified that

$$\begin{aligned}\mathrm{CE}^2(\widehat{V}) = &\left(\frac{1}{5} - 3\Delta^2(1-\Delta)^2\right) m^{-4} \\ &- \frac{2}{5}\Delta(1-\Delta)(1+\Delta-9\Delta^2+6\Delta^3)m^{-5} \\ &+ \frac{1}{5}\Delta^2(1-\Delta)^2(1+2\Delta-2\Delta^2)m^{-6},\end{aligned} \tag{5.2.6}$$

see the corresponding graph in Fig. 5.2.1(b). If $\Delta \sim \mathrm{UR}[0,1)$, then

$$\mathrm{Mean}\,\mathrm{CE}^2(\widehat{V}) = \frac{1}{10m^4} + \frac{1}{105m^6}, \tag{5.2.7}$$

which is $O(m^{-4})$, whereas for the straight line segment we had $O(m^{-2})$. A general explanation of this is given in Section 5.2.4 below.

5.2.2 Practical variance prediction formula

The target parameter is usually the volume $V > 0$ of a domain $Y \subset \mathbb{R}^3$. Consider an arbitrary sampling axis Ox, and a plane $L_2^3(x)$ normal to this axis at a point of abscissa x. Then, by Eq. (1.4.5),

$$V = \int_D A(x)\,\mathrm{d}x, \tag{5.2.8}$$

where $D \subset \mathbb{R}$ is also a domain (e.g., the orthogonal projection of the domain Y onto Ox), whereas $A(x) \equiv A(Y \cap L_2^3(x)), x \in D$, with $A(x) = 0, \notin D$, is the measurement function, namely the section area function in this case. The probe adopted to estimate V is a test system of Cavalieri planes a distance T apart and normal to the sampling axis Ox. Let $\{A_1, A_2, \ldots, A_n\}$ denote the corresponding sample of section areas, in sequential order, where A_1, A_n denote the first and the last non-zero section areas, respectively. By virtue of Eq. (2.25.10), the Cavalieri estimator

$$\widehat{V} = T \cdot (A_1 + A_2 + \cdots + A_n), \tag{5.2.9}$$

see also Eq. (4.8.1), is unbiased for V. A predictor of its variance, based on a single sample, is

$$\mathrm{var}(\widehat{V}) = \alpha(q) \cdot T^2 \cdot (3C_0 - 4C_1 + C_2), \; n \geq 3, \tag{5.2.10}$$

$$\mathrm{var}(\widehat{V}) = \frac{1}{6} \cdot T^2 \cdot (C_0 - C_1), \quad n = 2, \; q = 0, \tag{5.2.11}$$

$$C_k = \sum_{i=1}^{n-k} A_i A_{i+k}, \quad k = 0, 1, \ldots, n-1. \tag{5.2.12}$$

The coefficient $\alpha(q)$ depends on the smoothness constant q of $A(x)$, whose meaning is explained in the next section. Its expression is

$$\alpha(q) = \frac{\Gamma(2q+2)\zeta(2q+2)\cos(q\pi)}{(2\pi)^{2q+2}(1-2^{2q-1})}, \qquad q \in [0,1],\ q \neq 1/2, \tag{5.2.13}$$

and $\alpha(1/2) = \zeta(3)/(8\pi^2 \log 2)$, see Eq. (5.4.14) below, where $\Gamma(\cdot)$ is the Gamma function, whereas

$$\zeta(s) = \sum_{k=1}^{\infty} \frac{1}{k^s}, \qquad s > 1, \tag{5.2.14}$$

is Riemann's Zeta function, e.g. $\zeta(2) = \pi^2/6$, $\zeta(3) = 1.2020\ldots$, $\zeta(4) = \pi^4/90$, $\zeta(5) = 1.0369\ldots$, etc. The constant q may be estimated by the Kiêu–Souchet formula,

$$\widehat{q} = \frac{1}{2\log 2} \cdot \log\left(\frac{3C_0 - 4C_2 + C_4}{3C_0 - 4C_1 + C_2}\right) - \frac{1}{2}, \tag{5.2.15}$$

which requires a sample size $n \geq 5$.

5.2.3 Derivation of the variance prediction formula

The target is a volume see Eq. (5.2.8). Now, however, it is convenient to set $f(x) \equiv A(x)$. In general, f is of bounded variation and integrable in a domain $D \subset \mathbb{R}$, and $f(x) = 0$, $x \notin D$. The Cavalieri estimator of V becomes

$$v(z) = T \sum_{k \in \mathbb{Z}} f(z + kT), \qquad z \sim \mathrm{UR}[0, T). \tag{5.2.16}$$

The problem is to predict $\mathrm{Var}\{v(z)\}$ from a single Cavalieri sample. The preceding examples suggest that an exact approach is at best tedious, and inaccessible for an arbitrary object. The methods used in this chapter, based on the transitive theory of G. Matheron, aim at predicting a sort of 'mean' variance, analogous to the ones obtained in the examples, which excludes the oscillations, and it is called the 'extension term' of the true variance.

As a first try, bearing in mind that $\mathbb{P}(\mathrm{d}z) = \mathrm{d}z/T$, we have

$$\begin{aligned} \mathbb{E}(v^2(z)) &= T \int_0^T \sum_{i \in \mathbb{Z}} \sum_{j \in \mathbb{Z}} f(z + iT) \cdot f(z + jT)\, \mathrm{d}z \\ &\underset{j=i+k}{=} T \sum_{k \in \mathbb{Z}} g(kT), \end{aligned} \tag{5.2.17}$$

where

$$g(h) = \int_{\mathrm{R}} f(x) f(x+h)\, \mathrm{d}x, \qquad h \in \mathbb{R}, \tag{5.2.18}$$

is called the covariogram of $f(x)$. It can be verified that (i) $g(0) \geq 0$, (ii) $g(h) \leq g(0)$, (iii) $g(-h) = g(h)$, and (iv) $\int g(h)\,\mathrm{d}h = V^2$. Thus,

$$\mathrm{Var}\{v(z)\} = T \sum_{k \in \mathbb{Z}} g(kT) - \int_{\mathbb{R}} g(h)\,\mathrm{d}h. \tag{5.2.19}$$

Unfortunately the preceding identity cannot be used to estimate the variance directly, because the estimator of the integral is precisely the summation, hence their difference is identically zero. A useful approach runs as follows. Because $v(z)$ is periodic of period T, it can be expressed by the following Fourier series,

$$\begin{aligned} v(z) &= \sum_{k \in \mathbb{Z}} c_k \exp(2\pi i k T^{-1} z), \\ c_k &= \int_{\mathbb{R}} f(z) \exp(-2\pi i k T^{-1} z)\,\mathrm{d}z \equiv (\mathcal{F} f)(k/T), \end{aligned} \tag{5.2.20}$$

where $\mathcal{F} f$ denotes the Fourier transform of f. By Eq. (2.10.13) we have $g = f * \check{f}$, and by the convolution theorem, $\mathcal{F} g = (\mathcal{F} f)(\overline{\mathcal{F}} f) \geq 0$, where $\overline{\mathcal{F}}$ is the conjugate of $\mathcal{F}$. Now, using Parseval's theorem the second moment of $v(z)$ may be written as follows,

$$T^{-1} \int_0^T v^2(z)\,\mathrm{d}z = \sum_{k \in \mathbb{Z}} c_k \bar{c}_k = \sum_{k \in \mathbb{Z}} (\mathcal{F} g)(k/T). \tag{5.2.21}$$

Because $g(-h) = g(h)$,

$$G(t) \equiv (\mathcal{F} g)(t) = 2 \int_0^\infty g(h) \cos(2\pi t h)\,\mathrm{d}h, \tag{5.2.22}$$

which is real and non-negative. Moreover, by property (iv) above we have $G(0) = V^2$, whereby,

$$\mathrm{Var}\{v(z)\} = 2 \sum_{k=1}^{\infty} G(k/T). \tag{5.2.23}$$

Early applications of the preceding result required the knowledge of f, which is not practical. G. Matheron developed an effective approach noting that an important contribution of the variance is explained by the behaviour of g near the origin. In fact, by property (ii) above, the integrand of $G(t)$ satisfies $g(h)|\cos(2\pi t h)| \leq g(0)$, $h \in \mathbb{R}$. Thus, he proposed to model g near the origin by a polynomial of possibly fractional order. Later, K. Kiêu and S. Souchet related the order of the polynomial to the smoothness constant q of the measurement function f. If q is an integer, then q is the smallest order of the derivative of f which exhibits finite jumps. For instance, for the straight line segment studied above, $f(x) = 1_{[0,L]}(x)$, which exhibits jumps of $+1$ and -1 at 0 and L respectively, whereby $q = 0$. On the other hand, in the ball example, $f(x) = \pi(R^2 - x^2)$, $x \in [-R, R]$, which is continuous, but its first derivative exhibits a jump equal to $2\pi R$ at $-R$ and at R, whereby $q = 1$. It can be shown that if the smoothness constant of f is q (in which case we say that f is q-smooth), then

g is $(2q + 1)$-smooth. More recently, using fractional calculus, M. García-Fiñana extended the theory to infinite jumps, in which case q can take any real value in the interval $[0, 1]$. Consequently, a reasonable model for g near the origin is

$$g(h) = b_0 + b_{2q+1}|h|^{2q+1} + b_2 h^2, \qquad q \in [0, 1],\ q \neq 1/2, \tag{5.2.24}$$

where $\{b_0, b_{2q+1}, b_2\}$ are numerical coefficients which can be estimated from a Cavalieri sample. The idea of G. Matheron was to submit the preceding model to the rhs of Eq. (5.2.23), implicitly assuming that $h \in \mathbb{R}$. The Fourier transforms of b_0 and of $b_2 h^2$ are proportional to the Dirac delta function, hence their contribution is zero. The relevant expression of $(\mathcal{F}|h|^{2q+1})(\rho)$ is given by Eq. (A.3.3), whereby Eq. (5.2.23) yields the following extension term of the variance, which excludes the Zitterbewegung and higher-order terms, namely

$$\mathrm{Var}_E\{v(z)\} = -4b_{2q+1}T^{2q+2} \cdot \frac{\cos(q\pi)}{(2\pi)^{2q+2}} \cdot \Gamma(2q+2)\zeta(2q+2). \tag{5.2.25}$$

Given q, the coefficients $\{b_0, b_{2q+1}, b_2\}$ can be approximated from Eq. (5.2.24) if $g(h)$ is known at $h = 0, T, 2T$. The relevant coefficient becomes

$$b_{2q+1} = \frac{3g(0) - 4g(T) + g(2T)}{(2^{2q+1} - 4)T^{2q+1}}, \tag{5.2.26}$$

and its substitution into the rhs of Eq. (5.2.25) yields

$$\mathrm{Var}_E\{v(z)\} = \alpha(q) \cdot T \cdot (3g(0) - 4g(T) + g(2T)), \qquad q \in [0, 1], \tag{5.2.27}$$

where $\alpha(q)$ is given by Eq. (5.2.13). The estimator

$$\widehat{g}(kT) = TC_k, \quad C_k = \sum_{i=1}^{n-k} f_i f_{i+k}, \quad k = 0, 1, \ldots, n-1, \tag{5.2.28}$$

is unbiased for $g(kT)$. To see this, set $f_i \equiv f(z + iT)$, $i \geq 1$, compute $\mathbb{E}(C_k)$ with respect to $z \sim \mathrm{UR}[0, T)$, and use the definition of $g(h)$. Finally, replacing the covariogram terms with their estimators in the rhs of Eq. (5.2.27), the predictor of the error variance of $\widehat{V} \equiv v(z)$ given by Eq. (5.2.10) is obtained.

If $n = 2$, then we may replace the covariogram model given by Eq. (5.2.24) with one containing at most two unknown coefficients, such as

$$g(h) = b_0 + b_{2q+1}|h|^{2q+1}, \tag{5.2.29}$$

whereby,

$$b_{2q+1} = -\frac{g(0) - g(T)}{T^{2q+1}}, \tag{5.2.30}$$

which, substituted into the rhs of Eq. (5.2.25) yields a negative expression unless $q \in [0, 1/2)$, because $g(0) - g(T) \geq 0$ and $\cos(q\pi) < 0$ for $q \in (1/2, 1]$. The variance predictor given by Eq. (5.2.11) corresponds to $q = 0$.

To estimate the smoothness constant q, note that the extension term given by Eq. (5.2.25) with the period T replaced with $2T$ becomes 2^{2q+1} times the extension term with period T, whereby we obtain Eq. (5.2.15).

5.2.4 Examples

We consider, in turn, the two examples described in Section 5.2.1, in order to illustrate the meaning of Eq. (5.2.25).

Cavalieri estimator of segment length

For the segment $Y = [0, L]$ considered above the measurement function is the indicator function, hence its covariogram is the geometric covariogram, see Section 3.8.1, namely

$$g(h) = \begin{cases} L - |h| & \text{if } -L \leq h < L, \\ 0 & \text{otherwise}, \end{cases} \tag{5.2.31}$$

which is a special case of Eq. (5.2.24) with $q = 0$ and $b_1 = -1$. Thus Eq. (5.2.25) yields,

$$\operatorname{Var}_E(\widehat{L}) = \frac{1}{6}T^2, \tag{5.2.32}$$

which is the rhs of Eq. (5.2.4).

Cavalieri estimator of ellipsoid volume

For a ball of radius R centred at the origin the measurement function is the section area function $\pi(R^2 - x^2)$, $|x| \leq R$, hence

$$\begin{aligned} \pi^{-2}g(r) &= \int_{-R}^{R-r} (R^2 - x^2)(R^2 - (x+r)^2)\,\mathrm{d}x \\ &= \frac{16}{15}R^5 - \frac{4}{3}R^3r^2 + \frac{2}{3}R^2r^3 - \frac{1}{30}r^5, \quad r \equiv |h| \leq 2R, \end{aligned} \tag{5.2.33}$$

whose first three terms coincide with Eq. (5.2.24) with $q = 1$ and $b_3 = (2\pi^2/3)R^2$. Thus, the rhs of Eq. (5.2.25) divided by $V^2 = (4\pi/3)^2R^6$ yields,

$$\mathrm{CE}_E^2(\widehat{V}) = \frac{1}{10} \cdot \left(\frac{T}{2R}\right)^4, \tag{5.2.34}$$

which is the first term in the rhs of Eq. (5.2.7) because $2R/T = m$. If the last term in the rhs of Eq. (5.2.33) is incorporated into the model, then we obtain also the second term in the rhs of Eq. (5.2.7). These coincidences will hold whenever the relevant terms of the true covariogram coincide with those of the model given by Eq. (5.2.24).

Remark

As hinted in Section 5.2.3, using a model of $g(h)$ in Eq. (5.2.23) implicitly assumes that the range of $r \equiv |h|$ is $[0, \infty)$, whereas in the preceding example the true range of r is $[0, 2R]$. This is one of the reasons for obtaining Eq. (5.2.34) instead of the exact Eq. (5.2.6).

5.2.5 Application to discrete systematic sampling

Our purpose is to derive Eq. (4.4.10). The total parameter given by Eq. (4.3.6) is equivalent to

$$\gamma = \int_0^N f(x)\,\mathrm{d}x, \tag{5.2.35}$$

where the measurement function is a histogram function with unit class widths, defined as follows,

$$f(x) = \begin{cases} y_i, \; x \in (i-1, i], & i = 1, 2, \ldots, N, \\ 0, \;\; x \notin (0, N]. \end{cases} \tag{5.2.36}$$

The estimator of γ, see Eq. (4.4.9), may be written

$$\widehat{\gamma} = T \sum_{k \in \mathbb{Z}} f(z + kT), \quad z \sim \mathrm{UR}[0, T),\; T = 1, 2, \ldots, \tag{5.2.37}$$

and therefore Eq. (5.2.25) may be used with $q = 0$, because $f(x)$ has finite jumps. Thus,

$$\mathrm{Var}_E(\widehat{\gamma}) = -\frac{T^2}{6} \cdot g'(0). \tag{5.2.38}$$

For $h \in (0, 1)$ we have,

$$f(x)f(x+h) = \begin{cases} y_i^2, & x \in (i-1, i-h], \\ y_i y_{i+1}, & x \in (i-h, i], \end{cases} \tag{5.2.39}$$

whereby,

$$
\begin{aligned}
g(h) &= \sum_i \left\{ \int_{i-1}^{i-h} y_i^2 \, dx + \int_{i-h}^{i} y_i y_{i+1} \, dx \right\} \\
&= \sum_i y_i^2 - h \left\{ \sum_i y_i^2 - \sum_i y_i y_{i+1} \right\},
\end{aligned}
\tag{5.2.40}
$$

and now Eq. (5.2.38) yields Eq. (4.4.10).

5.2.6 Notes

1. Early developments

Eq. (5.2.19) was derived by Moran (1950), who noted that it could be used only in special cases. Eq. (5.2.23) appears in the same paper and in Kendall and Moran (1963). The breakthrough for the application to bounded sets came with the 'transitive theory' of Matheron (1965, 1971). In particular, an expression basically equivalent to Eq. (5.2.25), without a precise definition of the smoothness constant, appears already in Matheron (1965), p. 81. Up to the review of Cruz-Orive (1989a), however, Matheron's covariogram model, of which Eq. (5.2.24) is a special case, was applied only for $q = 0$. The explicit estimator in Eq. (5.2.10), with $\alpha(0) = 1/12$, was given by Gundersen and Jensen (1987).

2. Further refinements

In a study involving human brain tumours of approximately ellipsoidal shape, Neil Roberts (1992, personal communication) noticed that Eq. (5.2.10) with $\alpha(0) = 1/12$ overestimated $\mathrm{Var}(\widehat{V})$ by a factor of about 20. In order to verify this he computed $\mathrm{Var}(\widehat{V})$ empirically by a resampling technique we were developing at the time (Roberts, Cruz-Orive, Reid, Brodie, Bourne, and Edwards, 1993). The tumour material was reused by Roberts, Puddephat, and McNulty (2000), Fig. 1. This observation, together with the fact that the covariogram of the ball section area, see Eq. (5.2.33), lacks the first order term, prompted Cruz-Orive (1993) to use the model given by Eq. (5.2.24) with $q = 1$, thereby obtaining $\alpha(1) = 1/240 = \alpha(0)/20$. This attracted the interest of Kiên Kiêu and collaborators, who formalized the connection between $\mathrm{Var}(\widehat{V})$ and the smoothness properties of the measurement function (Souchet (1995); Kiêu (1997); Kiêu, Souchet, and Istas (1999); Gundersen, Jensen, Kiêu, and Nielsen (1999)). Later, García-Fiñana (2000), and García-Fiñana and Cruz-Orive (1998, 2000a, 2000b, 2004) extended the theory, reformulating in particular Eq. (5.2.25) and Eq. (5.2.10) for $q \in [0, 1]$.

3. Estimation of the smoothness constant

Eq. (5.2.15) was proposed by Kiêu et al. (1999). According to K. Kiêu (personal communication) the formula was borrowed from the area of economics, and did not have to be robust in the Cavalieri context. For fractional q, its estimator may vary with the number of sections, see Fig. 4 of García-Fiñana and Cruz-Orive (2004). It may be advisable to estimate q for each concrete organ of interest (e.g. brain, kidney, etc.) via Eq. (5.2.15) from at least 50 Cavalieri section areas measured as exactly as possible, then using the same estimate routinely. For a human brain, the latter authors obtained $\widehat{q} = 0.42$, whereby $\alpha(0.42) = 0.0275$ yielded reasonable variance predictions for different sample sizes. For details, see Section 4.9.4, Note 2. From a practical viewpoint, the constant q may be regarded as an auxiliary parameter to obtain a reasonable variance predictor under Cavalieri sampling.

4. Notes on the examples

The segment length example was considered by Moran (1950). For the ellipsoid, Eq. (5.2.6) was given by Cruz-Orive (1989a) using the equivalence with the ball (Cruz-Orive, 1985, Eq. 2.2). Analogous exact variance formulae for these, and for other geometric models, were also obtained by Kellerer (1989) and by Mattfeldt (1989).

5. The application to discrete sampling

The variance approximation given by Eq. (4.4.10) does not depend on the class widths, but only on the jumps of the measurement function (Cruz-Orive, 1987b, 1989b). It is a particular case of Eq. 9 from Kiêu, Souchet, and Istas (1999), or of Eq. 7 from García-Fiñana and Cruz-Orive (2004), with $m = 0$.

6. Modifications on the classical Cavalieri design

Baddeley, Dorph-Petersen, and Jensen (2006) considered the case in which the distance between sections is a random variable, a possibility hinted at in the Appendix from Pache et al. (1993). A number of further developments followed, see Stehr, Kiderlen, and Dorph-Petersen (2022) and references therein. The latter authors derive variance predictors for integer smoothness constant.

5.3 Cavalieri Section Areas Affected By Local Errors

5.3.1 Practical variance prediction formula with local errors

With the setup described in Section 5.2.2, here we consider the common case in which each section area A_i is not measured exactly, but estimated without bias by $\widehat{A}_i$, say, in which case the unbiased estimator of V becomes

$$\widetilde{V} = T \cdot (\widehat{A}_1 + \widehat{A}_2 + \cdots + \widehat{A}_n). \tag{5.3.1}$$

Set $\sigma_i^2 = \mathrm{Var}(\widehat{A}_i | A_i)$, the local error variance of each section area estimator. Then,

$$\mathrm{var}(\widetilde{V}) = \alpha(q) \cdot T^2 \cdot \{3(\widehat{C}_0 - v_n) - 4\widehat{C}_1 + \widehat{C}_2\} + T^2 v_n, \quad n \geq 3, \tag{5.3.2}$$

$$\mathrm{var}(\widetilde{V}) = \frac{1}{6} \cdot T^2 \cdot (\widehat{C}_0 - \widehat{C}_1 - v_2) + T^2 v_2, \quad n = 2,\ q = 0, \tag{5.3.3}$$

$$\widehat{C}_k = \sum_{i=1}^{n-k} \widehat{A}_i \widehat{A}_{i+k}, \quad k = 0, 1, \ldots, n-1, \tag{5.3.4}$$

$$v_n = \sigma_1^2 + \sigma_2^2 + \cdots + \sigma_n^2. \tag{5.3.5}$$

In turn, v_n has to be predicted as dictated by the subsampling design adopted in each case to estimate the $\{A_i\}$, see next for a common example.

5.3.2 Cavalieri section areas estimated by point counting

Suppose that each A_i is estimated by point counting with a square grid of side length $h > 0$. If the grid is UR, then $\widehat{A}_i = h^2 P_i$ is unbiased for A_i, see Eq. (4.6.1). Thus,

$$\widetilde{V} = T \cdot h^2 \cdot (P_1 + P_2 + \cdots + P_n) \tag{5.3.6}$$

is unbiased for V, see Eq. (4.8.2). Moreover, if the grid is IUR then a predictor of σ_i^2 is given by Eq. (5.11.2) below, namely,

$$\widehat{\sigma}_i^2 = 0.07283 \cdot \widehat{B}_i \cdot h^3, \quad i = 1, 2, \ldots, n, \tag{5.3.7}$$

where $\widehat{B}_i$ is a UE of the boundary length of the ith Cavalieri section, see Eq. (4.6.2). Often the shape factor

$$\varphi = \bar{B}/\sqrt{\bar{A}}, \tag{5.3.8}$$

where $\bar{B}$, $\bar{A}$ denote means obtained from a Cavalieri sample, tends to be rather stable for a given object, and it may be approximated using a systematic subset of three or four Cavalieri sections. In this case it is convenient to estimate v_n, see Eq. (5.3.5), as follows,

$$\widehat{v}_n = c \cdot h^3 \sum_{i=1}^{n} \widehat{B}_i = c \cdot h^4 \cdot \varphi \cdot \sqrt{nP}, \tag{5.3.9}$$

where $c = 0.07283$, and $P \equiv P_1 + P_2 + \cdots + P_n$. Then, dividing both sides of Eq. (5.3.2) by $\widetilde{V}^2 = T^2 h^4 P^2$, for $n \geq 3$ we obtain,

$$\mathrm{ce}^2(\widetilde{V}) = \frac{\alpha(q)}{P^2} \cdot \left[3\left(\widetilde{C}_0 - c\varphi\sqrt{nP}\right) - 4\widetilde{C}_1 + \widetilde{C}_2\right] + \frac{c\varphi\sqrt{n}}{P^{3/2}}, \tag{5.3.10}$$

$$\widetilde{C}_k = \sum_{i=1}^{n-k} P_i P_{i+k}, \quad k = 0, 1, \ldots, n-1, \tag{5.3.11}$$

which involves neither the sampling period T nor the grid size h. The first term in the rhs of Eq. (5.3.10) estimates the between sections contribution to the total square coefficient of error of $\widetilde{V}$, and the second the point counting contribution within sections.

The smoothness constant q may be estimated by means of Eq. (5.2.15) with C_0 replaced with $\widetilde{C}_0 - c\varphi\sqrt{nP}$ and C_k with $\widetilde{C}_k$, $k = 1, 2, 4$.

5.3.3 Derivation of the variance prediction formula with local errors

For convenience we replace the notation A, $\widehat{A}$ used above with f, $\widehat{f}$, respectively. The problem is to adapt Eq. (5.2.27) to this case. The model adopted is,

$$\widehat{f}(x) = f(x) + e(x), \quad x \in \mathbb{R}, \tag{5.3.12}$$

where $e(x)$ is a random error which vanishes outside a bounded domain and has the following properties: (i) $\mathbb{E}\{e(x)\} = 0$, (ii) $\mathrm{Var}\{e(x)\} = \sigma^2(x) \geq 0$, non-random, and (iii) $\mathrm{Cov}\{e(x), e(x+h)\} = 0$, $|h| \geq T$. The updated Cavalieri estimator of V becomes

$$\widetilde{V} = T \sum_{k \in \mathbb{Z}} \widehat{f}(z + kT) = \widehat{V} + T \sum_{k \in \mathbb{Z}} e(z + kT), \tag{5.3.13}$$

which is also unbiased for V. Now, by Eq. (A.1.44),

$$\begin{aligned}
\mathrm{Var}(\widetilde{V}) &= \mathrm{Var}\{\mathbb{E}(\widetilde{V}|z)\} + \mathbb{E}\{\mathrm{Var}(\widetilde{V}|z)\} \\
&= \mathrm{Var}(\widehat{V}) + T^2 \sum_{k \in \mathbb{Z}} \int_0^T \sigma^2(z + kT)\, \frac{\mathrm{d}z}{T} \\
&= \mathrm{Var}(\widehat{V}) + T \int_{-\infty}^{\infty} \sigma^2(x)\, \mathrm{d}x.
\end{aligned} \tag{5.3.14}$$

The mean value of the covariogram of the estimator $\widehat{f}$ of f is

$$\mathbb{E}g(h; \widehat{f}) = \int_{-\infty}^{\infty} \mathbb{E}\{\widehat{f}(x)\widehat{f}(x+h)\}\, \mathrm{d}x = g(h) + \mathbb{E}g(h; e), \tag{5.3.15}$$

where, by the properties of $e(x)$,

$$\mathbb{E}g(h;e) = \int_{-\infty}^{\infty} \mathbb{E}\{e(x)e(x+h)\}\,\mathrm{d}x$$
$$= \begin{cases} \int_{-\infty}^{\infty} \sigma^2(x)\,\mathrm{d}x & \text{if } h=0, \\ 0 & \text{if } |h| \geq T, \end{cases} \tag{5.3.16}$$

is the mean covariogram of the error function. From the preceding two results,

$$g(0) = \mathbb{E}g(0;\widehat{f}) - \int_{-\infty}^{\infty} \sigma^2(x)\,\mathrm{d}x, \tag{5.3.17}$$

$$g(h) = \mathbb{E}g(h;\widehat{f}), \quad |h| \geq T. \tag{5.3.18}$$

The first term in the rhs of Eq. (5.3.14) is approximated by Eq. (5.2.27) in which $g(0)$ is replaced with Eq. (5.3.17). On the other hand, the second term in the rhs of Eq. (5.3.14) is estimated by $T^2 v_n$. It is easy to verify that the estimator

$$\widehat{\mathbb{E}}g(h;\widehat{f}) = T\widehat{C}_k, \quad \widehat{C}_k = \sum_{i=1}^{n-k} \widehat{f}_i \widehat{f}_{i+k}, \quad k = 0, 1, \ldots, n-1, \tag{5.3.19}$$

where $\widehat{f}_i = f_i + e_i$ is the discretized version of Eq. (5.3.12), is unbiased for $\mathbb{E}g(h;\widehat{f})$. Substitution of the pertinent approximations into Eq. (5.3.14) yields Eq. (5.3.2).

5.3.4 Notes

1. Local errors

The contribution of local errors to $\mathrm{Var}(\widetilde{V})$ was called the 'nugget effect' by Matheron (1971), who developed the general concept mainly in a materials sciences context under the model-based approach. For the design-based case, on p. 12 he described in words the fact that, in the presence of the nugget effect, $\mathbb{E}g(0;\widehat{f}) > g(0)$, see Eq. (5.3.17).

2. The practical formula

Eq. (5.3.2), (with $\alpha(0) = 1/12$), was given in Cruz-Orive (1993). The derivation was improved by Kiêu (1997) via Eq. (5.3.14). The predictors Eq. (5.3.2) and Eq. (5.3.10) were revisited by Cruz-Orive (1999) and by Gundersen, Jensen, Kiêu, and Nielsen (1999). For an early application of Eq. (5.3.10) (with $\alpha(1) = 1/240$) see Roberts, Garden, Cruz-Orive, Whitehouse, and Edwards (1994).

5.4 Cavalieri Slabs Affected By Local Errors

5.4.1 Practical variance prediction formulae

A slab in $\mathbb{R}^3$ is a three-dimensional probe, and therefore the target parameter $\gamma(Y) > 0$ defined on a compact set $Y \subset \mathbb{R}^3$ may stand for N (which is the typical target in this context), L, S, or V. Let $L_t(x)$ denote a slab of thickness $t > 0$ normal to an arbitrary sampling axis at a point of abscissa x. Then, by Eq. (1.4.12),

$$\gamma(Y) = \frac{1}{t} \int_{\mathbb{R}} \gamma_t(x)\, \mathrm{d}x, \tag{5.4.1}$$

where

$$\gamma_t(x) \equiv \gamma(Y \cap L_t(x)), \quad x \in \mathbb{R}, \tag{5.4.2}$$

(with $\gamma_t(\emptyset) = 0$) is the measure of the slice contents (which has the same dimension as the target object). The probe consists of Cavalieri slabs of period $T \geq t > 0$, see Fig. 5.4.1. The primary data are the true slab contents,

$$\{\gamma_{tk} \equiv \gamma_t(z + kT),\ k = 1, 2, \ldots, n\}, \quad z \sim \mathrm{UR}[0, T). \tag{5.4.3}$$

Here we consider the general case in which the observed sample consists of unbiased estimators of the slab contents, which we renumber in the usual order as $\{\widehat{\gamma}_{t1}, \widehat{\gamma}_{t2}, \ldots, \widehat{\gamma}_{tn}\}$. By Eq. (2.25.20), a UE of $\gamma \equiv \gamma(Y)$ is

$$\widetilde{\gamma} = \frac{T}{t} \cdot (\widehat{\gamma}_{t1} + \widehat{\gamma}_{t2} + \cdots + \widehat{\gamma}_{tn}). \tag{5.4.4}$$

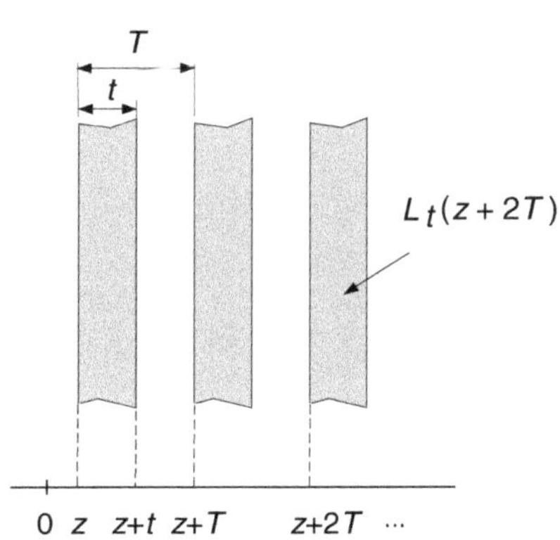

Fig. 5.4.1 Portions of three Cavalieri slab projections onto a normal plane.

A predictor of $\mathrm{Var}(\widetilde{\gamma})$ from a sample is

$$\mathrm{var}(\widetilde{\gamma}) = \alpha(q,\tau)\cdot\tau^{-2}\cdot\left[3\left(\widehat{C}_0 - \upsilon_n\right) - 4\widehat{C}_1 + \widehat{C}_2\right] + \tau^{-2}\upsilon_n, \tag{5.4.5}$$

where

$$\begin{aligned}
\tau &= t/T, \quad \tau \in [0,1],\\
\widehat{C}_k &= \sum_{i=1}^{n-k}\widehat{\gamma}_{ti}\cdot\widehat{\gamma}_{t,i+k}, \quad k = 0,1,\ldots,n-1,\\
\upsilon_n &= \sigma_1^2 + \sigma_2^2 + \cdots + \sigma_n^2,\\
\sigma_i^2 &= \mathrm{Var}(\widehat{\gamma}_{ti}|\gamma_{ti}).
\end{aligned} \tag{5.4.6}$$

As usual, the local error variances $\{\sigma_i^2\}$ have to be estimated in each particular case. The first term in the rhs of Eq. (5.4.5) estimates the slices component, and the second one the local error component. On the other hand,

$$\begin{aligned}
\alpha(q,\tau) &= \frac{\cos(\pi q)}{D(q,\tau)}\cdot\frac{8\Gamma(2q+4)}{(2\pi)^{2q+4}}\cdot\sum_{k=1}^{\infty}\frac{1-\cos(2\pi\tau k)}{k^{2q+4}}, \qquad (5.4.7)\\
D(q,\tau) &= 4(1+\tau)^{2q+3} + 4(1-\tau)^{2q+3} - (2+\tau)^{2q+3} - (2-\tau)^{2q+3}\\
&\quad - 6\tau^{2q+3} + 8(2^{2q+1}-1),
\end{aligned}$$

where $q \in [0,1]$ is the smoothness constant of a virtual measurement function $f(x)$ which is defined in the next section. The coefficient $\alpha(q,\tau)$ is plotted in Fig. 5.4.2 for $q \in [0,1]$ and for several values of τ. The summation in the rhs of Eq. (5.4.7) may be evaluated using the following identity,

$$\sum_{k=1}^{\infty}\frac{1-\cos(2\pi\tau k)}{k^s} = \zeta(s) - \mathrm{Re}(\mathrm{Li}_s(\mathrm{e}^{2\pi i\tau})), \tag{5.4.8}$$

where

$$\mathrm{Li}_s(z) = \sum_{k=1}^{\infty}\frac{z^k}{k^s}, \quad (|z| \le 1,\ s > 1), \tag{5.4.9}$$

is the polylogarithm function. The rhs of Eq. (5.4.8) may be computed with the aid of software packages such as Maple®, or Mathematica®.

Cases $q = 0, 1$

For integer q we have,

$$\alpha(q,\tau) = \frac{2}{q+2}\cdot\frac{B_{2q+4}(\tau) - B_{2q+4}(0)}{D(q,\tau)}, \quad q = 0,1,\ldots, \tag{5.4.10}$$

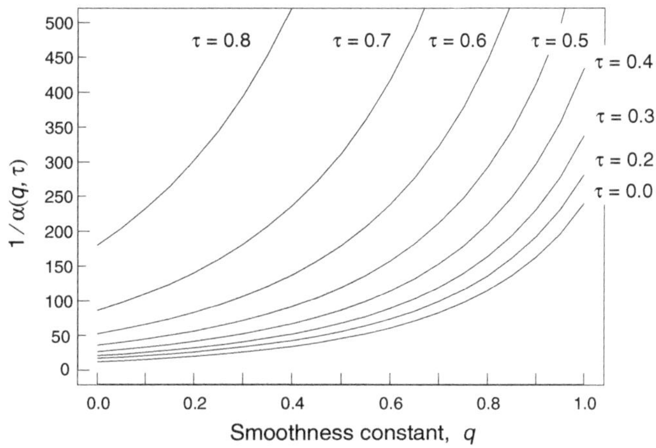

Fig. 5.4.2 Graph of the reciprocal of the coefficient given by Eq. (5.4.7). Reproduced from Cruz-Orive (2006), with permission of Wiley-Blackwell.

where $B_r(x)$ is the Bernoulli polynomial of integer order r. In particular,

$$\alpha(0,\tau) = \frac{1}{6}\cdot\frac{(1-\tau)^2}{2-\tau}, \tag{5.4.11}$$

$$\alpha(1,\tau) = \frac{1}{6}\cdot\frac{(1+2\tau-2\tau^2)(1-\tau)^2}{40-10\tau^2+3\tau^3}. \tag{5.4.12}$$

Cases $\tau = 0, 1$

If the slabs are replaced with planes, then the limit of the rhs of Eq. (5.4.7) as $\tau \to 0$ is the coefficient $\alpha(q)$ given by Eq. (5.2.13), as expected. On the other hand, $\alpha(q, 1) = 0$, also as expected.

Case $q = 1/2$

A passage to the limit in the rhs of Eq. (5.4.7) as $q \to 1/2$ yields,

$$\alpha(1/2,\tau) = \frac{3}{\pi^4 A(\tau)}\cdot\sum_{k=1}^{\infty}\frac{1-\cos(2\pi\tau k)}{k^5}, \tag{5.4.13}$$

$$\begin{aligned} A(\tau) = &-4(1+\tau)^4\log(1+\tau) - 4(1-\tau)^4\log(1-\tau) \\ &+ (2+\tau)^4\log(2+\tau) + (2-\tau)^4\log(2-\tau) \\ &- 6\tau^4\log(\tau) - 32\log 2. \end{aligned}$$

A further passage to the limit when $\tau \to 0$ in the preceding expression yields

$$\alpha(1/2, 0) = \frac{\zeta(3)}{8\pi^2 \log 2} \approx \frac{1}{45}. \qquad (5.4.14)$$

5.4.2 Derivations

A basic assumption underlying Eq. (5.4.5) is that there exists an integrable function $f : \mathbb{R} \to \mathbb{R}^+$ such that the slice contents can be expressed as follows,

$$\gamma_t(x) = \int_x^{x+t} f(y)\,\mathrm{d}y, \quad x \in \mathbb{R},\ t > 0. \qquad (5.4.15)$$

The parameter q is the smoothness constant of f. The observable function is the regularization of f by the indicator function of the segment $[0, t)$, which we denote by 1_t for short. Thus, by the analogue of Eq. (2.10.14),

$$f_t(x) = t^{-1} \cdot \gamma_t(x) = t^{-1} \cdot (f * \check{1}_t)(x), \qquad (5.4.16)$$

which shows that f_t is a regularization or filtering of f by the segment $[0, t)$. Now, by Eq. (5.4.1), the target parameter reads

$$\gamma(Y) = \int_{\mathbb{R}} f_t(x)\,\mathrm{d}x. \qquad (5.4.17)$$

Ignoring for the moment the local errors, the Cavalieri UE of $\gamma(Y)$ is

$$\widehat{\gamma} = T \cdot \sum_{k \in \mathbb{Z}} f_t(z + kT), \quad z \sim \mathrm{UR}[0, T), \qquad (5.4.18)$$

(see for instance Eq. (4.9.2)), and by analogy with Eq. (5.2.23),

$$\mathrm{Var}(\widehat{\gamma}) = 2 \sum_{k=1}^{\infty} G_t(k/T), \qquad (5.4.19)$$

where $G_t = \mathcal{F} g_t$ and g_t is the covariogram of f_t, which may be expressed as follows. Consider the geometric covariogram of the segment $[0, t)$, namely,

$$K(h) = (1_t * \check{1}_t)(h) = t - |h|, \quad -t \le h < t, \qquad (5.4.20)$$

see Eq. (5.2.31). Then,

$$g_t = f_t * \check{f}_t = t^{-1} f * \check{1}_t * t^{-1} \check{f} * 1_t = t^{-2} f * \check{f} * 1_t * \check{1}_t = t^{-2} g * K, \qquad (5.4.21)$$

and, by the convolution theorem,

$$G_t = t^{-2}\mathcal{F}g\mathcal{F}K. \tag{5.4.22}$$

In order to obtain the extension term $\mathrm{Var}(\widehat{\gamma})$, we adopt the model of g given by Eq. (5.2.24). Again, $(\mathcal{F}g)(\rho) = (\mathcal{F}|h|^{2q+1})(\rho)$, is given by Eq. (A.3.3), whereas

$$(\mathcal{F}K)(\rho) = \int_{-t}^{t}(t-|h|)\cdot\exp(-2\pi\mathrm{i}h)\,\mathrm{d}h = \frac{1-\cos(2\pi t\rho)}{2\pi^2\rho^2}. \tag{5.4.23}$$

Now Eq. (5.4.22) yields the model of G_t, and in turn Eq. (5.4.19) yields the extension term,

$$\begin{aligned}\mathrm{Var}_E(\widetilde{\gamma}) = -t^{-2}b_{2q+1}T^{2q+4}\cdot\frac{\cos(\pi q)}{2\pi^4}\\ \cdot\frac{\Gamma(2q+2)}{(2\pi)^{2q}}\cdot\sum_{k=1}^{\infty}\frac{1-\cos(2\pi(t/T)k)}{k^{2q+4}}.\end{aligned} \tag{5.4.24}$$

It remains to find an expression of b_{2q+1} analogous to Eq. (5.2.26). From Eq. (5.4.21),

$$g_t(r) = \frac{1}{t^2}\int_0^t (t-x)[g(r+x)+g(r-x)]\,\mathrm{d}x, \quad r>0, \tag{5.4.25}$$

which can be evaluated explicitly for the model given by Eq. (5.2.24). The relevant coefficient becomes,

$$\begin{aligned}b_{2q+1} &= -\frac{t^2}{T^{2q+3}}\cdot\frac{(2q+2)(2q+3)}{D(q,\tau)}\\ &\quad\cdot(3g_t(0)-4g_t(T)+g_t(2T)),\end{aligned} \tag{5.4.26}$$

see Note 1 in the next section. Finally, $g_t(kT)$ may be estimated without bias as follows,

$$\widehat{g}_t(kT) = \frac{T}{t^2}\cdot C_k, \quad C_k = \sum_{i=1}^{n-k}\gamma_{ti}\gamma_{t,i+k}, \quad k=0,1,\ldots,n-1. \tag{5.4.27}$$

Substituting the preceding results into the rhs of Eq. (5.4.24), and incorporating the local error contributions, the predictor given by Eq. (5.4.5) is obtained.

For integer q, Eq. (5.4.10) follows from Eq. (5.4.7) because, for $s=1,2,\ldots$,

$$\sum_{k=1}^{\infty}\frac{1-\cos(2\pi\tau k)}{k^{2s}} = (-1)^{s-1}\frac{(2\pi)^{2s}}{2\Gamma(2s+1)}\cdot(B_{2s}(0)-B_{2s}(\tau)), \tag{5.4.28}$$

where $B_{2s}(x)$ is a Bernoulli polynomial of even order, e.g., $B_2(x) = x^2 - x + 1/6$, $B_4(x) = x^4 - 2x^3 + x^2 - 1/30$, etc.

5.4.3 Notes

1. Variance prediction for Cavalieri slabs

Eq. (5.4.5) was derived by Cruz-Orive (2006), where some details omitted in Section 5.4.2 can be found. Some numerical values of $\alpha(q, \tau)$ can be obtained from Table 1 and Fig. 1 of that paper (the latter figure is reproduced here, see Fig. 5.4.2). Eq. (5.4.11) and Eq. (5.4.12) were first obtained by Gual-Arnau and Cruz-Orive (1998). For results involving Bernoulli polynomials, e.g. Eq. (5.4.28), see for instance Abramowitz and Stegun (1965).

2. The smoothness constant

It should be stressed that, if the target parameter is volume, i.e., if $\gamma \equiv V$, then the smoothness constant q corresponds to the measurement (area) function with $\tau = 0$. If $\gamma \equiv N$, then we may take $q = 0$.

3. Applications

Eq. (5.4.5) is applied in Section 4.9.3, with no local errors, namely with $\upsilon_n = 0$. In an animal model study on particle retention in hamster lungs, Cruz-Orive and Geiser (2004) used primary slab subsampling with physical Cavalieri disectors. The target was the number of particles (polystyrene microspheres) deposited in the hamster lung alveoli after inhalation. The local error contribution υ_n was estimated using two different approaches, namely the Poisson model for the spatial distribution of particles, and the double disector technique. See also Section 5.5.3, Note 2 below.

5.5 The Splitting Design for Cavalieri Slabs

5.5.1 Practical variance prediction formulae

The target parameter $\gamma(Y)$ is given by Eq. (5.4.1). The sampling design consists of the following two stages.

Stage 1. Intersect the object Y with a primary series of Cavalieri slabs of thickness t and period $T \geq t > 0$, namely,

$$\{L_t(z + kT),\ k \in \mathbb{Z}\}, \quad z \sim \mathrm{UR}[0, T). \tag{5.5.1}$$

The primary sampling fraction is

$$\tau = t/T, \quad 0 < \tau \leq 1. \tag{5.5.2}$$

Stage 2. For a given realization of z, the primary slab series is split into $n \geq 2$ systematic subsets. The ith subset is

$$\{L_t(z + (i-1)T + knT),\ k \in \mathbb{Z}\}, \quad i = 1, 2, \ldots, n,\ n = 2, 3, \ldots, \tag{5.5.3}$$

see Fig. 5.5.1 for an illustration with $n = 2$. The period of each subset is nT, and the corresponding sampling fraction is τ/n.

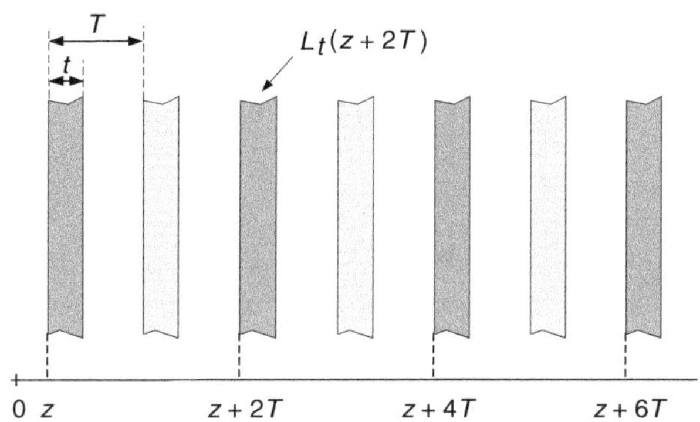

Fig. 5.5.1 Illustration of the splitting design with two subsets of Cavalieri slabs.

The primary sample total is,

$$Q \equiv \sum_{k \in \mathbb{Z}} \gamma_t(z + kT), \tag{5.5.4}$$

where γ_t is defined by Eq. (5.4.2). The corresponding UE of $\gamma(Y)$ is

$$\widehat{\gamma} = \frac{1}{\tau} \cdot Q. \tag{5.5.5}$$

For each z, the n possible subsample totals are

$$\left\{ Q_i \equiv \sum_{k \in \mathbb{Z}} \gamma_t(z + (i-1)T + knT),\ i = 1, 2, \ldots, n \right\}, \tag{5.5.6}$$

and clearly,

$$Q = Q_1 + Q_2 + \cdots + Q_n. \tag{5.5.7}$$

Suppose that each subtotal Q_i is estimated without bias by $\widehat{Q}_i$, namely,

$$\widehat{Q}_i = Q_i + e_i, \quad (i = 1, 2, \ldots, n), \tag{5.5.8}$$

$$\widehat{Q} = \sum_{j=1}^{n} \widehat{Q}_i = Q + \sum_{i=1}^{n} e_i, \tag{5.5.9}$$

where the $\{e_i\}$ are uncorrelated random local errors with zero mean and corresponding variances $\{\sigma_i^2\}$, with

$$v_n \equiv \sigma_1^2 + \sigma_2^2 + \cdots + \sigma_n^2. \tag{5.5.10}$$

The UE of $\gamma(Y)$ now is

$$\widetilde{\gamma} = \frac{1}{\tau} \cdot \widehat{Q}, \tag{5.5.11}$$

and a predictor of its variance is

$$\begin{aligned} \mathrm{var}(\widetilde{\gamma}) = & \frac{(1-\tau)^2}{\tau^2(n+1-2\tau)} \left[\frac{n}{n-1} \sum_{i=1}^{n} (\widehat{Q}_i - n^{-1}\widehat{Q})^2 - \widehat{v}_n \right] \\ & + \frac{1}{\tau^2} \cdot \widehat{v}_n. \end{aligned} \tag{5.5.12}$$

where $\widehat{v}_n$ is an estimator of v_n. This design is usually applied to estimate particle number, in which case the primary measurement function introduced in Section 5.4.2 is expected to exhibit finite jumps. Hence the preceding predictor assumes $q = 0$ for the underlying measurement function f defined via Eq. (5.4.15).

5.5.2 Derivations

If $\{Q_1, Q_2, \ldots, Q_n\}$ represents a UR permutation of the n subtotals, then the latter are identically distributed and therefore,

$$\begin{aligned} \mathbb{E}(Q_i|Q) &= \frac{1}{n} Q, \\ \mathbb{E}(Q_i) &= \frac{1}{n} \mathbb{E}(Q). \end{aligned} \tag{5.5.13}$$

(Here, and in the sequel, $i = 1, 2, \ldots, n$.) It follows that the second stage estimator

$$\widehat{\gamma}_i = \frac{n}{\tau} \cdot Q_i \tag{5.5.14}$$

is unbiased for $\gamma(Y)$ and the $\{\widehat{\gamma}_i\}$ are identically distributed. The estimator $\widehat{\gamma}$, see Eq. (5.5.5), comes from Cavalieri slices of thickness t and period T, whereas each estimator $\widehat{\gamma}_i$ comes from Cavalieri slices of thickness t and period nT. Taking Eq. (5.4.11) into account, for $q = 0$ the rhs of Eq. (5.4.24) yields, respectively,

$$\mathrm{Var}_E(\widehat{\gamma}) = -\frac{1}{6} \cdot T^2 (1-\tau)^2 b_1, \tag{5.5.15}$$

$$\mathrm{Var}_E(\widehat{\gamma}_i) = -\frac{1}{6} \cdot T^2 (n-\tau)^2 b_1. \tag{5.5.16}$$

Eliminating the coefficient b_1 we obtain,

$$\mathrm{Var}_E(\widehat{\gamma}_i) = \left(\frac{n-\tau}{1-\tau}\right)^2 \mathrm{Var}_E(\widehat{\gamma}). \tag{5.5.17}$$

Because the $\{Q_i\}$ are identically distributed, the preceding result implies the following model identity (henceforth the subscript 'E' is removed, for short),

$$\mathrm{Var}(Q_i) = \theta^2 \mathrm{Var}(Q), \tag{5.5.18}$$

$$\theta \equiv \frac{n-\tau}{n(1-\tau)}, \quad (\theta \geq 1). \tag{5.5.19}$$

The incorporation of local errors, see Eq. (5.5.8) and Eq. (5.5.9), yields,

$$\mathbb{E}(\widehat{Q}_i|\widehat{Q}) = \frac{1}{n}\widehat{Q}, \tag{5.5.20}$$

$$\mathrm{Var}(\widehat{Q}_i) = \mathrm{Var}(Q_i) + \sigma_i^2 = \theta^2\mathrm{Var}(Q) + \sigma_i^2, \tag{5.5.21}$$

$$\mathrm{Var}(\widehat{Q}) = \mathrm{Var}(Q) + v_n, \tag{5.5.22}$$

and eliminating $\mathrm{Var}(Q)$ from the last two equations we obtain,

$$\mathrm{Var}(\widehat{Q}_i) = \theta^2\mathrm{Var}(\widehat{Q}) + (\sigma_i^2 - \theta^2 v_n). \tag{5.5.23}$$

The last steps of the derivation of Eq. (5.5.12) run as follows. By Eq. (A.1.44), we have

$$\mathrm{Var}(\widehat{Q}_i) = \mathbb{E}\{\mathrm{Var}(\widehat{Q}_i|\widehat{Q})\} + \mathrm{Var}\{\mathbb{E}(\widehat{Q}_i|\widehat{Q})\}. \tag{5.5.24}$$

Using Eq. (5.5.20) and Eq. (5.5.23), and adding up from $i = 1$ to $i = n$, we obtain

$$\begin{aligned} n\theta^2\mathrm{Var}(\widehat{Q}) + v_n - n\theta^2 v_n &= \sum_{i=1}^{n} \mathbb{E}\{\mathrm{Var}(\widehat{Q}_i|\widehat{Q})\} + \frac{1}{n}\cdot\mathrm{Var}(\widehat{Q}) \\ &= \mathbb{E}\left\{\sum_{i=1}^{n}(\widehat{Q}_i - n^{-1}\widehat{Q})^2|\widehat{Q}\right\} + \frac{1}{n}\cdot\mathrm{Var}(\widehat{Q}). \end{aligned} \tag{5.5.25}$$

By Eq. (5.5.11) we have $\mathrm{Var}(\widetilde{\gamma}) = \tau^{-2}\mathrm{Var}(\widehat{Q})$, whereby,

$$\mathrm{var}(\widetilde{\gamma}) = \frac{\sum_{i=1}^{n}(\widehat{Q}_i - n^{-1}\widehat{Q})^2 + (n\theta^2 - 1)v_n}{\tau^2(n\theta^2 - 1/n)}. \tag{5.5.26}$$

Finally, replacing θ with the rhs of Eq. (5.5.19), then subtracting and adding the term $\tau^{-2}v_n$ in the rhs of the preceding expression, and replacing v_n with its estimator $\widehat{v}_n$, the variance predictor given by Eq. (5.5.12) is obtained.

5.5.3 Notes

1. The splitting design

Eq. (5.5.12) was derived in Cruz-Orive (2004). It constitutes a generalization of Eq. 15 from Cruz-Orive (1990), which was obtained for $n = 2$ and $\tau = 0$. The possibility of a generalization was suggested by Neil Roberts (personal communication). The derivation steps following Eq. (5.5.24) were suggested by Marta García-Fiñana (personal communication). The formula may be tentatively applied to fractionator designs such as that illustrated in Fig. 4.11.1, in which Cavalieri slices are replaced with systematic fragments from the target object.

2. Application

Besides the Cavalieri slices design, Cruz-Orive and Geiser (2004) also applied the splitting design to the same data. Fig. 8 of that paper suggests that the Cavalieri slices design with double disector subsampling yielded the lowest variance estimates, followed by the splitting design, also with double disector subsampling.

5.6 Precision of the Estimation of Particle Number in the Plane With Systematic Quadrats

5.6.1 Practical variance prediction formula

Let $Y \subset \mathbb{R}^2$ represent a bounded and finite set of $N \equiv N(Y)$ particles in the plane. The target parameter is N. For an illustration, see Section 4.10. The probe is a test system Λ_x of quadrats, see Eq. (2.25.28). The fundamental tile J_0 is a square of side length T and the fundamental probe T_0 is a square quadrat of side length t, where $0 < t \leq T$. Let Q denote the total, random number of particles captured by Λ_x, $x \sim \mathrm{UR}(J_0)$. If Y consists of point particles, then Q is their total number within the quadrats, otherwise the particles are captured according to an unbiased sampling rule such as the forbidden line rule. As a special case of Eq. (2.25.28), the estimator

$$\widehat{N} = \frac{1}{\tau^2} \cdot Q, \tag{5.6.1}$$

where $\tau = t/T \in (0, 1]$, is unbiased for N. In order to interpret the predictor of $\mathrm{Var}(\widehat{N})$ given below, the quadrat sample is regarded as a two-stage sample. The first stage sample consists of a set of n Cavalieri stripes of thickness t and period T encompassing the particle population Y. In the second stage, every stripe is subsampled by a perpendicular series of Cavalieri stripes with the same parameters (t, T). The result is a portion of Λ_x providing the following data.

- Q_{oi}, Q_{ei}: total numbers of particles captured by the odd and the even numbered quadrats, respectively, within the ith stripe.
- $Q_i = Q_{oi} + Q_{ei}$: total number of particles captured by all the quadrats within the ith stripe. Thus, $Q = \sum_{i=1}^{n} Q_i$.

The variance predictor reads as follows,

$$\operatorname{var}(\widehat{N}) = \frac{\alpha(0,\tau)}{\tau^4}\{3(C_0 - \widehat{v}_n) - 4C_1 + C_2\} + \frac{\widehat{v}_n}{\tau^4}, \quad n \geq 3, \tag{5.6.2}$$

$$C_k = \sum_{i=1}^{n-k} Q_i Q_{i+k}, \quad k = 0, 1, 2, \tag{5.6.3}$$

$$\widehat{v}_n = \sum_{i=1}^{n} \operatorname{var}(Q_i) = \frac{(1-\tau)^2}{3-2\tau}\sum_{i=1}^{n}(Q_{oi} - Q_{ei})^2. \tag{5.6.4}$$

The first term in the rhs of Eq. (5.6.2) is the contribution of the Cavalieri stripes. It is analogous to the first term in the rhs of Eq. (5.4.5) for Cavalieri slabs with $q = 0$, and the coefficient $\alpha(0,\tau)$ is given by Eq. (5.4.11). On the other hand, $\widehat{v}_n/\tau^4$ is the within stripes contribution, and $\widehat{v}_n$ is evaluated by the splitting design approach using Eq. (5.6.4). Each term in the rhs of Eq. (5.6.4) is a special case of Eq. (5.5.12) with $n = 2$ quadrat subsets (namely the odd and the even numbered ones) and no local errors, because Q_{oi}, Q_{ei} are directly observable for every stripe.

The variance predictor is first evaluated for a given direction of the stripes, and then for the perpendicular direction, the final predictor being the average of both results.

5.6.2 Derivation

The first term in the rhs of Eq. (5.6.2) bears a factor τ^{-4}, whereas the corresponding factor in Eq. (5.4.5) is τ^{-2}. To explain this, let N_i denote the total number of particles captured by the ith stripe. With the quadrat design, a UE of N_i is $\widehat{N}_i = \tau^{-1}Q_i$ In Eq. (5.4.6) the estimator $\widehat{\gamma}_{ti}$ is analogous to $\widehat{N}_i$, hence the $\{\widehat{C}_k\}$ in Eq. (5.4.6) correspond to τ^{-2} times the $\{C_k\}$ in Eq. (5.6.2).

To justify the factor τ^{-4} in the second term of Eq. (5.6.2), note that the estimator $\widehat{v}_n$ of v_n as defined in Eq. (5.4.6) is analogous to $\sum_{i=1}^{n} \operatorname{var}(\widehat{N}_i) = \tau^{-2}\sum_{i=1}^{n} \operatorname{var}(Q_i)$, whereas in Eq. (5.6.2), $\widehat{v}_n = \sum_{i=1}^{n} \operatorname{var}(Q_i)$.

5.6.3 Notes

1. Motivation

Eq. (5.6.2) was first proposed by Cruz et al. (2015) to predict the precision of population size estimation with systematic quadrats on aerial views, with applications in sociology and ecology. The method was extended to gigapixel images by Cruz and González-Villa (2021), who handled the perspective artifacts affecting the images – see also Cruz-Orive and Cruz (2020).

2. Double splitting design

Gómez et al. (2019) compared the performance of Eq. (5.6.2) against an alternative formula in which the Cavalieri component (first term in the rhs) is replaced with its splitting design version analogous to Eq. (5.6.4). The authors used automatic resampling on 26 point particle populations exhibiting a variety of patterns. Both formulae yielded nearly the same predictors of the mean variance, but the variance among predictors was lower for Eq. (5.6.2) than for the splitting alternative. The derivations given in Section 5.6.2 were outlined in the Appendix of the latter paper, and hints were also given to generalize Eq. (5.6.2) to higher dimensions.

5.7 Precision of the Estimation of Planar Curve Length With a Square Grid

5.7.1 Practical variance prediction formula

The target is a bounded, piecewise smooth curve $Y \subset \mathbb{R}^2$ of finite length $B > 0$, which is the target parameter. In Section 4.6, the curve is the boundary of the orthogonal projection of a leaf onto the observation plane. The probe is a square grid of test lines whose fundamental tile J_0 is a square of side length $T > 0$. The curve and the grid are IUR relative to each other, see Fig. 2.26.1. For $\phi \sim \mathrm{UR}[0, \pi/2)$, let l_1, l_2 denote the total orthogonal projected lengths of the curve onto two axes making angles ϕ and $\phi + \pi/2$ with a fixed axis Ox_1, respectively. Then, by Cauchy's formula, see Eq. (2.22.1), the estimator

$$\widehat{B} = \frac{\pi}{4}(l_1 + l_2) \tag{5.7.1}$$

is unbiased for B. Now, sample Cavalieri lines of period T along each projection axis. For the ith axis, let

$$\{I_{i1}, I_{i2}, \ldots, I_{in_i}\}, \quad i = 1, 2, \tag{5.7.2}$$

denote the total numbers of intersections determined in the curve by the n_i Cavalieri lines hitting the curve. Then, the estimator

$$\widehat{l_i} = T \sum_{j=1}^{n_i} I_{ij} \equiv T \cdot I_i, \quad i = 1, 2, \tag{5.7.3}$$

is unbiased for l_i. An equivalent design applies if the curve is equipped with an associated vector (x, ω), where $x \sim \mathrm{UR}(J_0)$ and $\omega \sim \mathrm{UR}[0, 2\pi)$, and independent. Consequently, the Buffon–Steinhaus estimator

$$\widetilde{B} = \frac{\pi}{4} \cdot (\widehat{l}_1 + \widehat{l}_2) = \frac{\pi}{4} \cdot T \cdot (I_1 + I_2), \tag{5.7.4}$$

is unbiased for B, see Section 4.6.

The length estimators $\widehat{B}$ and $\widetilde{B}$ may be regarded as one- and two-stage estimators, respectively.

A predictor of $\mathrm{Var}(\widetilde{B})$ from a single IUR superimposition of the grid and the curve is

$$\mathrm{var}(\widetilde{B}) = \frac{\pi^2}{96} \cdot \left[(\widehat{l}_1 - \widehat{l}_2)^2 - \widehat{v}_2 \right] + \frac{\pi^2}{16} \cdot \widehat{v}_2, \tag{5.7.5}$$

where

$$\widehat{l}_i = T \cdot I_i, \tag{5.7.6}$$

$$\widehat{v}_2 = \sum_{i=1}^{2} \mathrm{var}(\widehat{l}_i | l_i), \tag{5.7.7}$$

$$\mathrm{var}(\widehat{l}_i | l_i) = \frac{T^2}{12} \cdot (3C_{0i} - 4C_{1i} + C_{2i}), \quad n_i \geq 3, \tag{5.7.8}$$

$$\mathrm{var}(\widehat{l}_i | l_i) = \frac{T^2}{6} \cdot (C_{0i} - C_{1i}), \quad n_i = 2, \tag{5.7.9}$$

$$C_{ki} = \sum_{j=1}^{n_i - k} I_{ij} I_{i,j+k}, \quad k = 0, 1, \ldots, n_i - 1, \ i = 1, 2. \tag{5.7.10}$$

The first term in the rhs of Eq. (5.7.5) estimates the variance contribution due to the variability among curve projections, whereas the second one estimates the variance contribution due to the variability among the intersection counts within projections. In Section 4.6, see also Note 1 of Section 5.7.3 below, the former contribution was relatively negligible. Note that the former and the latter contributions are in fact predictors of $\mathrm{Var}\{\mathbb{E}(\widetilde{B}|\widehat{B})\} = \mathrm{Var}(\widehat{B})$ and of $\mathbb{E}\{\mathrm{Var}(\widetilde{B}|\widehat{B})\}$, respectively, whose sum is just $\mathrm{Var}(\widetilde{B})$.

5.7.2 Derivations

The target parameter considered so far was an integral of the form given by Eq. (5.2.8). In the present application, however, the curve length B may be expressed as in Eq. (1.19.7), namely,

$$B = \frac{1}{2} \int_0^{\pi} l(\omega) \, \mathrm{d}\omega, \tag{5.7.11}$$

where $l(\omega)$ is the measurement function. Here $l(\omega)$ is periodic of period π, that is, $l(\omega) = l(\omega + k\pi)$, $k \in \mathbb{Z}$. It follows that the corresponding covariogram,

$$g(h) = \int_0^{\pi} l(\omega) l(\omega + h)\, \mathrm{d}\omega, \quad h \in [0, \pi), \tag{5.7.12}$$

is also periodic of period π, namely $g(h) = g(h + k\pi)$, $k \in \mathbb{Z}$. The extension term of $\mathrm{Var}(\widehat{B})$, where $\widehat{B}$ is the first stage estimator of B given by Eq. (5.7.1) may still be predicted by Eq. (5.2.11), with the important difference that now, instead of Eq. (5.2.12), we have

$$C_k = \sum_{i=1}^{n} l_i l_{i+k}, \text{ with } l_{i+n} = l_i. \tag{5.7.13}$$

Thus,

$$C_0 = l_1^2 + l_2^2, \quad C_1 = l_1 l_2 + l_2 l_1, \tag{5.7.14}$$

$$\mathrm{var}(\widehat{B}) = \frac{1}{6} \cdot \left(\frac{\pi}{4}\right)^2 \cdot (l_1 - l_2)^2. \tag{5.7.15}$$

The application of Eq. (5.3.3) yields Eq. (5.7.5). The estimator $\widehat{l}_i$, see Eq. (5.7.6), is an ordinary Cavalieri estimator in which the measurement function is the number of intersections between the target curve and a Cavalieri test line. This measurement function is integer-valued, and it therefore exhibits finite jumps, whereby its smoothness constant is $q = 0$. Therefore, Eq. (5.7.8) and Eq. (5.7.9) are special cases of Eq. (5.2.10) and Eq. (5.2.11), respectively.

5.7.3 Notes

1. The Cauchy variance component

The problem of predicting the precision of the Buffon–Steinhaus estimator of curve length with a square grid is a classic with an abundant literature. Cruz-Orive and Gual-Arnau (2002), Section 8.2, estimated the Cauchy contribution $\mathrm{Var}(\widehat{B})$ using the theory developed by Gual-Arnau and Cruz-Orive (2000). Their approach was different from Matheron's, in that the covariogram was modelled entirely (that is, not just near the origin) by the periodic extension (of period π) of a positive definite polynomial $g_m(h)$ of degree $2m + 2$, $m \in \{0, 1\}$, satisfying the property $g_m(h) = g_m(\pi - h)$. With this approach (called the global covariogram approach), we obtain

$$\mathrm{var}_0(\widehat{B}) = (\pi^2/48) \cdot (l_1 - l_2)^2 \text{ for } m = 0, \tag{5.7.16}$$

$$\mathrm{var}_1(\widehat{B}) = (\pi^2/240) \cdot (l_1 - l_2)^2 \text{ for } m = 1. \tag{5.7.17}$$

In Section 9 of the former paper, a synthetic resampling experiment without local errors revealed unimportant differences in performance among the global and Matheron's approaches. This agreed with the fact, outlined in Remark 2.2 from Gual-Arnau and Cruz-Orive (2000), that the periodic covariogram for circular sampling, and the ordinary covariogram for Cavalieri sampling, yielded the same expressions of the exact variance in each case. The important point, however, is that, if the measurement function is periodic, then the $\{C_k\}$ have to be computed as in Eq. (5.7.13) under either approach. With this condition, we have adopted Matheron's approach to obtain Eq. (5.7.15) – note that the coefficient $\pi^2/96$ lies in between the corresponding coefficients in Eq. (5.7.16) and Eq. (5.7.17), respectively, which sounds reasonable. In retrospect, using the global approach does not seem worthwhile.

2. The total error variance predictor

By means of automatic replications of the estimator $\widetilde{B}$ on a set of nine digitized planar projections of DNA molecules of various shapes, Gómez, Cruz, and Cruz-Orive (2016) checked the performance of $\text{var}(\widetilde{B})$, adopting $\text{var}_1(\widehat{B})$, see Eq. (5.7.17), for the Cauchy contribution. The main conclusions were: (i) The true Cauchy contribution was practically negligible. Thus, $\text{Var}(\widetilde{B})$ was mainly explained by the local error variance component υ_2 due to the intersection counts. (ii) The predictive performance of $\text{var}(\widetilde{B})$ was satisfactory for practical purposes. (iii) The predictive performance of $\text{var}_1(\widehat{B})$ was poor.

5.8 IUR Test Systems to Estimate Planar Area, or Volume: Preliminary Comments

The Cavalieri designs considered in the preceding sections amount to systematic sampling along an axis of arbitrarily fixed direction. Here we introduce isotropy – thus, the test system and the target object are IUR relative to each other. An example is the Isotropic Cavalieri (ICav) design, see Section 2.26.3. As a result the variance prediction formulae are much simpler. The target parameter is planar area, or volume, and the target is a domain with piecewise smooth boundary – actually, the variance predictors involve the boundary measure (namely the perimeter length, or the surface area of the boundary of the target domain).

The basic technique used to derive the predictors, based on G. Matheron's transitive theory, is described by way of example in Section 5.9.2.

5.8.1 Notes

1. Remark on notation

The estimator given by Eq. (5.3.6), for instance, requires us to distinguish among the distance T between test planes and the distance h between test points. Here there are no such restrictions because we only deal with one-stage estimators, and therefore the basic length parameter of a test system (e.g. the distance between test planes, lines, or points) is always denoted by T.

2. References

With few changes, the material in Sections 5.10–5.17 below is taken from Cruz-Orive (2013), which is an expository paper based on some of the results of Kiêu and Mora (2006, 2009). The basic ideas and techniques emanate from Matheron (1965, 1971).

3. The software package pgs of R

Kiên Kiêu and Marianne Mora developed relevant theory, and specially advanced computing techniques, leading to their remarkable software package pgs (precision of geometric sampling) of R (`http://www.r-project.org/`), which is essential for the efficient computation of the numerical factors involved in the ensuing estimators. At present (2021) the pgs software, and its documentation, can be downloaded freely from the following URL: `https://mran.microsoft.com/snapshot/2017-02-04/web/packages/pgs/index.html`.

5.9 Isotropic Cavalieri Lines in the Plane

5.9.1 Practical variance prediction formula

The target parameter is the area $A > 0$ of a fixed domain $T \subset \mathbb{R}^2$ with piecewise smooth boundary ∂Y of finite length B. Consider an IUR test system $\Lambda_{z,\phi}$ of parallel test lines of period $T > 0$, see Eq. (2.26.8) and Fig. 5.9.1(c). The purpose is to predict the variance of the ICav area estimator

$$\widehat{A} = T \cdot L(Y \cap \Lambda_{z,\phi}) = T \sum_{i=1}^{n} L_i, \tag{5.9.1}$$

where L_1, L_n denote the first and the last non-zero intercept lengths, respectively. A linear intercept may consist of a finite number of separate segments if the target domain is not convex. The extension term of the variance reads,

$$\mathrm{Var}_E(\widehat{A}) = \frac{\zeta(3)}{2\pi^3} \cdot BT^3 \approx 0.01938\, BT^3, \tag{5.9.2}$$

and the practical predictor is obtained by replacing B with its estimator, based e.g. on Eq. (2.26.10).

5.9.2 Derivation

Consider an orthogonal frame Ox_1x_2 in which the axis Ox_1 makes an angle $\phi \in [0, \pi)$ with the axis OX_1 of a fixed orthogonal frame OX_1X_2. Referred to OX_1X_2, consider a test line $L_1^2(x_1, \phi)$. For a given ϕ (which is omitted in the subsequent notation, for simplicity), the measurement function is the intercept length $f(x_1) = L(Y \cap L_1^2(x_1, \phi))$. Now, referred to Ox_1x_2, we have

$$A = \int_{\mathbb{R}} \left[\int_{\mathbb{R}} 1_Y(x_1, x_2)\, \mathrm{d}x_2 \right] \mathrm{d}x_1 = \int_{\mathbb{R}} f(x_1)\, \mathrm{d}x_1, \tag{5.9.3}$$

where 1_Y denotes the indicator function of the domain Y. In G. Matheron's terminology, f is said to be a grading of order 1 of 1_Y, because a variable, namely x_2, has been integrated out. Then, for the given orientation, the analogue of Eq. (5.2.23) is

$$\mathrm{Var}(\widehat{A}) = 2 \sum_{k=1}^{\infty} G_1(k/T), \tag{5.9.4}$$

where $G_1 = \mathcal{F}_1 g_1$ and g_1 is the covariogram of f. Consider the bivariate geometric covariogram g_2 of 1_Y, see Eq. (3.8.1), namely,

$$g_2(h_1, h_2) = \int_{\mathbb{R}} \int_{\mathbb{R}} 1_Y(x_1, x_2) \cdot 1_Y(x_1 + h_1, x_2 + h_2)\, \mathrm{d}x_1\, \mathrm{d}x_2. \tag{5.9.5}$$

Then,

$$g_1(h_1) = \int_{\mathbb{R}} g_2(h_1, h_2)\, \mathrm{d}h_2, \tag{5.9.6}$$

which is itself a grading of order 1 of g_2. Now, set $G_2 = \mathcal{F}_2 g_2$, namely,

$$G_2(t_1, t_2) = \int_{\mathbb{R}^2} g_2(h_1, h_2) \exp\{-2\pi\mathrm{i}(t_1h_1 + t_2h_2)\}\, \mathrm{d}h_1\, \mathrm{d}h_2. \tag{5.9.7}$$

Then, $G_1 = \mathcal{F}_1 g_1$ is given by

$$G_1(t_1) = \int_{\mathbb{R}} g_1(h_1) \exp\{-2\pi \mathrm{i}(t_1 h_1)\} \,\mathrm{d}h_1 = G_2(t_1, 0). \tag{5.9.8}$$

If the angle ϕ is isotropic, then the mean of $g_2(h_1, h_2)$ with $\mathbb{P}(\mathrm{d}\phi) = \mathrm{d}\phi/\pi$ is the isotropic covariogram $g_2(r)$, $r = (h_1^2 + h_2^2)^{1/2}$, see Eq. (3.8.3), and G_2 is also isotropic, see Section A.3.2. Therefore,

$$\begin{aligned} G_2(t_1, t_2) &\equiv G(\rho), \quad \rho = (t_1^2 + t_2^2)^{1/2}, \\ G_1(t_1) &= G_2(t_1, 0) \equiv G(\rho), \quad \rho = t_1. \end{aligned} \tag{5.9.9}$$

The preceding identities suggest the following notational convention,

$$\mathcal{F}_1 g_1 = \mathcal{F}_2 g_2 = G(\rho), \tag{5.9.10}$$

which must be interpreted bearing in mind that the definition of ρ depends on the dimension j of $\mathcal{F}_j$. Eq. (5.9.10) allows us to compute a model of $\mathcal{F}_1 g_1$ given a model of $\mathcal{F}_2 g_2$. The latter is easy because g_2 is a geometric covariogram – see next – whereas $\mathcal{F}_1 g_1$ would in principle not be that easy because g_1 is not a geometric covariogram.

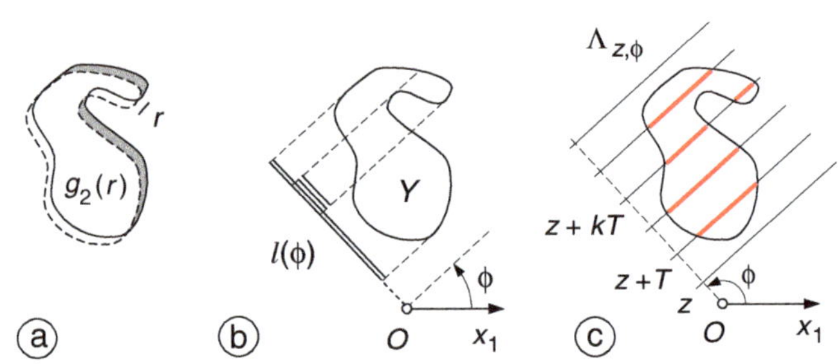

Fig. 5.9.1 (a) Geometric covariogram $g_2(r)$ of a planar domain Y. (b) Total orthogonal projection of the boundary of Y. (c) Linear intercepts (red segments) determined in Y by a test system of ICav lines, see Eq. (5.9.1). Modified from Cruz-Orive (2013), with permission of Springer.

The first Eq. (3.8.2) suggests the following approximation of $g_2(r, \phi)$ for small r, and for a given orientation $\phi \in [0, \pi)$, namely,

$$g_2(r, \phi) \approx A - r \cdot \frac{1}{2} l(\phi), \tag{5.9.11}$$

where $l(\phi)$ denotes the length of the total orthogonal projection of ∂Y onto an axis normal to the vector $h = (r, \phi)$, see Fig. 5.9.1(a,b). Under isotropy, Cauchy's projection formula yields $\mathbb{E}\{l(\phi)/2\} = B/\pi$, see Eq. (2.22.1), whereby the isotropic model of g_2 for small r becomes

$$g_2(r) = b_0 + b_1 r, \quad r \geq 0, \; r \text{ small}, \tag{5.9.12}$$

where $b_0 = A$ and $b_1 = -B/\pi$ in the present case. The next step is to compute the corresponding model version of $G_1 = \mathcal{F}_1 g_1$ via $\mathcal{F}_2 g_2$ using Eq. (5.9.10). Here, and in the sequel, we just need Eq. (A.3.4) for $k = 1$, namely,

$$(\mathcal{F}_d r)(\rho) = \frac{-\Gamma\left(\frac{d+1}{2}\right)}{2\pi^{(d+3)/2}} \cdot \frac{1}{\rho^{d+1}}, \quad (d = 1, 2, \ldots). \tag{5.9.13}$$

Thus,

$$G_1(\rho) = (\mathcal{F}_1 g_1)(\rho) = (\mathcal{F}_2 g_2)(\rho) = b_1 \cdot (\mathcal{F}_2 r)(\rho) = -\frac{b_1}{4\pi^2} \cdot \frac{1}{\rho^3}, \quad (\rho > 0), \tag{5.9.14}$$

which, submitted to the rhs of Eq. (5.9.4), yields Eq. (5.9.2).

5.10 Isotropic Cavalieri Planes

5.10.1 Practical variance prediction formula

The target parameter is the volume V of a fixed domain $Y \subset \mathbb{R}^3$ with piecewise smooth boundary ∂Y of finite area S. The test system $\Lambda_{z,u}$ consists of isotropic Cavalieri (ICav) planes of period $T > 0$, see Section 2.26.3. For each orientation $u \sim \mathrm{UR}(\mathbb{S}^2_+)$, and with $z \sim \mathrm{UR}[0, T)$, the UE of the ICav volume estimator is

$$\widehat{V} = T \cdot A(Y \cap \Lambda_{z,u}) = T \sum_{i=1}^{n} A_i, \tag{5.10.1}$$

and the extension term of its variance is,

$$\mathrm{Var}_E(\widehat{V}) = \frac{\pi}{360} S T^4 \approx 0.008727\, S T^4. \tag{5.10.2}$$

With the same probe, S may be estimated using Eq. (2.26.18).

5.10.2 Derivation

For each orientation $u \sim \mathrm{UR}(\mathbb{S}^2_+)$ of the IR sampling axis $L^3_1(0, u)$, adopt an orthogonal reference trihedron $Ox_1x_2x_3$, where Ox_1 is the sampling axis, whereby the Cavalieri planes are parallel to the plane Ox_2x_3. The analogue of Eq. (5.9.3) is

$$V = \int_{\mathbb{R}} \int_{\mathbb{R}} \int_{\mathbb{R}} 1_Y(x_1, x_2, x_3)\, dx_1\, dx_2\, dx_3 = \int_{\mathbb{R}} f(x_1)\, dx_1, \tag{5.10.3}$$

where $f(x_1) = A(Y \cap L_2^3(x_1, u))$ is the measurement function, namely the section area determined in the domain Y by a plane normal to the sampling axis at the point $(x_1, 0, 0)$. Here f is a grading of order 2 of 1_y because the two variables x_2, x_3 have been integrated out. In order to apply the analogue of Eq. (5.9.4) note that, for small r,

$$g_3(r, u) \approx V - r \cdot \frac{1}{2} a(u), \tag{5.10.4}$$

where $a(u)$ denotes the total projected area of the domain Y onto a plane normal to u. Under isotropy, $\mathbb{E}\{a(u)/2\} = S/4$, see Eq. (2.22.7), and therefore, the model for the isotropic geometric covariogram of 1_Y analogous to Eq. (5.9.12) is

$$g_3(r) = b_0 + b_1 r, \quad r \geq 0,\ r \text{ small}, \tag{5.10.5}$$

where $b_0 = V$ and $b_1 = -S/4$. By an analogous argument to that leading to Eq. (5.9.10) we now have $\mathcal{F}_1 g_1 = \mathcal{F}_3 g_3 = G(\rho)$ and applying Eq. (5.9.13) with $d = 3$ we obtain,

$$G_1(\rho) = b_1 \cdot (\mathcal{F}_3 r)(\rho) = -\frac{b_1}{2\pi^3} \cdot \frac{1}{\rho^4}, \quad (\rho > 0), \tag{5.10.6}$$

and Eq. (5.10.2) follows by applying Eq. (5.9.4), bearing in mind that $\zeta(4) = \pi^4/90$.

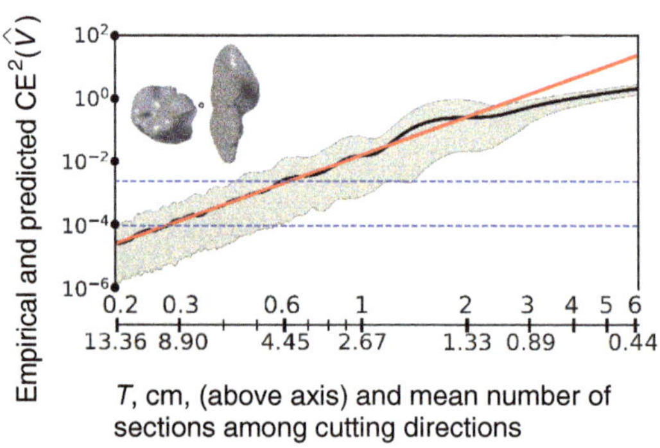

Fig. 5.10.1 Illustration of the performance of the variance predictor given by Eq. (5.10.2). The two dotted, horizontal lines, correspond to $\mathrm{CE}(\widehat{V}) = 0.01$ and 0.05. Modified from González-Villa et al. (2018), with permission of Image Analysis and Stereology.

5.10.3 Note

1. Error variance predictor for ICav planes

González-Villa et al. (2018) checked Eq. (5.10.2) by automatic Monte Carlo resampling of ICav test planes using the volume rendering of the union of two rat brain hemispheres of total volume V, see Fig. 5.10.1. The black curve represents the empirical $\mathrm{CE}^2(\widehat{V})$, whereas the red line corresponds to Eq. (5.10.2) divided by V^2. The prediction is seen to be quite reasonable for $T < 2\,\mathrm{cm}$, which corresponds to a mean number of sections greater than 1.33. For each value of T, the grey band contains 1250 ICav volume estimates corresponding to as many pseudosystematic IR orientations of the ICav planes.

5.11 Isotropic Square Grid of Test Points in the Plane

5.11.1 Practical variance prediction formula

The target domain, and the target parameter, are as in Section 5.9.1. The probe is an IUR square grid $\Lambda_{x,\omega}$ of test points, where $x = (x_1, x_2) \sim \mathrm{UR}(J_0)$, $\omega \sim \mathrm{UR}[0, 2\pi)$ are independent random variables. The fundamental tile J_0 is a square of side length $T > 0$. The UE of A is

$$\widehat{A} = T^2 \cdot P(Y \cap \Lambda_{x,\omega}), \tag{5.11.1}$$

and the extension term of its variance is

$$\mathrm{Var}_E(\widehat{A}) = \frac{Z(3,2)}{4\pi^3} \cdot BT^3 \approx 0.072837\, BT^3, \tag{5.11.2}$$

(compare with Eq. (5.9.2)), where $Z(s, d)$ is a special version of the Epstein Zeta function, namely,

$$Z(s,d) = \sum_{k_1 \in \mathbb{Z}} \sum_{k_2 \in \mathbb{Z}} \cdots \sum_{k_d \in \mathbb{Z}}{}' \frac{1}{(k_1^2 + k_2^2 + \cdots + k_d^2)^{s/2}}, \qquad s > d. \tag{5.11.3}$$

The $'$ following the d-fold summation indicates that the point $k_1 = k_2 = \cdots = k_d = 0$ is excluded. This function may be computed with the package pgs (see Section 5.8.1, Note 3) using the command

$$\texttt{Ezeta(s, RectLatd())}. \tag{5.11.4}$$

The coefficient of BT^3 in the rhs of Eq. (5.11.2) may be evaluated directly with the following pgs command,

$$\texttt{area.mse(PPRectLat2(1, 1, 1))}. \tag{5.11.5}$$

5.11.2 Derivation

Here the measurement function is the indicator $1_Y(x_1, x_2)$ itself, hence no grading is involved. Also, we need the two-dimensional version of Eq. (5.9.4), namely,

$$\operatorname{Var}(\widehat{A}) = \sum_{k_1\in\mathbb{Z}} \sum_{k_2\in\mathbb{Z}}{}' G_2(k_1/T, k_2/T), \tag{5.11.6}$$

where $G_2 = \mathcal{F}_2 g_2$, and g_2 is the geometric covariogram of 1_Y, see Eq. (5.9.5). Under isotropy, we adopt the model given by Eq. (5.9.12) and, because there is no grading, we have to use directly

$$G_2(\rho) = b_1 \cdot (\mathcal{F}_2 r)(\rho) = -\frac{b_1}{4\pi^2} \cdot \frac{1}{\rho^3}, \quad \rho = (t_1^2 + t_2^2)^{1/2}, \tag{5.11.7}$$

which, submitted to Eq. (5.11.6) yields,

$$\operatorname{Var}_E(\widehat{A}) = \frac{1}{4\pi^2} \cdot \frac{B}{\pi} \cdot T^3 \sum_{k_1\in\mathbb{Z}} \sum_{k_2\in\mathbb{Z}}{}' \frac{1}{(k_1^2 + k_2^2)^{3/2}}, \tag{5.11.8}$$

namely Eq. (5.11.2).

5.12 Isotropic Fakir Probe to Estimate Volume

5.12.1 Practical variance prediction formula

The target domain, and the target parameter, are as in Section 5.10.1. The probe is an IUR fakir probe $\Lambda_{z,u}$ see Section 2.26.4 and Fig. 5.12.1, namely a test system of test lines normal to an IR sampling plane. The test lines are drawn through the vertices of an IUR square grid lying on the sampling plane, and the fundamental tile J_0 of the grid is a square of side length $T > 0$. For any given orientation $u \in \mathbb{S}_+^2$, the fakir volume estimator is

$$\widehat{V} = T^2 \cdot L(Y \cap \Lambda_{z,u}) = T^2 \sum_{i=1}^{n} L_i, \tag{5.12.1}$$

where the $\{L_i\}$ is the set of renumbered positive linear intercept lengths. Further, under isotropy,

$$\operatorname{Var}_E(\widehat{V}) = \frac{Z(4,2)}{8\pi^3} \cdot ST^4 \approx 0.024296\, ST^4, \tag{5.12.2}$$

where S may be estimated using Eq. (2.26.20). Compare with Eq. (5.10.2).

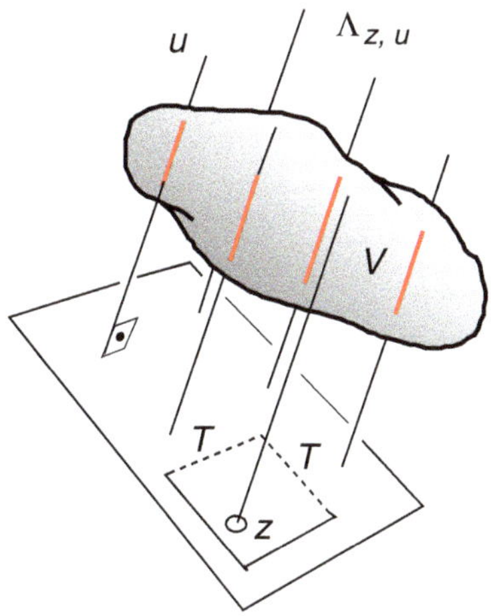

Fig. 5.12.1 Domain hit by an isotropic fakir probe. Modified from Cruz-Orive (2013), with permission of Springer.

5.12.2 Derivation

For each orientation $u \sim \mathrm{UR}(\mathbb{S}_+^2)$ of the sampling plane $L_2^3(0, u)$ adopt an orthogonal reference trihedron $Ox_1x_2x_3$, where Ox_1x_2 is the sampling plane, and the test lines are parallel to the axis Ox_3. The analogue of Eq. (5.10.3) is

$$V = \int_{\mathbb{R}} \int_{\mathbb{R}} \int_{\mathbb{R}} 1_Y(x_1, x_2, x_3)\, \mathrm{d}x_1\, \mathrm{d}x_2\, \mathrm{d}x_3 = \int_{\mathbb{R}} \int_{\mathbb{R}} f(x_1, x_2)\, \mathrm{d}x_1\, \mathrm{d}x_2, \tag{5.12.3}$$

where $f(x_1, x_2) = L(Y \cap L_1^3(x, u))$ is the measurement function, namely the intercept length determined in the domain Y by a test line normal to the sampling plane at the point $x = (x_1, x_2)$. Here f is a grading of order 1 of 1_Y because the variable x_3 has been integrated out. In order to compute the required Fourier transform $G_2 = \mathcal{F} g_2$ of the covariogram g_2 of f under isotropy, we adopt the model given by Eq. (5.10.5) for the geometric covariogram of 1_Y, and we apply the relationship $\mathcal{F}_2 g_2 = \mathcal{F}_3 g_3$, and Eq. (5.9.13). We obtain

$$G_2(\rho) = b_1 \cdot (\mathcal{F}_3 r)(\rho) = -\frac{b_1}{2\pi^3} \cdot \frac{1}{\rho^4}, \quad \rho = (t_1^2 + t_2^2)^{1/2}, \tag{5.12.4}$$

which is the same as Eq. (5.10.6) with a different definition of ρ. Now Eq. (5.11.6) yields

$$\mathrm{Var}_E(\widehat{V}) = \frac{1}{2\pi^3} \cdot \frac{S}{4} \cdot T^4 \sum_{k_1 \in \mathbb{Z}} \sum_{k_2 \in \mathbb{Z}}{}' \frac{1}{(k_1^2 + k_2^2)^2}, \tag{5.12.5}$$

which is Eq. (5.12.2).

5.13 Isotropic Cubic Grid of Test Points

5.13.1 Practical variance prediction formula

The target domain, and the target parameter, are as in Section 5.10.1. The probe is an IUR spatial grid $\Lambda_{x,u_2,T}$ of test points whose fundamental tile J_0 is a cube of side length $T > 0$, endowed with the IUR probability element given by Eq. (2.29.3). The unbiased volume estimator is

$$\widehat{V} = T^3 \cdot P(Y \cap \Lambda_{x,u_2,T}), \tag{5.13.1}$$

see Eq. (2.29.1), and the extension term of its variance is

$$\mathrm{Var}_E(\widehat{V}) = \frac{Z(4,3)}{8\pi^3} \cdot ST^4 \approx 0.066649\, ST^4, \tag{5.13.2}$$

compare with Eq. (5.12.2).

5.13.2 Derivation

The preceding result is a straightforward extension of the planar case (Section 5.11.2) to three dimensions. No grading is involved, and therefore we need

$$G_3(\rho) = b_1 \cdot (\mathcal{F}_3 r)(\rho) = -\frac{b_1}{2\pi^3} \cdot \frac{1}{\rho^4}, \quad \rho = (t_1^2 + t_2^2 + t_3^2)^{1/2}. \tag{5.13.3}$$

Application of the three-dimensional version of Eq. (5.11.6) yields

$$\mathrm{Var}_E(\widehat{V}) = \frac{1}{2\pi^3} \cdot \frac{S}{4} \cdot T^4 \sum_{k_1 \in \mathbb{Z}} \sum_{k_2 \in \mathbb{Z}} \sum_{k_3 \in \mathbb{Z}}{}' \frac{1}{(k_1^2 + k_2^2 + k_3^2)^2}, \tag{5.13.4}$$

which is Eq. (5.13.2).

5.14 Isotropic Cavalieri Stripes in the Plane

5.14.1 Practical variance prediction formula

The setup is that of Section 5.9.1, but now we use a system $\Lambda_{t,z,\phi}$ of IUR Cavalieri stripes of period $T > 0$ and thickness $0 < t \leq T$. For each orientation $\phi \sim \text{UR}[0, \pi)$, and with $z \sim \text{UR}[0, T)$, the corresponding UE of A is

$$\widehat{A} = \frac{T}{t} \cdot A(Y \cap \Lambda_{t,z,\phi}) \equiv \frac{T}{t} \cdot \sum_{i=1}^{n} A_{ti}, \tag{5.14.1}$$

where A_{ti} denotes the stripe area determined in the planar domain Y by the ith stripe, and A_{t1}, A_{tn} represent the first and the last non-zero areas, respectively. With the usual notation $\tau = t/T$, we have,

$$\text{Var}_E(\widehat{A}) = \frac{1}{4\pi^5\tau^2} \sum_{k=1}^{\infty} \frac{1 - \cos(2\pi\tau k)}{k^5} \cdot BT^3. \tag{5.14.2}$$

The factor of BT^3 is plotted in Fig. 5.14.1(a) as a function of τ, and it may also be computed via Eq. (5.4.8), or directly using the pgs command of Eq. (5.16.8) below.

As τ tends to zero the preceding formula reduces to Eq. (5.9.2) for IUR Cavalieri lines, as expected, because

$$\lim_{\tau \to 0} \frac{1 - \cos(2\pi\tau k)}{\tau^2} = 2\pi^2 k^2. \tag{5.14.3}$$

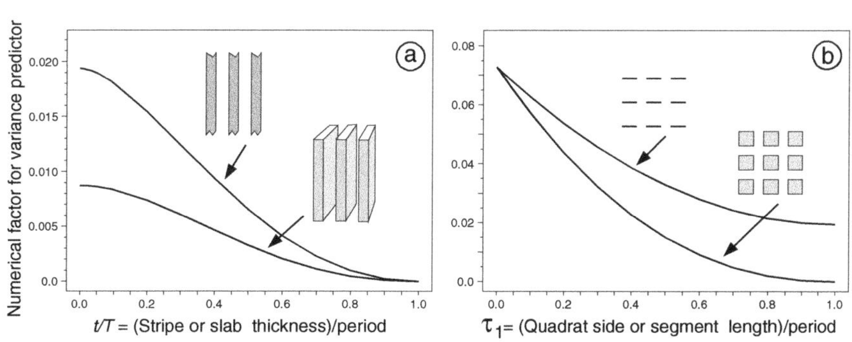

Fig. 5.14.1 (a) Graphs of the factor of BT^3 in Eq. (5.14.2) for ICav stripes, and of the factor of ST^4 in (5.15.2) for ICav slabs, respectively. (b) Graph of the coefficients $c_8(1, \tau_1)$ and $c_q(1, \tau_1, \tau_1)$ given by Eq. (5.17.3) and Eq. (5.16.5) for isotropic test systems of segments and of square quadrats, respectively, with a square fundamental tile in each case. Modified from Cruz-Orive (2013), with permission of Springer.

5.14.2 Derivation

In Section 5.4.2, set $\gamma(Y) \equiv A(Y)$. Then, for any orientation $\phi \sim \mathrm{UR}[0, \pi)$ of a Cavalieri stripe L_t, Eq. (1.4.11) yields

$$A = t^{-1} \int_{\mathbb{R}} A_t(x)\, \mathrm{d}x, \tag{5.14.4}$$

where

$$A_t(x) = A(Y \cap L_t(x)) = \int_x^{x+t} f(y)\, \mathrm{d}y, \tag{5.14.5}$$

is the stripe section area at $x \in \mathbb{R}$, and $f(y)$ is the intercept length defined by Eq. (5.9.3), namely a grading of order 1 of 1_Y. Define

$$f_t(x) = t^{-1} \cdot A_t(x) = t^{-1} \cdot (f * \check{1}_t)(x), \tag{5.14.6}$$

which is effectively the regularization or filtering of f by the segment $[0, t)$, see Eq. (5.4.16). By Eq. (5.14.4), the target parameter may now be expressed as

$$A = \int_{\mathbb{R}} f_t(x)\, \mathrm{d}x, \tag{5.14.7}$$

and therefore the UE of A may be written as follows,

$$\widehat{A} = T \sum_{k \in \mathbb{Z}} f_t(z + kT). \tag{5.14.8}$$

As a consequence Eq. (5.4.19) applies, that is,

$$\mathrm{Var}(\widehat{A}) = 2 \sum_{k=1}^{\infty} G_t(k/T), \tag{5.14.9}$$

where $G_t = \mathcal{F} g_t$ and g_t is the covariogram of f_t. By Eq. (5.4.22),

$$G_t = \mathcal{F} g_t = t^{-2} \mathcal{F} g \mathcal{F} K, \tag{5.14.10}$$

where g is the covariogram of f, and K is that of 1_t. We now adopt the isotropic covariogram model g_2 for 1_Y given by Eq. (5.9.12), whereby $\mathcal{F} g$ is given by Eq. (5.9.14). On the other hand, $\mathcal{F} K$ is given by Eq. (5.4.23). Substitution into the rhs of Eq. (5.14.10), and submitting the result to Eq. (5.14.9), we obtain

$$\mathrm{Var}(\widehat{A}) = -\frac{2}{4\pi^2 t^2} b_1 \sum_{k=1}^{\infty} \frac{1}{(k/T)^3} \cdot \frac{1 - \cos(2\pi(t/T)k)}{2\pi^2 (k/T)^2}, \tag{5.14.11}$$

which reduces to Eq. (5.14.2) after substituting $b_1 = -B/\pi$.

5.15 Isotropic Cavalieri Slabs

5.15.1 Practical variance prediction formula

The setup is the same as in Section 5.10, but now we use a system $\Lambda_{t,z,u}$ of ICav slabs of period $T > 0$ and thickness $0 < t \leq T$. For each isotropic orientation of the test system, the UE of V is

$$\widehat{V} = \frac{T}{t} \cdot V(Y \cap \Lambda_{t,z,\phi}) = \frac{T}{t} \cdot \sum_{i=1}^{n} V_{ti}, \tag{5.15.1}$$

where V_{ti} denotes the slice volume determined in the domain Y by the ith slab, with the usual notational convention. Now, setting $\tau = t/T$,

$$\mathrm{Var}_E(\widehat{V}) = \frac{\pi}{360}(1-\tau)^2(1+2\tau-2\tau^2)ST^4, \tag{5.15.2}$$

see Fig. 5.14.1(a).

For $\tau = 0$ we recover Eq. (5.10.2) for ICav planes, as expected.

5.15.2 Derivation

The path is an extension of that followed in Section 5.14.2 with $A_t(x)$ replaced with slice volume $V_t(x)$. Now we adopt the isotropic covariogram model g_3 given by Eq. (5.10.5) for 1_Y, whereby $\mathcal{F}g$ is given by Eq. (5.10.6) because f is a grading of order 2 of 1_Y, whereas $\mathcal{F}K$ is still given by Eq. (5.4.23). Now Eq. (5.14.10) and Eq. (5.14.9) yield,

$$\begin{aligned}\mathrm{Var}_E(\widehat{V}) &= -\frac{2}{2\pi^3 t^2} b_1 \sum_{k=1}^{\infty} \frac{1}{(k/T)^4} \cdot \frac{1-\cos(2\pi(t/T)k)}{2\pi^2(k/T)^2} \\ &= \frac{1}{8\pi^5\tau^2} \sum_{k=1}^{\infty} \frac{1-\cos(2\pi\tau k)}{k^6} \cdot ST^4. \end{aligned} \tag{5.15.3}$$

Application of Eq. (5.4.28) for $s = 3$ yields the closed form given by Eq. (5.15.2).

5.16 Isotropic Systematic Quadrats in the Plane

5.16.1 Practical variance prediction formula

As in Section 5.11, the target parameter is the area $A > 0$ of a fixed planar domain $Y \subset \mathbb{R}^2$. Here, however, the probe is an IUR test system $\Lambda_{x,\omega}$ of rectangular quadrats whose fundamental tile $J_0 = (0, T_1] \times (0, T_2]$ is a rectangle of side lengths

$$T_1 > 0, \quad T_2 = \tau T_1, \quad \tau \in (0, 1], \tag{5.16.1}$$

and the fundamental probe $T_0 \subset J_0$ is another rectangle of side lengths

$$s_1 = \tau_1 T_1, \quad s_2 = \tau_2 T_2, \quad \tau_1, \tau_2 \in [0, 1]. \tag{5.16.2}$$

Fig. 5.16.1(a) illustrates the case $T_1 = \tau^{-1}$, $T_2 = 1$. Also, $x = (x_1, x_2) \sim \mathrm{UR}(J_0)$, $\omega \sim \mathrm{UR}[0, 2\pi)$ and independent. As usual, it is equivalent to fix the test system and to associate an IUR vector (x, ω) to the domain Y, see Fig. 1.21.1. The corresponding UE of A is

$$\widehat{A} = \frac{T_1 T_2}{s_1 s_2} \cdot A(Y \cap \Lambda_{x,\omega}) = \frac{1}{\tau_1 \tau_2} \cdot A(Y \cap \Lambda_{x,\omega}), \tag{5.16.3}$$

and the extension term of its variance is,

$$\mathrm{Var}_E(\widehat{A}) = c_q(\tau, \tau_1, \tau_2) \cdot B T_2^3, \tag{5.16.4}$$

where

$$c_q(\tau, \tau_1, \tau_2) = \frac{1}{16\pi^7 (\tau_1 \tau_2)^2} \sum_{k_1 \in \mathbb{Z}} \sum_{k_2 \in \mathbb{Z}}{}' \frac{[1 - \cos(2\pi \tau_1 k_1)][1 - \cos(2\pi \tau_2 k_2)]}{k_1^2 k_2^2 (\tau^2 k_1^2 + k_2^2)^{3/2}}. \tag{5.16.5}$$

The preceding coefficient may be computed with the package pgs using the command

$$\texttt{area.mse(QRectLat2(h1} = \tau^{-1}, \texttt{h2} = 1, \texttt{hq} = \tau_1 \tau^{-1}, \texttt{vq} = \tau_2)), \tag{5.16.6}$$

where h1 (default: 1, unit lattice), h2 (default: h1, square lattice), denote the base and the height of J_0, and hq, vq those of T_0, respectively, see Fig. 5.16.1(a). Fig. 5.14.1(b) displays the plot of $c_q(1, \tau_1, \tau_2)$ against τ_1, corresponding to a square quadrat of side length $\tau_1 \in [0, 1]$ and a square fundamental tile of unit side length. For instance,

$$\texttt{area.mse(QRectLat2(1, 1, hq} = 0.4, \texttt{vq} = 0.4)), \tag{5.16.7}$$

yields $c_q(1, 0.4, 0.4) = 0.02272\ldots$.

For Cavalieri stripes it suffices to take $\tau_2 = 1$. Thus, the factor of BT^3 in the rhs of Eq. (5.14.2) may be evaluated with the command

$$\texttt{area.mse(QRectLat2(1, 1, hq} = \tau, \texttt{vq} = 1)). \tag{5.16.8}$$

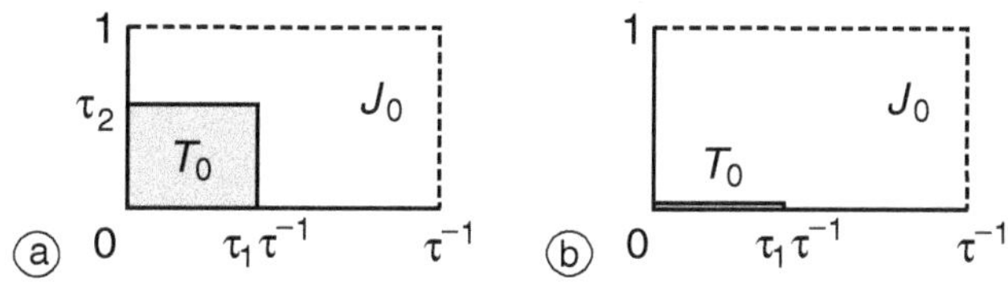

Fig. 5.16.1 (a,b) Definitions of the fundamental tile J_0 and probe $T_0 0$, see Sections 5.16.1 and 5.17.1 respectively.

Application of Eq. (5.14.3) shows that, as τ_1 and τ_2 tend to zero, and $\tau = 1$, Eq. (5.16.4) reduces to Eq. (5.11.2) for an IUR square grid of test points, as expected.

5.16.2 Derivation

For each orientation $\omega \sim \mathrm{UR}[0, 2\pi)$ of a quadrat T_0 relative to the domain Y, by Eq. (1.15.6) we have

$$A = \int_{\mathbb{R}^2} f_{T_0}(x)\, \mathrm{d}x, \tag{5.16.9}$$

where

$$f_{T_0}(x) = (s_1 s_2)^{-1} A(Y \cap T_0(x)) = (s_1 s_2)^{-1} 1_Y * \check{1}_{T_0}. \tag{5.16.10}$$

Now the UE of given by Eq. (5.16.3) may be written

$$\widehat{A} = T_2 T_2 \sum_{i \in \mathbb{Z}} \sum_{j \in \mathbb{Z}} f_{T_0}(x_1 + iT, x_2 + jT), \tag{5.16.11}$$

and

$$\mathrm{Var}(\widehat{A}) = \sum_{k_1 \in \mathbb{Z}} \sum_{k_2 \in \mathbb{Z}}{}' G_{T_0}(k_1/T_1, k_2/T_2), \tag{5.16.12}$$

where G_{T_0} denotes the Fourier transform of the covariogram g_{T_0} of f_{T_0}. Similarly as in Eq. (5.4.21), we have

$$g_{T_0} = f_{T_0} * \check{f}_{T_0} = (s_1 s_2)^{-2} g_2 * K_2, \tag{5.16.13}$$

where g_2 is the geometric covariogram of 1_Y, whereas K_2 is that of 1_{T_0}, namely,

$$K_2(h_1, h_2) = \begin{cases} (s_1 - |h_1|)(s_2 - |h_2|), & -s_i \le h_i \le s_i,\ i = 1, 2, \\ 0, & \text{otherwise}. \end{cases} \tag{5.16.14}$$

Application of the convolution theorem yields

$$G_{T_0} \equiv \mathcal{F}_2 g_{T_0} = (s_1 s_2)^{-2} \mathcal{F}_2 g_2 \mathcal{F}_2 K_2. \tag{5.16.15}$$

Under the isotropic model given by Eq. (5.9.12) for g_2, the corresponding Fourier transform $(\mathcal{F}_2 g_2)(\rho)$ is given by Eq. (5.11.7). On the other hand, recalling Eq. (5.4.23), the Fourier transform of K_2 is

$$(\mathcal{F}_2 K_2)(t_1, t_2) = \prod_{k=1}^{2} \frac{1 - \cos(2\pi s_k t_k)}{2\pi^2 t_k^2}. \tag{5.16.16}$$

Substitution into the rhs of Eq. (5.16.15), and then into the rhs of Eq. (5.16.12), yields Eq. (5.16.4).

5.17 Isotropic Grid of Straight Line Segments in the Plane

5.17.1 Practical variance prediction formula and its derivation

The target parameter is $A > 0$, as in the preceding section, but the fundamental probe $T_0 \subset J_0$ of the IUR test system $\Lambda_{x,\omega}$ is a straight line segment of length $s_1 = \tau_1 T_1$, $\tau_1 \in [0, 1]$ see Fig. 5.16.1(b). The corresponding UE of A is

$$\widehat{A} = \frac{T_1 T_2}{s_1} \cdot L(Y \cap \Lambda_{x,\omega}) = \frac{T_2}{\tau_1} \cdot L(Y \cap \Lambda_{x,\omega}), \tag{5.17.1}$$

and the extension term of its variance is

$$\mathrm{Var}_E(\widehat{A}) = c_s(\tau, \tau_1) \cdot B T_2^3, \tag{5.17.2}$$

where $c_s(\tau, \tau_1)$ is the limit of the rhs of Eq. (5.16.5) as τ_2 tends to zero, namely,

$$c_s(\tau, \tau_1) = \frac{1}{8\pi^5 \tau_1^2} \sum_{k_1 \in \mathbb{Z}} {\sum_{k_2 \in \mathbb{Z}}}' \frac{1 - \cos(2\pi \tau_1 k_1)}{k_1^2 (\tau^2 k_1^2 + k_2^2)^{3/2}}. \tag{5.17.3}$$

The preceding coefficient may be computed with the package pgs using the command

$$\texttt{area.mse(SRectLat2(h1} = \tau^{-1}, \texttt{h2} = 1, \texttt{end} = \texttt{c(}\tau^{-1}\tau_1, \texttt{0)))}, \tag{5.17.4}$$

see Fig. 5.16.1(b). Fig. 5.14.1(b) displays the plot of $c_q(1, \tau_1)$ against τ_1, corresponding to a segment of length $\tau_1 \in [0, 1]$ and a square fundamental tile of unit side length. For instance,

$$\texttt{area.mse(SRectLat2(1, 1, end = c(0.4, 0)))} \tag{5.17.5}$$

yields $c_s(1, 0.4) = 0.03871\ldots$.

5.17.2 Example

Let the target object $Y \subset \mathbb{R}^2$ be a disk of fixed area $A = \pi R^2$, and consider a test system of segments in which the fundamental tile is a square of side length $T > 0$, whereas the fundamental probe is a segment of length $t \equiv \tau T$, $\tau \in [0, 1]$. For $\tau = 0, 1/2, 1$ we have a square grid of test points, a grid of test segments of length $T/2$, and a grid of Cavalieri lines a distance T apart, respectively. In Fig. 5.17.1, the exact $\mathrm{CV}^2(\widehat{A})$ is plotted for each of the three test systems as functions of $\mathbb{E}(P) = A/T^2$, see also Section 5.18.4, Note 1. The corresponding predictor is always of the following form, (recall that $B = 2\pi R$ in this case),

$$\mathrm{CV}_E^2(\widehat{A}) = \frac{c \cdot BT^3}{A^2} = \frac{2\sqrt{\pi} \cdot c}{(\mathbb{E}P)^{3/2}}, \tag{5.17.6}$$

where c stands for each of the numerical factors of BT^3 in Eq. (5.11.2), Eq. (5.17.2) with $\tau = 1$, $\tau_1 = 1/2$ and Eq. (5.9.2), respectively. By Eq. (5.17.6), the log-log plots of the predictors are straight lines with a common slope equal to $-3/2$.

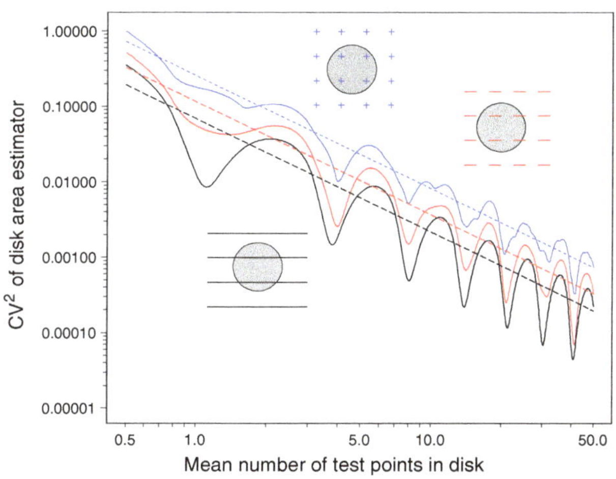

Fig. 5.17.1 See Section 5.17.2. Reproduced from Cruz-Orive (2013) with permission of Springer.

5.18 Apparent Paradoxes in Geometric Sampling

5.18.1 Test systems. The Rao–Blackwell theorem

Let $Y \subset \mathbb{R}^2$ be a domain of fixed area $A > 0$. Also, let Λ_t, Λ_0 be a test system of segments of a given length $t > 0$, and one of test points, respectively, both with a square fundamental tile of side length $T > t > 0$, see Fig. 5.17.1, upper insets. If Λ_t

is UR hitting Y, then by Eq. (2.25.3) a UE of A is

$$\widehat{A} = \frac{T^2}{t} \cdot l, \tag{5.18.1}$$

where $l \equiv L(Y \cap \Lambda_t)$ is the total intercept length determined in Y by Λ_t.

Now consider a given UR realization of Λ_t hitting Y, and suppose that Λ_0 is UR within Λ_t. Thus, the fundamental tiles of Λ_t and Λ_0 are equally oriented, and a test point of Λ_0 is UR within a test segment from the fixed realization of Λ_t. Note that no test points of Λ_0 lie outside Λ_t. Then, a UE of l is

$$\widehat{l} = t \cdot P, \tag{5.18.2}$$

where $P \equiv P(Y \cap \Lambda_t \cap \Lambda_0)$. Now, if in addition Λ_t is UR hitting Y, then,

$$\widetilde{A} = \frac{T^2}{t} \cdot \widehat{l} = T^2 P, \tag{5.18.3}$$

is also unbiased for A. Actually, $\widetilde{A}$ is formally identical to the UE of A obtained by hitting Y directly with a UR square grid Λ_0. Moreover,

$$\mathbb{E}(\widetilde{A}|\widehat{A}) = \frac{T^2}{t} \cdot l = \widehat{A}, \tag{5.18.4}$$

whereby the total variance identity yields

$$\begin{aligned} \mathrm{Var}(\widetilde{A}) &= \mathrm{Var}\{\mathbb{E}(\widetilde{A}|\widehat{A})\} + \mathbb{E}\{\mathrm{Var}(\widetilde{A}|\widehat{A})\} \\ &= \mathrm{Var}(\widehat{A}) + \mathbb{E}\{\mathrm{Var}(\widetilde{A}|\widehat{A})\} \\ &\geq \mathrm{Var}(\widehat{A}), \end{aligned} \tag{5.18.5}$$

which is a version of the Rao–Blackwell (R-B) theorem. Thus, because Λ_0 can 'slide' within Λ_t, sweeping it entirely, the latter test system is at least as precise as the former to estimate A. As observed also in Fig. 5.17.1, the test system Λ_T of Cavalieri lines is at least as precise as Λ_t for analogous reasons.

At first sight the preceding conclusion looks plausible because linear intercepts seem to bear more information than point hits. In reality, however, this argument is not warranted as far as error variance is concerned. If $\mathbb{E}(\widetilde{A}|\widehat{A}) \neq \widehat{A}$, then Eq. (5.18.5) does not need to hold. An example of this is illustrated in Fig. 5.18.1, in which the system of test points cannot sweep the one of segments remaining entirely within the segments, and therefore Eq. (5.18.4) does not apply.

In the example of Fig. 5.18.1, the fundamental tile of the test system Λ_0 of points is a square of side length T, hence $\widetilde{A} = T^2 P$. If Λ_0 is IUR hitting the domain Y then, by Eq. (5.11.2),

$$\mathrm{Var}_E(\widetilde{A}) \approx 0.072837\, BT^3. \tag{5.18.6}$$

On the other hand, the fundamental tile of the test system Λ_t of segments is a rectangle $J_0 = (0, 2T] \times (0, T]$, whereas the fundamental probe is the segment $T_0 = (0, T]$,

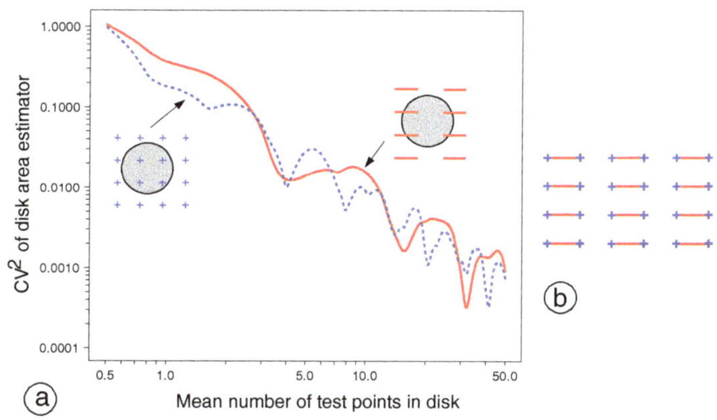

Fig. 5.18.1 As illustrated in (a), the test segments in (b) are not necessarily more precise than their endpoints to estimate disk area, see Section 5.18.1. Modified from Voss and Cruz-Orive (2009), with permission of Taylor & Francis.

whereby $\widehat{A} = 2Tl$. If Λ_t is IUR hitting Y, then the coefficient given by Eq. (5.17.3) can be computed by setting $\tau = 0.5$ in the command given by Eq. (5.17.4), namely,

$$\texttt{area.mse(SRectLat2(h1} = 2, \texttt{h2} = 1, \texttt{end} = \texttt{c(1,0)))}, \tag{5.18.7}$$

which yields,

$$\mathrm{Var}_E(\widehat{A}) \approx 0.084261\, BT^3. \tag{5.18.8}$$

Thus, contrary to intuition, $\mathrm{Var}_E(\widetilde{A}) < \mathrm{Var}_E(\widehat{A})$ in this case.

5.18.2 Single probes

Fig. 5.18.2(a) shows three bounded probes T_0, T_1, T_2, consisting of the union of two test points a distance l apart, a segment of length l, and one of length $2l$, respectively, hitting a disk Y of unit diameter isotropically and uniformly at random, with the purpose of estimating the area A of the disk. The joint probability element of each probe is given by Eq. (2.12.1), where, for convenience, $D_\oplus$ was in this case replaced with a reference disk, concentric with Y, of diameter $2+4l$ and area A_0. The test point probe T_0 can sweep the segment T_2 remaining entirely within it, but this does not work for T_1. If T_0 is UR within T_2, then we can estimate the intercept length $L(Y \cap T_2)$ by point counting, but this is not possible for $L(Y \cap T_1)$. As a consequence, the R-B theorem holds for T_0, T_2 but not for T_0, T_1, and in fact he graph confirms that estimating A with T_1 is not necessarily more precise than doing it with T_0, whereas T_2 is the most precise of the three probes for all values of $l > 0$.

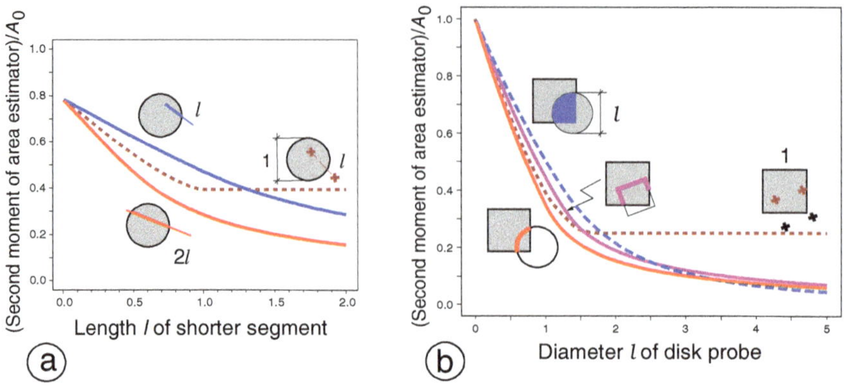

Fig. 5.18.2 Illustration of the apparent paradoxes described in Section 5.18.2. Modified from Voss and Cruz-Orive (2009), with permission of Taylor & Francis.

Fig. 5.18.3 illustrates the Jensen–Gundersen 'paradox': Unless the target disk is relatively small, estimating its area A by point counting using the four corners of an IUR square (bottom inset and blue curve) is more precise than doing it by measuring exactly the area itself of the intersection between the disk and the square (top inset and red curve). The apparent paradox is dispelled on realising that the two probes are in fact unrelated; the four corners cannot estimate the corresponding intersection area, and therefore the R-B theorem does not need to apply. By the usual argument, for the square to be always the more precise, its side should be twice as long.

Finally, Fig. 5.18.2(b) shows four different IUR probes aiming at estimating the area A of a fixed square Y. The R-B theorem applies only for the four corners of a square (T_0, say) compared with its circumscribed circle (T_1, say). In fact, T_0 can sweep T_1 entirely by rotating about a common centre, and therefore $L(Y \cap T_1)$ can be estimated without bias by point counting via $P(Y \cap T_1 \cap T_0)$. It follows that T_1 is more precise than T_0 to estimate A.

5.18.3 The Ohser paradox

Consider a motion-invariant process $Y \subset \mathbb{R}^2$ of straight lines, as described in Section 3.10.1, with length intensity B_A, which is the target parameter. As a probe we adopt a fixed, convex quadrat $T \subset \mathbb{R}^2$ of constant area $A(T)$ and boundary length $B(\partial T)$. By Eq. (3.2.3), the estimator

$$\widehat{B}_A = \frac{L(Y \cap T)}{A(T)} \equiv \frac{L}{A}, \tag{5.18.9}$$

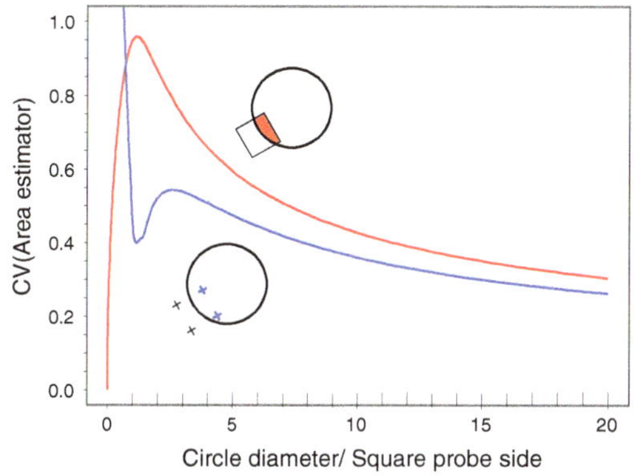

Fig. 5.18.3 Illustration of the Jensen–Gundersen 'paradox', see Sections 5.18.2 and 5.18.4, Note 1. Modified from Baddeley and Cruz-Orive (1995), with permission of the Applied Probability Trust.

namely the total intercept length captured by the quadrat, divided by the area of the quadrat, is a direct UE of B_A, see Fig. 5.18.4(a). On the other hand, by Eq. (3.3.10), the estimator

$$\widetilde{B}_A = \frac{\pi}{2} \cdot \frac{I(Y \cap \partial T)}{B(\partial T)} \equiv \frac{\pi}{2} \cdot \frac{I}{B}, \tag{5.18.10}$$

namely $\pi/2$ times the total number of intersections between the lines and the quadrat boundary, divided by the length of the latter, is also unbiased for B_A, see Fig. 5.18.4(b). At first sight the statistic I bears less information than L as far as line length is concerned, hence one could guess that $\widehat{B}_A$ is more precise than $\widetilde{B}_A$. To elucidate this, first note that the random number of hitting lines is almost surely (a.s.) equal to $I/2$, because a hitting line a.s. determines two intersections in ∂T. Second, by Section 3.10.1, property (ii), the mean intercept length determined by an invariant line hitting the convex quadrat T is given by Eq. (2.4.13), namely $\pi A/B$. Thus,

$$\mathbb{E}(L|I) = \frac{I}{2} \cdot \frac{\pi A}{B}, \tag{5.18.11}$$

whereby,

$$\mathbb{E}(\widehat{B}_A|\widetilde{B}_A) = \frac{\mathbb{E}(L|I)}{A} = \frac{I}{2} \cdot \frac{\pi A}{BA} = \frac{\pi}{2} \cdot \frac{I}{B} = \widetilde{B}_A, \tag{5.18.12}$$

and by the R-B theorem,

$$\text{Var}(\widehat{B}_A) \geq \text{Var}(\widetilde{B}_A). \tag{5.18.13}$$

An intuitive explanation of the preceding fact is that $\widehat{B}_A$ has two sources of variation, the first one due to the random number of hitting lines, and the second one due to the random intercept lengths within the quadrat, whereas $\widetilde{B}_A$ has only the first source of variation.

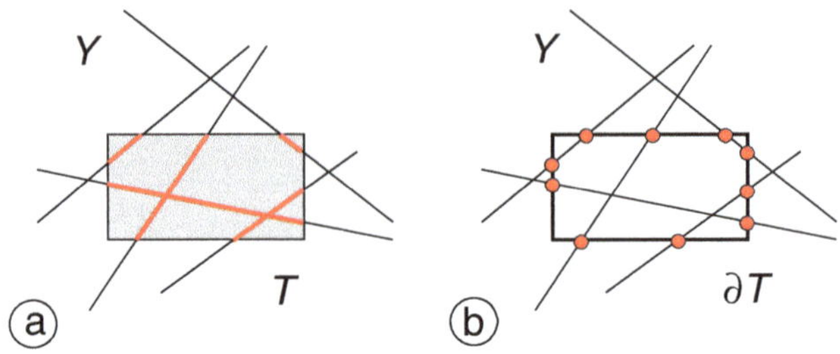

Fig. 5.18.4 Illustration of the Ohser 'paradox' (Section 5.18.3): to estimate the length intensity of a motion-invariant line process in the plane, measuring intercepts in a convex window, see (a), is less precise than counting their endpoints, see (b). Modified from Voss and Cruz-Orive (2009), with permission of Taylor & Francis.

5.18.4 Notes

1. The Jensen–Gundersen 'paradox'

This 'paradox', (Jensen & Gundersen, 1982), was the first one described in the stereology context. It caused some perplexity in the early 1980s until it was explained by Baddeley and Cruz-Orive (1995). The curves in Figs. 5.17.1, 5.18.1, and 5.18.2 correspond to exact expressions given by Voss and Cruz-Orive (2009). The Zitterbewegung is typical of test systems, and is usually not present when only single test probes are involved.

2. The Smit paradox

Such paradoxes are related to the so-called 'Smit paradox'. Suppose that Y is a second-order stationary random function with realizations on $\mathbb{R}$. Under some technical conditions, Smit (1961) showed that the mean of a given number n of equispaced observations of Y in a given interval $[0, T]$ can be a more precise estimator of the mean of Y than the integral of Y in $[0, T]$ divided by T. For related results, see Baddeley and Cruz-Orive (1995), Section 5.3.

3. The stereological R-B theorem

Baddeley and Cruz-Orive (1995), Sections 2 and 3, gave general conditions for the R-B theorem to work in the stereology context.

4. The Ohser paradox

The result in Eq. (5.18.13) was hinted at by Ohser (1990), p. 136.

Appendix

A.1 Prerequisites of Probability and Statistics

A.1.1 Basic mathematical concepts and notation

Numbers

A set X is a collection of elements. If x is an element of X, then we write $x \in X$, otherwise $x \notin X$. The set of real numbers $\mathbb{R} = (-\infty, \infty)$ is identified with the real line. The set of positive real numbers is denoted by $\mathbb{R}_+$. The set of integers is $\mathbb{Z} = \{\ldots, -1, 0, 1, \ldots\}$. The set $\mathbb{Z}_+$ of positive integers is the set $\mathbb{N}$ of natural numbers, namely, $\mathbb{Z}_+ = \mathbb{N} = \{1, 2, \ldots\}$. The set containing no elements is called the empty set and is denoted by $\emptyset$. Open, left-open, right-open and closed intervals are denoted by (a, b), $(a, b]$, $[a, b)$ and $[a, b]$ respectively.

Orders of magnitude

Suppose that the quantities u and v depend on a parameter which tends to some constant. Then, if $v \neq 0$,

$$\begin{cases} u = O(v) \text{ if } u/v \text{ remains bounded,} \\ u = o(v) \text{ if } u/v \to 0, \end{cases} \tag{A.1.1}$$

Set operations

If all the elements of a set A are contained in a set X, then A is a subset of X, and we write $A \subset X$. If B is another subset of X, then,

L. M. Cruz-Orive, *Stereology*, Interdisciplinary Applied Mathematics 59,
https://doi.org/10.1007/978-3-031-52451-6

$$\begin{aligned} A \cup B &= \{x \in X : \ x \in A \text{ or } x \in B\}, && \text{(union)}, \\ A \cap B &= \{x \in X : \ x \in A \text{ and } x \in B\}, && \text{(intersection)}, \\ A \backslash B &= \{x \in X : \ x \in A \text{ and } x \notin B\}, && \text{(difference)}. \end{aligned} \tag{A.1.2}$$

The complement A^c of A consists of all the elements of X which do not belong to A, namely,

$$A^c = X \backslash A. \tag{A.1.3}$$

The De Morgan laws read as follows,

$$A \cup B = (A^c \cap B^c)^c, \quad A \cap B = (A^c \cup B^c)^c. \tag{A.1.4}$$

The Cartesian product of two sets A, B is the following set of ordered pairs,

$$A \times B = \{(a, b) : \ a \in A \text{ and } b \in B\}. \tag{A.1.5}$$

Euclidean space

The plane – namely the two-dimensional Euclidean space $\mathbb{R}^2$ – may be regarded as the Cartesian product of two perpendicular real lines,

$$\mathbb{R}^2 = \mathbb{R} \times \mathbb{R}, \tag{A.1.6}$$

with the natural generalization to d-dimensional Euclidean space $\mathbb{R}^d$.

The Euclidean space $\mathbb{R}^d$ is a vector space over the field of real numbers. Thus,

$$\begin{aligned} &x, y \in \mathbb{R}^d \Rightarrow x + y \in \mathbb{R}^d, \\ &x \in \mathbb{R}^d, \ c \in \mathbb{R} \Rightarrow c \cdot x \in \mathbb{R}^d. \end{aligned} \tag{A.1.7}$$

An orthogonal reference frame Ox_1x_2 with a fixed origin O, axis of abscissas Ox_1, and perpendicular axis of ordinates Ox_2, is usually adopted for $\mathbb{R}^2$. A point, or vector, $x \in \mathbb{R}^2$ is thereby determined by its Cartesian coordinates (x_1, x_2), with a natural generalization to $\mathbb{R}^d$.

The multiplication, or homothety, of a set T with a real number c is equivalent to a change of scale without altering the shape of T, namely,

$$cT = \{c \cdot x : \ x \in T\}, \quad c \in \mathbb{R}, \ T \in \mathbb{R}^d. \tag{A.1.8}$$

For $c = -1$ we have a reflection of T, namely its symmetric about the origin,

$$\check{T} = -T = \{-x : \ x \in T\}. \tag{A.1.9}$$

A translation of a set T by a vector z is a change in the location of T without altering its size, shape and orientation, namely,

$$T_z = T + z = \{x + z : \ x \in T\}, \quad z \in \mathbb{R}^d, \ T \subset \mathbb{R}^d. \tag{A.1.10}$$

The rotated version of a vector $x \in \mathbb{R}^d$ about the origin is another vector $y = Rx$, where R is a $d \times d$ orthogonal matrix. The determinant of R is $|R| = +1$, and therefore $\|y\| = \|x\|$. The rotated version of a set T is the result of applying the same rotation to every point of T, namely,

$$RT = \{Rx : x \in T\}, \quad T \subset \mathbb{R}^d. \tag{A.1.11}$$

For instance, a counterclockwise rotation of the point $x \in \mathbb{R}^2$ by an angle $\omega \in [0, 2\pi)$ about the origin is the point $y \in \mathbb{R}^2$ given by

$$\begin{pmatrix} y_1 \\ y_2 \end{pmatrix} = R \begin{pmatrix} x_1 \\ x_2 \end{pmatrix}, \quad R = \begin{pmatrix} \cos\omega & -\sin\omega \\ \sin\omega & \cos\omega \end{pmatrix}. \tag{A.1.12}$$

A.1.2 Basic concepts of measure theory

While measure-theoretic considerations exceed the scope of this book, a brief, elementary overview of measure theory is useful because it underlies concepts of integral geometry, probability, and geometric sampling.

Measurable space

Consider a basic set Ω, and define a family $\mathcal{A}$ of subsets of Ω which is closed under the union and complement operations, that is,

$$(a) \quad \Omega \in \mathcal{A}, \tag{A.1.13}$$

$$\text{(b)} \quad \text{if } A \in \mathcal{A}, \text{ then } A^c \in \mathcal{A}, \tag{A.1.14}$$

$$\text{(c)} \quad \text{if } A_1, A_2, \ldots \in \mathcal{A}, \text{ then } A_1 \cup A_2 \cup \cdots \in \mathcal{A}. \tag{A.1.15}$$

By virtue of the De Morgan laws, the empty set $\emptyset$, and arbitrary intersections and differences among subsets of $\mathcal{A}$, also belong to $\mathcal{A}$. Such a family $\mathcal{A}$ is called a σ-algebra.

The pair $(\Omega, \mathcal{A})$ is called a measurable space. A measure on this space is a function $\mu : \mathcal{A} \to [0, \infty]$ which assigns a (possibly infinite) real value to each subset of $\mathcal{A}$, with the following two properties,

$$(1) \quad \mu(\emptyset) = 0, \tag{A.1.16}$$

$$(2) \quad \mu(A_1 \cup A_2 \cup \cdots) = \mu(A_1) + \mu(A_2) + \cdots, \tag{A.1.17}$$

for all subsets $A_1, A_2, \ldots \in \mathcal{A}$ which are mutually disjoint, namely $A_i \cap A_j = \emptyset$ whenever $i \neq j$. Eq. (A.1.17) means that a measure must be additive.

An important particular case of a measurable space is $(\mathbb{R}^d, \mathcal{B}^d)$, where $\mathcal{B}^d$ is the smallest σ-algebra of Borel sets on $\mathbb{R}^d$. The family $\mathcal{B}^d$ of Borel sets includes all conceivable geometric sets considered here. Of special interest in geometry is the

Lebesgue measure on $(\mathbb{R}^d, \mathcal{B}^d)$. For instance L, A, and V are Lebesgue measures representing interval length on the real line, area in the plane, and volume in space, respectively.

Measurable function

An $\mathcal{A}$-measurable function is a function $X : \Omega \longrightarrow \mathbb{R}$ defined on Ω such that for any Borel set $B \in \mathcal{B}^1$ of the real line (namely in any conceivable collection of intervals) we have $\{\omega \in \Omega : X(\omega) \in B\} \subset \mathcal{A}$.

A.1.3 Probability

The concept of probability is a prerequisite to understand sampling and estimation methods, and it allows scientific modelling in the presence of uncertainty.

A random experiment is a mechanism producing unpredictable results. Each individual result is called an elementary event. The set of all elementary events or states of nature is called the sample space Ω.

In the discrete case Ω is a finite, or countably infinite set of elementary events. One is often concerned with a finite number N of elementary events, namely,

$$\Omega = \{e_1, e_2, \ldots, e_N\}, \quad N \in \mathbb{N}. \tag{A.1.18}$$

For instance, if a roulette wheel with N numbers points to the ith number, then we say that the elementary event e_i has taken place.

In the continuous case Ω may be a finite interval,

$$\Omega = (0, h], \quad h > 0. \tag{A.1.19}$$

To illustrate a connection with the discrete case, partition Ω into N subintervals of equal length h/N. A mechanism chooses a point $X \in (0, h]$. If X lies in the ith subinterval, then one says that the ith event has occurred. Thus, in this case Ω may again be expressed as in Eq. (A.1.18). As N tends to infinity, the length of each subinterval becomes infinitesimal, and the corresponding elementary event is

$$X \in (x, x + \mathrm{d}x], \quad x \in (0, h]. \tag{A.1.20}$$

In any case, consider the σ-algebra $\mathcal{A}$ of subsets of Ω. The elements of $\mathcal{A}$ are combinations of elementary events, simply called events. For instance, in the discrete case an event may be: "the roulette wheel points to an even number". In the continuous case, "a point z lies in the interval $(0, h/2]$", etc. A probability is a measure $\mathbb{P}$

defined on the measurable space $(\Omega, \mathcal{A})$ with the special property

$$\mathbb{P}(\Omega) = 1. \tag{A.1.21}$$

This implies that the probability $\mathbb{P}(A)$ of any event $A \in \mathcal{A}$ will always lie in the interval $[0, 1]$. The triplet $(\Omega, \mathcal{A}, \mathbb{P})$ is called a probability space.

The events A and B are disjoint, or mutually exclusive, if $A \cap B = \emptyset$, in which case $\mathbb{P}(A \cap B) = 0$ and $\mathbb{P}(A \cup B) = \mathbb{P}(A) + \mathbb{P}(B)$. For any two events A and B, not necessarily disjoint, the total probability identity is

$$\mathbb{P}(A \cup B) = \mathbb{P}(A) + \mathbb{P}(B) - \mathbb{P}(A \cap B). \tag{A.1.22}$$

If $\mathbb{P}(B) > 0$, then the conditional probability of A given B, namely the probability that A occurs given that B occurs, is

$$\mathbb{P}(A|B) = \frac{\mathbb{P}(A \cap B)}{\mathbb{P}(B)}. \tag{A.1.23}$$

The events A, B are said to be independent if, and only if,

$$\mathbb{P}(A|B) = \mathbb{P}(A), \tag{A.1.24}$$

that is, the knowledge that B has occurred bears no information on the probability that A will occur. Equivalently,

$$\mathbb{P}(A \cap B) = \mathbb{P}(A) \cdot \mathbb{P}(B). \tag{A.1.25}$$

A.1.4 Random variables

Univariate random variables

A univariate random variable X is an $\mathcal{A}$-measurable function taking real values, that is, a measurable function that assigns a real value to each event of a sample space. A realization of X is a particular value x of X. The probability that X takes a value in the interval $(-\infty, x]$ is the distribution function (df) of X, that is,

$$F(x) = \mathbb{P}(X \leq x), \quad x \in \mathbb{R}. \tag{A.1.26}$$

A discrete random variable X takes discrete values with probabilities given by its probability function (pf):

$$p_i = \mathbb{P}(X = x_i), \ i \in \mathbb{Z}, \quad \sum_{i \in \mathbb{Z}} p_i = 1. \tag{A.1.27}$$

The expected value of any measurable function $\psi(X)$ of X is

$$\mathbb{E}\{\psi(X)\} = \sum_{i \in \mathbb{Z}} \psi(x_i) p_i. \tag{A.1.28}$$

A continuous random variable X takes discrete values with probability zero, that is, $\mathbb{P}(X = x) = 0$. In this case, non-zero probabilities are associated with intervals. We define the probability element of X as follows,

$$\mathbb{P}(\mathrm{d}x) := \mathbb{P}(x < X \leq x + \mathrm{d}x). \tag{A.1.29}$$

If X is absolutely continuous (henceforth simply called 'continuous'), then,

$$\mathbb{P}(\mathrm{d}x) = \mathrm{d}F(x) = F'(x)\,\mathrm{d}x = f(x)\,\mathrm{d}x, \tag{A.1.30}$$

where $f : \mathbb{R} \to [0, \infty)$ is an integrable function called the probability density function (pdf) of X. Thus,

$$\int_{\mathbb{R}} \mathbb{P}(\mathrm{d}x) = \int_{\mathbb{R}} f(x)\,\mathrm{d}x = 1. \tag{A.1.31}$$

The mean, or expected value, of a measurable function $\psi(X)$ of X reads,

$$\mathbb{E}\{\psi(X)\} = \int_{\mathbb{R}} \psi(x) f(x)\,\mathrm{d}x. \tag{A.1.32}$$

In particular, the variance of a univariate random variable is defined as

$$\mathrm{Var}(X) = \mathbb{E}\{X - \mathbb{E}(X)\}^2 = \mathbb{E}(X^2) - \{\mathbb{E}(X)\}^2. \tag{A.1.33}$$

The standard deviation is the square root of the variance, namely,

$$\mathrm{SD}(X) = \sqrt{\mathrm{Var}(X)}. \tag{A.1.34}$$

For a positive random variable, a useful descriptor is the coefficient of variation,

$$\mathrm{CV}(X) = \frac{\mathrm{SD}(X)}{\mathbb{E}(X)}. \tag{A.1.35}$$

Continuous bivariate random variable. Marginal and conditional distributions

A bivariate random variable (X_1, X_2) takes values in $\mathbb{R}^2$ and it has a joint distribution function

$$F(x_1, x_2) = \mathbb{P}(X_1 \leq x_1, X_2 \leq x_2), \quad (x_1, x_2) \in \mathbb{R}^2, \tag{A.1.36}$$

where $\mathbb{P}(A, B)$ stands for $\mathbb{P}(A \cap B)$. Each of the two components X_1, X_2 is in turn a univariate random variable.

If X_1 and X_2 are continuous, then their joint probability element is

$$\begin{aligned}\mathbb{P}(\mathrm{d}x_1, \mathrm{d}x_2) &= \mathbb{P}(x_1 < X_1 \le x_1 + \mathrm{d}x_1, x_2 < X_2 \le x_2 + \mathrm{d}x_2)\\ &= f(x_1, x_2)\,\mathrm{d}x_1\,\mathrm{d}x_2,\end{aligned} \tag{A.1.37}$$

where $f : \mathbb{R}^2 \to [0, \infty)$ is an integrable function called the joint pdf of (X_1, X_2). Thus,

$$\int_{\mathbb{R}^2} \mathbb{P}(\mathrm{d}x_1, \mathrm{d}x_2) = \int_{\mathbb{R}^2} f(x_1, x_2)\,\mathrm{d}x_1\,\mathrm{d}x_2 = 1. \tag{A.1.38}$$

In a Cartesian frame $Ox_1x_2x_3$ the graph of $x_3 = f(x_1, x_2)$ is a surface. By Eq. (A.1.38), the volume under the graph is equal to 1.

The mean value of a measurable function $\psi(X_1, X_2)$ reads

$$\mathbb{E}\{\psi(X_1, X_2)\} = \int_{\mathbb{R}^2} \psi(x_1, x_2) f(x_1, x_2)\,\mathrm{d}x_1\,\mathrm{d}x_2. \tag{A.1.39}$$

The marginal pdf $f_1(x_1)$ of X_1 is

$$f_1(x_1) = \int_{\mathbb{R}} f(x_1, x_2)\,\mathrm{d}x_2, \tag{A.1.40}$$

and similarly for the marginal pdf of X_2.

The conditional pdf of X_2 given a value $X_1 = x_1$ is

$$f_{2|1}(x_2|x_1) = \frac{f(x_1, x_2)}{f_1(x_1)}, \tag{A.1.41}$$

and similarly for $f_{1|2}(x_1|x_2)$.

Variance decomposition formula

The conditional mean $\mathbb{E}(X_2|X_1)$ and variance $\mathrm{Var}(X_2|X_1)$ of X_2 given X_1 are functions of X_1, and they are defined via Eq. (A.1.32) with $f_{2|1}(x_2|x_1)$ in the place of $f(x)$. It is easy to verify that

$$\mathbb{E}_1[\mathbb{E}(X_2|X_1)] = \mathbb{E}(X_2), \tag{A.1.42}$$

where $\mathbb{E}_1(\cdot)$ denotes expectation with respect to $f_1(x_1)$. More generally,

$$\mathbb{E}_1\{\mathbb{E}[\psi(X_2)|X_1]\} = \mathbb{E}[\psi(X_2)]. \tag{A.1.43}$$

The corresponding definitions for X_1 given X_2 are analogous. An important identity, relevant in sampling theory, is the following,

$$\mathrm{Var}(X_2) = \mathrm{Var}_1\{\mathbb{E}(X_2|X_1)\} + \mathbb{E}_1\{\mathrm{Var}(X_2|X_1)\}. \tag{A.1.44}$$

In fact,

$$\begin{aligned}\mathrm{Var}_1\{\mathbb{E}(X_2|X_1)\} &= \mathbb{E}_1\{\mathbb{E}(X_2|X_1)\}^2 - \{\mathbb{E}(X_2)\}^2,\\ \mathbb{E}_1[\mathrm{Var}(X_2|X_1)] &= \mathrm{E}(X_2^2) - \mathbb{E}_1\{\mathbb{E}(X_2|X_1)\}^2,\end{aligned} \tag{A.1.45}$$

and adding up the rhs's of the preceding two identities we get $\mathrm{Var}(X_2)$.

Independent and uncorrelated random variables

The random variables X_1, A_2 are independent if, and only if, the events $X_1 \le x_1$ and $X_2 \le x_2$ are independent for all $x_1, x_2 \in \mathbb{R}$. For continuous random variables a convenient, equivalent condition is

$$f(x_1, x_2) = f_1(x_1) \cdot f_2(x_2). \tag{A.1.46}$$

The covariance of X_1, X_2 is

$$\begin{aligned}\mathrm{Cov}(X_1, X_2) &= \mathbb{E}\{(X_1 - \mathbb{E}(X_1))(X_2 - \mathbb{E}(X_2))\}\\ &= \mathbb{E}(X_1 X_2) - \mathbb{E}(X_1) \cdot \mathbb{E}(X_2).\end{aligned} \tag{A.1.47}$$

If X_1 and X_2 are independent, then they are uncorrelated because, by independence, $\mathbb{E}(X_1 X_2) = \mathbb{E}(X_1) \cdot \mathbb{E}(X_2)$, hence $\mathrm{Cov}(X_1, X_2) = 0$. However the converse is not always true.

Remark on notation

In mathematical statistics it is customary to denote random variables by capital letters, and realizations of them by lower case letters. For simplicity, in this book a lower case letter may represent indistinctly a random variable, or a realization of it. For instance, a UR point in a domain may be denoted by z instead of Z. Random angles may be denoted, for instance, by ϕ, θ, ω instead of Φ, Θ, Ω, etc.

A.1.5 Uniform random variables

Uniform random (UR) variables play a central role in any sampling and simulation procedures. Here we concentrate on concepts relevant to geometric sampling.

Consider a set $D = \{x_1, x_2, \ldots, x_N\}$ of N fixed real numbers. The discrete random variable U is uniform in D, and we write $U \sim \mathrm{UR}(D)$ (where the symbol "$\sim$" means "is distributed as"), if

$$p_i = \mathbb{P}(U = x_i) = \frac{1}{N}, \quad i = 1, 2, \ldots, N. \tag{A.1.48}$$

If $D = \{0, 1, \ldots, 9\}$, then the UR variable U is usually called a random digit. Independent realizations of U are generated by suitable algorithms. Quality statistical software packages, such as R©, are equipped with random number generators that are reliable for most practical purposes.

The continuous UR variable $U \sim \mathrm{UR}[0, 1)$ has probability element

$$\mathbb{P}(\mathrm{d}u) = \begin{cases} \mathrm{d}u, & u \in [0, 1), \\ 0, & \text{otherwise}. \end{cases} \tag{A.1.49}$$

and a realization of it may be generated by means of a sequence of as many independent UR digits as required. The probability element of a UR variable X in an arbitrary interval of a fixed length $H > 0$ is

$$\mathbb{P}(\mathrm{d}x) = \frac{\mathrm{d}x}{H}, \quad x \in [a, a + H), \ a \in \mathbb{R}, \tag{A.1.50}$$

and we say that $X \sim \mathrm{UR}[a, a + H)$. Here, a realization of X may be generated as

$$X = a + UH, \quad U \sim \mathrm{UR}[0, 1). \tag{A.1.51}$$

A.1.6 Basic concepts of sampling theory

The purpose is to complement Sections 4.1–4.3 with a few technical details.

Sampling theory establishes precise rules to draw information from a population, which is a well-defined set of elements. Traditional sampling theory deals with finite populations of discrete elements, e.g., neurons of a given type in a well-defined brain compartment. The sampling frame defines the sampling units. A sampling unit may be a whole element, or an easily observable part of it, e.g., the nucleolus of a neuron. Each sampling unit has an associated measure, e.g. the neuron volume, in which case the working population is a set of neuron volumes $Y = \{V_1, V_2, \ldots, V_N\}$. The target parameter γ is a combination of such numbers, e.g. their arithmetic mean $\gamma = \mathbb{E}(V)$.

In practice the measurement of all the population units, and thereby the exact determination of γ, is inaccessible, and a sample, namely a subset of the population, has to be drawn. Sampling theory dictates the sampling protocol, called the sampling design, and, in addition, a formula, called an estimator $\widehat{\gamma}$ of γ, involving the values of the sampled units.

The sampling design dictates a random mechanism to draw the sample and, consequently, $\widehat{\gamma}$ is a random variable whose statistical distribution is called its sampling distribution. A concrete sample will yield a numerical value of $\widehat{\gamma}$, namely an estimate, and we do not know how close this estimate will be from the target γ. In other words, we do not know the magnitude of the random error $\widehat{\gamma} - \gamma$. Theory, however, can predict the performance of the sampling design, namely the statistical behaviour of such an error. An unbiased design, which is usually very easy to implement in stereology, will warrant a priori that, apart from uncontrolled artifacts which may arise in practice, the bias of $\widehat{\gamma}$, namely the mean error,

$$\text{Bias}(\widehat{\gamma}) = \mathbb{E}(\widehat{\gamma}) - \gamma, \tag{A.1.52}$$

over its sampling distribution, is equal to zero. We then say that $\widehat{\gamma}$ is a (design) unbiased estimator (UE) of γ.

Unbiasedness alone does not warrant that $\widehat{\gamma}$ is a satisfactory estimator, because the error may vary too much among all possible samples. It is therefore desirable that the mean square error,

$$\text{MSE}(\widehat{\gamma}) = \mathbb{E}(\widehat{\gamma} - \gamma)^2, \tag{A.1.53}$$

is as small as desired. This may be achieved by increasing the sample size n if $\widehat{\gamma}$ is unbiased and $\text{Var}(\widehat{\gamma}) = O(n^{-\alpha})$, $\alpha > 0$.

A small variance $\text{Var}(\widehat{\gamma})$ means that $\widehat{\gamma}$ is precise, namely that its mean square deviation with respect to its own mean $\mathbb{E}(\widehat{\gamma})$, is small – this is the popular saying that $\widehat{\gamma}$ is "reproducible" – but, unfortunately, this will only ensure that $\widehat{\gamma}$ concentrates away from the target γ if $\widehat{\gamma}$ is biased, see Fig. 4.1.1. We rather want that $\widehat{\gamma}$ is as accurate as desired, namely that $\text{MSE}(\widehat{\gamma})$ is as small as desired. The identity

$$\text{MSE}(\widehat{\gamma}) = \text{Var}(\widehat{\gamma}) + \{\text{Bias}(\widehat{\gamma})\}^2 \tag{A.1.54}$$

illustrates that, to be accurate, a "reproducible" estimator has to be also unbiased. Unbiasedness can only be warranted by the theory underlying the sampling design, because the target γ is unknown. For instance, the estimator $\widehat{A}$ in Eq. (4.1.2) is unbiased because the identity in Eq. (4.1.1) holds, and the identities leading to the UEs given in this book stem in turn from the theory developed in Chapters 1 and 2, and not from numerical replications.

In applied science, if $\widehat{\gamma}$ is an estimator, then $\text{Var}(\widehat{\gamma})$ is called the error variance of $\widehat{\gamma}$, whereas $\text{SD}(\widehat{\gamma})$ is called the standard error, and denoted by $\text{SE}(\widehat{\gamma})$. Also, $\text{CV}(\widehat{\gamma})$ is called the coefficient of error of $\widehat{\gamma}$, and denoted by $\text{CE}(\widehat{\gamma})$.

For simplicity, the preceding example assumes that the neuron volumes sampled at a first stage are directly observable. In practice, however, each sampled volume has to be estimated at a second sampling stage in which the sampling frame depends on the preferred design. For instance, if the Cavalieri method is used, then for each sampled neuron an artificial population is conceived consisting of the infinitely many planar sections normal to a given sampling axis, and the sample will consist of a few Cavalieri sections. Thus, the scenario is discrete at the first sampling stage, but continuous at the second stage. In the example, the estimator $\widehat{\gamma}$ is the combined outcome of the two stages constituting the sampling design. Section 4.22.1 illustrates a two-stage design in which the nucleator is used at the second stage.

A.1.7 Notes

For an introduction to probability theory and mathematical statistics see, for instance, Feller (1968), and Grimmett and Stirzaker (2001). For an introduction to statistics see, for instance, Lindgren (1993). A classical reference on sampling theory is Cochran (1977).

A.2 Approximate Mean and Variance of Non-Linear Functions of Random Variables

A.2.1 Univariate case

Let X denote a random variable with finite mean μ and variance σ^2, and consider a smooth function $f(X)$ which admits a Taylor expansion near μ. Then,

$$\mathbb{E}\{f(X)\} \approx f(\mu) + \frac{1}{2} f''(\mu) \cdot \sigma^2, \tag{A.2.1}$$

$$\mathrm{Var}\{f(X)\} \approx (f'(\mu))^2 \sigma^2. \tag{A.2.2}$$

The preceding expressions are exact if $f(X) = a + bX$, (a, b constants), because the limited Taylor expansion used includes up to quadratic terms, see Section A.2.2 below.

Special case: logarithmic transformation

Suppose that the random variable of interest is $X > 0$, but we work with its logarithm of base Δ, namely,

$$Y := f(X) = \log_\Delta(X), \quad \Delta > 0. \tag{A.2.3}$$

The problem is to recover, at least approximately, the original parameters $\mu := \mathbb{E}X$ and $\sigma^2 := \mathrm{Var}(X)$ from observations of Y. Noting that

$$f(X) = \log X / \log \Delta, \tag{A.2.4}$$

application of Eq. (A.2.1) and Eq. (A.2.2) yields,

$$\mathbb{E}X \approx \Delta^{\mathbb{E}(Y)+\frac{1}{2}(\log\Delta)\mathrm{Var}(Y)},$$
$$\mathrm{CV}^2(X) \approx (\log\Delta)^2\mathrm{Var}(Y), \tag{A.2.5}$$

respectively, where $\log(\cdot)$ denotes natural logarithm.

A.2.2 Case of two random variables

Let (X_1, X_2) denote a pair of random variables with finite moments, with means $\mu \equiv (\mu_1, \mu_2)$, variances (σ_1^2, σ_2^2), and covariance σ_{12}, and let $f \equiv f(X_1, X_2)$ denote a bounded, real, smooth function of X_1 and X_2. We have,

$$\mathbb{E}(f) \approx f(\mu) + \frac{1}{2}\{f_{11}(\mu)\sigma_1^2 + 2f_{12}(\mu)\sigma_{12} + f_{22}(\mu)\sigma_2^2\}, \tag{A.2.6}$$
$$\mathrm{Var}(f) \approx f_1^2(\mu)\sigma_1^2 + 2f_1(\mu)f_2(\mu)\sigma_{12} + f_2^2(\mu)\sigma_2^2, \tag{A.2.7}$$

which involve the definitions $f_i(\mu) = [\partial f/\partial X_i]_\mu$, $f_{ii}(\mu) = [\partial^2 f/\partial X_i^2]_\mu$, and $f_{12}(\mu) = [\partial^2 f/(\partial X_1 \partial X_2)]_\mu$.

Eq. (A.2.1) and Eq. (A.2.2) follow by setting $\sigma_2^2 = \sigma_{12} = 0$ in the rhs of the preceding two equations, respectively.

Derivation

Up to second-order terms, the limited Taylor expansion of f near μ reads

$$\begin{aligned} f &= f(\mu) + (X_1 - \mu_1)f_1(\mu) + (X_2 - \mu_2)f_2(\mu) \\ &\quad + \frac{1}{2}\Big\{(X_1 - \mu_1)^2 f_{11}(\mu) + 2(X_1 - \mu_1)(X_2 - \mu_2)f_{12}(\mu) \\ &\quad + (X_2 - \mu_2)^2 f_{22}(\mu)\Big\}. \end{aligned} \tag{A.2.8}$$

Taking expectations on both sides of the preceding identity, Eq. (A.2.6) is obtained. On the other hand, raising both sides of Eq. (A.2.8) to the power two, taking expectations, and subtracting $(\mathbb{E}f)^2$, the approximation given by Eq. (A.2.7) is obtained.

A.2.3 Cochran's formula for a ratio estimator

Consider a pair of random variables (Y, X), $X > 0$, and define the ratio

$$R = \mathbb{E}(Y)/\mathbb{E}(X). \tag{A.2.9}$$

The problem is to estimate R from a sample of n independent observations $\{(Y_i, X_i),\ i = 1, 2, \ldots, n\}$. A ratio-unbiased estimator is

$$\widehat{R} = \sum_{i=1}^{n} Y_i \Big/ \sum_{i=1}^{n} X_i. \tag{A.2.10}$$

A ratio estimator $\widehat{R}$ is 'ratio-unbiased' if the expected value of the numerator, divided by the expected value of the denominator, is equal to R.

By Eq. (A.2.6), the approximate coefficient of bias of $\widehat{R}$ becomes

$$\begin{aligned} \mathrm{CB}(\widehat{R}) :&= \frac{\mathbb{E}(\widehat{R}) - R}{R} \\ &\approx \frac{1}{n}\{\mathrm{CV}^2(X) - \rho \cdot \mathrm{CV}(Y) \cdot \mathrm{CV}(X)\}, \end{aligned} \tag{A.2.11}$$

where ρ denotes the correlation coefficient of (X, Y). On the other hand, by Eq. (A.2.7),

$$\begin{aligned} \mathrm{CE}^2(\widehat{R}) :&= \frac{\mathrm{Var}(\widehat{R})}{R^2} \\ &\approx \frac{1}{n}\{\mathrm{CV}^2(Y) + \mathrm{CV}^2(X) - 2\rho \cdot \mathrm{CV}(Y) \cdot \mathrm{CV}(X)\}. \end{aligned} \tag{A.2.12}$$

Replacing the items in the rhs of the preceding approximation with the corresponding estimators, and rearranging, we obtain,

$$\mathrm{ce}^2(\widehat{R}) = \frac{n}{n-1}\left(\frac{\sum Y_i^2}{(\sum Y_i)^2} + \frac{\sum X_i^2}{(\sum X_i)^2} - 2\frac{\sum Y_i X_i}{\sum Y_i \sum X_i}\right), \tag{A.2.13}$$

which is known as Cochran's formula. The summations run from 1 to n.

A.2.4 Goodman's formula for a product, or a ratio, of independent estimators

Suppose that a parameter γ of interest can be expressed as the product of two parameters, namely,

$$\gamma = \gamma_1 \gamma_2, \tag{A.2.14}$$

and let $\widehat{\gamma_1}$, $\widehat{\gamma_2}$ represent independent estimators of γ_1, γ_2, respectively. Then,

$$\widehat{\gamma} = \widehat{\gamma_1}\widehat{\gamma_2}, \tag{A.2.15}$$

estimates γ and applying Eq. (A.2.7), $\mathrm{CE}^2(\widehat{\gamma})$ may be predicted approximately by Goodman's formula,

$$\mathrm{ce}^2(\widehat{\gamma}) = \mathrm{ce}^2(\widehat{\gamma}_1) + \mathrm{ce}^2(\widehat{\gamma}_2). \tag{A.2.16}$$

Now, suppose that

$$\gamma = \gamma_1/\gamma_2, \quad \gamma_2 \neq 0. \tag{A.2.17}$$

Then,

$$\widehat{\gamma} = \widehat{\gamma}_1/\widehat{\gamma}_2, \tag{A.2.18}$$

and, by independence, Eq. (A.2.12) may be applied with $\rho = 0$. Thus, Eq. (A.2.16) applies also in this case.

A.2.5 Notes

For further details and applications of Sections A.2.1 and A.2.2 see, for instance, Chatfield (1983). For ratio estimation and Cochran's formula, see Cochran (1977), and for Goodman's formula, see Goodman (1960).

A.3 The Fourier Transform of r^ω

Eq. (A.3.3) and Eq. (A.3.4) below are used to obtain Eq. (5.2.25) and Eq. (5.9.13), respectively.

A.3.1 Basic formulae

Consider the vector $x = (x_1, x_2, \ldots, x_d)' \in \mathbb{R}^d$, $d \geq 1$, of modulus $r \equiv \|x\| = (x'x)^{1/2}$, and a real constant $w > -d$, $w \neq 2k$, and $w \neq -d - 2k$ for $k = 1, 2, \ldots$. We want to show that the Fourier transform of r^w has the following expression,

$$\begin{aligned}(\mathcal{F}_d r^w)(t) &= \int_{\mathbb{R}^d} r^w \exp(-2\pi \mathrm{i} t'x)\,\mathrm{d}x \\ &= \frac{1}{\pi^{w+d/2}} \cdot \frac{\Gamma\left(\dfrac{w+d}{2}\right)}{\Gamma\left(-\dfrac{w}{2}\right)} \cdot \frac{1}{\rho^{w+d}},\end{aligned} \tag{A.3.1}$$

where $t = (t_1, t_2, \ldots, t_d)'$, and $\rho \equiv \|t\| = (t't)^{1/2}$ is the modulus of t. Thus, $(\mathcal{F}_d r^w)(t)$ depends on t via ρ only. A convenient alternative expression is

$$(\mathcal{F}_d r^w)(\rho) = -\frac{\sin(w\pi/2)}{\pi^{w+d/2+1}} \cdot \Gamma\left(\frac{w+d}{2}\right)\Gamma\left(\frac{w}{2}+1\right) \cdot \frac{1}{\rho^{w+d}}, \tag{A.3.2}$$

which may be obtained by applying Euler's reflection formula to $\Gamma(-w/2)$.

For $d = 1$ and $w = 2q + 1$, $q \in [0, 1]$, application of Legendre's duplication formula yields

$$(\mathcal{F}_1 r^{2q+1})(\rho) = -2\cos(q\pi)\frac{\Gamma(2q+2)}{(2\pi)^{2q+2}} \cdot \frac{1}{\rho^{2q+2}}. \tag{A.3.3}$$

Moreover, for $w = 2k - 1$, $k = 1, 2, \ldots$, Eq. (A.3.2) becomes

$$(\mathcal{F}_d r^{2k-1})(\rho) = \frac{(-1)^k}{\pi^{2k+d/2}}\Gamma\left(k + \frac{d-1}{2}\right)\Gamma\left(k + \frac{1}{2}\right)\frac{1}{\rho^{2k+d-1}}. \tag{A.3.4}$$

A.3.2 Derivations

The first step is to show that the Fourier transform $F(t) \equiv (\mathcal{F}_d r^w)(t)$ of r^w depends on t via ρ only, and it has the following form,

$$F(t) = c_d(w) \cdot \rho^{-w-d}, \tag{A.3.5}$$

where $c_d(w)$ is a constant which depends on w and d only. First, we show that $F(t)$ is homogeneous of degree $-w - d$. In fact, for any real number $\alpha > 0$, set $\alpha x = y$, whereby $x = \alpha^{-1}y$, $r = \alpha^{-1}\|y\|$, $\mathrm{d}x = \mathrm{d}x_1\,\mathrm{d}x_2 \cdots \mathrm{d}x_d = \alpha^{-d}\,\mathrm{d}y$, and we have

$$\begin{aligned} F(\alpha t) &= \int_{\mathbb{R}}^{d} \alpha^{-w}\|y\|^w \exp(-2\pi \mathrm{i}\alpha t'\alpha^{-1}y)\alpha^{-d}\,\mathrm{d}y \\ &= \int_{\mathbb{R}^d} \alpha^{-w-d}\|y\|^w \exp(-2\pi \mathrm{i}t'y)\,\mathrm{d}y \\ &= \alpha^{-w-d}F(t). \end{aligned} \tag{A.3.6}$$

Moreover, $F(t)$ is rotation-invariant. In fact, for any $d \times d$ orthogonal matrix A, set $x = Ay$, whereby $r = \|y\|$, $\mathrm{d}x = \mathrm{d}y$, and therefore $F(At) = F(t)$. This implies that $F(t)$ depends on ρ only. From the preceding two results, Eq. (A.3.5) follows.

To proceed with the second step, define $\Phi = \mathcal{F}\varphi$ and $\Psi = \mathcal{F}\psi$, with arguments in $\mathbb{R}^d$. Using the inverse transform

$$\varphi(x) = \int_{\mathbb{R}^d} \Phi(t)\exp(2\pi \mathrm{i}t'x)\,\mathrm{d}t, \tag{A.3.7}$$

and applying Fubini's theorem, we obtain Parseval's formula,

$$\begin{aligned}\int_{\mathbb{R}^d} \varphi(x)\psi(x)\,\mathrm{d}x &= \int_{\mathbb{R}^d} \psi(x)\,\mathrm{d}x \int_{\mathbb{R}^d} \Phi(t)\exp(2\pi\mathrm{i}t'x)\,\mathrm{d}t \\ &= \int_{\mathbb{R}^d} \Phi(t)\,\mathrm{d}t \int_{\mathbb{R}^d} \psi(x)\exp(2\pi\mathrm{i}t'x)\,\mathrm{d}x \\ &= \int_{\mathbb{R}^d} \Phi(t)\Psi(-t)\,\mathrm{d}t. \end{aligned} \tag{A.3.8}$$

Now, set $\varphi(x) = r^w$, whereby $\Phi(t) = c_d(w)\rho^{-w-d}$, as we have seen. On the other hand, set $\psi(x) = \exp(-r^2/2)$, whereby

$$\Psi(t) = (2\pi)^{d/2}\exp(-2\pi^2\rho^2), \tag{A.3.9}$$

which is well known from the Normal distribution theory. In the identity resulting from the application of Parseval's formula, a change to spherical coordinates yields $\mathrm{d}x = r^{d-1}\,\mathrm{d}r\,\mathrm{d}u_{d-1}$ and $\mathrm{d}t = \rho^{d-1}\,\mathrm{d}\rho\,\mathrm{d}u_{d-1}$, where $\mathrm{d}u_{d-1}$ is the area element of the unit sphere $\mathbb{S}^{d-1}$. After diving both sides of the formula by the area of $\mathbb{S}^{d-1}$, the integrals become unidimensional, namely,

$$\int_0^\infty r^{w+d-1}\mathrm{e}^{-r^2/2}\,\mathrm{d}r = c_d(w)(2\pi)^{d/2}\int_0^\infty \rho^{-w-1}\mathrm{e}^{-2\pi^2\rho^2}\,\mathrm{d}\rho. \tag{A.3.10}$$

The preceding integrals are expressible as Gamma functions and, after simplification, $c_d(w)$ is the factor of ρ^{-w-d} in the rhs of Eq. (A.3.1).

A.3.3 Notes

The preceding proof is adapted from Gel'fand and Shilov (1964), pp. 192–194. Matheron (1965), p. 279, also used related formulae to study the precision of systematic sampling, see Section 5.2.6, Note 1.

List of Notation

- $\emptyset$: Empty set.
- $\oplus$: Minkowski addition, Eq. (1.16.1).
- $1_Y(x)$: Indicator function, Eq. (1.4.3).
- 2D, 3D stand for two- and three-dimensional Euclidean space, namely $\mathbb{R}^2$, $\mathbb{R}^3$, respectively.
- $x \in A$: x is an element of the set A.
- $B \subset A$: B is a subset of (or is contained in) A.
- $A \uparrow B$: The set A hits the set B. Equivalently, $A \cap B \neq \emptyset$.
- $A \equiv B$: A and B are identical. Also: A is an alternative notation for B, or vice versa.
- $A := B$: A is defined by B.
- $\alpha(\cdot)$, α: Section measure, Eq. (1.3.4).
- $A(\cdot)$, A: Area of a planar domain.
- a: Area of a fundamental tile, Section 2.25.2.
- a_0: Area of a quadrat probe, Section 2.25.6.
- AP: Associated point, Section 1.13.1.
- AV: Associated vector, Section 1.13.1.
- $B(\cdot)$, B: Curve, or boundary length, in the plane, Eq. (1.3.7). Number of bridges, Fig. 1.12.1.
- b: Essentially linear curve element, Eq. (1.3.7).
- $C(\cdot)$, C: Total curvature of a planar curve, Eq. (1.10.2).
- $C(\cdot)$: Scalar covariance, Eq. (3.6.7), Eq. (3.6.8).
- c_1, c_{10}: Constants, Section 1.5.5.
- c_2, c_{20}: Constants, Section 1.16.4.
- $\mathrm{CV}(\cdot)$: Coefficient of variation of a random variable, Eq. (A.1.35).
- $\mathrm{CE}(\cdot)$: Coefficient of variation of an estimator, Section A.1.6.
- $\mathrm{cv}(\cdot)$, $\mathrm{ce}(\cdot)$: Estimators of $\mathrm{CV}(\cdot)$, $\mathrm{CE}(\cdot)$, respectively.
- D: Reference set, usually a cube, or a ball, Section 2.1.2.
- $D_\oplus(u_d)$, $D_\oplus$: Section 2.11.1.
- $\mathbb{E}$: Expected value or true mean, Eq. (A.1.28), Eq. (A.1.32).
- $\mathbb{E}_W(\gamma)$: W-weighted mean of a parameter γ, Eq. (2.32.1).

L. M. Cruz-Orive, *Stereology*, Interdisciplinary Applied Mathematics 59,
https://doi.org/10.1007/978-3-031-52451-6

- FUR: Uniform random with a fixed orientation, Sections 2.4.1 and 2.5.1.
- $\gamma(\cdot)$, γ: Measure of a compact set, Section 1.3.3, Eq. (1.3.4).
- $\widehat{\gamma}$: Estimator of a parameter γ, Eq. (4.1.2), Section A.1.6.
- γ_V: Ratio γ/V, Sections 2.6.2 and 4.2.3.
- $G(\cdot)$: Gaussian curvature, Section 1.12.1. Fourier transform of the covariogram $g(\cdot)$, Eq. (5.2.23).
- $G_{r,g-r}$: Grassmannian, Section 1.5.5.
- G_d: Special group of motions in $\mathbb{R}^d$, Section 1.13.2.
- $G_{d[x]}$: Special group of rotations in $\mathbb{R}^d$ about a fixed point $x \in \mathbb{R}^d$, Sections 1.13.2 and 1.16.4.
- $g(\cdot)$: Covariogram, Eq. (5.2.18). Geometric covariogram, Eq. (3.8.1).
- H: Side length of a reference cube (Section 2.2.1), or diameter of a reference ball, Eq. (2.4.2), Eq. (2.5.4). Number of holes, Fig. 1.12.1.
- $H(\cdot)$: Caliper length of a domain in a given direction, see Fig. 1.10.1(a) for the planar case.
- H_K: Support set, or flower, of a planar convex set K, Fig. 1.8.2.
- $h(\cdot)$: Hitting measure, Eq. (1.19.1), Eq. (2.1.6).
- h: Length parameter of a fundamental tile, Fig. 4.6.1, Fig. 4.16.1(d). Disector thickness, Eq. (4.14.3).
- HP: Horizontal plane, Fig. 1.2.6.
- I: Number of intersections, Eq. (1.5.4). Number of islands, Fig. 1.12.1.
- ICav: Isotropic Cavalieri, Section 2.26.3.
- IR: Isotropic random, Sections 2.4.1 and 2.5.1.
- IUR: Isotropic uniform random, Sections 2.4.1 and 2.5.1.
- J_0: Fundamental tile of a test system, Eq. (1.21.1), Fig. 1.21.1.
- $\kappa(\cdot)$: Local curvature of a curve, Eq. (1.10.1).
- $K(\cdot)$: K-function, Eq. (3.6.13).
- Λ_{x,u_d}: Test system of bounded probes, Fig. 1.21.1, Eq. (1.21.8).
- $\Lambda_{z,u}$: Test system of unbounded probes, Fig. 1.21.2, Eq. (1.21.13).
- $L(\cdot)$, L: Linear intercept length, Eq. (1.4.4), Eq. (1.4.6). Length of a curve in space, Section 1.5.4.
- L_r^d: r-plane in $\mathbb{R}^d$, Fig. 1.2.1 and Fig. 1.2.3.
- L_2^*: Oriented plane, Section 1.13.3.
- $L_{r[0]}^d$: plane in $\mathbb{R}^d$ through a fixed point, Fig. 1.2.1 and Fig. 1.2.3.
- $L_{2\cdot v}^3$, $L_{1\cdot v}^2$: Vertical plane, and sine-weighted line in a vertical plane, respectively, Section 1.2.6.
- $L_{1(z)}^2$: A priori weighted test line for the invariator, Section 1.2.8 and Fig. 1.2.7.
- $L_{r,t}^d$: Slab with faces of dimension r and thickness t in $\mathbb{R}^d$, Fig. 1.2.5.
- $L_t^-(\cdot)$, L_t^-: Slab disector of thickness t, Fig. 2.15.2(b).
- $L_t^*(u_2)$: Oriented slab probe, Section 1.7.4.
- L_{1+}^2, L_{2+}^3: Ray emanating from a fixed point, and half-plane emanating from a fixed axis, respectively, Section 1.2.7.
- lhs: Left-hand side (of equation).
- l: Length of a bounded test curve, Eq. (1.15.1), Eq. (1.16.5).

- $l(\cdot)$: Total orthogonal projected length of a bounded curve onto an axis at a given direction, Fig. 1.19.1(b).
- $\mu(\cdot)$: Measure, Section A.1.2.
- $M(\cdot)$: Integral of mean curvature of a bounded surface, Eq. (1.11.7).
- M: Final linear magnification, Section 4.2.1, Eq. (4.6.1).
- $\text{MSE}(\cdot)$: Mean square error, Eq. (A.1.53).
- $\nu(\cdot)$: Measure of a bounded test probe. Section 1.14, Eq. (1.14.1).
- $\nabla_{2[0]}$, ∇_2: Triangle areas involved in Eq. (1.2.32) and Eq. (1.2.34), respectively. The symbol '∇' reads 'nabla'.
- $N(\cdot)$, N: Number, Section 1.1.
- O_k: Surface area of the k-dimensional unit sphere, Eq. (1.5.16).
- $P(\cdot)$, P: Number of test points hitting a domain, Eq. (1.4.2), Eq. (4.6.1).
- $\mathbb{P}(\cdot)$: Probability, Section A.1.3.
- $\mathbb{P}(\mathrm{d}x)$: Probability element of a continuous random variable, Eq. (A.1.29).
- $Q(\cdot)$, Q: Number of transects between a plane and a spatial curve, Section 1.5.4. Number of point particles captured by a slab, Sections 1.7.4, 2.15.1.
- $Q^-(\cdot)$, Q^-: Number of particles captured by a disector, Section 2.15.2.
- q: Smoothness constant, paragraph following Eq. (5.2.23).
- $\mathbb{R}^d$: d-dimensional Euclidean space.
- rhs: Right-hand side (of equation).
- $S(\cdot)$, S: Surface area, Eq. (1.3.9).
- $\mathbb{S}$, $\mathbb{S}^2$: Unit circle and unit sphere, respectively.
- $\mathbb{S}_+$, $\mathbb{S}^2_+$: Unit half-circle and unit hemisphere, respectively.
- srs: Simple random sampling, Section 4.4.2.
- $\tau(\cdot)$, τ: Membrane thickness, Section 3.14.1, Eq. (4.13.1), and rotation angle involved in the kinematic density, Eq. (1.13.4), respectively. Also, $\tau = t/T$, sampling fraction, Eq. (5.5.2).
- T: Bounded test probe, abbreviation of T_r^d, Section 1.20.1. Period of Cavalieri planes, Eq. (2.25.9). Size parameter of a test system, Section 5.8.1, Note 1.
- T_r^d: r-dimensional bounded test probe in $\mathbb{R}^d$, see particular cases in Section 1.13.
- t: Slab thickness, Fig. 1.2.5.
- u: Axial direction defined on the unit hemisphere, Fig. 1.2.2(b), see also Section 1.5.5.
- u_d: In $\mathbb{R}^d$, $u_d \in G_{d[0]}$, Section 1.16.4. Also, $u_d = (u_{d-1}, u_{d-2}, \ldots, u_1)$, where $u_k \in \mathbb{S}^k$, $k = 1, 2, \ldots, d-1$, see Section 1.13.2 for $d = 3$.
- UR: Uniform random, Section A.1.5.
- $V(\cdot)$, V: Volume, Eq. (1.4.5), Eq. (1.4.6).
- v, $\bar{v}_N$, $\bar{v}_V$: Estimator of individual particle volume, of the number-weighted mean particle volume (Eq. (4.22.1)), and of the volume-weighted mean particle volume (Section 4.21.1, respectively.
- $\text{Var}(\cdot)$: Variance of a random variable, Eq. (A.1.33).
- $\text{var}(\cdot)$: Estimator of $\text{Var}(\cdot)$.
- VA: Vertical axis, Section 1.2.6.
- VCav: Vertical Cavalieri, Section 2.28.1.
- $W(\cdot)$: Vertical section functional, Eq. (1.17.5).

- $\chi(\cdot)$, χ: Euler–Poincaré characteristic, Sections 1.10.2 and 1.12.2.
- x: A point or vector, usually in $\mathbb{R}^d$.
- Y: Target compact set, Section 1.3.1, or target random process, Section 3.1.1.
- ∂Y, Y°: Boundary and interior, respectively, of a domain Y, Section 1.3.1. Thus, $Y = Y^\circ \cup \partial Y$.
- $Y'(\cdot)$: Orthogonal projection of a domain Y onto an axis at a given direction. Its length is $H(\cdot)$, Fig. 1.19.1(a).
- $\zeta(\cdot)$: Riemann's Zeta function, Eq. (5.2.14).
- $Z(\cdot, \cdot)$: Epstein Zeta function, Eq. (5.11.3).
- $Z(\cdot)$: Vertical projection functional, Eq. (1.19.29).
- Z_r^d: r-cylinder in $\mathbb{R}^d$, see Eq. (1.13.11) for $d = 3$.
- z: A point or vector, usually in an r-plane.

References

Abercrombie, M. (1946). Estimation of nuclear population from microtome sections. *Anat Rec*, *94*, 239-47.

Abramowitz, M., & Stegun, I. A. (1965). *Handbook of Mathematical Functions*. New York: Dover.

Andersen, I. T., Hahn, U., & Jensen, E. B. V. (2015). Optimal PPS sampling with vanishing auxiliary variables – with applications in microscopy. *Scand J Statist*, *42*, 136-48.

Artacho-Pérula, E., Roldán-Villalobos, R., & Cruz-Orive, L. M. (1999). Application of the fractionator and vertical slices to estimate total capillary length in skeletal muscle. *J Anat*, *195*, 429–37.

Auneau, J., & Jensen, E. V. B. (2010). Expressing intrinsic volumes as rotational integrals. *Adv Appl Math*, *45*, 1-11.

Bach, G. (1963). Gründung einer internationalen Gesellschaft für Stereologie. *Z wiss Mikrosk*, *65*, 190-195.

Baddeley, A. J. (1984). Vertical sections. In W. Weil & R. V. Ambartzumian (Eds.), *Stochastic Geometry and Stereology, Oberwolfach 1983*. Berlin: Teubner.

Baddeley, A. J. (1985). An anisotropic sampling design. In N. Nagel (Ed.), *Geobild '85* (p. 92-97). Jena: Friedrich-Schiller-Universität.

Baddeley, A. J. (1999). A crash course in stochastic geometry. In O. E.Barndorff-Nielsen, W. S. Kendall, & M. N. M. van Lieshout (Eds.), *Stochastic Geometry: Likelihood and Computation* (p. 1-35). London: Chapman and Hall.

Baddeley, A. J., & Cruz-Orive, L. M. (1995). The Rao–Blackwell theorem in stereology and some counterexamples. *Adv Appl Prob*, *27*, 2-19.

Baddeley, A. J., Dorph-Petersen, K. A., & Jensen, E. B. V. (2006). A note on the stereological implications of irregular spacing of sections. *J Microsc*, *222*, 177-81.

Baddeley, A. J., Gundersen, H. J. G., & Cruz-Orive, L. M. (1986). A note on the stereological implications of irregular spacing of sections. *J Microsc*, *142*, 259-76.

L. M. Cruz-Orive, *Stereology*, Interdisciplinary Applied Mathematics 59, https://doi.org/10.1007/978-3-031-52451-6

Baddeley, A. J., Howard, C. V., Boyde, A., & Reid, S. (1987). Three-dimensional analysis of the spatial distribution of particles using the tandem-scanning reflected light microscope. *Acta Stereologica*, *6*, 87-100.

Baddeley, A. J., & Jensen, E. B. V. (2005). *Stereology for Statisticians*. Chapman and Hall/CRC.

Baddeley, A. J., Moyeed, R. A., Howard, C. V., & Boyde, A. (1993). Analysis of three-dimensional point pattern with replication. *Appl Statist*, *42*, 641-68.

Baddeley, A. J., Rubak, E., & Turner, R. (2016). *Spatial Point Patterns: Methodology and Applications With R*. Boca Raton: Chapman and Hall/CRC.

Balanzat, M. (1940). Généralisation de quelques formules de géometrie intégrale. *C R Acad Sci Paris*, *210*, 596-598.

Barbier, E. (1860). Note sur le problème de l'aiguille et le jeu du joint couvert. *J Math Pures Appl*, *5*, 273-86.

Batra, S., König, M. F., & Cruz-Orive, L. M. (1995). Unbiased estimation of capillary length from vertical slices. *J Micosc*, *178*, 152-59.

Bendtsen, T. F., & Nyengaard, J. R. (1989). Unbiased estimation of particle number using sections – an historical perspective with special reference to the stereology of glomeruli. *J Microsc*, *153*, 93-102.

Beneš, V., & Rataj, J. (2004). *Stochastic Geometry: Selected Topics*. Boston: Kluwer.

Bertrand, J. (1889). *Calcul des probabilités*. Paris: Gauthier-Villars.

Bhanu Prasad, P., & Jernot, J. P. (1991). Topological description of the densification of a granular medium. *J Microsc*, *163*, 211-20.

Bhanu Prasad, P., Lantuejoul, C., Jernot, J. P., & Chermant, J. (1989). Unbiased estimation of the Euler–Poincaré characteristic. proc. ecs v freiburg 1989. *Acta Stereol*, *8/2*, 101-6.

Blaschke, W. (1935). Integralgeometrie 2: Zur Ergebnissen von M.W. crofton. *Bull Soc Roum Sci*, *37*, 3-11.

Blaschke, W. (1936–1937). *Vorlesungen uber Integralgeometrie*. Hamburg: Hamburger Math Einzelschr 20 und 22. [Vol 1 and 2, 2nd ed. New York: Chelsea (1949).].

Boyce, R. W., & Gundersen, H. J. G. (2018). The automatic proportionator estimator is highly efficient for estimation of total number of sparse cell populations. *Front Neuroanat*, *12*, Article 19. Retrieved from doi.org/10.3389/fnana.2018.00019

Braendgaard, H., & Gundersen, H. J. G. (1986). The impact of recent stereological advances on quantitative studies of the nervous system. *J Neurosci Meth*, *18*, 39-78.

Brüngger, A., & Cruz-Orive, L. M. (1987). Nuclear morphometry of nuclear malignant melanoma and benign nevocytic nevy. *Arch Dermatol Res*, *279*, 412-14.

Buffon, G. (1777). Essai d'arithmétique morale. In *Supplément à l'Histoire Naturelle* (Vol. 4, p. 46-123). Paris: Imprimérie Royale.

Bur, S., Bachofen, H., Gehr, P., & Weibel, E. R. (1985). Lung fixation by airway instillation: effects on capillary hematocrit. *Exper Lung Res*, *9*, 57-66.

Burri, P. H., & Weibel, E. R. (1971). Morphometric estimation of pulmonary diffusion capacity. II. Effect of environmental O2 on the growing lung. *Resp Physiol*, *11*, 247-64.

Cahn, J. W. (1967). The significance of average mean curvature and its determination by quantitative metallography. *Trans Metall Soc AIME*, *239*, 610-16.

Calka, P. (2003). An explicit expression for the distribution of the number of sides of the typical Poisson–Voronoi cell. *Adv Appl Prob*, *35*, 863-870.

Cauchy, A. (1832). *Mémoire sur la rectification des courbes et la quadrature des surfaces courbes.* Paris (lithograph), 11 pp [In: *Mem Acad Sci Paris* 22(1850):3–15; *Oeuvres complétes* 1/II, Paris: Gauthier-Villars (1908), 167–77].

Cavalieri, B. (1635). *Geometria Indivisibilibus Continuorum.* Bononi: Typis Clementis Ferronij. (Reprinted 1966 as: Geometria degli Indivisibili. Torino: Unione Tipografico-Editrice Torinese.)

Chatfield, C. (1983). *Statistics for Technology, 3rd edn.* London: Chapman and Hall.

Cochran, W. G. (1977). *Sampling Techniques, 3rd edn.* New York: J. Wiley.

Cox, D. R. (1962). *Renewal Theory*. London: Methuen.

Cox, D. R., & Miller, H. D. (1965). *The Theory of Stochastic Processes*. London: Methuen.

Crofton, M. W. (1868). On the theory of local probability. *Phil Trans Roy Soc London*, *158*, 181-99.

Crofton, M. W. (1869). Sur quelques théorèmes du calcul intégral. *C R Acad Sci Paris*, *68*, 1469-1470.

Cruz, M., Gómez, D., & Cruz-Orive, L. M. (2015). Efficient and unbiased estimation of population size. *PLoS ONE*, *10*(11), e0141868.

Cruz, M., & González-Villa, J. (2018). Simplified procedure for efficient and unbiased population size estimation. *PLoS ONE*, *13*(10), e0206091.

Cruz, M., & González-Villa, J. (2021). Unbiased population size estimation on still gigapixel images. *Sociol Meth Res*, *50*, 627-48.

Cruz-Orive, L. M. (1979). Estimation of sheet thickness distributions from linear and plane sections. *Biom J*, *21*, 717-30.

Cruz-Orive, L. M. (1980a). Best linear unbiased estimators for stereology. *Biometrics*, *36*, 595-605.

Cruz-Orive, L. M. (1980b). On the estimation of particle number. *J Microsc*, *120*, 15-27.

Cruz-Orive, L. M. (1982). The use of quadrats and test systems in stereology, including magnification corrections. *J Microsc*, *125*, 89-102.

Cruz-Orive, L. M. (1983). Distribution-free estimation of sphere size distributions from slabs showing overprojection and truncation, with a review of previous methods. *J Microsc*, *131*, 265-90.

Cruz-Orive, L. M. (1985). Estimating volumes from systematic hyperplane sections. *J Appl Prob*, *22*, 518-30.

Cruz-Orive, L. M. (1987a). Particle number can be estimated using a disector of unknown thickness: the selector. *J Microsc*, *145*, 121-42.

Cruz-Orive, L. M. (1987b). Precision of stereological estimators from systematic probes. Proc. ICS VII Caen 1987. *Acta Ster*, *6/III*, 153-8.
Cruz-Orive, L. M. (1987c). Stereology: historical notes and recent evolution. In E. R. Miles (Ed.), *Iss commemorative-memorial volume, acta ster 6/ii* (p. 43-56).
Cruz-Orive, L. M. (1987d). Stereology: recent solutions to old problems and a glimpse into the future. Proc. ICS VII Caen 1987. *Acta Ster*, *6/III*, 3-18.
Cruz-Orive, L. M. (1989a). On the precision of systematic sampling: a review of Matheron's transitive methods. *J Microsc*, *153*, 315-33.
Cruz-Orive, L. M. (1989b). Precision of systematic sampling on a step function. In A. Hübler, W. Nagel, B. D. Ripley, & G. Werner (Eds.), *Geobild '89* (p. 185-93). Berlin: Akademie-Verlag.
Cruz-Orive, L. M. (1989c). Second-order stereology: estimation of second moment volume measures. Proc. ECS V Freiburg 1989. *Acta Ster*, *8/2*, 641-46.
Cruz-Orive, L. M. (1990). On the empirical variance of a fractionator estimate. *J Microsc*, *160*, 89-95.
Cruz-Orive, L. M. (1993). Systematic sampling in stereology. Proc 49th Session, Florence. *Bull Intern Statis Inst*, *55*(2), 451-68.
Cruz-Orive, L. M. (1997). Stereology of single objects. *J Microsc*, *186*, 93-107.
Cruz-Orive, L. M. (1999). Precision of Cavalieri sections and slices with local errors. *J Microsc*, *193*, 182-98.
Cruz-Orive, L. M. (2000). *How long is a tree?* The 20th International Biometric Conference, Vol II, Invited Program, 189-98.
Cruz-Orive, L. M. (2002). Stereology: meeting point of integral geometry, probability, and statistics. In memory of Professor Luis A. Santaló (1911–2001). *Mathematicae Notae*, *41*, 49-98. (Special issue (Homenaje a Santaló))
Cruz-Orive, L. M. (2004). Precision of the fractionator from Cavalieri designs. *J Microsc*, *213*, 205-11.
Cruz-Orive, L. M. (2005). A new stereological principle for test lines in 3D. *J Microsc*, *219*, 18-28.
Cruz-Orive, L. M. (2006). A general variance predictor for Cavalieri slices. *J Microsc*, *222*, 158-65.
Cruz-Orive, L. M. (2008). Comparative precision of the pivotal estimators of particle size. *Image Anal Stereol*, *27*, 17-22.
Cruz-Orive, L. M. (2009a). The pivotal tessellation. *Image Anal Stereol*, *28*, 63-67.
Cruz-Orive, L. M. (2009b). Stereology: old and new. In V. Capasso, G. Aletti, & A. Micheletti (Eds.), *Proc 10th Eur Congr Stereol Image Anal.* Bologna: Esculapio.
Cruz-Orive, L. M. (2011). Flowers and wedges for the stereology of particles. *J Microsc*, *243*, 86-102.
Cruz-Orive, L. M. (2012). Uniqueness properties of the invariator, leading to simple computations. *Image Anal Stereol*, *31*, 87-96.
Cruz-Orive, L. M. (2013). Variance predictors for isotropic geometric sampling, with applications in forestry. *Stat Methods Appl*, *22*, 3-31.

Cruz-Orive, L. M. (2017). Stereology: a historical survey. *Image Anal Stereol*, *36*, 153-177.

Cruz-Orive, L. M., & Cruz, M. (2020). Transformation of a grid of quadrats to cope with perspective artifacts. *Image Anal Stereol*, *39*, 47-52.

Cruz-Orive, L. M., & Geiser, M. (2004). Estimation of particle number by stereology: an update. *J Aerosol Med*, *17*, 197-212.

Cruz-Orive, L. M., Gelšvartas, J., & Roberts, N. (2014). Sampling theory and automated simulations for vertical sections, applied to human brain. *J Microsc*, *253*, 119-50.

Cruz-Orive, L. M., & Gual-Arnau, X. (2002). Precision of circular systematic sampling. *J Microsc*, *207*, 225-42.

Cruz-Orive, L. M., & Gual-Arnau, X. (2015). The invariator design: an update. *Image Anal Stereol*, *34*, 147-59.

Cruz-Orive, L. M., & Gual-Arnau, X. (2020). Stereology with cylinder probes. *Image Anal Stereol*, *39*, 213-218.

Cruz-Orive, L. M., & Howard, C. V. (1991). Estimating the length of a bounded curve in three dimensions using total vertical projections. *J Microsc*, *163*, 101-13.

Cruz-Orive, L. M., & Howard, C. V. (1995). Estimation of individual feature surface area with the vertical spatial grid. *J Microsc*, *178*, 146-51.

Cruz-Orive, L. M., & Hunziker, E. B. (1986). Stereology for anisotropic cells: application to growth cartilage. *J Microsc*, *143*, 47-80.

Cruz-Orive, L. M., Insausti, A., Insausti, R., & Crespo, D. (2004). A case study from neuroscience involving stereology and multivariate analysis. In S. M. Evans, A. M. Janson, & J. M. Nyengaard (Eds.), *Quantitative Methods in Neuroscience* (p. 16-64). Oxford: Oxford University Press.

Cruz-Orive, L. M., Ramos-Herrera, M. L., & Artacho-Pérula, E. (2010). Stereology of isolated objects with the invariator. *J Microsc*, *240*, 94-110.

Cruz-Orive, L. M., & Roberts, N. (1993). Stereology of isolated objects with the invariator. *J Microsc*, *170*, 25-33.

Cruz-Orive, L. M., & Weibel, E. R. (1981). Sampling designs for stereology. *J Microsc*, *122*, 235-72.

Czuber, E. (1884). *Geometrische Wahrscheinlichkeiten und Mittelwerte*. Leipzig: Teubner.

Davy, P. J., & Miles, R. E. (1977). Sampling theory for opaque spatial specimens. *J Roy Statist Soc B*, *39*, 56-65.

Debrunner, H. E., Mani, P., Meyer, C., & Rätz, J. (1982). Professor Dr. Hugo Hadwiger, 1908–1981. *Elemente der Mathematik*, *37*, 65-78.

DeHoff, R. T. (1967). The quantitative estimation of mean surface curvature. *Trans Metall Soc AIME*, *239*, 617-21.

DeHoff, R. T. (1987). Use of the disector to estimate the Euler characteristic of three dimensional microstructures. In R. E. Miles (Ed.), *The Commemorative-Memorial Volume. Twenty-five years of stereology. Acta Stereol 6/ii* (p. 133-40).

DeHoff, R. T. (2004). Curvature and the topological properties of interconnected phases. In R. T. DeHoff & F. N. Rhines (Eds.), *Quantitative Microscopy* (p. 291-325). New York: McGraw-Hill.

DeHoff, R. T., & Rhines, F. N. (Eds.). (1968). *Quantitative Microscopy*. New York: McGraw-Hill.

Delesse, A. (1847). Procédé mécanique pour déterminer la composition des roches. *C R Acad Sci Paris*, *25*, 544-5.

De-lin, R. (1994). *Topics in Integral Geometry*. Singapore: World Scientific.

Deltheil, R. (1926). *Probabilités géométriques*. Paris: Gauthier-Villars.

Dieck, T. (2008). *Algebraic Topology*. Zürich: European Mathematical Society Publishing House.

Diggle, P. J. (1986). Displaced amacrine cells in the retina of a rabbit: analysis of a bivariate spatial point pattern. *J Neurosci Meth*, *18*, 115-25.

Diggle, P. J. (2003). *Statistical Analysis of Spatial Point Patterns. 2nd edn*. London: Arnold.

Do Carmo, M. P. (1976). *Differential Geometry of Curves and Surfaces*. Upper Saddle River, NJ: Prentice-Hall.

Dorph-Petersen, K. A., Nyengaard, J. R., & Gundersen, H. J. G. (2001). Tissue shrinkage and unbiased stereological estimation of particle number and size. *J Microsc*, *204*, 232-46.

Eggli, P. S., Hunziker, E. B., & Schenk, R. K. (1998). Quantitation of structural features characterizing weight- and less-weight-bearing regions in articular cartilage: a stereological analysis of medial femoral condyles in young adult rabbits. *Anat Record*, *222*, 217-27.

Elias, H. (1962). Stereology: scope and limits. *Stereologia*, *1*, 1-2.

Elias, H. (1963). The birth of the stereological society. In *Proc First Intern Congress for Stereology* (p. 2/1-2/3). Vienna: Vienna medical academy for postgraduated medical training.

Enns, E. G., & Ehlers, P. F. (1978). Random paths through a convex region. *J Appl Prob*, *15*, 144-52.

Evans, S. M., & Gundersen, H. J. G. (1989). Estimation of spatial distributions using the nucleator. Proc. ECS V Freiburg 1989. *Acta Stereol*, *8/2*, 395-400.

Feller, W. (1968). *An Introduction to Probability Theory and its Applications. Vol I, 3rd edn*. New York: J. Wiley.

Fernández-Viadero, C., González-Mandly, A., Verduga, R., Crespo, D., & Cruz–Orive, L. M. (2008). La estereología como herramienta de cuantificación del volumen y la atrofia cortical en el cerebro del anciano con demencia. *Rev Esp Geriatr Gerontol*, *43*, 32-43.

Fullman, R. L. (1953). Measurement of particle sizes in opaque bodies. *J Metals*, *5*, 447-52.

García-Fiñana, M. (2000). *Muestreo systemático en $\mathbb{R}$. Aplicaciones a la Estereología* (Unpublished doctoral dissertation). Universidad de Cantabria.

García-Fiñana, M. (2006). Confidence intervals in Cavalieri sampling. *J Microsc*, *222*, 146-57.

García-Fiñana, M., & Cruz-Orive, L. M. (1998). Explanation of apparent paradoxes in Cavalieri sampling. *Acta Stereologica*, *17*, 293-302.

García-Fiñana, M., & Cruz-Orive, L. M. (2000a). Fractional trend of the variance in Cavalieri sampling. *Image Anal Stereol*, *19*, 71-79.

García-Fiñana, M., & Cruz-Orive, L. M. (2000b). New approximations for the variance in Cavalieri sampling. *J Microsc*, *199*, 224-38.

García-Fiñana, M., & Cruz-Orive, L. M. (2004). Improved variance prediction for systematic sampling on $\mathbb{R}$. *Statistics*, *38*, 243-72.

García-Fiñana, M., Cruz-Orive, L. M., Mackay, C. E., Pakkenberg, E. B., & Roberts, N. (2003). Comparison of MR imaging against physical sectioning to estimate the volume of human cerebral compartments. *NeuroImage*, *18*, 505-16.

Gardi, J. E., Nyengaard, J. R., & Gundersen, H. J. G. (2008). Automatic sampling for unbiased and efficient stereological estimation using the proportionator in biological studies. *J Microsc*, *230*, 108-20.

Gehr, P., Mwangi, D. K., Ammann, A., Maloiy, G. M. O., Taylor, C. R., & Weibel, E. R. (1981). Design of the mammalian respiratory system. v. scaling morphometric pulmonary diffusing capacity to body mass: wild and domestic mammals. *Resp Physiol*, *44*, 61-86.

Gel'fand, I. M., & Shilov, G. E. (1964). *Generalized Functions. Vol I. Properties and Operations*. New York: Academic Press.

Giger, H., & Hadwiger, H. (1968). Über Treffzahlwahrscheilichkeiten im Eikörperfeld. *Z Wahr verw Gebiete*, *10*, 329-34.

Glagolev, A. A. (1933). On the geometrical methods of quantitative mineralogical analysis of rocks [in russian]. *Trans Inst Econ Min, Moskva*, *59*, 1-47.

Gokhale, A. M. (1990). Unbiased estimation of curve length in 3D using vertical slices. *J Microsc*, *159*, 133-41.

Gokhale, A. M. (1992). Estimation of length density l_v from vertical slices of unknown thickness. *J Microsc*, *167*, 1-8.

Gokhale, A. M., & Beneš, V. (1998). Estimation of average particle size from vertical projections. *J Microsc*, *191*, 195-200.

Gokhale, A. M., Evans, R. A., Mackes, J. L., & Mouton, P. R. (2004). Design-based estimation of surface area in thick tissue sections of arbitrary orientation using virtual cycloids. *J Microsc*, *216*, 25-31.

Gómez, A. I., Cruz, M., & Cruz-Orive, L. M. (2016). On the precision of curve length estimation in the plane. *Image Anal Stereol*, *35*, 1-14.

Gómez, A. I., Cruz, M., & Cruz-Orive, L. M. (2019). Variance prediction for population size estimation. *Image Anal Stereol*, *38*, 131-9.

González-Villa, J., Cruz, M., & Cruz-Orive, L. M. (2017). On the precision of the nucleator. *Image Anal Stereol*, *36*, 123-34.

González-Villa, J., Cruz, M., & Cruz-Orive, L. M. (2018). On the precision of the isotropic Cavalieri design. *Image Anal Stereol*, *37*, 225-32.

Goodman, L. A. (1960). On the exact variance of products. *J Amer Statist Ass*, *55*, 708-13.

Grimmett, G. R., & Stirzaker, D. R. (2001). *Probability and Random Processes, 3rd ed.* Oxford: Oxford University Press.

Gual-Arnau, X., & Cruz-Orive, L. M. (1998). Variance prediction under systematic sampling with geometric probes. *Adv Appl Prob*, *30*, 889-903.

Gual-Arnau, X., & Cruz-Orive, L. M. (2000). Systematic sampling on the circle and on the sphere. *Adv Appl Prob*, *32*, 628-47.

Gual-Arnau, X., & Cruz-Orive, L. M. (2002). Variance prediction for pseudosystematic sampling on the sphere. *Adv Appl Prob*, *34*, 469-83.

Gual-Arnau, X., & Cruz-Orive, L. M. (2009). A new expression for the density of totally geodesic submanifolds in space forms, with stereological applications. *Diff Geom Appl*, *27*, 124-28.

Gual-Arnau, X., & Cruz-Orive, L. M. (2016). New rotational integrals in space forms, with an application to surface area estimation. *Appl Math*, *61*, 489-501.

Gual-Arnau, X., & Cruz-Orive, L. M. (2020). Stereology with cylinder probes. *Image Anal Stereol*, *39*, 213-218.

Gual-Arnau, X., Cruz-Orive, L. M., & Nuño-Ballesteros, J. J. (2010). A new rotational integral formula for intrinsic volumes in space forms. *Adv Appl Math*, *44*, 298-308.

Gundersen, H. J. G. (1977). Notes on the estimation of the numerical density of arbitrary profiles: the edge effect. *J Microsc*, *111*, 219-23.

Gundersen, H. J. G. (1979). Estimation of tubule or cylinder l_v, s_v and v_v on thick sections. *J Microsc*, *117*, 333-45.

Gundersen, H. J. G. (1986). Stereology of arbitrary particles. a review of unbiased number and size estimators and the presentation of some new ones, in memory of William R. Thompson. *J Microsc*, *143*, 3-45.

Gundersen, H. J. G. (1988). The nucleator. *J Microsc*, *151*, 3-21.

Gundersen, H. J. G. (1992). Stereology: the fast lane between neuroanatomy and brain function – or still only a tightrope? *Acta Neurol Scand*, *137*, 8-13.

Gundersen, H. J. G. (2002a). Stereological estimation of tubular length. *J Microsc*, *207*, 155-60.

Gundersen, H. J. G. (2002b). The smooth fractionator. *J Microsc*, *207*, 191-210.

Gundersen, H. J. G., Bagger, P., Bendtsen, T. F., Evans, S. M., Korbo, L., Marcussen, N., . . . West, M. J. (1988). The new stereological tools: Disector, fractionator, nucleator and point sampled intercepts and their use in pathological research and diagnosis. *APMIS*, *96*, 857-81.

Gundersen, H. J. G., Boyce, R. W., Nyengaard, J. R., & Odgaard, A. (1993). The conneulor: unbiased estimation of connectivity using physical disectors under projection. *Bone*, *14*, 217-22.

Gundersen, H. J. G., & Jensen, E. B. (1983). Particle sizes and their distributions estimated from line- and point-sampled intercepts. Including graphical unfolding. *J Microsc*, *131*, 291-310.

Gundersen, H. J. G., & Jensen, E. B. (1985). Stereological estimation of the volumeweighted mean volume of arbitrary particles observed on random sections. *J Microsc*, *138*, 127-42.

Gundersen, H. J. G., & Jensen, E. B. (1987). The efficiency of systematic sampling in stereology and its prediction. *J Microsc*, *147*, 229-63.

Gundersen, H. J. G., Jensen, E. B., & Østerby, R. (1978). Distribution of membrane thickness determined by lineal analysis. *J Microsc*, *113*, 27-43.

Gundersen, H. J. G., Jensen, E. B. V., Kiêu, K., & Nielsen, J. (1999). The efficiency of systematic sampling in stereology – reconsidered. *J Microsc*, *193*, 199-211.

Gundersen, H. J. G., & Østerby, R. (1981). Optimizing sampling efficiency of stereological studies in biology: or 'do more less well!'. *J Microsc*, *121*, 65-73.

Hadwiger, H. (1957). *Vorlesungen über Inhalt, Oberfläche und Isoperimetrie*. Berlin: Springer.

Hall, P. (1988). *Introduction to the Theory of Coverage Processes*. New York: J. Wiley.

Hansen, L. V., Nyengaard, J. R., Andersen, J. B., & Jensen, E. B. V. (2011). The seniautomatic nucleator. *J Microsc*, *242*, 206-215.

Harding, E. F., & Kendall, D. G. (Eds.). (1974). *Stochastic Geometry. A Tribute to the Memory of Rollo Davidson*. London: J. Wiley.

Haug, H. (1985). Gibt es Nervenzellverluste während der Alterung in der menschlichen Hirnrinde? Ein morphometrischer Beitrag zu dieser Frage. *Nervenheilkunde*, *4*, 103-9.

Horvitz, D. G., & Thompson, D. J. (1952). A generalization of sampling without replacement from a finite universe. *J Amer Statist Ass*, *47*, 663-85.

Hostinský, B. (1925). Sur les probabilités géométriques. *Publ Fac Sci Univ Masaryk (Brno)*, *50*, 1-26.

Howard, C. V. (1986). Rapid nuclear volume estimation in malignant melanoma using point sampled intercepts in vertical sections: a review of the unbiased quantitative analysis of particles of arbitrary shape. In J. Y. Mary & J. P. Rigaut (Eds.), *Quantitative Image Analysis in Cancer Cytology and Histology* (p. 245-56). Amsterdam: Elsevier.

Howard, C. V., Cruz-Orive, L. M., & Yaegashi, H. (1992). Estimating neuron dendritic length in 3D from total vertical projections and from vertical slices. *Acta Neurol Scand Suppl*, *137*, 14-19.

Howard, C. V., & Reed, M. G. (2005). *Unbiased Stereology. Three-dimensional Measurement in Microscopy, 2nd ed.* Oxford: Bios/Taylor & Francis.

Howard, C. V., Reid, S., Baddeley, A. J., & Boyde, A. (1985). Unbiased estimation of particle density in the tandem scanning reflected light microscope. *J Microsc*, *138*, 203-12.

Hykšová, M., Kalousková, A., & Saxl, I. (2012). Early history of geometric probability and stereology. *Image Anal Stereol*, *31*, 1-16.

Insausti, A. M., Cruz-Orive, L. M., Jáuregui, L., Manrique, M., & Insausti, R. (1999). Stereological assessment of the glial reaction to chronic deafferentation of the cochlear nuclei in the macaque monkey (Macaca fascicularis). *J Comp Neurol*, *414*, 485-94.

Insausti, A. M., Insausti, R., Cruz-Orive, L. M., & Manrique, M. (2000). Stereological analysis of the cochlear nuclei of monkey (Macaca fascicularis) after deafferentation. *Image Anal Stereol*, *19*, 133-37.

Jensen, E. B. (1987). Design- and model-based stereological analysis of arbitrarily shaped particles. *Scand J Statist*, *14*, 161-80.

Jensen, E. B. (1991). Recent developments in the stereological analysis of particles. *Ann Inst Statist Math*, *43*, 455-68.

Jensen, E. B. (1998). *Local Stereology*. Singapore: World Scientific.

Jensen, E. B., & Gundersen, H. J. G. (1982). Stereological ratio estimation based on counts from integral test systems. *J Microsc*, *125*, 51-66.

Jensen, E. B., & Gundersen, H. J. G. (1983). On the estimation of the volume-weighted distribution of particle sizes. In E. B. Jensen & H. J. G. Gundersen (Eds.), *Proceedings of the Second International Workshop on Stereology and Stochastic Geometry. Memoirs No. 6* (p. 81–103). Department of Theoretical Statistics, Institute of Mathematics, University of Aarhus.

Jensen, E. B., & Gundersen, H. J. G. (1985). The stereological estimation of moments of particle volume. *J Appl Prob*, *22*, 82-98.

Jensen, E. B., & Gundersen, H. J. G. (1987). Stereological estimation of surface area of arbitrary particles. *Acta Stereol, 6, Suppl. III*, 25-30.

Jensen, E. B., & Gundersen, H. J. G. (1989). Fundamental stereological formulae based on isotropically orientated probes through fixed points with applications to particle analysis. *J Microsc*, *153*, 249-67.

Jensen, E. B., & Gundersen, H. J. G. (1993). The rotator. *J Microsc*, *170*, 35-44.

Jensen, E. B., Gundersen, H. J. G., & Østerby, R. (1979). Determination of membrane thickness distribution from orthogonal intercepts. *J Microsc*, *115*, 19-33.

Jensen, E. B., & Kiderlen, M. (Eds.). (2017). *Tensor Valuations and Their Applications in Stochastic Geometry and Imaging*. Berlin: Springer.

Jensen, E. B., & Sørensen, F. B. (1991). A note on stereological estimation of the volume-weighted second moment of particle volume. *J Microsc*, *164*, 21-27.

Jensen, E. B., & Sundberg, R. (1986a). Generalized assocoated point methods for sampling planar objects. *J Microsc*, *144*, 55-70.

Jensen, E. B., & Sundberg, R. (1986b). Statistical models for stereological inference about spatial structures: on the applicability of best linear unbiased estimators in stereology. *Biometrics*, *42*, 735-51.

Jolly, G. M. (1979). Sampling of large objects. In R. M. Cormack, G. P. Patil, & D. S. Robson (Eds.), *Sampling Biological Populations* (p. 193-201). Fairland MD: International Co-operative Publishing House.

Karlsson, L. M., & Cruz-Orive, L. M. (1997). Estimation of mean particle size from single sections. *J Microsc*, *186*, 121-32.

Karlsson, L. M., & Liljeborg, A. (1994). Second-order stereology for pores in translucent alumina studied by confocal scanning laser microscopy. *J Microsc*, *175*, 186-94.

Kellerer, A. M. (1983). On the number of clumps resulting fromthe overlap of randomly placed fiigures in a plane. *J Appl Prob*, *20*, 126-135.

Kellerer, A. M. (1986). The variance of a poisson process of domains. *J Appl Prob*, *23*, 307-21.

Kellerer, A. M. (1989). Exact formulae for the precision of systematic sampling. *J Microsc*, *153*, 285-300.

Kendall, M. G., & Moran, P. A. P. (1963). *Geometrical Probability*. London: Ch. Griffin.

Kiêu, K. (1997). *Three Lectures on Systematic Geometric Sampling. Memoirs No. 13/1997*. Aarhus: Department of Theoretical Statistics, University of Aarhus.

Kiêu, K., & Jensen, E. B. V. (1993). Stereological estimation based on isotropic slices through fixed points. *J Microsc*, *170*, 45-51.

Kiêu, K., & Mora, M. (2006). Precision of stereological planar area predictors. *J Microsc*, *222*, 201-11.

Kiêu, K., & Mora, M. (2009). Advances on the precision of several stereological volume estimators. In V. Capasso, G. Aletti, & A. Micheletti (Eds.), *Proc 10th Eur Congr Stereol Image Anal.* Bologna: Esculapio.

Kiêu, K., Souchet, S., & Istas, J. (1999). Precision of systematic sampling and transitive methods. *J Statist Plan Inf*, *77*, 263-79.

Kubínová, L., & Janáček, J. (2001). Confocal microscopy and stereology: estimating volume, number, surface area and length by virtual test probes applied to three-dimensional images. *Microsc Res Techn*, *53*, 425-35.

Kubínová, L., Mao, X. M., Janáček, J., & Archembeau, J. O. (2003). Stereology techniques in radiation biology. *Radiation Research*, *160*, 110-19.

Larsen, J. O. (1998). Stereology of nerve cross sections. *J Neurosci Meth*, *85*, 107-18.

Larsen, N. Y., Ziegel, J. F., Nyengaard, J. R., & Jensen, E. B. V. (2019). Stereological estimation of particle shape from vertical sections. *J Microsc*, *275*, 183-94.

Lindgren, B. W. (1993). *Statistical Theory, 4th edn.* New York: Rouledge.

Lockwood, E. H. (1961). *A Book of Curves*. Cambridge: Cambridge University Press.

Mack, C. (1954). The expected number of clumps when convex laminae are placed at random and with random orientation on a plane area. *Proc Camb Phil Soc*, *50*, 581-85.

Maletti, G. M., & Wulfsohn, D. (2006). Evaluation of variance models for fractionator sampling of trees. *J Microsc*, *222*, 228-41.

Matérn, B. (1960). Spatial variation. *Meddelanden från Statens Skogsforskningsinstitut*, *49*(5), 1-114. ([2nd ed. Berlin: Springer, 1986])

Matheron, G. (1965). *Les variables régionalisées et leur estimation*. Paris: Masson et Cie.

Matheron, G. (1967). *éléments pour une théorie des milieux poreux*. Paris: Masson et Cie.

Matheron, G. (1971). *The theory of regionalized variables and its applications*. Les Cahiers du Centre de Morphologie Mathématique de Fontainebleau, No. 5. Fontainebleau: École Nationale Supérieure des Mines de Paris. ([Matheron's theory of regionalised variables, Pawlowsky-Glahn, Serra J, eds., Oxford: Oxford University Press, 2019])

Matheron, G. (1975). *Random Sets and Integral Geometry*. New York: J. Wiley.

Matheron, G. (1976). La formule de Crofton pour les sections épaisses. *J App Prob*, *13*, 707-13.

Mattfeldt, T. (1989). The accuracy of one-dimensional systematic sampling. *J Microsc*, *153*, 301-13.

Mattfeldt, T., Frei, H., & Rose, C. (1993). Second-order stereology of benign and malignant alterations of the human mammary gland. *J Microsc*, *171*, 143-51.

Mayhew, T. M., & Cruz-Orive, L. M. (1974). Second-order stereology of benign and malignant alterations of the human mammary gland. *J Microsc*, *102*, 195-207.

McMillan, A. M., & Sørensen, F. B. (1992). The efficient and unbiased estimation of nuclear size variability using the 'selector'. *J Microsc*, *165*, 433-37.

McNulty, V., Cruz-Orive, L. M., Roberts, N., Holmes, C. J., & Gual-Arnau, X. (2000). Estimation of brain compartment volume from MR Cavalieri slices. *J Comp Assist Tomography*, *24*, 466-77.

Mecke, J. (1981). Stereological formulas for manifold processes. *Prob Math Statist*, *2*, 31-35.

Michel, R. P., & Cruz-Orive, L. M. (1988). Application of the Cavalieri principle and vertical sections method to lung: estimation of volume and pleural surface area. *J Microsc*, *150*, 117-36.

Miles, R. E. (1964). Random polygons determined by random lines in a plane I, II. *Proc Nat Acad Sci*, *52*, 901-7, 1157-60.

Miles, R. E. (1971). Isotropic random simplices. *Adv Appl Prob*, *3*, 353-382.

Miles, R. E. (1972). Multidimensional perspectives in stereology. *J Microsc*, *95*, 181-95.

Miles, R. E. (1974). On the elimination of edge effects in planar sampling. In E. F. Harding & D. G. Kendall (Eds.), *Stochastic Geometry. A Tribute to the Memory of Rollo Davidson* (p. 228-47). London: J. Wiley.

Miles, R. E. (1976). Estimating aggregate and overall characteristics from thick sections in transmission microscopy. *J Microsc*, *107*, 227-33.

Miles, R. E. (1978a). The importance of proper model specification in stereology. In R. E. Miles & J. Serra (Eds.), *Geometrical Probability and Biological Structures: Buffon's 200th Anniversary. Proceedings, Paris 1977* (p. 115-36). Lect Not Biomath No. 23. Berlin: Springer.

Miles, R. E. (1978b). The sampling, by quadrats, of planar aggregates. *J Microsc*, *113*, 257-67.

Miles, R. E. (1979). Some new integral geometric formulae, with stochastic applications. *J Appl Prob*, *16*, 592-606.

Miles, R. E. (1981). Stereological formulae based upon planar curve section of surfaces in space. *J Microsc*, *121*, 21-27.

Miles, R. E. (1983). Stereology for embedded aggregates of not-necessarily convex particles. In E. B. Jensen & H. J. G. Gundersen (Eds.), *Proceedings of the Second International Workshop on Stereology and Stochastic Geometry. Memoirs No. 6* (p. 121-47). Department of Theoretical Statistics, Institute of Mathematics, University of Aarhus.

Miles, R. E. (1985). A comprehensive set of stereological formulae for embedded aggregates of not-necessarily-convex particles. *J Microsc*, *138*, 115-25.

Miles, R. E. (Ed.). (1987). *The Commemorative-Memorial Volume. Twenty-five Years of Stereology*. Acta Stereol 6/II.

Miles, R. E., & Davy, P. J. (1976). Precise and general conditions for the validity of a comprehensive set of stereological fundamental formulae. *J Microsc*, *107*, 211-26.

Miles, R. E., & Davy, P. J. (1977). On the choice of quadrats in stereology. *J Microsc*, *110*, 27-44.

Miles, R. E., & Serra, J. (Eds.). (1978). *Geometrical Probability and Biological Structures: Buffon's 200th Anniversary. Proceedings, Paris 1977*. Lect Not Biomath No. 23. Berlin: Springer.

Moran, P. A. P. (1944). Measuring the surface area of a convex body. *Ann Math*, *45*, 793-99.

Moran, P. A. P. (1950). Numerical integration by systematic sampling. *Proc Camb Phil Soc*, *46*, 111-15.

Moran, P. A. P. (1966). Measuring the length of a curve. *Biometrika*, *53*, 359-64.

Mosekilde, L. (1990). Measuring the length of a curve. *Bone and Mineral*, *10*, 13-35.

Mouton, P. R., Gokhale, A. M., Ward, N. L., & West, M. J. (2002). Stereological length estimation using spherical probes. *J Microsc*, *206*, 54-64.

Murthy, N. N. (1967). *Sampling Theory and Methods*. Calcutta: Statistical Publishing Society.

Naveira, A. M., & Reventós, A. (Eds.). (2009). *Luis Antonio Santaló. Selected Works*. Berlin: Springer.

Noorhafshan, A., Omidi, A., Karbalay-Doust, S., Aliabadi, E., & Dehghani, F. (2011). Effects of curcumin on the dorsal root ganglion structure and functional recovery after sciatic nerve crush in rat. *Micron*, *42*, 449-55.

Nyengaard, J. R., & Gundersen, H. J. G. (1992). The isector: a simple and direct method for generating isotropic, uniform random sections from small specimens. *J Microsc*, *165*, 427-31.

Ogbuihi, S., & Cruz-Orive, L. M. (1990). Estimating the total number of lymphatic valves in infant lungs with the fractionator. *J Microsc*, *158*, 19-30.

Ohser, J. (1990). *rundlagen und praktische Möglichkeiten der Charakterisierung struktureller Inhomogeneitäten von Werkstoffen. doctor scientiae technicarum*. Freiberg (Sachsen); Bergakademie Freiberg.

Ohser, J., & Nagel, W. (1996). The estimation of the euler–poincaré characteristic from observations on parallel sections. *J Microsc*, *184*, 117-26.

Ohser, J., & Schladitz, K. (2009). *3D images of materials structures. Processing and Analysis*. Weinheim: Wiley-VCH.

Pache, J. C., Roberts, N., Zimmerman, A., Vock, P., & Cruz-Orive, L. M. (1993). Vertical LM sectioning and parallel CT scanning designs for stereology: application to human lung. *J Microsc*, *170*, 9-24.

Pawlas, Z., Nyengaard, J. R., & Jensen, E. B. V. (2009). Particle sizes from sectional data. *Biometrics*, *65*, 216-24.

Petkantschin, P. (1936). Zusammenhänge zwischen den Dichten der linearen Unterräume im *n*-dimensionalen Raum. *Abh Math Sem Hamburg*, *11*, 249-310.

Poincaré, H. (1912). *Calcul des probabilités, 2nd ed.* Paris: Gauthier-Villars.

Rafati, A. H., Ziegel, J. F., Nyengaard, J. R., & Jensen, E. B. V. (2016). Stereological estimation of particle shape and orientation from volume tensors. *J Microsc*, *263*, 229-37.

Rey-Pastor, J., & Santaló, L. A. (1951). *Geometría integral.* Madrid: Espasa-Calpe.

Ripley, B. D. (1977). Modelling spatial patterns. *J Roy Statist Soc B*, *39*, 172-212.

Ripley, B. D. (1981). *Spatial Statistics.* New York: J. Wiley.

Robbins, H. E. (1944). On the measure of a random set. *Ann Math Statist*, *15*, 70-74.

Roberts, N., Cruz-Orive, L. M., Bourne, M., Herfkens, R. J., Karwoski, R. A., & Whitehouse, G. H. (1997). Analysis of cardiac function by MRI and stereology. *J Microsc*, *187*, 31-42.

Roberts, N., Cruz-Orive, L. M., Reid, N. M. K., Brodie, D. A., Bourne, M., & Edwards, R. H. T. (1993). Unbiased estimation of human body composition by the Cavalieri method using magnetic resonance imaging. *J Microsc*, *171*, 239-53.

Roberts, N., Garden, A. S., Cruz-Orive, L. M., Whitehouse, G. H., & Edwards, R. H. T. (1994). Estimation of fetal volume by MRI and stereology. *Br J Radiol*, *67*, 1067-94.

Roberts, N., Howard, C. V., Cruz-Orive, L. M., & Edwards, R. H. T. (1991). The application of total vertical projections for the unbiased estimation of the length of blood vessels and other structures by magnetic resonance imaging. *Magn Res Imaging*, *9*, 917-25.

Roberts, N., Puddephat, M. J., & McNulty, V. (2000). The benefit of stereology for quantitative radiology. *Br J Radiol*, *73*, 679-697.

Roberts, N., Reed, M., & Nesbitt, G. (1997). Estimation of the connectivity of a synthetic porous medium. *J Microsc*, *187*, 110-18.

Rosiwal, A. K. (1898). Über geometrische Gesteinsanalysen. Ein einfacher Weg zur ziffermässigen Feststellung des Quantitäsverhältnisses der Mineralbestandtheile gemengter Gesteine. *Verhandlungen KK Geol Reichsanstalt*, *5-6*, 143-75.

Sagaseta, M. (2005). *Estereología: Metodología básica y ejemplos. Trabajo dirigido en estadística y computación.* Internal Report. Santander: Facultad de Ciencias, Universidad de Cantabria.

Saltykov, S. A. (1946). The method of intersections in metallography. (In Russian). *Zavodskaya Laboratorija*, *12*(9-10), 816-25.

Saltykov, S. A. (1958). *Stereometric Metallography, 2nd ed. (3rd ed. 1970).* State Publishing House for Metals and Sciences, Moscow, (in Russian). ((German translation of 3rd edition: VEB Deutscher Verlag für Grundstoffindustrie, Leipzig, 1974))

Sandau, K. (1987). How to estimate the area of a surface using the spatial grid. *Acta Stereol*, *6/III*, 31-6.

Sandau, K., & Hahn, U. (1994). Some remarks on the accuracy of surface area estimation using the spatial grid. *J Microsc*, *173*, 67-72.

Santaló, L. A. (1936a). Geometría integral 4: Sobre la medida cinemática en el plano. *Abh Math Sem Univ Hamburg*, *11*, 222-236.

Santaló, L. A. (1936b). Geometría integral 7: Nuevas aplicaciones del concepto de medida cinemática en el plano y en el espacio. tesis doctoral. *Rev Acad Ci Madrid*, *33*, 451-504. ([Reprinted in *Luis Antonio Santaló. Selected Works*, A. M. Naveira & A. Reventós A, eds. (2009), pp. 167–212.])

Santaló, L. A. (1936c). *Über das kinematische Mass im Raum*. Paris: Hermann & Cie. Éditeurs.

Santaló, L. A. (1939). Geometría integral de figuras ilimitadas. *Publ Inst Mat Univ Litoral*, *I/2*, 1-58.

Santaló, L. A. (1942). Integral formulas in Crofton's style on the sphere and some inequalities referring to spherical curves. *Duke Math J*, *9*, 707-22. ([Reprinted in *Luis Antonio Santaló. Selected Works*, A. M. Naveira & A. Reventós A, eds. (2009), pp. 253-68.])

Santaló, L. A. (1943). Sobre la distribución probable de corpúsculos en un cuerpo, deducida de la distribución en sus secciones y problemas análogos. *Rev Unión Mat Argentina*, *9*, 145-64. ([Reprinted in *Luis Antonio Santaló. Selected Works*, A. M. Naveira & A. Reventós A, eds. (2009), pp. 705-24.])

Santaló, L. A. (1953). *Introduction to Integral Geometry*. Paris: Hermann & Cie. Éditeurs.

Santaló, L. A. (1956). Sur la mesure des espaces linéaires qui coupent un corps convexe et problèmes qui s'y rattachent. In *Colloque sur les questions de réalité en géométrie, liège 23–26 mai 1955* (p. 177-90). Paris: Masson.

Santaló, L. A. (1976). *Integral Geometry and Geometric Probability*. Addison-Wesley, Reading, Massachusetts.

Scherle, W. (1970). A simple method for volumetry of organs in quantitative stereology. *Mikroskopie*, *26*, 57-60.

Schneider, R., & Weil, W. (2008). *Stochastic and Integral Geometry*. Berlin: Springer.

Serra, J. (1982). *Image Analysis and Mathematical Morphology*. London: Academic Press.

Shay, J. (1975). Economy of effort in electron microscope morphometry. *Amer J Pathol*, *81*, 503-12.

Smit, J. C. (1961). Estimation of the mean of a stationary stochastic process by equidistant observations. *Trab Estadist*, *12*, 35-45.

Smith, C. S., & Guttman, L. (1953). Measurement of internal boundaries in three-dimensional structures by random sectioning. *Trans Metall Soc AIME*, *197*, 81-7.

Snedecor, G. W., & Cochran, W. G. (1980). *Statistical Methods. 8th ed.* Ames, Iowa: The Iowa State University Press.

Sommerville, D. M. (1958). *An Introduction to the Geometry of n Dimensions*. New York: Dover.

Sørensen, F. B. (1989). Objective histopathologic grading of cutaneous malignant melanomas by stereologic estimation of nuclear volume. *Cancer*, *63*, 1784-98.

Sørensen, F. B. (1991). Stereological estimation of mean and variance of nuclear volume from vertical sections. *J Microsc*, *162*, 203-29.

Souchet, S. (1995). *Précision de l'estimateur de Cavalieri.* Rapport de stage, D.E.A. de Statistiques et Modèles Aléatoires Appliqués à la Finance, Université Paris-VII. Laboratoire de Biométrie, INRA-Versailles.

Stark, A. K., Gundersen, H. J. G., Gardi, J. E., Pakkenberg, B., & Hahn, U. (2011). The saucor, a new stereological tool for analysing the spatial distributions of cells, exemplified by human neocortical neurons and glial cells. *J Microsc*, *242*, 132-47.

Stehr, M., Kiderlen, M., & Dorph-Petersen, K. A. (2022). Improving Cavalieri volume estimation based on non-equidistant planar sections: The trapezoidal estimator. *J Microsc*, *288*, 40-53.

Steiner, J. (1840). Über parallele Flächen. *Monatsber preuss Akad Wiss, Berlin*, 114-118. ([In: *Gesammelte Werke 2*, 171–76. Berlin: Reimer (1882)])

Steinhaus, H. (1930). Zur Praxis der Rektifikation und zum Längenbegriff. *Ber Akad Wiss Leipzig*, *82*, 120-30.

Sterio, D. C. (1984). The unbiased estimation of number and sizes of arbitrary particles using the disector. *J Microsc*, *134*, 127-36.

Stoyan, D., Kendall, W. S., & Mecke, J. (1995). *Stochastic Geometry and Its Applications, 2nd. ed.* Chichester: J. Wiley.

Struik, D. J. (1961). *Lectures on Classical Differential Geometry, 2nd edn.* New York: Dover.

Tandrup, T. (1993). A method for unbiased and efficient estimation of number and mean volume of specified neuron subtypes in rat dorsal root ganglion. *J Compar Neurol*, *329*, 269-76.

Thioulouse, J., Mathy, B., & Ploye, H. (1985). Estimation d'un volume a partir de coupes sériées: sous-echantillonnage, covariogramme transitif et calcul de précision. *Mikroskopie*, *42*, 215-24.

Thompson, W. R. (1932). The geometric properties of microscopic congurations. I. General aspects of projectometry. *Biometrika*, *24*, 21-6.

Thompson, W. R., Hussey, R., Matteis, J. T., Meredith, W. C., Wilson, G. C., & Tracy, F. E. (1932). The geometric properties of microscopic congurations. II. Incidence and volume of islands of Langerhans in the pancreas of a monkey. *Biometrika*, *24*, 27-38.

Thomson, E. (1930). Quantitative microscopic analysis. *J Geol*, *38*, 193-222.

Thórisdóttir, O., & Kiderlen, M. (2014). The invariator principle in convex geometry. *Adv Appl Math*, *58*, 63-87.

Thórisdóttir, O., Rafati, A. H., & Kiderlen, M. (2014). Estimating the surface area of nonconvex particles from central planar sections. *J Microsc*, *255*, 49-64.

Tinajero-Bravo, M., Eslava-Gómez, G., & Cruz-Orive, L. M. (2014). Conditions for exact Cavalieri estimation. *Image Anal Stereol*, *33*, 219-24.

Tobias, R. T. (1981). Volume: some historical perspectives. *Math Gazette*, *65*, 261-71.

Underwood, E. E. (1970). *Quantitative Stereology.* Reading, Mass.: Addison-Wesley.

Varga, O. (1935). Integralgeometrie 3. Croftons Formeln für den Raum. *Math Z*, *40*, 387-405.

Vesterby, A., Kragstrup, J., Gundersen, H. J. G., & Melsen, F. (1987). Unbiased stereologic estimation of surface density in bone using vertical sections. *Bone*, *8*, 13-17.

Von Neumann, J. (1951). Various techniques used in connection with random digits. Monte Carlo methods. *Nat. Bureau Standards*, *12*, 36-38.

Voss, F., & Cruz-Orive, L. M. (2009). Second moment formulae for geometric sampling with test probes. *Statistics*, *43*, 329-65.

Weibel, E. R. (1963). *Morphometry of the Human Lung*. New York: Academic Press.

Weibel, E. R. (1967). Structure in space and its appearance on sections. In H. Elias (Ed.), *Stereology* (p. 15-26). Berlin-Heidelberg: Springer.

Weibel, E. R. (1969). Stereological principles for morphometry in electron microscopic cytology. *Int Rev Cytol*, *26*, 235-302.

Weibel, E. R. (1979). *Stereological Methods. Vol. 1: Practical Methods for Biological Morphometry*. London: Academic Press.

Weibel, E. R. (1980). *Stereological Methods. Vol. 2: Theoretical Foundations*. London: Academic Press.

Weibel, E. R., Gehr, P., Cruz-Orive, L. M., Müller, A. E., Mwangi, D. K., & Haussener, V. (1981). Design of the mammalian respiratory system. IV. Morphometric estimation of pulmonar diffusing capacity; critical evaluation of a new sampling method. *Resp Physiol*, *44*, 39-59.

Weibel, E. R., & Knight, B. W. (1964). A morphometric study on the thickness of the pulmonary air-blood barrier. *J Cell Biol*, *21*, 367-84.

West, M. J. (2012). *Basic Stereology for Biologists and Neuroscientists*. New York: CSH Laboratory Press.

Wicksell, S. D. (1925). The corpuscle problem. A mathematical study of a biometric problem. *Biometrika*, *17*, 84-9.

Wirjadi, O., Schladitz, K., Easwaran, P., & Ohser, J. (2016). Estimating fibre direction distributions of reinforced composites from tomographic images. *Image Anal Stereol*, *35*, 167-79.

Wulfsohn, D., Aravena Zamora, F., Potin Téllez, C., Zamora Lagos, I., & García-Fiñana, M. (2012). Multilevel systematic sampling to estimate total fruit number for yield forecasts. *Precision Agric*, *12*, 256-75.

Wulfsohn, D., Gundersen, H. J. G., Jensen, E. B. V., & Nyengaard, J. R. (2004). Multilevel systematic sampling to estimate total fruit number for yield forecasts. *J Microsc*, *215*, 111-20.

Wulfsohn, D., Knust, J., Ochs, M., Nyengaard, J. R., & Gundersen, H. J. G. (2010). Stereological estimation of the total number of ventilatory units in mice lungs. *J Microsc*, *238*, 75-89.

Wulfsohn, D., Nyengaard, J. R., Gundersen, H. J. G., Cutler, A. J., & Squires, T. M. (1999). Non-destructive, stereological estimation of plant root lengths, branching pattern and diameter distribution. *Plant and Soil*, *214*, 15-26.

Wulfsohn, D., Sciortino, M., Aaslyng, J. M., & García-Fiñana, M. (2010). Non-destructive, stereological estimation of canopy surface area. *Biometrics*, *66*, 159-68.

Youngs, T. A., Boyce, T. W., Mosekilde, L., Søgaard, C. H., Paddock, C. L., & Gundersen, H. J. G. (1994). Direct stereological estimation of 3–D connectivity density in human illiac cancellous bone: the effect of age and sex. Proc 6ECS Prague 1993. *Acta Stereol*, *13/1*, 55-60.

Ziegel, J. F., Nyengaard, J. R., & Jensen, E. B. V. (2015). Estimating particle shape and orientation using volume tensors. *Scand J Stat*, *42*, 813-31.

Author Index

L. M. Cruz-Orive, *Stereology*, Interdisciplinary Applied Mathematics 59,
https://doi.org/10.1007/978-3-031-52451-6

Subject Index

L. M. Cruz-Orive, *Stereology*, Interdisciplinary Applied Mathematics 59,
https://doi.org/10.1007/978-3-031-52451-6

GPSR Compliance
The European Union's (EU) General Product Safety Regulation (GPSR) is a set of rules that requires consumer products to be safe and our obligations to ensure this.

If you have any concerns about our products, you can contact us on

ProductSafety@springernature.com

In case Publisher is established outside the EU, the EU authorized representative is:

Springer Nature Customer Service Center GmbH
Europaplatz 3
69115 Heidelberg, Germany

www.ingramcontent.com/pod-product-compliance
Ingram Content Group UK Ltd.
Pitfield, Milton Keynes, MK11 3LW, UK
UKHW022217051026
15624UKWH00001B/199
9783031524509